Municipal Wastewater Treatment

Municipal Wastewater Treatment

Guest Editors

Yung-Tse Hung
Hamidi Abdul Aziz
Issam A. Al-Khatib

Basel • Beijing • Wuhan • Barcelona • Belgrade • Novi Sad • Cluj • Manchester

Guest Editors

Yung-Tse Hung
Civil and Environmental
Engineering
Cleveland State University
Cleveland
USA

Hamidi Abdul Aziz
Civil Engineering
Universiti Sains Malaysia
Nibong Tebal
Malaysia

Issam A. Al-Khatib
Institute of Environmental
and Water Studies
Birzeit University
Birzeit
Palestine

Editorial Office
MDPI AG
Grosspeteranlage 5
4052 Basel, Switzerland

This is a reprint of the Special Issue, published open access by the journal *International Journal of Environmental Research and Public Health* (ISSN 1660-4601), freely accessible at: https://www.mdpi.com/journal/ijerph/special_issues/municipal_wastewater.

For citation purposes, cite each article independently as indicated on the article page online and as indicated below:

Lastname, A.A.; Lastname, B.B. Article Title. *Journal Name* **Year**, *Volume Number*, Page Range.

ISBN 978-3-7258-3531-7 (Hbk)
ISBN 978-3-7258-3532-4 (PDF)
https://doi.org/10.3390/books978-3-7258-3532-4

Contents

About the Editors . . . . **ix**

Preface . . . . **xi**

Noudeng Vongdala, Hoang-Dung Tran, Tran Dang Xuan, Rolf Teschke and Tran Dang Khanh
Heavy Metal Accumulation in Water, Soil, and Plants of Municipal Solid Waste Landfill in
Vientiane, Laos
Reprinted from: *Int. J. Environ. Res. Public Health* **2019**, *16*, 22,
https://doi.org/10.3390/ijerph16010022 . . . . **1**

Junyuan Guo, Yuling Zhou, Yijin Yang, Cheng Chen and Jiajing Xu
Effects of Hydraulic Loading Rate on Nutrients Removal from Anaerobically Digested Swine
Wastewater by Multi Soil Layering Treatment Bioreactor
Reprinted from: *Int. J. Environ. Res. Public Health* **2018**, *15*, 2688,
https://doi.org/10.3390/ijerph15122688 . . . . **14**

**Yaohui Wu, Wen Liu, Yonghong Wang, Xinjiang Hu, Zhengping He, Xiaoyong Chen and Yunlin
Zhao**
Enhanced Removal of Antibiotic in Wastewater Using Liquid Nitrogen-Treated Carbon Material:
Material Properties and Removal Mechanisms
Reprinted from: *Int. J. Environ. Res. Public Health* **2018**, *15*, 2652,
https://doi.org/10.3390/ijerph15122652 . . . . **24**

Malwina Tytła
The Effects of Ultrasonic Disintegration as a Function of Waste Activated Sludge Characteristics and
Technical Conditions of Conducting the Process—Comprehensive Analysis
Reprinted from: *Int. J. Environ. Res. Public Health* **2018**, *15*, 2311,
https://doi.org/10.3390/ijerph15102311 . . . . **38**

Hamidi Abdul Aziz, Nur Nasuha Ahmad Puat, Motasem Y. D. Alazaiza and Yung-Tse Hung
Poultry Slaughterhouse Wastewater Treatment Using Submerged Fibers in an Attached Growth
Sequential Batch Reactor
Reprinted from: *Int. J. Environ. Res. Public Health* **2018**, *15*, 1734,
https://doi.org/10.3390/ijerph15081734 . . . . **60**

Magdalena Lebiocka, Agnieszka Montusiewicz and Agnieszka Cydzik-Kwiatkowska
Effect of Bioaugmentation on Biogas Yields and Kinetics in Anaerobic Digestion of Sewage Sludge
Reprinted from: *Int. J. Environ. Res. Public Health* **2018**, *15*, 1717,
https://doi.org/10.3390/ijerph15081717 . . . . **72**

Ruizhu Hu, Tinglin Huang, Aofan Zhi and Zhangcheng Tang
Full-Scale Experimental Study of Groundwater Softening in a Circulating Pellet Fluidized Reactor
Reprinted from: *Int. J. Environ. Res. Public Health* **2018**, *15*, 1592,
https://doi.org/10.3390/ijerph15081592 . . . . **88**

Junjun Ma, Bing Li, Lincheng Zhou, Yin Zhu, Ji Li and Yong Qiu
Simple Urea Immersion Enhanced Removal of Tetracycline from Water by Polystyrene Microspheres
Reprinted from: *Int. J. Environ. Res. Public Health* **2018**, *15*, 1524,
https://doi.org/10.3390/ijerph15071524 . . . . **99**

Krzysztof Poszytek, Joanna Karczewska-Golec, Anna Ciok, Przemyslaw Decewicz, Mikolaj Dziurzynski, Adrian Gorecki, et al.
Genome-Guided Characterization of *Ochrobactrum* sp. POC9 Enhancing Sewage Sludge Utilization—Biotechnological Potential and Biosafety Considerations
Reprinted from: *Int. J. Environ. Res. Public Health* **2018**, *15*, 1501,
https://doi.org/10.3390/ijerph15071501 . . . . **114**

Qinglin Fang, Wenlai Xu, Gonghan Xia and Zhicheng Pan
Effect of C/N Ratio on the Removal of Nitrogen and Microbial Characteristics in the Water Saturated Denitrifying Section of a Two-Stage Constructed Rapid Infiltration System
Reprinted from: *Int. J. Environ. Res. Public Health* **2018**, *15*, 1469,
https://doi.org/10.3390/ijerph15071469 . . . . **131**

Xianze Wang, Zhongmou Liu, Zhian Ying, Mingxin Huo and Wu Yang
Adsorption of Trace Estrogens in Ultrapure and Wastewater Treatment Plant Effluent by Magnetic Graphene Oxide
Reprinted from: *Int. J. Environ. Res. Public Health* **2018**, *15*, 1454,
https://doi.org/10.3390/ijerph15071454 . . . . **144**

Amin Mojiri, Akiyoshi Ohashi, Noriatsu Ozaki, Ahmad Shoiful and Tomonori Kindaichi
Pollutant Removal from Synthetic Aqueous Solutions with a Combined Electrochemical Oxidation and Adsorption Method
Reprinted from: *Int. J. Environ. Res. Public Health* **2018**, *15*, 1443,
https://doi.org/10.3390/ijerph15071443 . . . . **157**

Iveta Pandová, Anton Panda, Jan Valíček, Marta Harničárová, Milena Kušnerová and Zuzana Palková
Use of Sorption of Copper Cations by Clinoptilolite for Wastewater Treatment
Reprinted from: *Int. J. Environ. Res. Public Health* **2018**, *15*, 1364,
https://doi.org/10.3390/ijerph15071364 . . . . **172**

Eduard Rott, Bertram Kuch, Claudia Lange, Philipp Richter and Ralf Minke
Influence of Ammonium Ions, Organic Load and Flow Rate on the UV/Chlorine AOP Applied to Effluent of a Wastewater Treatment Plant at Pilot Scale
Reprinted from: *Int. J. Environ. Res. Public Health* **2018**, *15*, 1276,
https://doi.org/10.3390/ijerph15061276 . . . . **184**

Allisen N. Okeyo, Nolonwabo Nontongana, Taiwo O. Fadare and Anthony I. Okoh
Vibrio Species in Wastewater Final Effluents and Receiving Watershed in South Africa: Implications for Public Health
Reprinted from: *Int. J. Environ. Res. Public Health* **2018**, *15*, 1266,
https://doi.org/10.3390/ijerph15061266 . . . . **204**

Eduard Rott, Bertram Kuch, Claudia Lange, Philipp Richter, Amélie Kugele and Ralf Minke
Removal of Emerging Contaminants and Estrogenic Activity from Wastewater Treatment Plant Effluent with UV/Chlorine and UV/H$_2$O$_2$ Advanced Oxidation Treatment at Pilot Scale
Reprinted from: *Int. J. Environ. Res. Public Health* **2018**, *15*, 935,
https://doi.org/10.3390/ijerph1505935 . . . . **221**

Qinglin Fang, Wenlai Xu, Zhijiao Yan and Lei Qian
Effect of Potassium Chlorate on the Treatment of Domestic Sewage by Achieving Shortcut Nitrification in a Constructed Rapid Infiltration System
Reprinted from: *Int. J. Environ. Res. Public Health* **2018**, *15*, 670,
https://doi.org/10.3390/ijerph1504670 . . . . **239**

Xiao Bian, Hui Gong and Kaijun Wang
Pilot-Scale Hydrolysis-Aerobic Treatment for Actual Municipal Wastewater: Performance and
Microbial Community Analysis
Reprinted from: *Int. J. Environ. Res. Public Health* **2018**, *15*, 477,
https://doi.org/10.3390/ijerph15030477 . . . . **250**

Xudong Chen, Zhongwen Xu, Liming Yao and Ning Ma
Processing Technology Selection for Municipal Sewage Treatment Based on a Multi-Objective
Decision Model under Uncertainty
Reprinted from: *Int. J. Environ. Res. Public Health* **2018**, *15*, 448,
https://doi.org/10.3390/ijerph15030448 . . . . **258**

Junping Lv, Xuechun Wang, Wei Liu, Jia Feng, Qi Liu, Fangru Nan, et al.
The Performance of a Self-Flocculating Microalga *Chlorococcum* sp. GD in Wastewater with Different
Ammonia Concentrations
Reprinted from: *Int. J. Environ. Res. Public Health* **2018**, *15*, 434,
https://doi.org/10.3390/ijerph15030434 . . . . **276**

Yihuan Deng and Andrew Wheatley
Mechanisms of Phosphorus Removal by Recycled Crushed Concrete
Reprinted from: *Int. J. Environ. Res. Public Health* **2018**, *15*, 357,
https://doi.org/10.3390/ijerph15020357 . . . . **289**

About the Editors

Yung-Tse Hung

Prof. Dr. Yung Tse Hung, Ph.D., P.E., DEE, Fellow-ASCE, served as a Professor of Civil Engineering at Cleveland State University from 1981 to 2024. He earned his B.S. and M.S. in Civil Engineering from Cheng Kung University, Taiwan, and his Ph.D. from the University of Texas at Austin. Prof. Hung has taught at 16 universities across eight countries and started the public health engineering program at the University of Canterbury, New Zealand, in 1972. He has served on the faculties of numerous universities globally, including in New Zealand, USA, Hong Kong, UAE, Singapore, Australia, Russia, and Kyrgyzstan. Prof. Hung's research focuses on biological wastewater treatment, industrial water pollution control, and municipal wastewater treatment. He has published approximately 40 books, 242 book chapters, 198 refereed publications, and 352 other scholarly works, totaling around 811 publications and presentations. He is a Fellow of ASCE, a Diplomate of AAEE, a Fellow of the Ohio Academy of Science, a member of AEESP, and a Life Member of WEF. He serves as Editor-in-Chief for several international journals and books and is a registered professional engineer in Ohio and North Dakota.

Hamidi Abdul Aziz

Prof. Dr. Hamidi Abdul Aziz is a distinguished professor in Environmental Engineering at the School of Civil Engineering, Universiti Sains Malaysia. He earned his Ph.D. in Civil Engineering (Environmental) from the University of Strathclyde in 1992. Currently, he heads the Solid Waste Management Cluster (SWAM) at Universiti Sains Malaysia. With 29 years of experience in teaching and researching in environmental engineering, Professor Aziz has made significant contributions in areas such as solid waste management, landfill technology, water and wastewater treatment, leachate treatment, bioremediation, pollution control, and environmental impact assessment. To date, he has mentored over 100 Ph.D. and MSc students and has published over 200 ISI papers and several books. Additionally, he serves as an Editor and Editorial Board Member for several international journals. In recognition of his outstanding research contributions, the Malaysia Academy of Sciences named him a Top Research Scientist of Malaysia in 2012. In 2020, he was listed among the top 2% of scientists in his field globally by the prestigious Stanford University.

Issam A. Al-Khatib

Prof. Dr. Issam A. Al-Khatib is a faculty member at the Institute of Environmental and Water Studies, Birzeit University, Palestine. His expertise spans the topics of water resource management, environmental assessment, wastewater management, and climate change, with a focus on environmental health and sustainable development. Dr. Al-Khatib has led numerous research and evaluation projects on topics such as medical waste management, solid waste, water and sanitation, and urban environments. He is dedicated to promoting public environmental awareness through training, workshops, and community campaigns. He has contributed to national committees on environmental standards, including swimming pool water quality regulations. Dr. Al-Khatib has supervised over 40 MSc theses, focusing on the above-listed critical environmental issues, and is an expert in academic quality evaluation. Currently, he is working on a project focused on evaluating the attitudes, perceptions, and behaviors of Birzeit University students toward the management of single-use plastic waste. As of early 2025, Prof. Al-Khatib is ranked among the top 0.5% of researchers globally according to Scholar GPS, recognized for his exceptional contributions to environmental science, monitoring, and waste management.

Preface

Water has become a scarce resource in the world. The major objective of municipal wastewater treatment is to remove pollutants from wastewater before effluent is discharged back to the environment. Treated effluent can be utilized for various types of water reuse and for resource recovery. This reprint, including 21 papers, addresses diverse aspects of wastewater treatment, encompassing both municipal wastewater and industrial wastewater. One study explores the use of submerged fibers in an attached growth sequential batch reactor for effective treatment of poultry slaughterhouse wastewater. Another study focuses on the performance and microbial community analysis of pilot-scale hydrolysis–aerobic treatment for municipal wastewater. A separate study develops a multi-objective decision model to select appropriate municipal sewage treatment technologies under uncertainty. A different study investigates phosphorus removal mechanisms using recycled crushed concrete. Further studies examine the impact of C/N ratio and potassium chlorate on nitrogen removal in two-stage-constructed rapid infiltration systems. One particular study focuses on removing nutrients from anaerobically digested swine wastewater using a multi-soil layering treatment bioreactor. There is also a full-scale experimental study on groundwater softening using a circulating pellet fluidized reactor. Another analysis focuses on bioaugmentation effects on biogas yields and kinetics in the anaerobic digestion of sewage sludge. These studies collectively highlight innovative and effective methods for improving the efficiency of municipal wastewater treatment, as well as the treatment of other wastewater, and reducing water pollution.

Yung-Tse Hung, Hamidi Abdul Aziz, and Issam A. Al-Khatib
Guest Editors

International Journal of
*Environmental Research
and Public Health*

Article

Heavy Metal Accumulation in Water, Soil, and Plants of Municipal Solid Waste Landfill in Vientiane, Laos

Noudeng Vongdala [1], Hoang-Dung Tran [2], Tran Dang Xuan [1,*], Rolf Teschke [3] and Tran Dang Khanh [4]

1   Graduate school for International Development and Cooperation, Hiroshima University, Hiroshima 739-8529, Japan; vongdala1986@gmail.com
2   Nguyen Tat Thanh University, Ho Chi Minh City 702000, Vietnam; tranhoangdung1975@yahoo.com
3   Department of Internal Medicine II, Division of Gastroenterology and Hepatology, Klinikum Hanau, D-63450 Hanau, Germany; rolf.teschke@gmx.de
4   Agricultural Genetics Institute, Pham Van Dong, Tu Liem, Hanoi 123000, Vietnam; tdkhanh@vaas.vn
*   Correspondence: tdxuan@hiroshima-u.ac.jp; Tel.: +81-82-424-6927

Received: 20 November 2018; Accepted: 17 December 2018; Published: 21 December 2018

Abstract: The municipal solid waste (MSW) landfill in Vientiane, Laos, which receives > 300 tons of waste daily, of which approximately 50% is organic matter, has caused serious environmental problems. This study was conducted to investigate the accumulated levels of heavy metals (HMs) (cadmium (Cd), chromium (Cr), copper (Cu), nickel (Ni), lead (Pb), and zinc (Zn)) in water (surface and groundwater), soil, and plants between dry and wet seasons according to the standards of the Agreement on the National Environmental Standards of Laos (ANESs), Dutch Pollutant Standards (DPSs), and the World Health Organization (WHO), respectively. Although no impact of pollution on the surface water was observed, the levels of Cr and Pb in the groundwater significantly exceeded the basics of ANESs and WHO in both seasons. The pollution caused by Cd and Cu reached the eco-toxicological risk level in the landfill soils and its vicinity. The vegetable *Ipomoea aquatica*, which is consumed by the nearby villagers, was seriously contaminated by Cr, Pb, Cu, and Zn, as the accumulation of these toxic metals was elevated to much greater levels as compared to the WHO standards. For the grass *Pennisetum purpureum* (elephant grass), the quantities of HMs in all plant parts were extreme, perhaps due to the deeper growth of its rhizome than *I. aquatica*. This study is the first to warn of serious HM pollution occurring in the water, soil, and plants in the MSW landfill of Vientiane, Laos, which requires urgent phytoremediation. The indication of what sources from the MSW principally cause the pollution of HMs is needed to help reduce the toxicological risks on Lao residents and the environment in Vientiane as well.

Keywords: municipal solid waste; landfill; heavy metals; soil; plants; water; pollution; health risk

1. Introduction

The management of municipal solid waste (MSW) is a challenge for the urban environment in many developing countries and is attributed to rapid population growth, unsatisfactory urbanisation, and undesireable economic growth. Thus, open dumping and unsanitary waste landfills are a pressing issue. MSW comprises household, healthcare, and industrial waste, but they are not segregated and are all disposed of into the same landfill [1]. The landfill is the principal place for solid waste dumping, which has resulted in serious environmental pollution and the spread of disease [2,3]. Leachate migration in open dumping sites is a dominant source of heavy metals (HMs) in surface and groundwater, soil, and plants [4–6]. If plants uptake HMs from polluted soil, there is a high possibility of HMs transferring to the human food chain through the consumption of vegetation or animals [7,8].

The most problematic HMs include cadmium (Cd), chromium (Cr), copper (Cu), lead (Pb), nickel (Ni), and zinc (Zn) [9,10]. Contaminated wastewater has resulted in the contamination of food crops [11,12].

When wastewater contains HMs at low concentrations, they may be useful for increasing productivity in agricultural production, as they are essential to living organism growth. However, high concentrations of HMs have negative effects on the environment [13,14]. Many studies have documented that both individual and combined HMs simultaneously expose humans and other organisms to toxic effects but are mediated by leached doses, duration of exposure, and genetic factors [15]. There are many compounds with hazardous effects leaching from MSW landfills that can harmfully influence human health and the environment. In addition, several hazardous compounds have also been found in the sediments but not in leachates. However, the leachates from MSW are a fundamental matrix of toxic contaminants to freshwater aquatic organisms [16,17].

Plants growing in an MSW landfill and its vicinity cycle both trace elements and contaminated HMs that can affect the food chain and can accumulate trace elements, especially HMs. They receive HMs from soils and partly from water and air [18]. However, some plants, such as *Myriophylhum aquaticum* (parrot feather), *Ludwigina palustris* (creeping primrose), and *Mentha aquatic* (water mint), have shown excellent phytoremediation strength for contaminated soils, groundwater, and wastewater by absorption [19,20]. In addition, the phytoremediation from plants is an efficient treatment technique for HMs due to its low cost and easy operation and maintainance [21,22].

The MSW landfill in Vientiane, Laos receives around 300 tons of waste daily, of which 40–50% is organic matter. The sources of solid waste include residential, commercial, institutional, and industrial activities [23]. When hazardous waste from hospitals and industry is disposed of into the landfill, it causes groundwater and surface water pollution [24]. In the wet season, surface water runs off from the landfill [25]. Therefore, analyses for HMs should be conducted on samples collected during both the dry and wet seasons. Residents in areas close to the MSW landfill who work at the dump, such as waste pickers, may have serious health problems. However, an analysis of chemical leachates in soils, water, and plants growing in the MSW has not yet been evaluated. This research was thus conducted to assess HM concentrations in water, soil, and plants in the landfill and surrounding area to understand the pollution situation and to help establish an appropriate management strategy. In addition, this information can help develop policies for site remediation and urban environmental quality management.

2. Materials and Methods

2.1. Description of Research Area

The MSW landfill is located in a suburban area of Vientiane, the capital of Laos (latitude 18°4′45.86″ N, longitude 102°50′57.18″ E, approximately 32 km from the urban centre). The landfill operation is open dumping in a total area of 100 ha of land space and has been operating since 2007. The area has a tropical climate characterized by wet season rainfall from May to September and a dry season from October to April. The MSW landfill is a flat field adjacent to agricultural land, including rice fields. It includes landfill management offices, a recycling factory, wastewater treatment ponds, wetlands, and a temporary hut for waste pickers. Residents in the surrounding areas of the MSW landfill have never received public health information. Many plants in the landfill such as *Ipomoea aquatica* (water spinach) are collected and sold in the nearby local markets for human consumption. The landfill location is also connected to small substreams, and wastewater runoff in the rainy season overflows to the rivers. Also, the residents catch fish from the polluted ponds and nearby streams, which are sold to local markets without concern for possible HM pollution from MSW.

In this study, samples from surface water were collected upstream to downstream from wetlands in the landfill site (Table 1). Groundwater and soil samples were gathered from inside and outside of the landfill. In addition, two plant species—*I. aquatica* and the grass *Pennisetum purpureum* (elephant grass)—were also collected from the landfill and surrounding area (see Figure 1 or Table 1).

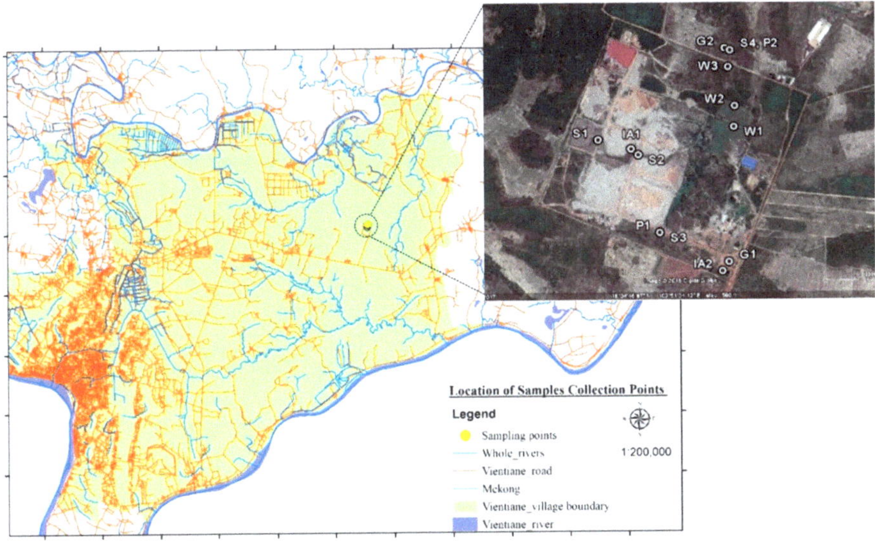

Figure 1. Map and satellite image of sampling locations in the landfill site.

2.2. Sample Collection

Three samples from the surface water in the landfill wetlands, from an average water depth of 1–1.5 m, were collected between each station from upstream to downstream. Good quality samples were randomly taken each season, and each station included two subsamples. They were stored in 1-L polyethylene bottles and subsequently adjusted by HNO_3 to obtain pH < 2 [26,27]. The values of pH, temperature (°C), electrical conductivity (EC), and dissolved oxygen (DO) were measured by HORIBA (U-50 Multiparameter Water Quality Meter, Kyoto, Japan) at field sites. Furthermore, two groundwater samples were collected from wells inside and in the vicinity of the landfill. The G1 (well) was used for the landfill management office at the landfill, and the G2 (well) was far from the landfill (70 m away) and it was frequently exploited by nearby farmers. The characterized flow of groundwater (G2) was unknown. The water samples were taken in both wet and dry seasons in 2017, and they were preserved in a way similar to the surface water samples. In detail, pH values of 8.32 and 7.53, temperatures of 19.14 and 25 °C, DO values of 12.03 and 0.52 mg/L, and EC values of 2.77 and 6.93 µs/cm were recorded in wet and dry seasons, respectively. The results showed that these parameters fitted the standards of the permissible limits of the National Environmental Standards of Laos, except the DO in the dry season.

Soil samples were randomly collected at 0–0.25-m depth at an area of 50 cm^2 in the landfill and locations at 60 m from the landfill according to a method described in Gworek et al. [28]. There were 48 samples gathered in both wet and dry seasons in 2017. The locations of soil samples inside the dumpsite were collected from the nearest recharge canals. At each sampling location, six subsamples were randomly collected to make a composite sample. The soils were placed in sterilized plastic bags, while the sampled sites were recorded by GPS (GARMIN GPS 62$_{sc}$ (surface and groundwater, soil and plant samples) (Table 1). The soils were air-dried at room temperature after collection. They were finely ground by an agate mortar, filtered through a 0.2-mm stainless-steel sieve to remove coarse debris, and stored in sealed containers at 4 °C for further analysis.

I. aquatica (water spinach) was collected in two ponds in the landfill, including leachate and fish ponds. The samples were collected in both wet and dry seasons (total of 24 samples). The roots, stems, and leaves were separated and stored in zipped polyethylene bags. The samples were washed thoroughly with tap water and rinsed with distilled water for 1 min to clear them of periphyton and detritus [28]. The *P. purpureum* samples were collected in two locations: inside and outside the landfill. The samples were dried at 40 °C for 2 days in an oven until a consistent weight was maintaned. The samples were ground into a fine powder by a mortar and kept in the dark at 5 °C for further analysis.

Table 1. Description of water, soil, and plant sampling sites (inside and outside the landfill).

Site	Latitude	Longitude	Description of Location
Surface water sampling sites			
W1	18°4′50.45″	102°51′13.20″	Leachate and wastewater were runoff from the landfill to the wetland (upstream)
W2	18°4′53.85″	102°51′13.35″	Leachate and wastewater were runoff from the landfill to wetland (middle stream)
W3	18°5′0.20″	102°51′12.26″	Leachate and wastewater were runoff from the landfill to wetland (downstream)
Groundwater sampling sites			
G1	18°4′28.40″	102°51′12.31″	Available groundwater (well) inside landfill used for the landfill management's office and waste pickers
G2	18°5′3.26″	102°51′11.70″	Available well used for domestic purposes of farmers was outside the landfill about 70 m away
Soil sampling sites			
S1	18°4′48.32″	102°50′50.46″	Random samples of soils in the landfill were near recharge canals
S2	18°4′45.62″	102°50′57.60″	Random samples of soils in the landfill were near recharge canals
S3	18°4′33.17″	102°51′0.81″	Random samples of soils in the landfill were near recharge canals
S4	18°5′2.90″	102°51′12.59″	Random samples of soils were outside the landfill site about 60 m away
Plant sampling sites			
IA1	18°4′46.82″	102°50′55.98″	Random samples of *Ipomoea aquatica* in the wastewater or leachate area (inside landfill)
IA2	18°4′26.91″	102°51′11.23″	Random samples of *I. aquatica* in the fish pond was near the landfill's office.
P1	18°4′33.17″	102°51′0.81″	Random samples of grass in the landfill
P2	18°5′2.90″	102°51′12.59″	Random samples of grass were outside the landfill site about 70 m away

2.3. Chemical Analysis for Heavy Metals

The surface and groundwater samples were filtered by filter papers (Qualitative, circle, 90 mm Ø, Whatman™, New Jersey, US) to obtain a 100-mL solution, to which 1 mL of HNO_3 (65%) was added and then heated for 2 h without boiling at 80–90 °C. The samples were cooled to room temperature and then filtered with 0.2-μm syringe filters [29]. The waters were analyzed using a multitype inductively coupled plasma emission spectrometer (ICP-ES) and ICPE-9000 (Shimadzu, Tokyo, Japan).

Five grams of soil were added to 20 mL of HNO_3 (7 mol/L), stirred for 1 h, and then placed in an autoclave at 120 °C for 30 min. The mixture was cooled to room temperature, filtered by filter papers, and diluted by deionized water [30]. The samples were filtered again by a 0.2-μm syringe filter and transferred to tubes for analyses by ICPE-9000 (Shimadzu, Tokyo, Japan).

The roots, stems, and leaves of the plant samples were dried and ground into powder. An amount of 0.5 g of each sample was added to 6 mL of concentrated HNO_3 (65%) and 2 mL of concentrated HCl (30%) and stood until the reaction was completed. The mixture was moved to an autoclave for 66 min at 132 °C for digestion [30]. The plant samples were analyzed by ICPE-9000 (Shimadzu, Tokyo, Japan).

2.4. Statistical Analysis

The statistical analyses were implemented by Minitab® 17.3.0 (Copyright © 2016 by Minitab Inc., Shanghai, China). Mean and standard deviation values were evaluated using one-way analysis of variance (ANOVA). Tukey's test was applied to compare pairs among seasons, treatments, and data of the standards at a *p*-value of $p < 0.05$, which indicated the significant differences by 95% confidence.

3. Results

3.1. Concentration of Heavy Metals in Surface and Groundwater

Table 2 shows that several HM contents were below the detectable limit of the analyzed instrument and were noted as not detected (ND). The results in Table 2 indicate that no trace of cadmium (Cd) and zinc (Zn) was detectable in surface and groundwater of both the wet and dry seasons. The accumulation of chromium (Cr), copper (Cu), nickel (Ni), and lead (Pb) was observed, but none of them exceeded the standard levels of the Agreement on the National Environmental Standards of Laos (ANESs). Similarly, no presence of Cd and Zn was found in G1 (groundwater inside the landfill) and G2 (groundwater outside the landfill) in both the wet and dry seasons (Table 2; Figure 2). However, Ni was not detected in both G1 and G2 in the dry season, whilst Pb was not found in the wet season in G2. In addition, no accumulation of Cu was observed in G2 in both the wet and dry seasons (Table 2; Figure 2). In the groundwater inside the MSW landfill, the levels of Cr and Pb significantly exceeded the standards of both ANESs and the World Health Organization (WHO). On the other hand, traces of Ni and Cu were below the permissible limit of ANESs and WHO (Table 2). In locations outside the landfill (G2), although Cr, Ni, and Pb were detected, only the contents of Pb in the dry season were much higher than those of ANESs and WHO standards (Table 2). The levels of HMs in the dry season were higher than the wet season. The results suggest that seasonal variation significantly influences heavy metal concentrations.

Table 2. Heavy metal concentrations between the wet and dry seasons in surface and groundwater (mg/L).

Site	Heavy Metal	Wet Season	Dry Season	Standards (mg/L)	
		Mean $\pm$ SD	Mean $\pm$ SD	ANESs	WHO
W1-3	Cd	ND	ND	0.03	-
	Cr	0.05 ± 0.015 [b]	0.19 ± 0.057 [a]	0.5	-
	Cu	0.01 ± 0.005 [b]	0.05 ± 0.018 [a]	0.5	-
	Ni	0.01 ± 0.001 [b]	0.04 ± 0.005 [a]	0.2	-
	Pb	0.02 ± 0.008 [b]	0.17 ± 0.042 [a]	0.2	-
	Zn	ND	ND	1.0	-
G1	Cd	ND	ND	0.003	0.003
	Cr	0.06 ± 0.01 [a]	0.08 ± 0.020 [a]	0.05	0.05
	Cu	0.01 ± 0.011 [a]	0.004 ± 0.01 [a]	1.50	2.00
	Ni	0.01 ± 0.010	ND	0.02	0.07
	Pb	0.05 ± 0.020 [a]	0.06 ± 0.013 [a]	0.01	0.01
	Zn	ND	ND	5.00	3.00
G2	Cd	ND	ND	0.003	0.003
	Cr	0.02 ± 0.019 [a]	0.04 ± 0.022 [a]	0.05	0.05
	Cu	ND	ND	1.50	2.00
	Ni	0.001 ± 0.001 [b]	ND	0.02	0.07
	Pb	ND	0.04 ± 0.014 [a]	0.01	0.01
	Zn	ND	ND	5.00	3.00

SD: standard deviation; ND: not detected; W: station of surface water sampling, including three main stations/points (W1–3); G1: groundwater in the landfill; G2: groundwater outside the landfill; WHO: World Health Organization, Guidelines for Drinking-Water Quality (2011); ANESs: Agreement on the National Environmental Standards of Laos (2009); values in a row with similar superscript letters are not significantly different ($p < 0.05$).

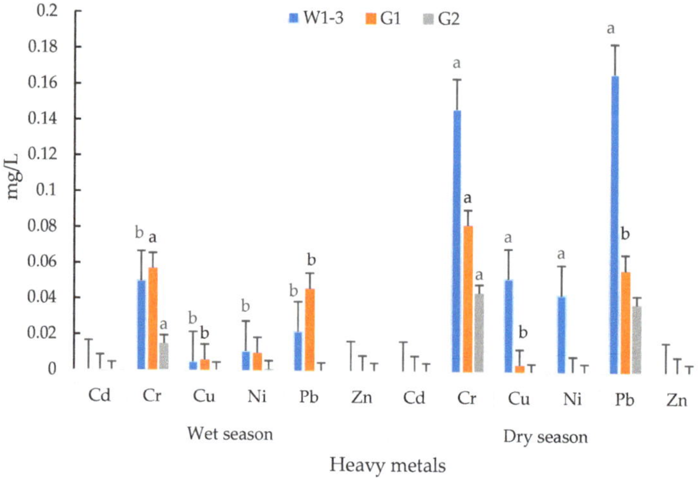

Figure 2. Seasonal variation of heavy metals in water. Bar values are mean ± standard error (SE); bars with similar letters are not significantly different ($p < 0.05$).

3.2. Concentration of Heavy Metals in Soil

Table 3 shows that the HMs from soils in the landfill were generally in much greater quantities than those of soils from the nearby location of the MSW landfill. The amount of HMs differed, and Zn was the greatest (>17-fold), whilst Cd and Cr were the least (>3-fold) (Figure 3). On the other hand, the quantities between the wet and dry seasons varied among the HMs. Amounts of Cd and Cu in the landfill exceeded the target value of the Dutch Pollutant Standards. In addition, the level of Cd overcame the eco-toxicological risk (outside landfill) (Table 3).

Table 3. Accumulation of heavy metals in soils compared with the Dutch standards (mg/kg).

Sites	Heavy Metal	Wet Season	Dry Season	Dutch Standards	
		Mean ± SD	Mean ± SD	Tv	Iv
	Cd	3.76 ± 0.33 [a]	3.73 ± 1.12 [a]	0.8	12.0
	Cr	39.67 ± 3.78 [a]	48.08 ± 13.67 [a]	100.0	380.0
S1–3	Cu	66.82 ± 27.52 [a]	54.06 ± 20.99 [a]	36.0	190.0
	Ni	19.43 ± 0.84 [a]	19.94 ± 4.91 [a]	35.0	210.0
	Pb	80.17 ± 19.33 [a]	67.99 ± 19.07 [a]	85.0	530.0
	Zn	77.46 ± 57.88 [a]	52.48 ± 34.59 [a]	140.0	720.0
	Cd	1.02 ± 0.64 [b]	1.06 ± 0.05 [b]	0.8	12.0
	Cr	10.02 ± 4.24 [b]	19.33 ± 1.95 [b]	100.0	380.0
S4	Cu	7.96 ± 2.79 [b]	11.70 ± 0.50 [b]	36.0	190.0
	Ni	5.65 ± 1.72 [b]	9.61 ± 0.06 [b]	35.0	210.0
	Pb	16.03 ± 5.40 [b]	21.47 ± 0.42 [b]	85.0	530.0
	Zn	4.39 ± 2.68 [b]	4.79 ± 0.56 [b]	140.0	720.0

SD: standard deviation; S: stations of soil sampling in landfill, including three main stations (S1–3); S4: the main station of soil sampling outside the landfill; Tv: target value (values > Tv: eco-toxicological risk); Iv: intervention value (values > Iv: environmental risk); values in columns with similar superscript letters are not significantly different ($p < 0.05$).

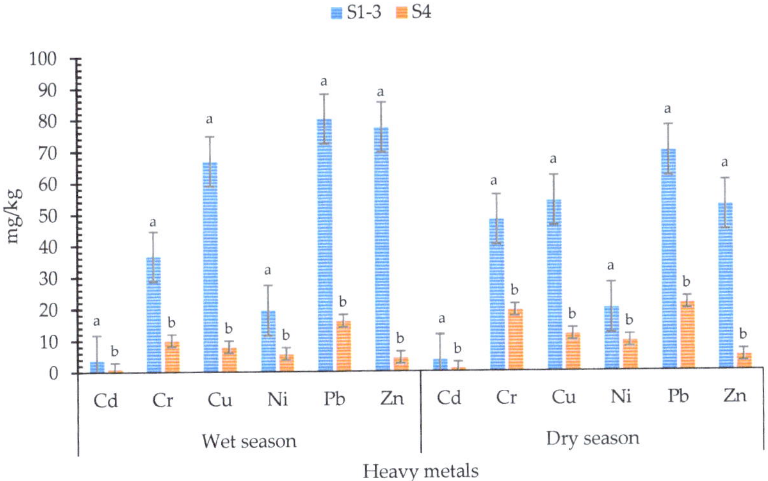

Figure 3. Significant difference in two locations and seasonal variation of heavy metals in soils. Bar values are mean ± SE; bars with similar letters are not significantly different ($p < 0.05$).

3.3. Concentrations of Heavy Metals in Plants

Table 4 shows that the roots of *I. aquatica* accumulated higher quantities of HMs than those allowed by WHO standards in both the wet and dry seasons. The maximal pollution levels were recorded for Cd, Cr, Pb, and Zn (8–56-fold higher than the WHO standards) in the stems and roots (Table 4). The leaves of this vegetable were not affected by Cd, Cu, Ni, Cd, and Ni in the wet and dry seasons, respectively, whilst the HM levels in stems did not exceed the standards of WHO in Cd, Cu, and Ni in the wet and dry seasons, respectively (Table 4). Considering that the villagers commonly consume leaves and stems of *I. aquatica*, Cr, Pb, Cu, and Zn in this vegetable presented in much higher quantities than the WHO standards (Table 4).

For *P. purpureum*, the quantity of Cd in its roots was > 50-fold greater than that of *I. aquatica* in the landfill, although the accumulated level of Cd was > 8-fold lower than the plants growing outside the MSW (Table 4). In the dry season, the amount of Cd in the rhizome of *P. purpureum* was also reduced by > 4-fold compared to the wet season. The accumulation of Cd in the stems and leaves of *P. purpureum* was similar to its roots. This evidence indicates that in the wet season, *P. purpureum* absorbs a much greater amount of Cd than in the dry season (Table 4). Furthermore, Cr, Cu, Ni, Pb, and Zn in different plant parts of *P. purpureum* were analogous to that of Cd. These chemicals were accreted in much higher quantities than that of *I. aquatica* and all of them exceeded the WHO standards, except for Ni in the leaves of this grass (Table 4). The accumulation of these HMs was significantly reduced in the dry season. Except for Ni, the presence of Cd, Cr, Cu, Pb, and Zn was found in all plant parts of *P. purpureum* growing outside the landfill in the dry season. This revealed that the pollution caused by Ni was less problematic than other HMs (Table 4). The amount of HMs detected in *P. purpureum* may be explained by the deeper roots of the grass, as *I. aquatica* roots commonly spread either on the surface of the soil or water in the landfill.

Table 4. Levels of heavy metals (mg/kg) in plants between the wet and dry seasons.

Plant Name/Site	Samples	Cd	Cr	Cu	Ni	Pb	Zn
				Wet Season			
I. aquatica	Leaves	ND	17.00 ± 4.27 [a]	8.30 ± 5.12 [a]	0.24 ± 0.34 [a]	13.26 ± 7.06 [a]	1.54 ± 0.64 [a]
	Stem	ND	18.63 ± 6.10 [a]	9.63 ± 9.75 [a]	ND	7.48 ± 7.11 [a]	8.54 ± 11.97 [a]
	Roots	0.16 ± 0.21 [a]	26.62 ± 1.54 [a]	36.79 ± 13.32 [a]	11.01 ± 9.74 [a]	38.94 ± 15.85 [a]	26.27 ± 6.46 [a]
Grass/P1	Leaves	ND	11.32 ± 2.05 [b]	21.10 ± 2.36 [b]	0.25 ± 0.24 [b]	2.85 ± 2.69 [b]	4.27 ± 7.39 [c]
	Stems	3.07 ± 1.57 [b]	40.47 ± 25.75 [b]	55.50 ± 37.50 [b]	15.93 ± 13.81 [b]	47.30 ± 36.84 [b]	38.47 ± 6.12 [ab]
	Roots	8.24 ± 3.13 [a]	164.33 ± 50 [a]	193.7 ± 30.02 [a]	71.33 ± 22.08 [a]	181 ± 49 [a]	55.03 ± 8.09 [a]
Grass/P2	Leaves	ND	11.14 ± 2.75 [b]	21 ± 14.53 [b]	0.55 ± 0.08 [b]	3.31 ± 2.97 [b]	0.33 ± 0.57 [c]
	Stems	ND	12.03 ± 2.29 [b]	63.2 ± 17.34 [b]	0.08 ± 0.14 [b]	2.83 ± 2.72 [b]	15.86 ± 11.12 [bc]
WHO standards	Roots	1.32 ± 1.72 [b]	20.22 ± 24.25 [b]	29.97 ± 34.42 [b]	5.13 ± 8.89 [b]	21.07 ± 26.92 [b]	35.27 ± 16.64 [ab]
		0.02	1.30	10.00	10.00	2.00	0.60
				Dry Season			
I. aquatica	Leaves	ND	15.37 ± 1.27 [a]	21.60 ± 7.68 [a]	0.39 ± 0.29 [a]	9.66 ± 1.64 [a]	27.60 ± 24.00 [a]
	Stems	0.11 ± 0.15 [a]	16.68 ± 5.44 [a]	35.98 ± 17.04 [a]	0.65 ± 0.30 [a]	9.12 ± 2.55 [a]	8.25 ± 3.02 [a]
	Roots	1.12 ± 1.16 [a]	26.10 ± 5.94 [a]	49.23 ± 0.47 [a]	5.79 ± 3.13 [a]	34.66 ± 4.05 [a]	29.18 ± 24.45 [a]
Grass/P1	Leaves	0.96 ± 0.93 [a]	35.37 ± 4.65 [a]	218.7 ± 9.07 [a]	2.45 ± 0.85 [a]	16.27 ± 3.04 [a]	37.8 ± 23.19 [a]
	Stems	0.07 ± 0.13 [a]	7.52 ± 2.56 [c]	8.42 ± 0.5 [d]	0.40 ± 0.70 [a]	0.38 ± 0.66 [b]	8.52 ± 9.59 [a]
	Roots	2.07 ± 1.83 [a]	18.7 ± 4.06 [b]	48 ± 13.42 [c]	4.57 ± 1.91 [a]	8.91 ± 3.85 [ab]	33.1 ± 23.31 [a]
Grass/P2	Leaves	0.63 ± 0.60 [a]	42.47 ± 4.61 [a]	159 ± 6.24 [b]	1.9 ± 2.11 [a]	13.36 ± 7.58 [a]	22.9 ± 31.02 [a]
	Stems	0.09 ± 0.15 [a]	7.07 ± 2.74 [e]	7.55 ± 2.44 [d]	0.72 ± 1.25 [a]	0.25 ± 0.27 [b]	7.07 ± 12.24 [a]
WHO standards	Roots	1.05 ± 1.82 [a]	13.5 ± 2.98 [be]	39.4 ± 18.71 [c]	2.37 ± 2.64 [a]	5.37 ± 6.21 [ab]	17.7 ± 24.37 [a]
		0.02	1.30	10.00	10.00	2.00	0.60

Values are mean ± SD (standard error); WHO: World Health Organization; P1 and P2: sampling stations of *Pennisetum purpureum* in and outside the landfill, respectively; ND: not detected. Columns with similar superscript letters are not significantly different at $p < 0.05$. P1 and P2: locations of sampling in and outside the landfill, respectively.

4. Discussion

In this study, the surface water showed values lower than the standards (data not shown); therefore, it did not affect the environment. In contrast, the groundwater in the wet and dry seasons had high concentrations of Pb and Cr that exceeded the permissible values of ANESs [31] and WHO [32,33]. High doses of Pb and Cr detected in groundwater might be due to the leachate of HM contamination from MSW [34,35]. However, the amounts of Cd, Cu, Ni, and Zn in both seasons were lower than the permissible value of the regulations [31,32]. Cd and Zn in the groundwater were not detected, perhaps because of the low content of Cd and Zn in the MSW [36], which might be reduced by plant absorption [37].

In general, the dry season showed higher HM concentrations than the wet season. Probably, in the wet season, low-strength leachate was generated; however, during the dry season, the reduced percolation and enhanced evaporation might have increased the leachate strength [38]. Although the accumulation of Cd, Cu, Ni, and Zn was lower than Pb and Cr in the groundwater (Table 2), it was reported that the long-term oxidation of residual organic matters, together with sulfur, nitrogen, and iron in MSW, may lead to the release of greater amounts of HMs [39].

The Cd and Cu in the soil samples exceeded the levels of Dutch Pollutant Standards to reach an eco-toxicological risk [34]. The higher amount of Cd than the target value [34] in each station might result in broader contamination by leachate migration. Vegetation that absorbs an excessive amount of Cd and other HMs (Table 4) may adversely influence the health of people and animals living in the landfill and in its vicinity [40]. The tremendous amount of Cd and Cu may be related to the high quantities of these toxic metals in waste compositions that are disposed of in the landfill [41]. Kitchen, ash, plastic, and industrial wastes are the primary sources of metals in MSW landfills [42,43]. In contrast, the accumulation of Pb, Cr, Ni, and Zn in soil was found to be lower than the target value standards [34]. However, the quantities of these HMs may be elevated by long-term accumulation [39]. It was reported that HM concentrations were generally lower during the rainy season and higher in dry season [44]. In this study, the sampling stations were close to the leachate canals of the landfill, where the polluted water flowed to and contaminated the surface soil. In contrast, the samples from outside landfill had higher concentrations in the dry season [44]. HM contamination in soils depends on the dose in the MSW and environmental effects [45].

High doses of Cd, Cr, Pb, and Zn were detected in the edible parts (leaves and stems) of *I. aquatica*. Those HMs exceeded the permissible limit of WHO standards (5–86-fold). Although Cr, Cd, and Zn are acknowledged as essential elements to plants, higher concentrations of these HMs can be toxic [46–48]. The accumulation of Pb, Cd, and Cr in *I. aquatica* presents health problems for people living in the landfill as well as nearby villagers. This vegetable is commonly collected and sold in local markets and is also used to feed the residents' animals (pigs).

It was found that the grass *P. purpureum* absorbed a much higher quantity of Cd as well as other HMs than *I. aquatica* in the landfill (Table 4), perhaps due to the deeper roots of the grass grown in soils, which elevates its ability to absorb heavy metals [49]. Several plants, such as *Pistia stratiotes*, *Eichhornia crassipess*, *Hydrocotyle umbellatta*, *Lemna minor*, *Tyhpa latifolia*, *Scirpus acutus* [50], *Micranthemum ubrosum* [51], and *I. aquatica* [52], play a significant role for phytoremediation, as they are one of the best and cheapest cleanup technologies for contaminated soils, groundwater, and wastewater. However, the high contamination of HMs in *P. purpureum* apparently affected the health of cattle such as cows and buffalos, something which requires further investigation. People living near the MSW landfill in Vientiane may be exposed to HM toxicity at high levels, which would probably lead to health effects [8,53]. Although some metals are essential for biological systems in humans and animals, acting as structural and catalytic components of proteins and enzymes, in higher concentrations, they can be toxic [54,55]. The findings of this study highlight the environmental risks posed by HMs in the MSW landfill of Vientiane.

This study is the first to examine the HM contamination of water, soil, and plants in the MSW landfill in Vientiane, Laos. It provided detailed information on the polluted levels of HMs in underground water, soil, and vegetation in the MSW landfill. The most problematic HMs included Pb, Cr, Cd, Cu, and Zn (Tables 2–4) that may seriously affect the health of local residents near the MSW landfill and environment. Similar investigations of MSW landfills in other big cities of Laos, such as Savannakhet, Pakxe, and Thakhek, should be conducted. Analyzing how much HMs from MSW has penetrated into the environment is necessary for health protection in developing countries. In neighboring countries of Laos, such as Thailand and Vietnam, although the analysis of HMs has been well organized [56,57], MSW pollution has been unstoppable due to the mismanagement of waste classifications. The results of this study should be also submitted to the authorities in Vientiane, Laos, as they are relevant to environmental and health protection. Lawmakers should address the level of toxic HM penetration into the water, soil, and plants in the MSW landfill.

5. Conclusions

This study provided evidence of the serious pollution of the water, soils, and plants growing in the MSW landfill in Vientiane, Laos. Except for Ni, the polluted levels of Cd, Cr, Cu, Pb, and Zn exceeded the standards of ANESs, WHO, and Dutch Pollutant Standards, VROM (2000) to reach levels of eco-toxicological risk. The high accumulation of HMs in the edible plant parts of the widely consumed vegetable *I. aquatica* may cause a dilemma for the villagers living in and around the MSW landfill. The leachates by these HMs from MSW to the surrounding area may enlarge the extent of environmental pollution and health problems, thus requiring immediate phytoremediation and settlement. Frequent monitoring of the surface water, groundwater, and soil quality is necessary to determine the pollution levels and possibly initiate remedial measures. Education and legislation on landfill waste management must be academic and strict, from elementary school to university levels, throughout the country. In addition, the government should pay attention to improving landfill systems such as wastewater and leachate treatment systems. The MWS should be classified and analyzed to understand what sources in the MSW landfill principally caused the leaches of HMs in groundwater, soil, and vegetation. This may help to prevent the pollution by HMs from the MWS, therefore reducing the health problems of Laotian residents in the MSW landfill as well as in Vientiane.

Author Contributions: N.V. and T.D.X. wrote the paper as well as designing the experiments, collecting samples, and analyzing all of the water, soil, and plant samples. H.-D.T., T.D.K. and R.T. revised the manuscript.

Funding: The authors appreciate Japanese Grant Aid for Human Resource Development Scholarship (JDS) to provide Noudeng Vongdala a scholarship. Thanks are also due to Nguyen Tat Thanh University for a partial support to this research.

Acknowledgments: The authors thank the Analysis Center, Hiroshima University for their assistance in analyses of HMs.

Conflicts of Interest: The authors declare no conflict of interest.

References

1. United Nations Environment Programme (UNEP). Waste Management in ASEAN Countries. 2017. Available online: https://wedocs.unep.org/bitstream/handle/20.500.11822/21134/waste_mgt_asean_summary.pdf?sequence=1&isAllowed=y (accessed on 30 June 2018).
2. Bakis, R.; Tuncan, A. An investigation of heavy metal and migration through groundwater from the landfill area of Eskisehir in Turkey. *Environ. Monit. Assess.* **2011**, *176*, 87–98. [CrossRef] [PubMed]
3. Giusti, L. A review of waste management practices and their impact on human health. *Waste Manag.* **2009**, *29*, 2227–2239. [CrossRef] [PubMed]
4. Esakku, S.; Palanivelu, K.; Joseph, K. Assessment of heavy metals in a municipal solid waste dumpsite. In Proceedings of the Workshop on Sustainable Landfill Management, Chennai, India, 3–5 December 2003; Volume 35, pp. 139–145.
5. Kanmani, S.; Gandhimathi, R. Assessment of heavy metal contamination in soil due to leachate migration from an open dumping site. *Appl. Water Sci.* **2012**, *13*, 193–205. [CrossRef]

6. Slack, R.J.; Gronow, J.R.; Voulvoulis, N. Household hazardous waste in municipal landfills: Contaminants in leachate. *Sci. Total Environ.* **2005**, *337*, 119–137. [CrossRef] [PubMed]

7. Chary, N.S.; Kamala, C.; Raj, D.S. Assessing risk of heavy metals from consuming food grown on sewage irrigated soils and food chain transfer. *Ecotoxicol. Environ. Safety* **2008**, *69*, 513–524. [CrossRef] [PubMed]

8. Nica, D.V.; Bura, M.; Gergen, I.; Harmanescu, M.; Bordean, D. Bioaccumulative and conchological assessment of heavy metal transfer in a soil-plant-snail food chain. *Chem. Cent. J.* **2012**, *15*, 1–15. [CrossRef] [PubMed]

9. Jaishankar, M.; Tseten, T.; Anbalagan, N.; Mathew, B.B.; Beeregowda, K.N. Toxicity, mechanism and health effects of some heavy metals. *Interdisciplinary Toxicol.* **2014**, *13*, 60–72. [CrossRef] [PubMed]

10. Muddarisna, N.; Krisnayanti, B.D.; Utami, S.R.; Utami, E.; Handayanto, E. Phytoremediation of mecury-contaminated soil using three wild plant species and its effect on maize growth. *Applied Ecol. Environ. Sci.* **2013**, *1*, 27–32.

11. Arora, M.; Kiran, B.; Rani, S.; Rani, A.; Kaur, B.; Mittal, N. Heavy metal accumulation in vegetables irrigated with water from different sources. *Food Chem.* **2008**, *111*, 811–815. [CrossRef]

12. Chen, L.; Zhou, S.; Shi, Y.; Wang, C.; Li, B.; Li, Y.; Wu, S. Heavy metals in food crops, soil, and water in the Lihe River Watershed of the Taihu Region and their potential health risks when ingested. *Sci. Total Environ.* **2018**, *615*, 141–149. [CrossRef]

13. Mertz, W. The essential trace elements. *Science* **1981**, *213*, 1332–1338. [CrossRef] [PubMed]

14. Muchuweti, M.; Birkett, J.; Chinyanga, E.; Zvauya, R.; Scrimshaw, M.; Lester, J. Heavy metal content of vegetables irrigated with mixtures of wastewater and sewage sludge in Zimbabwe: Implications for human health. *Agric. Ecosyst. Environ.* **2006**, *112*, 41–48. [CrossRef]

15. Tchounwou, P.B.; Yedjou, C.G.; Patlolla, A.K.; Sutton, D.J. Heavy metals toxicity and the environment. In *Molecular, Clinical and Environmental Toxicology*; Springer: Basel, Switzerland, 2012; Volume 101, pp. 133–164.

16. Öman, C.B.; Junestedt, C. Chemical characterization of landfill leachates-400 parameters and compounds. *Waste Manag.* **2008**, *28*, 1876–1891. [CrossRef]

17. Clarke, B.O.; Anumol, T.; Barlaz, M.; Snyder, S.A. Investigating landfill leachate as a source of trace organic pollutants. *Chemosphere* **2015**, *127*, 269–275. [CrossRef] [PubMed]

18. Kabata-Pendias, A.; Pendias, H. *Trace Elements in Soils and Plants*, 4th ed.; CRC press: Boca Raton, FL, USA, 2010; Volume 548, pp. 93–118, INBN-13: 978-1-4200-9370.

19. Hinchman, R.R.; Negri, M.C.; Gatliff, E.G. Phytoremediation: Using green plants to clean up contaminated soil, groundwater and wastewater. In Proceedings of the International Topical Meeting on Nuclear and Hazardous Waste Management, Seattle, WA, USA, 18–23 August 1996; Volume 96, pp. 1–13.

20. Kamal, M.; Ghaly, A.E.; Mahmoud, N.; Cote, R. Phytoaccumulation of heavy metals by aquatic plants. *Environ. Int.* **2004**, *29*, 1029–1039. [CrossRef]

21. Kivaisi, A.K. The potential for constructed wetlands for wastewater treatment and reuse in developing countries: A review. *Ecol. Eng.* **2001**, *16*, 545–560. [CrossRef]

22. Madera-Parra, C.A.; Peña-Salamanca, E.J.; Peña, M.R.; Rousseau, D.P.; Lens, P.N. Phytoremediation of landfill leachate with *Colocasia esculenta, Gynerum sagittatumand*, and *Heliconia psittacorumin* constructed wetlands. *Int. J. Phytoremediat.* **2014**, *17*, 16–24. [CrossRef]

23. Climate and Clean Air Coalition Municipal Solid Waste Initiative (CCAC). Solid Waste Management City Profile, Vientiane Capital, LAO People's Democratic Republic. 2015. Available online: http://www.waste.ccacoalition.org/sites/default/files/files/vientiane_city_profile_vientiane_capital_lao.pdf (accessed on 30 June 2018).

24. Vodyanitskii, Y.N. Biochemical processes in soil and groundwater contaminated by leachates from municipal landfills (mini review). *Ann. Agrar. Sci.* **2016**, *14*, 249–256. [CrossRef]

25. Wuana, R.A.; Okieimen, F.E. Heavy metals in contaminated soils: A review of sources, chemistry, risks and best available strategies for remediation. *Isrn Ecol.* **2011**, *2011*. [CrossRef]

26. Ontario Ministry of the Environment and Climate Change (MOECC). Protocol for the Sampling and Analysis of Industrial/Municipal Wastewater. 2016. Available online: http://www.downloads.ene.gov.on.ca/envision/env_reg/er/documents/2016/011--7834_Protocol.pdf (accessed on 5 May 2018).

27. Government of Western Australia. Field Sampling Guideline: A Guideline for Field Sampling for Surface Water Quality Monitoring Programs. 2009. Available online: https://www.water.wa.gov.au/__data/assets/pdf_file/0020/2936/87154.pdf (accessed on 5 May 2018).

28. Gworek, B.; Dmuchowski, W.; Koda, E.; Marecka, M.; Baczewska, A.H.; Brągoszewska, P.; Osiński, P. Impact of the municipal solid waste Łubna landfill on environmental pollution by heavy metals. *Water* **2016**, *8*, 470. [CrossRef]

29. SHIMADZU. Environmental Analyses-Shimadzu Analysis Guidebook. 2005. Available online: https://applicationstation.ssi.shimadzu.com/sites/default/files/environmental-analyses-guidebook.pdf (accessed on 5 May 2018).

30. Andersen, K.J.; Kisser, M.I. *Digestion of Solid Matrices–Desk Study Horizontal*; Eurofins A/A: Kwai Chung, Denmark, 2004; Volume 59, pp. 25–33.

31. Ministry of Natural Resources and Environment, Laos, MONEL. *Agreement on National Environmental Standards of Laos (ANESs)*; MONREL: Vientiane, Laos, 2009.

32. WHO (World Health Organization). *Guidelines for Drinking-Water Quality*, 4th ed.; WHO: Geneva, Switzerland, 2011; Volume 554, pp. 327–433. ISBN 978-92-4-1548151.

33. WHO (World Health Organization). *Permissible Limits of Heavy Metals in Soil and Plants*; World Health Organization: Geneva, Switzerland, 1996.

34. Bird, G.; Brewer, P.A.; Macklin, M.G.; Balteanu, D.; Driga, B.; Serban, M.; Zaharia, S. The solid-state partitioning of contaminant metals and as in river channel sediments of the mining affected Tisa drainage basin, northwestern Romania and eastern Hungary. *Appl. Geochem.* **2003**, *18*, 1583–1595. [CrossRef]

35. Van Ryan Kristopher, R.G.; Parilla, R. Analysis of heavy metals in Cebu city sanitary landfill, Philippines. *J. Environ. Sci. Manag.* **2014**, *17*, 50–59.

36. Kar, D.; Sur, P.; Mandai, S.K.; Saha, T.; Kole, R.K. Assessment of heavy metal pollution in surface water. *Int. J. Environ. Sci. Tech.* **2008**, *5*, 119–124. [CrossRef]

37. Engin, M.S.; Uyanik, A.; Kutbay, H.G. Accumulation of heavy metals in water, sediments and wetland plants of Kizilirmak Delta (Samsun, Turkey). *Int. J. Phytoremediat.* **2015**, *17*, 66–75. [CrossRef] [PubMed]

38. Tatsi, A.A.; Zouboulis, A.I. A field investigation of the quantity and quality of leachate from a municipal solid waste landfill in a Mediterranean climate (Thessaloniki, Greece). *Adv. Environ. Res.* **2002**, *6*, 207–219. [CrossRef]

39. Kjeldsen, P.; Barlaz, M.A.; Rooker, A.P.; Baun, A.; Ledin, A.; Christensen, T.H. Present and long-term composition of MSW landfill leachate: A review. *Crit. Rev. Eniron. Sci. Tech.* **2002**, *41*, 297–336. [CrossRef]

40. Arao, T.; Ishikawa, S.; Murakami, M.; Abe, K.; Maejima, Y.; Makino, T. Heavy metal contamination of agricultural soil and countermeasures in Japan. *Paddy Water Environ.* **2010**, *8*, 247–257. [CrossRef]

41. Calace, N.; Liberatori, A.; Petronio, B.M.; Pietroletti, M. Characteristics of different molecular weight fractions of organic matter in landfill leachate and their role in soil sorption of heavy metals. *Environ. Pollut.* **2001**, *113*, 331–339. [CrossRef]

42. Awasthi, A.K.; Zeng, X.; Li, J. Environmental pollution of electronic waste recycling in India: A critical review. *Environ. Pollut.* **2016**, *211*, 259–270. [CrossRef]

43. Long, Y.Y.; Shen, D.S.; Wang, H.T.; Lu, W.J.; Zhao, Y. Heavy metal source analysis in municipal solid waste (MSW): Case study on Cu and Zn. *J. Hazard. Mater.* **2011**, *186*, 1082–1087. [CrossRef]

44. Olafisoye, O.B.; Adefioye, T.; Osibote, O.A. Heavy metals contamination of water, soil, and plants around an electronic waste dumpsite. *Environ. Study* **2013**, *22*, 1431–1439.

45. Samadder, S.R.; Prabhakar, R.; Khan, D.; Kishan, D.; Chauhan, M.S. Analysis of the contaminants released from municipal solid waste landfill site: A case study. *Sci. Total Environ.* **2017**, *580*, 593–601. [CrossRef] [PubMed]

46. Cakmak, I.; Marschner, H. Effect of zinc nutritional status on activities of superoxide radical and hydrogen peroxide scavenging enzymes in bean leaves. In *Plant Nutrition-from Genetic Engineering to Field Practice*; Springer: Dordrecht, The Netherlands, 1993; pp. 133–137.

47. Athar, M.; Vohora, S.B. *Heavy Metals and Environment*, 1st ed.; K.K. Gupta for New Age International: New Delhi, India, 1995; Volume 224, pp. 58–63. ISBN 81-224-0769-2.

48. Rai, U.N.; Sinha, S. Distribution of metals in aquatic edible plants: *Trapa natans* (Roxb.) Makino and *Ipomoea aquatica* Forsk. *Environ. Monit. Assess.* **2001**, *70*, 241–252. [CrossRef] [PubMed]

49. Durowoju, O.S.; Odiyo, J.O.; Ekosse, G.I.E. Variations of heavy metals from geothermal spring to surrounding soil and *Mangifera indica*–Siloam village, Limpopo province. *Sustainability* **2016**, *8*, 60. [CrossRef]

50. Farid, M.; Irshad, M.; Fawad, M.; Ali, Z.; Eneji, A.E.; Aurangzeb, N.; Ali, B. Effect of cyclic phytoremediation with different wetland plants on municipal wastewater. *Int. J. Phytoremediat.* **2014**, *16*, 572–581. [CrossRef] [PubMed]
51. Islam, M.S.; Ueno, Y.; Sikder, M.T.; Kurasaki, M. Phytofiltration of arsenic and cadmium from the water environmnt using *Micranthemum umbrosum* (JF Gmel) SF Blake as a hyperaccumulator. *Int. J. Phytoremediat.* **2013**, *15*, 1010–1021. [CrossRef]
52. Chanu, L.B.; Gupta, A. Phytoremediation of lead using *Ipomoea aquatica* Forsk. in hydroponic solution. *Chemosphere* **2016**, *156*, 407–411. [CrossRef]
53. Gbaruko, B.C.; Friday, O.V. Bioaccumulation of heavy metals in some fauna and flora. *Int. J. Environ. Sci. Tech.* **2007**, *6*, 197–202. [CrossRef]
54. Liu, J.; Cao, L.; Dou, S. Bioaccumulation of heavy metals and health risk assessment in three benthic bivalves along the coast of Laizhou Bay, China. *Mar. Pollut. Bull.* **2017**, *117*, 98–110. [CrossRef]
55. Nazir, R.; Khan, M.; Masab, M.; Rehman, H.U.; Rauf, N.U.; Shahab, S.; Shaheen, Z. Accumulation of heavy metals (Ni, Cu, Cd, Cr, Pb, Zn, Fe) in the soil, water and plants and analysis of physico-chemical parameters of soil and water collected from Tanda Dam Kohat. *J. Pharm. Sci. Res.* **2015**, *7*, 89–97.
56. Schneider, P.; Anh, L.H.; Wagner, J.; Reichenbach, J.; Hebner, A. Solid waste management in Ho Chi Minh City, Vietnam: Moving towards a circular economy? *Sustainability* **2017**, *9*, 286. [CrossRef]
57. Yukalang, N.; Clarke, B.; Ross, K. Barriers to effective municipal solid waste management in a rapidly urbanizing area in Thailand. *Int. J. Environ. Res. Public Health* **2017**, *14*, 1013. [CrossRef] [PubMed]

International Journal of
*Environmental Research
and Public Health*

Article

Effects of Hydraulic Loading Rate on Nutrients Removal from Anaerobically Digested Swine Wastewater by Multi Soil Layering Treatment Bioreactor

Junyuan Guo *, Yuling Zhou, Yijin Yang, Cheng Chen and Jiajing Xu

College of Resources and Environment, Chengdu University of Information Technology, Chengdu 610225, China; yuzijiang626@163.com (Y.Z.); jinyiyang@163.com (Y.Y.); 3160202006@cuit.edu.cn (C.C.); xujiajingkk@163.com (J.X.)
* Correspondence: gjy@cuit.edu.cn; Tel./Fax: +86-288-596-6913

Received: 31 August 2018; Accepted: 12 October 2018; Published: 29 November 2018

Abstract: A multi soil layering (MSL) treatment bioreactor was developed aiming at nutrients removal from anaerobically digested swine wastewater (ADSW). The start-up of the MSL bioreactor and its performance in nutrients removal at different hydraulic loading rate (HLR) were investigated. Results showed that the MSL bioreactor was successfully started up after operation for 28 days, and at this time, the removal efficiencies of ammonia-N, total nitrogen (TN) and total phosphorus (TP) in the ADSW reached 63.6%, 58.5%, and 46.5%, respectively. The MSL bioreactor showed a stable performance during the whole working process with varying HLR from 80 to 200 L/(m^2·day). Maximum removal efficiencies of ammonia-N, TN and TP were obtained at 160 L/(m^2·day), and was appeared as 94.2%, 94.4%, and 92.5%, respectively. It was worth noting that iron scraps were the key factor that enhanced the independent capability of the MSL bioreactor in TP removal, because there was only 21.4–25.8% of the TP was removed when the MSL bioreactor run with no iron addition.

Keywords: multi soil layering treatment (MSL); anaerobically digested swine wastewater (ADSW); ammonia-N removal; total nitrogen (TN) removal; total phosphorus (TP) removal

---

1. Introduction

Nowadays, swine wastewater was considered as one of the biggest culprits for the severe agricultural non-point pollution, because of its high concentration of ammonia, organic pollutants, and phosphorus were not managed properly [1,2]. Although anaerobic digestion was thoroughly investigated and applied technology in the treatment of the swine wastewater all over the world, residual nutrients were still considerable in the liquid named "ADSW" (anaerobically digested swine wastewater) [3]. In recent years, land treatment systems were becoming more and more popular in the treatment of the ADSW [4].

The MSL (multi soil layering) bioreactor was known as an effective soil-based technology in the conventional poorly functioning sewage treatment via the enhancing inherent ability of soil [5]. This system has been tested to remove pollutants from polluted river water [6] and domestic wastewater [7], but there was no research related to nutrients removal from swine wastewater. Why can nutrients be effectively removed by the MSL bioreactor from ADSW? As known, the MSL bioreactor was a biofilm reactor that filled with two kinds of media—one was the soil mixture block (SMB) composed of soil and some organic materials, such as sawdust and charcoal, with a ratio of 85:15 by dry weight, and another was the zeolite and iron layers. In case of natural ventilation or manual intervention ventilation, aerobic environments were formed in the zeolite layers due to its porous structures, when the ADSW flowed through the MSL bioreactor, ammonia-N was adsorbed by the zeolite and the nitrifying bacteria were then grew fast. At the same time, anoxic environments were formed in the SMB after who was soaked within the ADSW, and the denitrifying bacteria were gradually grown [7]. Therefrom, ammonia-N was transformed to NO_3^- first in the aerobic environments and then to gaseous nitrogen in the anoxic environments [8]. In the transformation process of NO_3^- to gaseous nitrogen in the anoxic environments in SMB, the organic materials (e.g., sawdust and charcoal) can be used as extra carbon sources for denitrifying bacteria [9]. The role of the iron that was added in the MSL bioreactor was enhancing TP (total phosphorus) removal by the chemical reaction between Fe^{3+} and PO_4^{3-} [10]. It can clearly be seen that the MSL bioreactor will be an option in the removal of nutrients from the ADSW.

Aims of this study were (1) to construct a MSL bioreactor to remove nutrients simultaneously from the ADSW, (2) to investigate the effects of HLR (hydraulic loading rate) on the removal of nutrients, and (3) to discuss the mechanisms that the MSL bioreactor can remove nutrients during its working period.

2. Materials and Methods

2.1. Swine Wastewater

Swine wastewater for detecting the performance of MSL bioreactor in nutrients removal was obtained from a pig farm at Chengdu city, Sichuan Province, China. In this pig farm, the joint of Up-flow Anaerobic Sludge Bed (USAB) and Sequencing Batch Reactor (SBR) was applied to treat swine wastewater. The swine wastewater mainly including pig manure, pig urine, and pig house flushing water. Concentrations of COD (chemical oxygen demand), ammonia-N,TN (total nitrogen) and TP of the swine wastewater were 5683.4, 645.8, 752.2, and 26.5 mg/L, respectively, which appeared as 1130.5, 682.6, 761.8, and 22.8 mg/L, respectively after treated by the UASB. The pH values of the swine wastewater and the correspondingly ADSW were almost the same, 6.7 and 6.8, respectively.

2.2. Experimental Apparatus

According to our previous study, the experimental MSL bioreactor was constructed by using a lidless acrylic box filled with different media [11]. Dimensions the MSL bioreactor and the layout of the filter media were showed in Figure 1. The acrylic box was measured as 450 mm × 250 mm × 700 mm, whose bottom was drilled with an aperture area rate of 28%. The aperture was 0.8 mm. From the bottom of the acrylic box, a pebble layer was firstly filled, and then the alternately filling of zeolite layers, iron layers, and SMB layers. Numbers of pebble layers, zeolite layers, iron layers, and soil layers were 1, 4, 3, and 3, respectively. Natural zeolite was selected to build the zeolite layers, and the SMB with a size of 220 mm × 110 mm × 80 mm was the mixture of clayey soil and sawdust (with a ratio of 85:15 by dry weight). The iron scraps were lathe iron cutting scraps obtained from the metal technology practice base of Chengdu University of Information Technology. After being treated by the MSL bioreactor, the water was allowed to naturally flow into the collection tank through the aperture bottom.

2.3. ADSW Treatment Process

To start up the MSL bioreactor, firstly, the mixture of the ADSW and the sludge from the UASB of the pig farm, with a volume ratio of 2:1, was pumped and dispersed into the filter media to seed the MSL bioreactor. After one week, the mixture of the ADSW and the sludge was replaced by the same volume of ADSW alone to continuously seed the MSL bioreactor. During the start-up process, the influent HLR was kept at approximately 50 L/(m^2·day). After start up the MSL bioreactor successfully, effects of the HLR on the MSL bioreactor in nutrients removal were investigated. During the experiment, the mean HLR was set at 80, 120, 160, and 200 L/(m^2·day) in steps, and at each HLR condition, after 7 day's stable run of the MSL bioreactor at 50 L/(m^2·day), the HLR was began changed and water samples were continuously taken and analyzed for 14 days. In addition, the effects of inlet ammonia-N concentration on ammonia-N removal by the MSL were investigated as well.

2.4. Analytic Methods

Analysis methods of COD, ammonia-N, NO_3^-, TN, and TP were summarized as follows: COD using the potassium dichromate method, ammonia-N using the Nessler's reagent colorimetric method, NO_3^- by the ultraviolet spectrophotometry method, TN by the potassium persulfate oxidation-ultraviolet spectrophotometry method, and TP using the potassium persulfate digestion colorimetric method [12]. Water pH was measured using pH meter (PHS-3C). The microstructure of the zeolites before and after start-up of the MSL bioreactor was characterized with environmental scanning electron microscopy (Quanta 200, FEI Ltd., Eindhoven, The Netherlands).

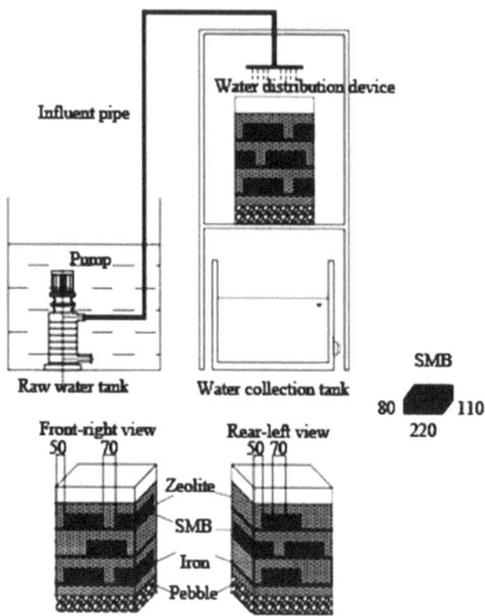

Figure 1. Experimental device and detailed structure of the MSL (multi soil layering treatment) bioreactor.

3. Results and Discussion

3.1. Start-Up of the MSL

As shown in Figure 2a, before start-up of the MSL bioreactor, the raw zeolite in the MSL showed an irregular void structure, which was beneficial to the growth of microorganisms and formed the biofilm. After seven days of start-up, a change of the zeolite color from grayish white to brown was discovered by the naked eye, and suggested that there were microorganisms growing on the surface of the zeolite. As shown in Figure 2b, after 14 days of start-up, a velvet biofilm that had been apparently growing on the surface of the zeolite indicated an efficient biofilm formation in the MSL bioreactor, which can adsorb and degrade ammonia-N and organic pollutants in ADSW [7].

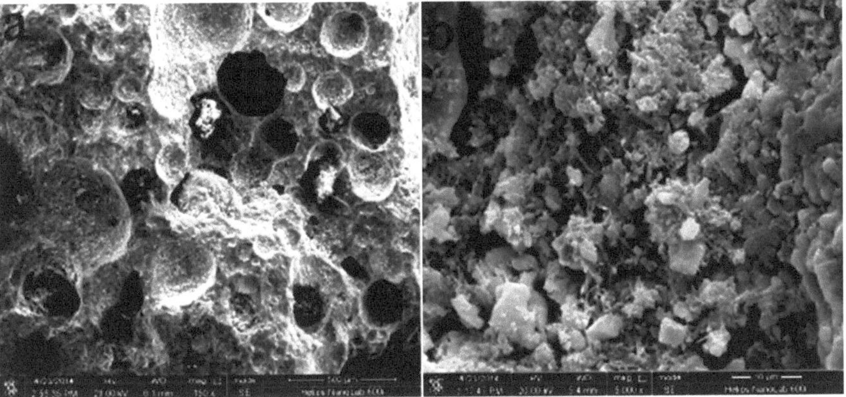

Figure 2. SEM images of the zeolites in the MSL (multi soil layering treatment) bioreactor before start-up (**a**) and after 14 days of start-up (**b**).

From the 15th day onwards, the water qualities in and out the MSL were continuously collected and monitored, which were shown in Figure 3. The removal efficiencies of COD, ammonia-N, TN, and TP was increased from 21.2%, 33.5%, 29.8%, and 20.5% at 15 days to 48.8%, 63.6%, 58.5%, and 46.5% at 28 days, respectively, along with the prolonged time. This is due to the continuous ingestion of organic matter, nitrogen and phosphorus in the ADSW by microorganisms in the MSL bioreactor [11].

At 28–35 days, the removal efficiencies of COD, ammonia-N, TN, and TP can reach 50.5%, 65.4%, 60.7%, and 48.5%, respectively (the corresponding effluent concentrations were lower than 559.6, 236.2, 299.4, and 11.7 mg/L, respectively). The relative deviations of the two monitoring results of COD, ammonia-N, TN, and TP in the effluent were less than 5%, indicated that the MSL bioreactor reached stable operation. This state further illustrated that the biofilm was successfully formed in the MSL bioreactor [11].

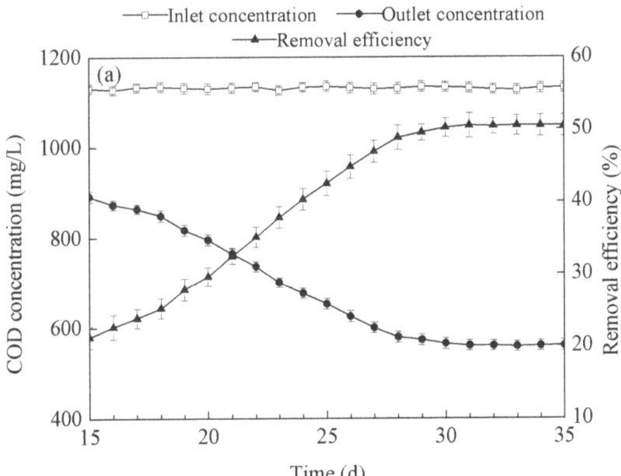

Figure 3. *Cont.*

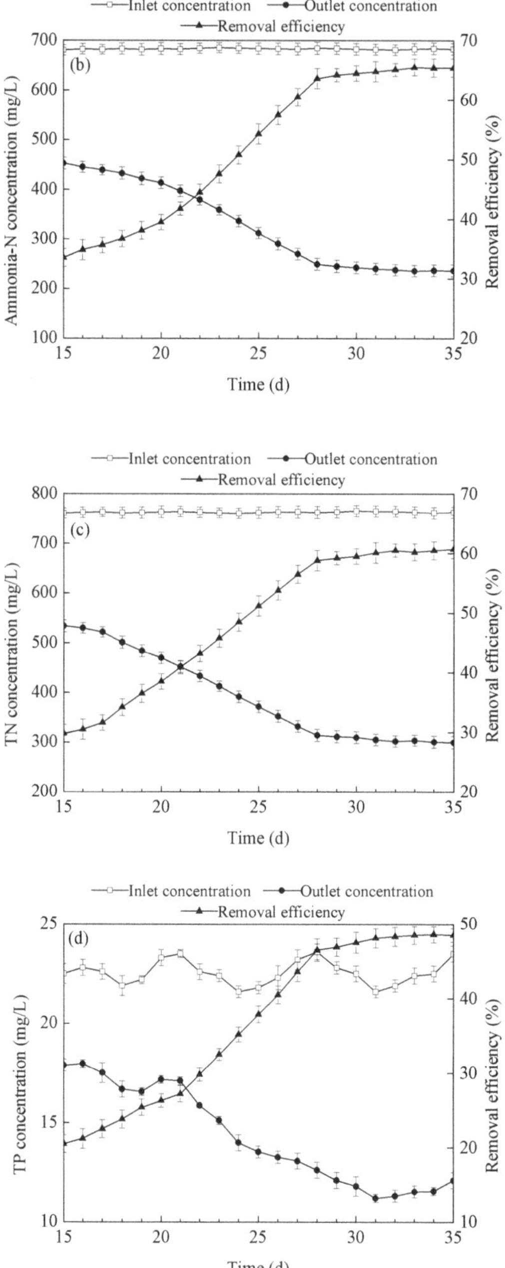

Figure 3. Concentrations and removal efficiencies of COD (chemical oxygen demand) (**a**), ammonia-N (**b**), TN (total nitrogen) (**c**) and TP (total phosphorus) (**d**) during the start-up period of the MSL bioreactor.

3.2. Effects of Hydraulic Loading Rate (HLR) on Nutrients Removal

3.2.1. Phosphorus Removal

As shown in Figure 4a, under different HLR from 80 to 200 L/(m^2·day), 87.6–94.4% of TP on average can be removed from the ADSW, with the final effluent TP concentration kept in the range of 1.3–2.8 mg/L, meaning that the TP in the ADSW can be effectively removed by the MSL bioreactor. The literature has reported that the TP removal in the MSL bioreactor was mainly due to the chemical reaction between Fe^{3+} and PO_4^{3-}, and the adsorption by $Fe(OH)_3$ [7,11].

$$Fe - 2e^- \rightarrow Fe^{2+} \tag{1}$$

$$Fe^{2+} - e^- \rightarrow Fe^{3+} \tag{2}$$

$$O_2 + 2H_2O + 4e^- \rightarrow 4OH^- \tag{3}$$

$$Fe^{3+} + 3OH^- \rightarrow Fe(OH)_3 \qquad K^{\theta}_{sp} = 4.0 \times 10^{-38} \tag{4}$$

$$Fe^{3+} + PO_4^{3-} \rightarrow FePO_4 \qquad K^{\theta}_{sp} = 1.3 \times 10^{-22} \tag{5}$$

As shown in Equations (1) and (2), in aerobic environment, iron was transformed into Fe^{2+}, and then to Fe^{3+}, which aids in TP removal through forming $FePO_4$ precipitate, at the same time, the reaction $O_2 + 2H_2O + 4e^- \rightarrow 4OH^-$ (Equation (3)) occurred, a small amount of $Fe(OH)_3$ was formed, which can enhance TP removal through coagulation. TP was mainly removed through these two ways.

According to the equilibriums (Equations (4) and (5)), $K_{sp}{}^{\theta}$ of the $FePO_4$ precipitation equilibrium was higher than Fe^{3+} hydrolysis equilibrium, thus, Fe^{3+} reacted with PO_4^{3-} prior to OH^-, thus far, chemical reaction between Fe^{3+} and PO_4^{3-} was the key way for TP removal [13].

In order to certify the important role of iron in TP removal, a control MSL bioreactor that was run without an iron layer in the same procedure. As shown in Figure 4b, a relatively poor TP removal efficiency of about 21.4–25.8% was obtained at different HLR, which further illustrated that the TP was mainly removed due to the chemical reaction between Fe^{3+} and PO_4^{3-}, and the precipitate would subsequently be adsorbed and/or intercepted by the filter media in the hybrid system. A similar conclusion was reported by Sato et al. (2005) [10].

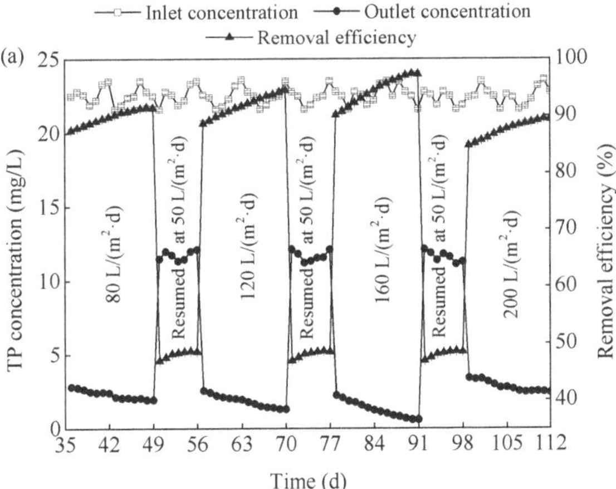

Figure 4. *Cont.*

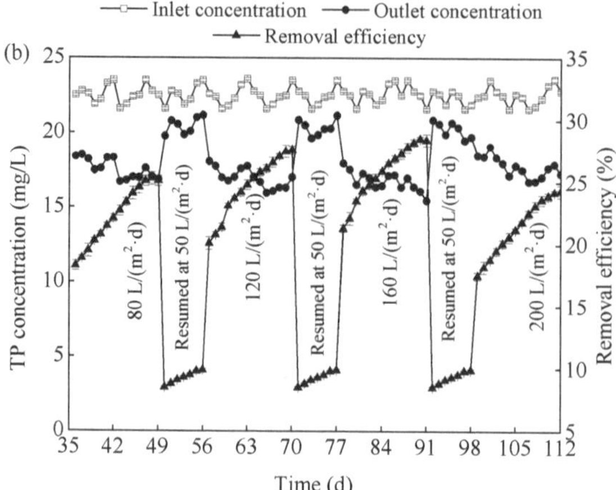

Figure 4. Effects of HLR (hydraulic loading rate) on TP (total phosphorus) removal by the MSL (multi soil layering) with (**a**) and without (**b**) Fe addition.

3.2.2. Ammonia-N Removal

Figure 5 clearly showed that the MSL bioreactor can remove 92.6% of ammonia-N from the ADSW at HLR of 120 L/(m^2·day), and 94.2% at 160 L/(m^2·day), whereas it was decreased to 88.1% when the HLR was adjusted to 200 L/(m^2·day). The main reason for the reduction of ammonia-N removal efficiency was that the nitrifying bacteria were at disadvantage in the fierce competition (by the HLR up to 200 L/(m^2·day)) of living space with heterotrophic bacteria [14]. Apart from this, HLR as high as 200 L/(m^2·day) shortened the hydraulic retention time (HRT) of the wastewater, which decreased the ammonia-N adsorption by zeolite and the nitrification by biofilm in the MSL bioreactor, and further decreased ammonia-N removal from the ADSW. A similar conclusion was reported by Luo et al. (2014) [7].

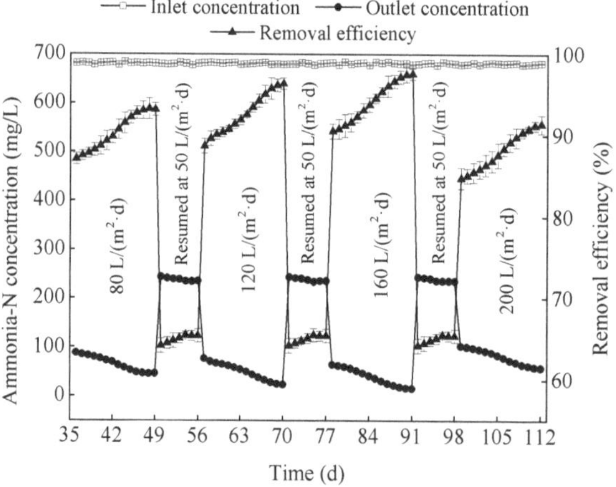

Figure 5. Effects of HLR (hydraulic loading rate) on ammonia-N removal by the MSL (multi soil layering).

As shown in Figure 6, the trend of NO$_3^-$ production rate was similar with that of ammonia-N removal rate at different HLR in the MSL bioreactor, both increased with the increasing HLR from 80 to 160 L/(m^2·day), indicating the occurrence of nitrification during the ammonia-N removal process.

In fact, under the HLR condition of 80, 120 and 160 L/(m^2·day), ammonia-N removal rate was appeared as 49.2, 75.7, and 102.7 g NH$^+_4$-N/(m^2·day), respectively, and the correspondingly NO$_3^-$ production rate was showed as 16.8, 27.9, and 37.8 g NO$_3^-$-N/(m^2·day), respectively. However, both of ammonia-N removal rate and NO$_3^-$ production rate were slightly decreased when the HLR was continue increased to 200 L/(m^2·day), and they were shown as 94.2 g NH$^+_4$-N/(m^2·day) and 34.8 g NO$_3^-$-N/(m^2·day), respectively. Some other research reported quite different information that the nitrification rate was dropped obviously with the excessive increasing HLR [15,16]. On the one hand, due to the large specific surface area and ion exchange function of the zeolite in the MSL bioreactor, ammonia-N can be adsorbed more effectively than the traditional materials. On the other hand, the air layer was allowed to exist around the biofilm due to the application of media with a relatively large particle size in the MSL bioreactor, which promoted the air flow and oxygen diffusion to the biofilm, and further accelerated the mass transfer and degradation of ammonia-N [8].

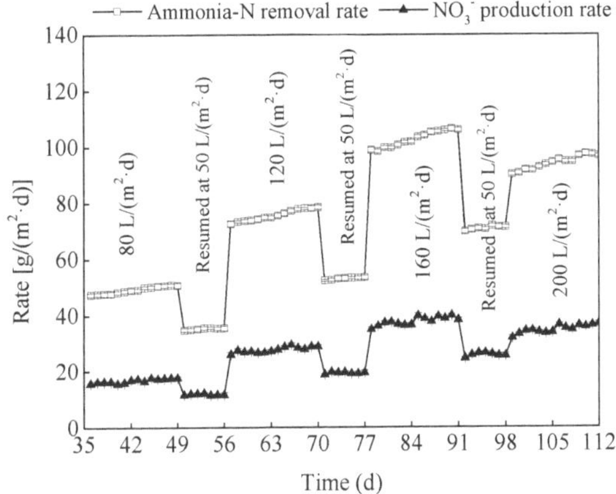

Figure 6. Ammonia-N removal and NO$_3^-$ production rates under different HLR (hydraulic loading rate) in the MSL (multi soil layering).

3.2.3. TN Removal

Figure 7 showed that about 52.5–80.4% of TN was removed from the ADSW by the MSL bioreactor at different HLR, better than that by the zeolite biofilter [17], indicated that the SMB was beneficial for denitrification and thus improved TN removal from the ADSW. Denitrification was proofed by the researches on biofilm reactor as the dominant mechanism that NO$_3^-$ was converted to nitrogen gases to escape from wastewater [13].

In fact, TN removal efficiency reached 85.9%, 89.2%, and 92.5% on average at HLR of 80, 120, and 160 L/(m^2·day), respectively, whereas 83.6% was obtained when the HLR was up to 200 L/(m^2·day). At low HLR, the nitrogen and organic loading was low, the ammonia-N was nitrified into NO$_3^-$ and then drained out of the MSL bioreactor fast, which led to poor TN removal. This was due to the lack of carbon sources that the NO$_3^-$ could not be denitrified in time [11]. With the increasing HLR from 80 to 160 L/(m^2·day), the denitrification rate was improved, and the TN removal efficiency was significantly increased. However, the MSL bioreactor did not perform well in TN removal from the ADSW at the excessive HLR of 200 L/(m^2·day), due to the short HRT and the quickly shedding biofilm [18]. From the effects of HLR on TN removal, it seems that some improvements can still be made to enhance the performance of the MSL bioreactor in denitrification. Schipper et al. (2010) reported two ways to enhance the denitrification in the MSL bioreactor, one was the continuous supply of carbon source to the denitrifying bacteria, and the other was the maintenance of adequate saturated hydraulic conductivity [19]. So, alternatives for more effective denitrification of the MSL bioreactor may include: adding effective solid carbon sources into the SMBs, decreasing the particle size of the filter media in the MSL to hold more water.

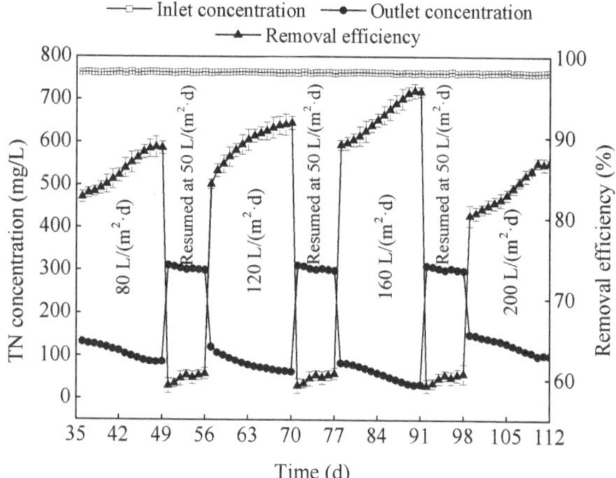

Figure 7. Effects of HLR (hydraulic loading rate) on TN (total nitrogen) removal by the MSL (multi soil layering).

4. Conclusions

A MSL bioreactor was constructed to remove nutrients from the ADSW, whose performance in nutrients removal was strongly affected by the HLR. Removal efficiencies of ammonia-N, TN and TP was increased with the increasing HLR from 80 to 160 L/(m²·day), and reached the peak values of 94.2%, 94.4%, and 92.5%, respectively, whereas the excessive HLR reduced the independent capability of the MSL bioreactor in ADSW treatment. The increasing HLR accelerated the water flow rate in the MSL, thereby accelerated the mass transfer process between the liquid phase and the biofilm phase, which can improve the degradation of pollutants, while the excessive HLR shortened the HRT and led to the quickly shedding biofilm, which limited the pollutants removal. In the ADSW treatment process, iron scraps played an important role in the TP removal, and there was only 21.4–25.8% of the TP was removed when the MSL was run with no iron addition. All in all, the MSL was effective in nutrients removal from the ADSW under different HLR.

Author Contributions: Formal analysis, Y.Y.; investigation, J.G., Y.Z., and J.X.; project administration, J.G.; supervision, J.G.; writing—review and editing, C.C.

Funding: This research was funded by the National Natural Science Foundation of China (Grant No. 51508043), Basic Projects of Science and Technology Department of Sichuan Provincial (2016JY0015; 2014JY0245), and Projects (J201515) supported by CUIT.

Conflicts of Interest: The authors declare no conflict of interest.

References

1. Vanotti, M.B.; Dube, P.J.; Szogi, A.A.; García-gonzález, M.C. Recovery of ammonia and phosphate minerals from swine wastewater using gas-permeable membranes. *Water Res.* **2017**, *112*, 137–146. [CrossRef] [PubMed]
2. Wen, S.; Liu, H.Y.; He, H.J.; Luo, L.; Li, X.; Zeng, G.M.; Zhou, Z.L.; Lou, W.; Yang, C.P. Treatment of anaerobically digested swine wastewater by Rhodobacter blasticus and Rhodobacter capsulatus. *Bioresour. Technol.* **2016**, *222*, 33–38. [CrossRef] [PubMed]
3. Luo, L.; He, H.J.; Yang, C.P.; Wen, S.; Zeng, G.M.; Wu, M.J.; Zhou, Z.L.; Lou, W. Nutrient removal and lipid production by Coelastrella sp. in anaerobically and aerobically treated swine wastewater. *Bioresour. Technol.* **2016**, *216*, 135–141. [CrossRef] [PubMed]

4. Zhao, B.; Li, J.; Leu, S.Y. An innovative wood-chip-framework soil infiltrator for treating anaerobic digested swine wastewater and analysis of the microbial community. *Bioresour. Technol.* **2014**, *173*, 384–391. [CrossRef] [PubMed]

5. Sato, K.; Iwashima, N.; Wakatsuki, T.; Masunaga, T. Quantitative evaluation of treatment processes and mechanisms of organic matter, phosphorus, and nitrogen removal in a multi-soil-layering system. *Soil Sci. Plant Nutr.* **2011**, *57*, 475–486. [CrossRef]

6. Masunaga, T.; Sato, K.; Mori, J.; Shirahama, M.; Kudo, H.; Wakatsuki, T. Characteristics of wastewater treatment using a multi-soil-layering system in relation to wastewater contamination levels and hydraulic loading rates. *Soil Sci. Plant Nutr.* **2007**, *53*, 215–223. [CrossRef]

7. Luo, W.; Yang, C.P.; He, H.J.; Zeng, G.M.; Yan, S.; Cheng, Y. Novel two-stage vertical flow biofilter system for efficient treatment of decentralized domestic wastewater. *Ecol. Eng.* **2014**, *64*, 415–423. [CrossRef]

8. Guan, Y.D.; Chen, X.; Zhang, S.; Luo, A.C. Performance of multi-soil-layering system (MSL) treating leachate from rural unsanitary landfills. *Sci. Total Environ.* **2012**, *420*, 183–190.

9. Wang, L.; Zheng, Z.; Luo, X.; Zhang, J. Performance and mechanisms of a microbial-earthworm eco filter for removing organic matter and nitrogen from synthetic domestic wastewater. *J. Hazard. Mater.* **2011**, *195*, 245–253. [CrossRef] [PubMed]

10. Sato, K.; Masunaga, T.; Wakatsuki, T. Water movement characteristics in a multi-soil-layering system. *Soil Sci. Plant Nutr.* **2005**, *51*, 75–82. [CrossRef]

11. Guo, J.Y.; Zhou, Y.L.; Jiang, S.L.; Chen, C. Feasibility investigation of a multi soil layering bioreactor for domestic wastewater treatment. *Environ. Technol.* **2018**. [CrossRef] [PubMed]

12. American Public Health Association; American Water Works Association; Water Environment Federation. *Standard Methods for the Examination of Water and Wastewater*, 20th ed.; American Public Health Association: Washington, DC, USA, 1998.

13. Zhang, Y.; Cheng, Y.; Yang, C.P.; Luo, W.; Zeng, G.M.; Lu, L. Performance of system consisting of vertical flow trickling filter and horizontal flow multi-soil-layering reactor for treatment of rural wastewater. *Bioresour. Technol.* **2015**, *193*, 424–432. [CrossRef] [PubMed]

14. Krasnits, E.; Beliavsky, M.; Tarre, S.; Green, M. PHA based denitrification: Municipal wastewater vs. acetate. *Bioresour. Technol.* **2013**, *132*, 28–37. [CrossRef] [PubMed]

15. Hu, B.; Wheatley, A.; Ishtchenko, V.; Huddersman, K. The effect of shock loads on SAF bioreactors for sewage treatment works. *Chem. Eng. J.* **2011**, *166*, 73–80. [CrossRef]

16. Van den Akker, B.; Holmes, M.; Pearce, P.; Cromar, N.J.; Fallowfield, H.J. Structure of nitrifying biofilms in a high-rate trickling filter designed for potable water pre-treatment. *Water Res.* **2011**, *45*, 3489–3498. [CrossRef] [PubMed]

17. Guo, J.Y.; Zhou, M.J.; Gan, P.F.; Tan, X.D.; Guo, Z.H.; Fu, L.; Huang, W.Y.; Bai, X. Performance of zeolite trickling filter in treatment of domestic wastewater and characteristics of the biofilm. *China Environ. Sci.* **2016**, *36*, 3601–3609.

18. Healy, M.G.; Rodgers, M.; Mulqueen, J. Denitrification of a nitrate-rich synthetic wastewater using various wood-based media materials. *J. Environ. Sci. Health Part A* **2006**, *41*, 779–788. [CrossRef] [PubMed]

19. Schipper, L.A.; Robertson, W.D.; Gold, A.J.; Jaynes, D.B.; Cameron, S.C. Denitrifying bioreactors—An approach for reducing nitrate loads to receiving waters. *Ecol. Eng.* **2010**, *36*, 1532–1543. [CrossRef]

International Journal of
Environmental Research and Public Health

Article

Enhanced Removal of Antibiotic in Wastewater Using Liquid Nitrogen-Treated Carbon Material: Material Properties and Removal Mechanisms

Yaohui Wu [1,2], Wen Liu [1], Yonghong Wang [1], Xinjiang Hu [1], Zhengping He [1], Xiaoyong Chen [1,3,*] and Yunlin Zhao [1,2,*]

[1]   College of Life Science and Technology, Central South University of Forestry and Technology, Changsha 410004, China; wyh752100@163.com (Y.W.); 15197704556@163.com (W.L.); bionano@163.com (Y.W.); xjhu@csuft.edu.cn (X.H.); 15673775866@163.com (Z.H.)
[2]   Hunan Research Center of Engineering Technology for Utilization of Environmental and Resources Plant, Central South University of Forestry and Technology, Changsha 410004, China
[3]   College of Arts and Sciences, Governors State University, University Park, Illinois, IL 60484, USA
*   Correspondence: ChenXyXyXyXy@163.com (X.C.); zyl8291290@163.com (Y.Z.)

Received: 21 October 2018; Accepted: 21 November 2018; Published: 26 November 2018

Abstract: Antibiotic residues in the aquatic environment have become a global problem posing a serious threat to the environment and an inherent health risk to human beings. In this study, experiments were carried to investigate the use of carbon material modified by liquid nitrogen treatment (CM1) and carbon material unmodified by liquid nitrogen treatment (CM2) as adsorbents for the removal of the antibiotic ampicillin from aqueous solutions. The properties of the CMs (CM1 and CM2) and the effects of variations of the key operating parameters on the removal process were examined, and kinetic, isothermal and thermodynamic experimental data were studied. The results showed that CM1 had larger specific surface area and pore size than CM2. The ampicillin adsorption was more effective on CM1 than that on CM2, and the maximum adsorption capacity of ampicillin onto CM1 and CM2 was 206.002 and 178.423 mg/g, respectively. The kinetic data revealed that the pesudo-second order model was more suitable for the fitting of the experimental kinetic data and the isothermal data indicated that the Langmuir model was successfully correlated with the data. The adsorption of ampicillin was a spontaneous reaction dominated by thermodynamics. In synthetic wastewater, CM1 and CM2 showed different removal rates for ampicillin: 92.31% and 86.56%, respectively. For an adsorption-based approach, carbon material obtained by the liquid nitrogen treatment method has a stronger adsorption capacity, faster adsorption, and was non-toxic, therefore, it could be a promising adsorbent, with promising prospects in environmental pollution remediation applications.

Keywords: liquid nitrogen; modified carbon; ampicillin; adsorption

---

1. Introduction

In recent years, antibiotic residues in the aquatic environment have become a persistent, bioaccumulation problem, potentially toxic in the medium to long term. China has a large consumption of, estimated to account to 1/4 of the total amount of antibiotics consumed globally every year [1]. Unfortunately however, due to poor adsorption in the gut of the animals, elimination of the antibiotics occurs and the majority are excreted unchanged in feces and urine. In addition, some antibiotics are dropped directly into the water by fish farmers, leading to masses of antibiotics present as the unmodified parent compound [2]. In the past years, it was recognized that antibiotics represent a new source of water contaminants [3]. Antibiotics have been frequently detected in wastewater treatment plants (WWTPs) and in rivers [4]. β-Lactams, as a low cost class of antibiotics with broad-spectrum applicability are commonly used around the world [5], and have already caused some environmental problems. The occurrence of β-lactam antibiotic-resistant bacteria in surface water, animal waste water and poultry farms was already detected [6]. Moreover, β-lactam antibiotic-resistant bacteria are also found in the bodies of organisms, such as in humans [7] and fishes [8]. Considering the treat that residual β-lactam antibiotics pose to the environment and the risk to human beings [9], the development of efficient removal techniques is becoming a matter of significance to be addressed.

A number of methods were reported in the literature for the removal of β-lactam antibiotics from water. Arslan-Alaton and Caglayan demonstrated that ozone produced by an advanced oxidation method had positive effects on penicillin removal from formulation effluent [10]. Giraldo-Aguirre and collaborators evidenced that the β-lactam antibiotics could be removed efficiently from pharmaceutical industry wastewater by a photo-Fenton method at near neutral pH [11]. Biodegradation can be used as a common method to degrade β-lactam antibiotics, for example, degrading ampicillin in aqueous solutions using biofilms [12]. Limousy et al. efficiently eliminated amoxicillin from aqueous solutions with an adsorption method, and the elimination rate was shiwn to be as high as 93% [13]. Among all these methods, adsorption represents an ideal method due to its easy operation, high efficiency, and lack of formation of toxic byproducts [14]. Carbon is widely used as adsorbent, and its adsorption capacity could be contributed to two major factors: surface groups and specific surface area. Previous studies showed that the adsorptivity of carbon was closely related to its surface groups [15]. In order to increase the adsorption capacity, many methods have been used to produce various activated carbons by changing the surface groups of the carbon. Moussavi et al. [16] compared the adsorption of β-lactam antibiotics on NH_4Cl treated activated carbon (NAC) and standard activated carbon (SAC). Over 99% of 50 mg/L β-lactam antibiotics were adsorbed using NAC at 0.4 g/L, while under similar experimental conditions only about 55% of β-lactam antibiotics was adsorbed by SAC. Although there have been some achievements in this area, many shortcomings still exist, such as the use of non-environmentally friendly denaturants (H_2SO_4, NH_4Cl, $(NH_4)_2S_2O_8$), complex processes and high cost. Also it was well known that the adsorption properties of carbons depended heavily on their specific surface area. Due to its large surface specific area and uniform aperture, mesoporous carbon materials are promising adsorbents for the removal of antibiotics from aqueous solution [17]. Therefore, how to increase the specific surface area has become a research hotspot. Liang et al. [18] reported that the specific surface area of mesoporous carbon materials was not only related to the chemical composition of the raw precursor, but also to the preparation conditions. They increased the specific surface area by adjusting the ratio of water and ethanol, the amount of hydrochloric acid, and the carbonization temperature. However, in the high temperature sintering process, the carbon frameworks tended to shrink, which reduced the pore size and the surface area [19]. Recently, it was found that liquid nitrogen could be applied as passivator during the preparation of $Ti\text{-}Al_2O_3$ composites by forming a passivation layer (Ti-N) which protected the original structure [20], but a possible protective effect of liquid nitrogen on the carbon structure in the preparation of mesoporous carbon has yet to be reported.

The present study aimed at directly comparing adsorptive ampicillin removal with liquid nitrogen treated mesoporous carbon material and that obtained by a common method. The primary goals were to: (1) investigate the influence of liquid nitrogen on the carbon structure and (2) reveal the mechanism(s) of ampicillin removal by carbon materials.

2. Materials and Methods

2.1. Preparation and Characteristics of Carbon Materials (CMs)

Two types of CMs were synthesized by a soft template method [21]. The preparation process consisted of the following steps: Pluronic F127 and phloroglucin were successively added into a beaker with ethanol. Then HCl was gradually added to the mixture with vigorous stirring for 90 min at 25 °C. As the reaction proceeded, the homogeneous reactant mixture was separated into two layers.

The upper clear fluid was poured out and the sediment was put in a crucible for drying 120 min at 60 °C in an oven (WGL-230B, TAISITE, Tianjin, China). The dried products were subsequently processed in two different ways. CM1 was obtained by placing the mixture in liquid nitrogen for half an hour and then calcining the mixture in a muffle furnace (SXL-1200M, SIOMM, Shanghai, China) at 800 °C for 3 h. CM2 was obtained by calcining the mixture in a muffle furnace at 800 °C for 3 h.

The specific surface area by the Brunauere-Emmette-Teller (BET) method, average pore diameter and total pore volume were detected by a surface and porosity analyzer (ASAP2020, Micromeritics Co., Shanghai, China). The morphology of the CMs was obtained by scanning electron microscopy (SEM, S-3400N, Hitachi, Tokyo, Japan). Changes of functional groups before and after ampicillin adsorption were determined by Fourier transform infrared spectroscopy (FT-IR, ALPHA, Beijing, China).

2.2. Detection of Ampicillin

The concentration of ampicillin in the aqueous phase was detected by a UV-spectrophotometry method (Nanodrop 2000/2000c, Thermo, Hong Kong, China). Ampicillin solution was mixed with imidazole solution [22] for 25 min at 60 °C, and then the UV-Vis absorption at 342 nm was recorded by a Nanodrop 2000 UV-Vis spectrophotometer (Nanodrop 2000/2000c, Thermo, Hong Kong, China). The absorption value corresponds to the concentration of ampicillin. Deionized-water produced by a pure water preparation system (FDY1001-UV-P, FLOM, Qingdao, China) was used in all experiments. The concentration of ampicillin in synthetic wastewater (SW) was detected at 220 nm using a High Performance Liquid Chromatography (HPLC) system (Agilent 1200, California, USA) [23]. A C_{18} column (5 μm particle size, 4.6 mmol/L × 250 mmol/L) was utilized. The mobile phase consisted of 0.02 mol/L potassium dihydrogen phosphate buffer and acetonitrile (92:8, *v:v*) delivered at a flow rate of 1.0 mL/min at 25 °C. It was filtered through a 0.45 μm size membrane filter (HDG-1A, Huadan, Beijing, China) and degassed before use.

2.3. Influence of pH on Adsorption of Ampicillin

The initial pH is regarded as the most important control parameter in the adsorption process. To determine the influence of pH on the adsorption of ampicillin, the pH in the experiments was adjusted to 2–9 using 1 mol/L NaOH or 1 mol/L HCl solution, using 1 mL solution of ampicillin with an initial concentration of 3 mmol/L and 5 mg adsorbent mixed in a 1.5 mL Eppendorf (EP) tube for 0.5 h at 25 °C.

2.4. Kinetic, Equilibrium and Thermodynamic Studies for Ampicillin Adsorption

To examine the adsorption of ampicillin on each carbon material, kinetic, equilibrium and thermodynamic studies were performed in batch experiments using 1.5 mL EP tubes. The experiments were conducted in triplicate at pH 7. Unless stated otherwise, after the adsorbent was completely mixed with the solution, the tubes were left standing still for a certain time before further use.

The kinetic study was performed to observe the effect of the reaction time on ampicillin adsorption. The influence of the adsorption time was confirmed by applying 1 mL solution of ampicillin with an initial concentration of 3 mmol/L and 5 mg adsorbent, mixed at 25 °C and pH 7 for 1–10 min. Samples were collected after the required reaction time. The results were analyzed by pseudo first-order and pseudo second-order kinetics.

The equilibrium study was conducted to investigate the influence of the equilibrium concentrations of ampicillin on the adsorption results. The experiments were performed at an equilibrium concentration of ampicillin of 0.5–3 mmol/L with carbon material doses of 5 mg in 1 mL of solution. The tubes were sampled after 0.5 h. The results were analyzed by the models of Langmuir and Freundlich.

The thermodynamic study was performed to examine the impact of temperature on the adsorption. The effect was determined by utilizing 1 mL solution of ampicillin with an initial concentration of 3 mmol/L and adsorbent amount of 5 mg in 1 mL solution mixing for 0.5 h at pH 7. A HH-2 digital electric-heated thermostatic water bath (HH-2, Guohua, Jiangsu, China) was used for controlling the reaction temperature.

2.5. Ampicillin Adsorption in SW

A typical SW was prepared according to the composition of the effluent from a sewage treatment plant in Hong Kong using glucose, ammonium chloride, urea, sodium nitrate and potassium dihydrogen phosphate [24]. The concentration of each compound was: 60 mg/L total organic carbon (TOC), 51 mg/L total nitrogen (TN), 5 mg/L total phosphorus (TP), and of these, the total nitrogen was composed of 40 mg/L ammonia nitrogen, 10 mg/L organic nitrogen and 1 mg/L nitrate nitrogen. In this experiment, 5 mg CM1 and 1 mmol/L ampicillin was added into SW and pure water, and then the effect of SW on the adsorption was studied.

2.6. Renew and Reuse of CMs

After the adsorption experiments, the used CMs were immersed in pure water for 30 s with vortex mixer, centrifuged (10,000 r/min, 5 min), and then collected. The desorption procedures were repeated four times before the CMs were gathered and dried at 60 °C. Thus, recovered CMs have been obtained. In the reuse of CMs experiments, the recovered CMs were applied to ampicillin solutions at the natural pH. The adsorption/desorption procedures were repeated ten times to evaluate the reusability of CMs.

3. Results

3.1. Characteristics of the Carbon Material

The N_2 adsorption-desorption isotherms of the tested CMs are provided in Figure 1, and the pore size distribution of the CMs is shown in the inset graph of Figure 1. Textural properties, e.g., specific surface area (S_{BET}, m^2/g), total pore volume (V_{Total}, cm^3/g), and micropore volume (V_{Mes}, cm^3/g) of the carbonaceous materials are listed in Table 1.

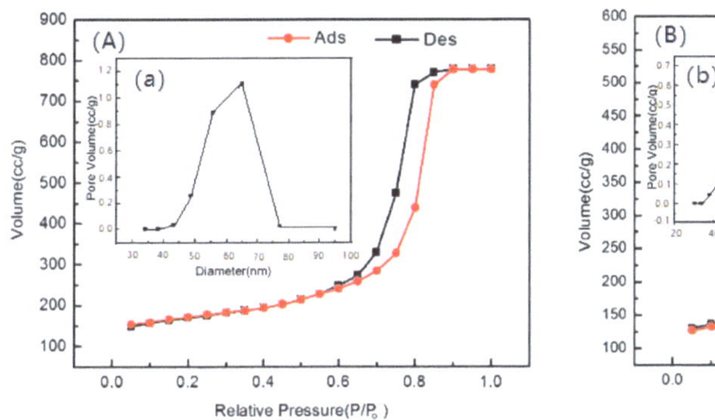
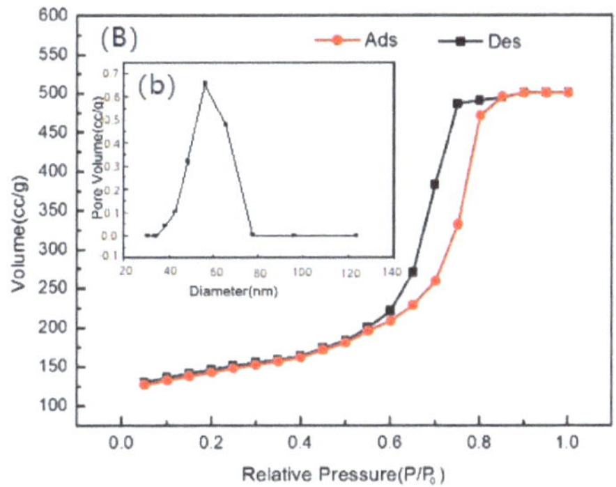

Figure 1. N_2 adsorption-desorption of CMs: (**A**) CM1; (**B**) CM2. The inset graph shows pore size distribution of CMs: (**a**) CM1; (**b**) CM2.

Table 1. Characteristics of synthesized (CM1, CM2) mesoporous carbons.

Characteristic	CM1	CM2
BET surface area (m^2/g)	550.52	422.93
Average pore width (nm)	62.98	45.83
Total pore volume (cm^3/g)	1.20	0.84
Mesopore volume (Vp) (cm^3/g)	0.84	0.71
Macropore volume (cm^3/g)	0.36	0.13

The porous structure, presented in Figure 1, revealed the nitrogen adsorption and desorption isotherms could be categorized as type IV isotherms. The nitrogen adsorption-desorption curve was related to the volume and size of the pores, so the textural properties of CMs could be known by analyzing the adsorption and desorption isotherms. According to the nitrogen adsorption curve and the nitrogen desorption curve, the nitrogen adsorption-desorption curve can be divided into five types: type I, type II, type III, type IV and type V [25]. The porous structure presented in Figure 1, revealed the nitrogen adsorption and desorption isotherms were categorized as type IV isotherm which indicates a mesoporous material [26]. As shown in Table 1, the effect of the liquid nitrogen treatments [27] on the textural properties of CM1 could also be observed. CM1 provided a higher surface area (550.52 m^2/g) and total pore volume (1.20 cm^3/g) than those provided by CM2 (S_{BET} = 422.93 m^2/g and V_{Total} = 45.83 cm^3/g), and the mean pore sizes of CM1 and CM2 were 62.98 nm and 45.83 nm, respectively. Figure 2 depicts two successive magnifications ($\times$50 k and $\times$100 k) of the tested CMs. As shown, it was clear from the $\times$100 k magnification, that for CM1, the pore structure and the surface were looser and smoother compared with CM2, which exhibited a compact and irregular surface. Thus, CM1 with wider interstitial spaces displayed a larger surface area and bigger pore width. These results were in agreement with the previous N$_2$ adsorption-desorption results of the CMs.

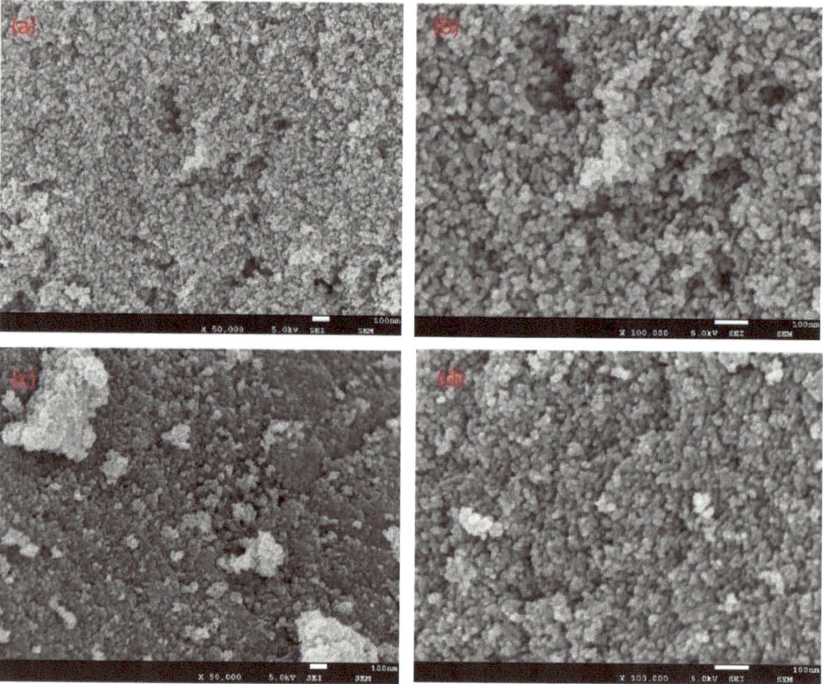

Figure 2. Analysis of CMs using SEM: (**a**,**b**) CM1; (**c**,**d**) CM2.

FTIR spectrum analysis helps determine the chemical compositions of materials. The FTIR spectra of the CMs before and after the adsorption of ampicillin in the wavelength range from 650 to 4000 cm^{-1} are shown in Figure 3.

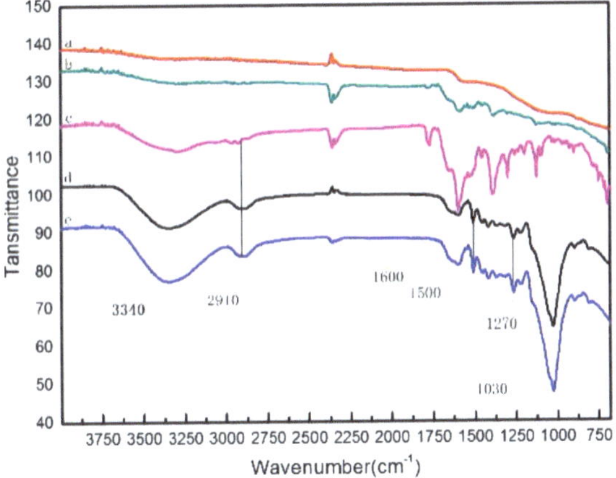

Figure 3. FTIR spectra of CM1 and CM2 before and after the adsorption of ampicillin: (a) CM2; (b) CM1. (c) ampicillin; (d) CM2 + ampicillin; (e) CM1 + ampicillin.

As shown, the FTIR spectra of the CMs were the same. However, the functional groups of the CMs before and after the adsorption of ampicillin were rather distinct. After adsorption, there were new peaks at 1270 cm^{-1} which are characteristic of N-H stretching vibrations, suggesting the presence of the secondary amide. The absorption band at 1600 cm^{-1} in all the FTIR spectra and 1500 cm^{-1} was assigned to C-H stretching vibrations of the benzene ring. The medium peak at 2910 cm^{-1} attributed to C-H stretching vibrations, indicating the presence of alkane/alkene groups on the activated carbons surface. The absorption band at 3340 cm^{-1} is usually ascribed to -OH stretching vibrations of alcohol, carboxylate, ester and/or ether groups [28]. The small peak at 1030 cm^{-1}, probably due to the infrared spectrum of C-S displacement toward low frequency by the effects of hydrogen bonds and ionic bonds.

3.2. Effect of pH

The effect of initial solution pH on the extent of adsorption of ampicillin is shown in Figure 4. Within the pH range studied (pH 2–9), both CMs reached the maximum adsorption when the pH was 3. Ampicillin (pKa = 2.96 for its free carboxylic acid groups and pKa = 7.22 for its free amino groups) is ionized in solution. At pHs between 2.96 and 7.22, ampicillin presents a zwitterionic structure [29]. As the pH increases, the structure of ampicillin could be strongly affected by the pH of the solution and the activated carbon become negatively charged.

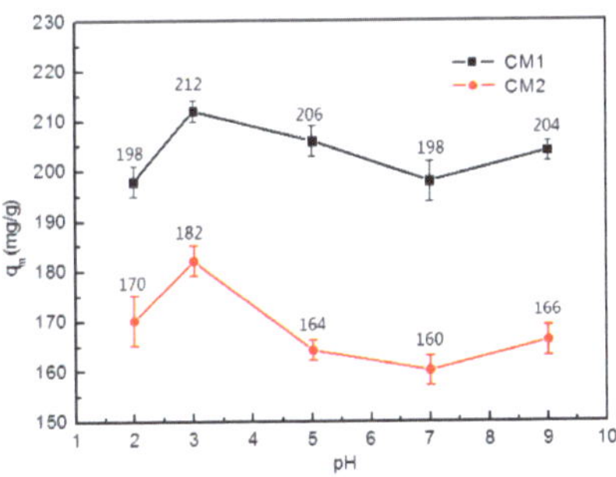

Figure 4. Effect of initial solution pH on the extent of adsorption of ampicillin (concentration of CMs = 5 g/L, concentration of ampicillin = 3 mmol/L, contact time = 30 min, at 25 °C).

3.3. Ampicillin Adsorption: Kinetic, Isotherm and Thermodynamic Analysis

The adsorption kinetics of ampicillin on the carbon materials was evaluated. Both 1st order and 2nd order adsorption kinetics were determined using Equations (1) and (2), respectively.

The pseudo-first-order expression is:

$$q_t = q_e \left(1 - e^{-k_1 t}\right) \tag{1}$$

where q_t (mg/g) is the concentration of ampicillin adsorbed mass per unit of the adsorbent at time t (min). k_1 (1/min) is the pseudo-first-order adsorption rate constant, q_e (mg/g) is the adsorption capacity counted by the pseudo-first-order model.

The pseudo-second-order equation is as follows:

$$q_t = \frac{q_e^2 k_2 t}{1 + q_e k_2 t} \tag{2}$$

where q_t (mg/g) is the concentration of ampicillin adsorbed mass per unit of the adsorbent at time t (min). k_2 (g/mg/min) is the pseudo-second-order adsorption rate constant, and q_e (mg/g) is the adsorption capacity computed by the pseudo-second-order model.

The parameters found for the different models are presented in Table 2, Some kinetic curve fitting results are shown in Figure 5.

Table 2. Kinetic parameters for the adsorption of ampicillin on CMs (concentration of CMs = 5 g/L, concentration of ampicillin = 3 mmol/L, pH = 7, at 25 °C).

CMs	the Pseudo-First-Order			the Pseudo-Second-Order		
	k_1 (1/min)	q (mg/g)	R^2	k_2 (g/mg/min)	q (mg/g)	R^2
CM1	36.632	192.257	0.961	0.019	206.002	0.999
CM2	6.036	155.728	0.885	0.010	178.423	0.998

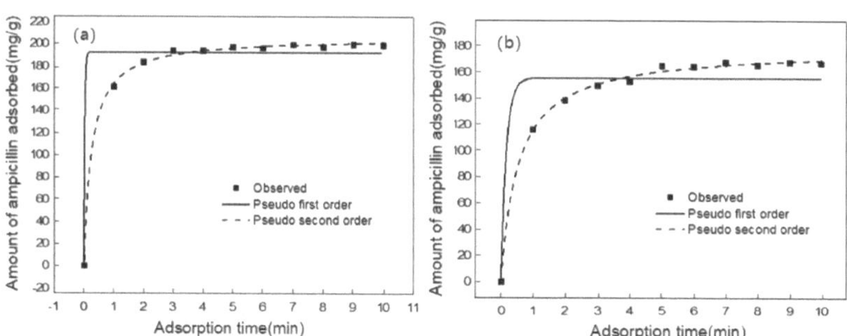

Figure 5. Kinetic modeling of ampicillin adsorption onto CMs: (**a**) CM1; (**b**) CM2. (concentration of CMs = 5 g/L, concentration of ampicillin = 3 mmol/L, pH = 7, at 25 °C).

The influence of contact time on the adsorption of ampicillin by the tested CMs is depicted in Figure 5a,b for CM1 and CM2, respectively. First, an exponential increase in the adsorption was registered within the first 3 min for CM1 and within 5 min for CM2, and the maximum adsorption capacities of ampicillin on CM1 and CM2 was 206.002 mg/g and 178.423 mg/g, respectively. Such a tendency confirms the significant effect of the CMs' properties (surface area and interstitial porosity) on the removal efficiency. These further proved that CM1 has a larger specific surface area and larger apertures than CM2, which could enhance the adsorption capacity of carbon materials and accelerate the adhesion of ampicillin to the carbon material [30]. The same result was previously reported in Table 1 regarding the surface area and pore volume. According to the values of R^2 (Table 2), it was easily obtained that the pseudo-second order model was more appropriate for depicting both CMs' kinetic data.

The Langmuir model (Equation (3)), and Freundlich model (Equation (4)) were used to further investigate the adsorption process and mechanism, with the results shown in Figure 6 and Table 3.

Langmuir isotherm:

$$q_e = \frac{q_L k_L C_e}{1 + k_L C_e} \tag{3}$$

where q_e (mg/g) is the ampicillin adsorption capacity at equilibrium, q_L (mg/g) is the maximum ampicillin adsorption capacity of CMs, and k_L (L/mg) is the constant that related to the rate of adsorption.

Freundlich isotherm:

$$q_e = K_F C_e^{1/n} \tag{4}$$

where q_e (mg/g) is the quantity of ampicillin adsorbed on CMs, C_e (mg/L) is the equilibrium concentration of ampicillin, K_f and n are the Freundlich constants which are correspond to adsorption capacity and intensity, respectively.

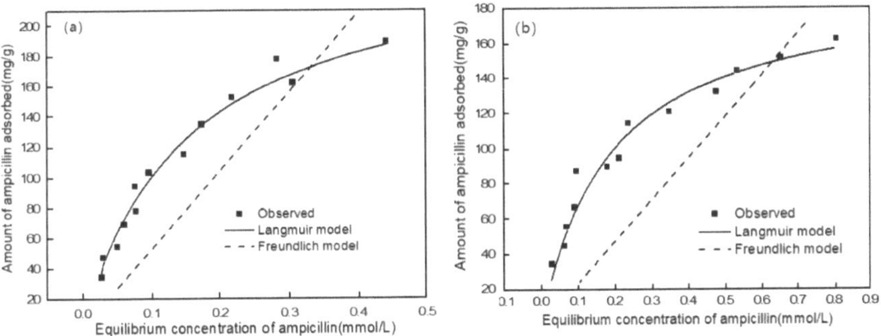

Figure 6. Equilibrium isotherm model of ampicillin adsorption onto CMs: (**a**) CM1; (**b**) CM2. (concentration of CMs = 5 g/L, contact time = 30 min, pH = 7, at 25 °C).

Table 3. Adsorption isotherm parameters for the adsorption of ampicillin on CMs (concentration of CMs = 5 g/L, contact time = 30 min, pH = 7, at 25 °C).

CMs	Langmuir			Freundlich		
	K_L (L/mg)	q_{max} (mg/g)	R^2	K_F (L/mg)	n	R^2
CM1	6.475	252.819	0.992	1.962	0.004	0.388
CM2	5.428	193.274	0.988	1.884	0.007	0.387

As shown in Table 3, both C_e/q_e versus C_e curve and R^2 values indicate that the Langmuir isotherm model had a better fit than the Freundlich model. From the Langmuir model, the maximum adsorption capacity (q_m) of ampicillin was calculated to be 252.819 mg/g for CM1, which was higher than that for CM2 (193.274 mg/g). This could be explained by the fact that CM1 had a larger specific surface area with more surface active sites after liquid nitrogen treatment. The mechanisms involved in the process of ampicillin adsorption and the adsorption capacity of CMs were determined by obtaining the corresponding adsorption isotherms. Figure 6 shows that the adsorption isotherms of ampicillin on the CMs could be classified as L-2 type, according to the Giles classification [31], suggesting that ampicillin molecules are adsorbed in parallel to the carbon surface and there is no major competition between ampicillin and water molecules for the active sites on the carbon.

Thermodynamic parameters, such as entropy change ($\triangle S°$), enthalpy change ($\triangle H°$) and Gibbs free energy change ($\triangle G°$) were computed with Equation (5):

$$\ln k_d = \frac{\Delta S°}{R} - \frac{\Delta H°}{RT} \tag{5}$$

where K_d is the adsorption distribution constant (L/mg), R is the universal gas constant (8.314 J/mol K), and T is the absolute temperature (K). The adsorption distribution constant K_d is computed using the following Equation (6):

$$K_d = \frac{C_{e,adsorption}}{C_{e,solution}} \tag{6}$$

where $C_{e,adsorption}$ and $C_{e,solution}$ are the equilibrium concentrations of ampicillin (mg/L) on the adsorbent (CMs) and in the solution, respectively.

The thermodynamic analysis is presented in Table 4. The maximum adsorption capacity was obtained at 55 °C, and the adsorption of ampicillin on CMs was affected as the temperature varied between the ranges of 35–55 °C. The negative values of Gibbs free energy change ($\triangle G°$) (-5.812––3.392 kJ/mol) at all temperatures indicated that ampicillin adsorption on CMs was a spontaneous reaction. CM1 and CM2 had a positive value of entropy ($\triangle S°$), which indicated that the entropy increased at the solid-solution interface. The distribution of enthalpy ($\triangle H°$) indicated that the adsorption process was endothermic and was more favorable at high temperatures, and might be caused by the changes in hydrogen bonding, protonation or Van der Waals forces.

Table 4. Thermodynamic parameters for ampicillin adsorption on CMs (concentration of CMs = 5 g/L, concentration of ampicillin = 3 mmol/L, contact time = 30 min, pH = 7).

CMs	Temp. (°C)	$\triangle$H (KJ/mol)	$\triangle$S (KJ/mol)	$\triangle$G (KJ/mol)
	35	12.471	49.884	−3.392
CM1	45	12.471	49.884	−3.89
	55	12.471	49.884	−4.612
	35	8.314	43.232	−5.001
CM2	45	8.314	43.232	−5.434
	55	8.314	43.232	−5.812

3.4. Ampicillin Adsorption in SW

In order to assess the feasibility of the CMs' practical application, the CMs were used to remove ampicillin from SW to study the ampicillin removal rate. As shown in Figure 7, in SW, both CM could not completely remove the ampicillin, and the removal ratio of 1 mmol/L ampicillin was 92.31% for CM1 and 86.56 for CM2, respectively.

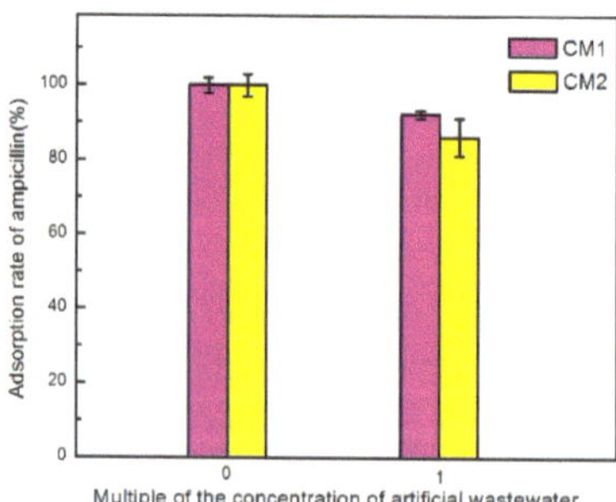

Figure 7. Removal rate of ampicillin in synthetic wastewater (concentration of CMs = 5 g/L, contact time = 30 min, at 25 °C).

3.5. Renew and Reuse of CMs

The experimental results of adsorption material regeneration were presented in Figure 8. Figure 8a shows the regeneration of CMs in pure water, and Figure 8b demonstrates the regeneration of the CMs in SW. As shown in Figure 8, the CMs could basically be regenerated after being washed only three times in pure water, however, CM2 needed four washes to reach the same goal in SW. This indicated that SW changed the interaction between sorbent and ampicillin, and decreased the effect of regeneration. The regeneration of CM1 under these two conditions were better than that of CM2.

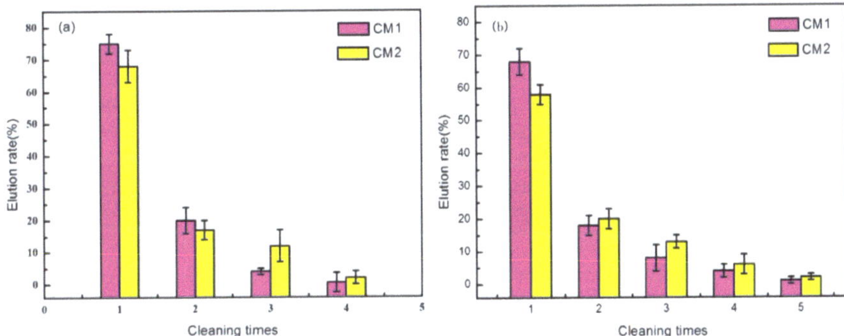

Figure 8. (**a**) Ampicillin elution from CMs in water; (**b**) Ampicillin elution from CMs in synthetic wastewater.

The reuse results of the CMs are shown in Figure 9, which indicates that after ten adsorption/desorption recycles, the ampicillin removal ratio with CM1 was higher than that with CM2. CM1's removal ratio was above 80.56%, and that of CM2's was nearly 68.57%. The good regeneration ability of CM1 made it suitable to capture ampicillin in sewage treatment, indicating that the modified method established in this work could be useful for practical application.

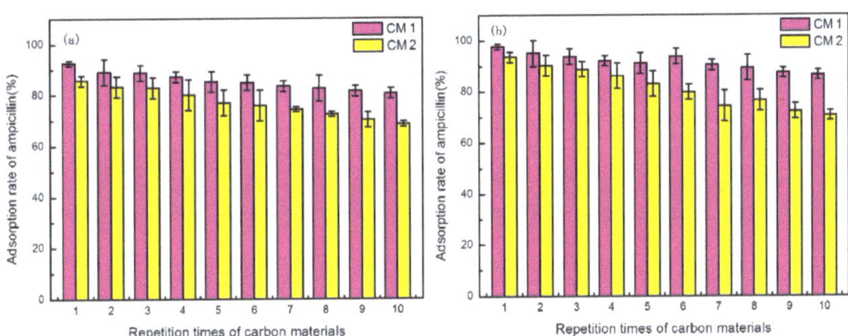

Figure 9. (**a**) Reuse of CMs in SW solution; (**b**) Reuse of CMs in normal SW (concentration of CMs = 5 g/L, contact time = 30 min, at 25 °C).

4. Discussion

Antibiotics are a global concern as emerging contaminants. In this study, ampicillin was removed from aqueous solutions using a novel carbon material. The textural properties of the carbon materials were characterized by N_2 adsorption-desorption isotherms, SEM and FTIR and the ampicillin removal mechanism was explored by kinetic, equilibrium and thermodynamic studies. The carbon materials were further used to remove ampicillin from SW.

Phenolic resin is widely used for the preparation of carbon materials due to the fact it is rich in hydroxyl groups and could form hydrogen bonds with the PEO segment of the triblock copolymer, which was a driving force for forming a carbon material structure. In previous studies, the main method for improving carbon materials was to change the catalyst during the preparation process [32,33]. However, no similar reports have been reported to prevent sintering during the carbonization process. In the process of synthesizing Ti-Al_2O_3, Liu et al. used liquid nitrogen passivated Ti nanopowder to prevent sintering, and to prepare a material with uniform phase dispersion [20].

In this research, a new absorbent (CM1) had been obtained by modified liquid nitrogen treatment, which was utilized to remove ampicillin from solution in comparison with CM2 (obtained without liquid nitrogen treatment). Combined with N_2 adsorption-desorption isotherms and SEM results, it was clear obtained that CM1 has a larger specific area and bigger pores than CM2. We confirm that liquid nitrogen could obviously increase specific surface area and aperture of CMs by forming a passivation layer as mentioned before [19]. According to the results of kinetic, equilibrium and thermodynamic studies, CM1 removed ampicillin more efficiently. Our results were consistent with others' research, as Álvarez has indicated that the surface area was increased from 352 to 391 $m^2\,g^{-1}$ by increasing the carbonization temperature of the carbon material, and the average pore width was changed from 0.86 nm to 1.31 nm. which led to an increase in the amount of tetracycline adsorbed from 53.9 to 58.1 $mg\,g^{-1}$ [34].

Regarding the effect of contact time, concentration and temperature, the results for the case of ampicillin demonstrated that the kinetic data were successfully correlated with the pseudo-second order model and the isothermal data were well-fitted to the Langmuir isotherm model. The high accuracy in the prediction of experimental data by the pseudo-second order model has been considered in many research works [35] as evidence that chemical bonds are involved in the adsorption mechanism of ampicillin onto CMs. This assumption has been confirmed by the result of the study of the FTIR spectra of the adsorbed phase onto CMs (Figure 3), which displayed new chemical bonds between both the adsorbed phase and the adsorbent. The Langmuir isotherm model results was in agreement with previous studies [36], which indicated that adsorption was a monolayer adsorption and CMs have homogeneous solid surfaces.

The experimental pH data also revealed a feeble change between pH 2 and 9 and the adsorption reached a maximum at pH 3, indicating that the structure of ampicillin had a major effect on the adsorption of ampicillin on CMs [28].

According to our desorption and the renew-recycle experiments, CM1 could basically be regenerated in water after being washed only three times, however, with synthetic wastewater the adsoption by CM1 was still up to 86.56%, but down to 68.57% with CM2. Ncibi et al. have reported that carbon materials with larger pore sizes were easier to reuse as it was difficult to wash substances out from small sized pores [30]. According to a previous study [37], other substances present in the synthetic wastewater can compete for the sorption sites with ampicillin, so the removal ratio was decreased.

These experiments have some limitations. Phenolic resin is relatively expensive compared to other carbon raw materials (bioresidue and coal based), so the liquid nitrogen modification method should be further applied to other activated carbons forms.

The liquid nitrogen treatment method is a new green and low cost physical modification method. Compared with other chemical modification methods, it could produce carbon materials with larger specific surface area and more adsorption sites to increase the adsorption capacity without changing the chemical properties. If combined with other modification methods, a new modified material with better adsorption ability would be achieved. Moreover, the large surface area of carbon materials could improve the properties of batteries [38] and increase hydrogen storage capacity [39], so the liquid nitrogen treatment method for carbon material modification may have broader application prospect in sewage treatment and industrial manufacture.

5. Conclusions

In this research, liquid nitrogen was used as a passivating agent to modify carbon materials, and a new absorbent (CM1) was obtained, which was utilized to remove ampicillin from aqueous solutions. Compared with CM2, the specific area and aperture of CM1 was larger, which can remove ampicillin more efficiently. Kinetic data were successfully correlated to the pseudo-second order model. Moreover, the thermodynamic results were consistent with the Langmuir model. According to the model, the adsorption capacities of CM1 and CM2 were 212 mg/g and 182 mg/g, respectively. The whole adsorption process was an endothermic reaction. The adsorption capacity was affected by the solution pH, and it reached a maximum at pH 3. The ampicillin removal ratio of CM1 was higher than that of CM2, however, it decreased with increasing concentration of SW. In the renew-recycle experiments, the removal ratio of ampicillin by CM2 was still up to 80.56%, but down to 68.57% with CM1.

Author Contributions: Y.W. (Yaohui Wu) conceived and designed the study; W.L., Y.W. (Yonghong Wang), and X.H. analyzed the data and discussed the results; X.C. discussed the results; W.L. wrote the paper; Y.Z. and Z.H. revised the paper.

Funding: This research was funded by the Major Science and Technology Program of Hunan province (2017NK1010), the Research Foundation of Education Bureau of Hunan Province (Grant No. 16A227), and Central South University of Forestry and Technology Graduate Technology Innovation Fund (20183015).

Acknowledgments: We thank the Nanomaterials Laboratory of Central South University of Forestry and Technology for measuring the carbon material textural properties.

References

1. Kümmerer, K. Antibiotics in the aquatic environment—A review—Part II. *Chemosphere* **2009**, *75*, 435–441. [CrossRef] [PubMed]
2. Sarmah, A.K.; Meyer, M.T.; Boxall, A.B. A global perspective on the use, sales, exposure pathways, occurrence, fate and effects of veterinary antibiotics (VAs) in the environment. *Chemosphere* **2006**, *65*, 725–759. [CrossRef] [PubMed]
3. Ma, Y.; Li, M.; Wu, M.; Li, Z.; Liu, X. Occurrences and regional distributions of 20 antibiotics in water bodies during groundwater recharge. *Sci. Total Environ.* **2015**, *518–519*, 498–506. [CrossRef] [PubMed]
4. Zhou, L.J.; Ying, G.G.; Liu, S.; Zhao, J.L.; Yang, B.; Chen, Z.F.; Lai, H.J. Occurrence and fate of eleven classes of antibiotics in two typical wastewater treatment plants in South China. *Sci. Total Environ.* **2013**, *452–453*, 365–376. [CrossRef] [PubMed]
5. Homem, V.; Santos, L. Degradation and removal methods of antibiotics from aqueous matrices—A review. *J. Environ. Manag.* **2011**, *92*, 2304–2347. [CrossRef] [PubMed]
6. Taučer-Kapteijn, M.; Hoogenboezem, W.; Heiliegers, L.; Bolster, D.D.; Medema, G. Screening municipal wastewater effluent and surface water used for drinking water production for the presence of ampicillin and vancomycin resistant enterococci. *Int. J. Hyg. Environ. Health* **2016**, *219*, 437–442. [CrossRef] [PubMed]
7. Kohlenberg, A.; Schwab, F.; Rüden, H. Wide dissemination of extended-spectrum beta-lactamase (ESBL)-producing Escherichia coli and Klebsiella spp. in acute care and rehabilitation hospitals. *Epidemiol. Infect.* **2012**, *140*, 528–534. [CrossRef] [PubMed]
8. Chao, S.; Cong, Z.; Fan, L.; Qiu, L.; Wei, W.; Meng, S.; Hu, G.; Kamira, B.; Chen, J. Occurrence of antibiotics and their impacts to primary productivity in fishponds around Tai Lake, China. *Chemosphere* **2016**, *161*, 127–135.
9. Baquero, F.; Martínez, J.L.; Cantón, R. Antibiotics and antibiotic resistance in water environments. *Curr. Opin. Biotechnol.* **2008**, *19*, 260–265. [CrossRef] [PubMed]
10. Arslan-Alaton, I.; Caglayan, A.E. Ozonation of Procaine Penicillin G formulation effluent Part I: Process optimization and kinetics. *Chemosphere* **2005**, *59*, 31–39. [CrossRef] [PubMed]
11. Giraldo-Aguirre, A.L.; Serna-Galvis, E.A.; Erazo-Erazo, E.D.; Silva-Agredo, J.; Giraldo-Ospina, H.; Flórez-Acosta, O.A.; Torres-Palma, R.A. Removal of β-lactam antibiotics from pharmaceutical wastewaters

using photo-Fenton process at near-neutral pH. *Environ. Sci. Pollut. Res. Int.* **2018**, *25*, 20293–20303. [CrossRef] [PubMed]

12. Liang, S.; Yu, L.; Xu, H.L. Treatment of ampicillin-loaded wastewater by combined adsorption and biodegradation. *J. Chem. Technol. Biotechnol.* **2010**, *85*, 814–820.

13. Limousy, L.; Ghouma, I.; Ouederni, A.; Jeguirim, M. Amoxicillin removal from aqueous solution using activated carbon prepared by chemical activation of olive stone. *Environ. Sci. Pollut. Res. Int.* **2017**, *24*, 9993–10004. [CrossRef] [PubMed]

14. Akhtar, J.; Amin, N.A.S.; Shahzad, K. A review on removal of pharmaceuticals from water by adsorption. *Desalin. Water Treat.* **2016**, *57*, 12842–12860. [CrossRef]

15. Cong, Q.; Yuan, X.; Qu, J. A review on the removal of antibiotics by carbon nanotubes. *Water Sci. Technol.* **2013**, *68*, 1679–1687. [CrossRef] [PubMed]

16. Moussavi, G.; Alahabadi, A.; Yaghmaeian, K.; Eskandari, M. Preparation, characterization and adsorption potential of the NH_4Cl-induced activated carbon for the removal of amoxicillin antibiotic from water. *Chem. Eng. J.* **2013**, *217*, 119–128. [CrossRef]

17. Peng, X.; Hu, F.; Lam, F.L.; Wang, Y.; Liu, Z.; Dai, H. Adsorption behavior and mechanisms of ciprofloxacin from aqueous solution by ordered mesoporous carbon and bamboo-based carbon. *J. Colloid Interface Sci.* **2015**, *460*, 349–360. [CrossRef] [PubMed]

18. Liang, C.; Dai, S. Synthesis of mesoporous carbon materials via enhanced hydrogen-bonding interaction. *J. Am. Chem. Soc.* **2006**, *128*, 5316–5317. [CrossRef] [PubMed]

19. Dang, X.; Wei, C.; Liu, H.; Lv, J.; Pan, D.; Li, X. Preparation, Morphology, and Structure of Thermotropic Liquid Crystalline Polyester-imide/Phenol-formaldehyde Resin Blends. *J. Macromol. Sci. Part A Chem.* **2012**, *49*, 378–384. [CrossRef]

20. Liu, Q.; Wang, Z.; Junyan, W.U. Effect of Raw Materials Cryomilled in Liquid Nitrogen on Mechanical Properties of Sintered $Ti-Al_2O_3$ Composites. *J. Univ. Jinan* **2018**, *281*, 285–290.

21. Meng, Y.; Gu, D.; Zhang, F.; Shi, Y.; Cheng, L.; Feng, D.; Wu, Z.; Chen, Z.; Wan, Y.; Stein, A.; Zhao, D. A Family of Highly Ordered Mesoporous Polymer Resin and Carbon Structures from Organic-Organic Self-Assembly. *Chem. Mater.* **2006**, *18*, 4447–4464. [CrossRef]

22. Arikan, O.A. Degradation and metabolization of chlortetracycline during the anaerobic digestion of manure from medicated calves. *J. Hazard. Mater.* **2008**, *158*, 485–490. [CrossRef] [PubMed]

23. Li, X.; Yu, C.; Cai, Y.; Liu, G.; Jia, J.; Wang, Y. Simultaneous determination of six phenolic constituents of danshen in human serum using liquid chromatography/tandem mass spectrometry. *J. Chromatogr. B Analyt. Technol. Biomed. Life Sci.* **2005**, *820*, 41–47. [CrossRef] [PubMed]

24. Wu, Y.; Tam, N.F.Y.; Wong, M.H. Effects of salinity on treatment of municipal wastewater by constructed mangrove wetland microcosms. *Mar. Pollut. Bull.* **2008**, *57*, 727–734. [CrossRef] [PubMed]

25. Kim, M.; Sohn, K.; Na, H.B.; Hyeon, T. Synthesis of Nanorattles Composed of Gold Nanoparticles Encapsulated in Mesoporous Carbon and Polymer Shells. *Nano Lett.* **2002**, *2*, 1383–1387. [CrossRef]

26. Chandra, D.; Bhaumik, A. Photoluminescence behavior of new mesoporous titanium-composites synthesized by using bidentate structure directing agents. *Microporous Mesoporous Mater.* **2007**, *101*, 348–354. [CrossRef]

27. Liu, S.; Guo, C.; Zhi, D.; Liang, X. Comparative proteomics reveal the mechanism of Tween80 enhanced phenanthrene biodegradation by Sphingomonas sp. GY2B. *Ecotoxicol. Environ. Saf.* **2017**, *137*, 256–264. [CrossRef] [PubMed]

28. Wang, G.; Tao, W.; Li, Y.; Sun, D.; Yan, W.; Huang, X.; Zhang, G.; Liu, R. Removal of ampicillin sodium in solution using activated carbon adsorption integrated with H_2O_2 oxidation. *J. Chem. Technol. Biotechnol. Biotechnol.* **2012**, *87*, 623–628. [CrossRef]

29. Alekseev, V.G.; Volkova, I.A. Acid-Base Properties of Some Penicillins. *Russ. J. Gen. Chem.* **2003**, *73*, 1616–1618. [CrossRef]

30. Ncibi, M.C.; Sillanpää, M. Optimized removal of antibiotic drugs from aqueous solutions using single, double and multi-walled carbon nanotubes. *J. Hazard. Mater.* **2015**, *298*, 102–110. [CrossRef] [PubMed]

31. Giles, C.H.; Macewan, T.H.; Nakhwa, S.N.; Smith, D. 786. Studies in adsorption. Part XI. A system of classification of solution adsorption isotherms, and its use in diagnosis of adsorption mechanisms and in measurement of specific surface areas of solids. *J. Chem. Soc.* **1960**, *111*, 3973–3993. [CrossRef]

32. Liu, J.; Qiao, S.Z.; Liu, H.; Chen, J.; Orpe, A.; Zhao, D.; Lu, G.Q. Extension of the Stöber method to the preparation of monodisperse resorcinol-formaldehyde resin polymer and carbon spheres. *Angew. Chem.* **2011**, *123*, 6069–6073. [CrossRef]
33. Zhou, H.; Xu, S.; Su, H.; Wang, M.; Qiao, W.; Ling, L.; Long, D. Facile preparation and ultra-microporous structure of melamine-resorcinol-formaldehyde polymeric microspheres. *Chem. Commun.* **2013**, *49*, 3763–3765. [CrossRef] [PubMed]
34. Álvarez-Torrellas, S.; Ribeiro, R.S.; Gomes, H.T.; Ovejero, G.; García, J. Removal of antibiotic compounds by adsorption using glycerol-based carbon materials. *Chem. Eng. J.* **2016**, *296*, 277–288. [CrossRef]
35. Ho, Y.S.; Mckay, G. The kinetics of sorption of divalent metal ions onto sphagnum moss peat. *Water Res.* **2000**, *34*, 735–742. [CrossRef]
36. Carabineiro, S.A.C.; Thavorn-Amornsri, T.; Pereira, M.F.R.; Serp, P.; Figueiredo, J.L. Comparison between activated carbon, carbon xerogel and carbon nanotubes for the adsorption of the antibiotic ciprofloxacin. *Catal. Today* **2012**, *186*, 29–34. [CrossRef]
37. Rahardjo, A.K.; Susanto, M.J.J.; Kurniawan, A.; Indraswati, N.; Ismadji, S. Modified Ponorogo bentonite for the removal of ampicillin from wastewater. *J. Hazard. Mater.* **2011**, *190*, 1001–1008. [CrossRef] [PubMed]
38. Peng, H.J.; Hao, G.X.; Chu, Z.H.; Lin, Y.W.; Lin, X.M.; Cai, Y.P. Porous carbon with large surface area derived from a metal–organic framework as a lithium-ion battery anode material. *RSC Adv.* **2017**, *7*, 34104–34109. [CrossRef]
39. Cai, J.; Li, L.; Lv, X.; Yang, C.; Zhao, X. Large surface area ordered porous carbons via nanocasting zeolite 10X and high performance for hydrogen storage application. *ACS Appl. Mater. Interfaces* **2014**, *6*, 167–175. [CrossRef] [PubMed]

 International Journal of
Environmental Research and Public Health

Article

The Effects of Ultrasonic Disintegration as a Function of Waste Activated Sludge Characteristics and Technical Conditions of Conducting the Process—Comprehensive Analysis

Malwina Tytła

Institute of Environmental Engineering, Polish Academy of Sciences, 34 M. Skłodowskiej-Curie St., 41-819 Zabrze, Poland; malwina.tytla@ipis.zabrze.pl; Tel.: +48-32-271-6481 (ext. 135)

Received: 11 September 2018; Accepted: 18 October 2018; Published: 20 October 2018

Abstract: A comprehensive analysis of the effects obtained in the process of ultrasonic disintegration (UD) of waste activated sludge (WAS), was conducted. Sludge samples were collected periodically from Central Wastewater Treatment Plant (WWTP) in Gliwice (Poland) and disintegrated in the two ultrasonic devices of different construction and technical parameters, i.e., WK-2010 (A) and ultrasonic washer (B). The experiments were performed under a constant energy supply per sludge volume $E_V = 160$ kWh·m^{-3}. The direct and technological effects, i.e., after UD and anaerobic digestion (AD) were investigated, respectively. Statistical analysis showed that characteristics and parameters of the WAS, which affects the magnitude of the direct effects create the following sequence: TS (total solids), VS (volatile solids), ΔT (temperature increase) > EPS (extracellular polymeric substances) > SCOD (soluble chemical oxygen demand) > CST (capillary suction time) > N_{TOT} (total nitrogen), P_{TOT} (total phosphorus) > pH. Whereas, in the case of technological effects, the above sequence was as follows: TS, VS > CST > N_{TOT}, P_{TOT} > pH. Ultrasonic disintegration of WAS prior to AD increased total biogas production (from 13.0% to 19.7%) and reduced the content of TS (from 4.1% to 8.2%) and VS (5.8% to 9.5%) in comparison to the control sample. This confirms the usefulness of ultrasonic disintegration as an effective method of sludge digestion intensification. The obtained results showed that changes in the characteristics of WAS have a significant impact on the magnitude of the effects of ultrasonic disintegration, especially TS, VS, ΔT, EPS, SCOD and CST. Concluding, it can be inferred that the most promising conditions for ultrasonic pretreatment conducted under constant energy supply per sludge volume, are: low power, long sonication time, large surface area of the emitter, and high increase of sludge temperature while conducting the process.

Keywords: waste activated sludge; ultrasonic disintegration; disintegration effects; sludge characteristics; wastewater treatment plant

---

1. Introduction

The development of new technologies and growing effectiveness of biological wastewater treatment observed in recent years, can be mainly accounted to the implementation of the European Council Directive [1] concerning urban wastewater treatment, which led to an increase in the amount of sewage sludge production [2,3]. The significance of the abovementioned issue is even more pressing due to the hazards which sludge may pose to the environment and the economy. It is estimated that by the year 2020, the amount of sludge produced in European Union (EU) countries will reach 12,997,000 Mg$_{TS}$, in which 950,000 Mg$_{TS}$ would be in Poland alone [4]. Moreover, the costs associated with sludge treatment and its disposal may constitute as much as 65% of the total operating costs of Wastewater Treatment Plant (WWTP) [5]. Therefore, it is necessary to introduce methods for intensifying anaerobic digestion, which is the most commonly applied process in sludge treatment, that enable sludge stabilization and mass reduction, improve its dewaterability and ensure hygienization [6,7]. Among these methods, we can distinguish disintegration. Nowadays, it is one of the most important environmental issues.

There are different types of disintegration techniques, i.e., mechanical (ultrasonic disintegration, homogenizations), thermal (hydrolysis—low and high temperature), chemical (hydrolysis with oxygen, ozone, sodium hydroxide) and biological (with application of enzymes) [8,9]. Currently, mechanical methods are the most effective and commonly used, especially ultrasonic disintegration, which is based on the cavitation phenomenon [2,10]. Ultrasonic disintegration shortens the hydrolytic phase, i.e., the rate—limiting step of anaerobic digestion (AD), and increases the efficiency of the process [2,6]. Moreover, ultrasonic disintegration (UD) increases the solubility of the organic compounds (by disrupting the sludge floc and cells), leading to a release of intracellular materials available for living organisms, which can be used as a substrate in the subsequent steps of anaerobic digestion [8,11]. It must be emphasized, that ultrasonic disintegration is one of the most secure and environmentally friendly methods of anaerobic sludge digestion intensification, to which mainly waste activated sludge (WAS) is subjected due to the difficulty in its decomposing [8]. The advantages of sludge UD are: lack of byproducts or use of additional reagents and possibility to intervene during the process conducting [11,12]. We can distinguish two types of ultrasonic disintegration effects: direct and technological. The direct effects are observed after ultrasonic pretreatment, whereas the technological during further sludge treatment, i.e., anaerobic digestion [11,13]. The obtained effects are described by appropriate indicators of disintegration, i.a. disintegration degree (DD_{COD}) [14], etc. The direct effects are monitored based on the physicochemical changes of sludge characteristics (parameters) before and after pretreatment, i.a. pH [11,15], concentration of soluble chemical oxygen demand (SCOD) [3,16], biogenic substances [17,18] and extracellular polymeric substances [9,19] in sludge supernatant, capillary suction time (CST) [7,20] or microscopy examination of a flocs disruption [5,21]. In the case of anaerobic digestion the expected effects are: increase in biogas production, total solids (TS) and volatile solids (VS) reduction, as well as dewatering ability improvement [10,13]. The effects of sludge pretreatment are also influence by operating parameters of the process conducting, i.a. technical construction of ultrasonic device, frequency (f), the amount of energy supplied per sludge volume (E_V) or total solids content (E_S), ultrasonic density (U_D) and intensity (U_I), sonication time (t) and process temperature (T) [18,22,23].

Therefore, taking into account the number of factors, which may have an impact on a magnitude of disintegration effects, a comprehensive analysis and selection of the most important sludge characteristics (parameters) and technical conditions of the process conducting are necessary. This is particularly important due to the fact that it is still unknown why in the similar conditions of sludge sonication, different effects are obtained. However, it is important to emphasize that there are no two identical sludges—their characteristics change over time. Therefore, it is important to carry out the ultrasonic disintegration in a cyclical manner, i.e., at an appropriate time intervals, which will allow observing changes in sludge characteristics and indicate the main parameters, that have the greatest impact on the obtained effects.

The aims of this study were: (a) to evaluate direct and technological effects based on the values of selected indicators and optical microscopy examination; (b) to determine sludge characteristics (parameters) and technical conditions of the process conducting, having the greatest impact on the direct and technological effects; (c) to conduct a comprehensive analysis of the disintegration effects, with including periodically changes of WAS characteristics and technical conditions of conducting the process, using selected statistical tests.

2. Materials and Methods

2.1. Sludge Collection and Analysis

The WAS samples were collected once a month over a 7-months period of time, starting from July 2013 up to January 2014. Sludge samples after mechanical thickening were collected from the advanced biological Central Wastewater Treatment Plant in Gliwice (Poland, Central Europe). The digested sludge (inoculum) was taken from a full scale anaerobic digester (mesophilically operated) in the same WWTP and used only in order to conduct anaerobic digestion. The collected sludge samples were stored in polypropylene tubes at 4 °C before further analysis. The operational parameters of the WWTP are reported in Table 1.

Table 1. The operational parameters of the Wastewater Treatment Plant.

Parameter [1]	Unit	Value
Population equivalent (PE)	-	155,009
Average flow (Q)	$m^3 \cdot d^{-1}$	31,913
Hydraulic retention time (HRT) of sludge in anaerobic digester	d	33

[1] Data for the year 2017 obtained from the Central WWTP in Gliwice.

In order to determine the changes of sludge samples characteristics—before and after UD/AD, selected parameters were considered, i.e., pH, SCOD, N_{TOT} (total nitrogen), P_{TOT} (total phosphorus), proteins and carbohydrates concentrations, CST, flocs disruption (microscopic examination; at $100\times$ magnification), TS and VS content, as well as volume and composition of the evolved biogas. Prior and after UD or AD, sludge samples were centrifuged for 30 min at 20,000 rpm (18 °C) and subjected to vacuum filtration throughout membrane filters (0.45 μm; Chemland, Gyeonggi-do, Korea) [24]. Sludge prepared in the following manner was measured for: SCOD, N_{TOT} and P_{TOT} concentrations. In order to determine the concentration of extracellular polymeric substances the thermally extraction was conducted. At the beginning, WAS sample was centrifuged at $2000\times$ g for 20 min. The residual sludge pellet was resuspended in the distilled water right to the original volume and placed into the water bath (at 80 °C) for 1 h. After incubation sludge was separated from extract by centrifugation at $2000\times$ g and $4500\times$ g over 20 min each time, respectively [25,26]. In the obtained extracts the concentration of proteins and carbohydrates, were measured. The albumin bovine (Acros Organics) and glucose (POCH), were used as a standard solutions. Moreover, the measurements of: pH, TS, VS, CST and sludge temperature increase (ΔT), as well as microscopic examination, were determined only in the sludge samples. Whereas, the biogas volume and composition (CH_4, CO_2, H_2S) were measured during sludge anaerobic digestion. The list of methods/devices used for sludge samples analysis, before and after UD and/or AD processes (including type of the obtained effects) are shown in Table 2.

Table 2. The list of methods used in this study.

Parameter	Methods/Devices	Reference	Type of Effects
pH	Electrometric method; Multi HQ40D (Hach Lange)	PN–EN 12176:2004 [27]	Direct; Technological
TS	Weight method (at 105 °C)	PN–EN 12880:2004 [28]	Technological;
VS	Weight method (at 550 °C)	PN–EN 12879:2004 [29]	Technological
$SCOD_0$; $SCOD_{UD}$	Potassium dichromate method; measurement tests LCI 400, LCK 014 (Hach Lange); UV–VIS DR 5000	ISO 15705:2002 [30]	Direct
N_{TOT}	Potassium oxidation method measurement tests LCK 238 (Hach Lange); UV–VIS DR 5000	ISO 11905-1:1997 [31]	Direct, Technological

Table 2. *Cont.*

Parameter	Methods/Devices	Reference	Type of Effects
P_{TOT}	Ammonium molybdate method; measurement tests LCK 350 (Hach Lange); UV–VIS DR 5000	ISO 6878:2004 [32]	Direct, Technological
Proteins	Folina—Ciocalteau reagent; UV–VIS DR 5000	Lowry et al., 1951 [33]	Direct
Carbohydrates	Phenol-Sulphuric acid reaction method; UV–VIS DR 5000	Dubois et al., 1956 [34]	Direct
CST	Quantity measurement; Capillary suction timer (Envolab)	PN–EN 14701–1:2007 [35]	Direct, Technological
ΔT	Quantity measurement; Digi—Sense (Cole—Parmer)	-	Direct
Flocs disruption	Optical microscopy; Optical microscope (MOTIC BA400)	-	Direct
Biogas production	Quantitative measurement; MULTITEC 540 gauge (Sewerin)	-	Technological
Biogas composition	Qualitative measurement; CH_4 and CO_2 (% vol.), H_2S (ppm); MULTITEC 540 gauge (Sewerin)	-	-

TS—total solids; VS—volatile solids; $SCOD_0$—SCOD of the supernatant of the original sludge; $SCOD_{UD}$—SCOD value of the supernatant of the disintegrated sludge; TN—total nitrogen; TP—total phosphorus; CST—capillary suction time; ΔT—temperature increase.

2.2. Experimental Design and Operating Conditions

2.2.1. Ultrasonic Disintegration

Two different experimental ultrasonic devices were applied for WAS ultrasonic disintegration. The first device (A) consists of disintegrator of high power disintegrator WK-2010, equipped with "sandwich" head and "lens" emitter (designed and manufactured by SemiInstruments, Zabrze, Poland). This device is also equipped with a gauge allowing to read the frequency, as wells as with a steel chamber where the UD of sludge samples takes place. The second device (B) consists of ultrasonic washer which is a rectangular chamber, equipped with a single "flat" emitter of "sandwich" type placed in the bottom of the chamber (designed and manufactured by ZUT Intersonic S.C., Olsztyn, Poland). The technical characteristics and operating conditions of the experimental devices are shown in Table 3.

Table 3. Technical characteristics and operating conditions of the experimental ultrasonic devices.

Parameter	Symbol	Unit	WK-2010	Ultrasonic Washer
Power	P	W	650	90
Frequency	f	kHz	25	25
Number of emitters	-	-	1	1
Emitter surface area	A_E	cm^2	78.5	19.6
Emitter diameter	d_E	cm	10	5
Emitter position	h_E	cm	1 [1]	built
Chamber dimensions	idc × dc	cm	14 × 7 [2]	15 × 13.7 × 20 [3]
Chamber volume	V_C	mL	1000	2000

[1] emitter position relative to the sludge mirror; [2] dimension of chamber for sludge UD (ultrasonic disintegration) in relation to WK-2010 (internal diameter, depth); [3] internal dimensions of the ultrasonic washer (front, side, depth).

The WAS samples pretreatment were performed under a constant energy supply (over time) per sludge volume, i.e., E_V = 160 kWh·m^{-3} (E_S—in the range of 10,868–23,226 kJ·kg^{-1} TS; app.

15,111 kJ·kg^{-1} TS). Sludge samples were disintegrated for: 267 s (WK-2010) and 1920 s (ultrasonic washer). The sample volume was constant and equal V = 0.3 L. The ultrasonic density (U$_D$) for each of experimental device was as follows: 2.2 W·cm^{-3} (WK-2010) and 0.3 W·cm^{-3} (ultrasonic washer). Whereas, the ultrasonic intensity (U$_I$) was: 2.6 W·cm^{-2} (WK-2010) and 1.9 W·cm^{-2}, respectively. The above conditions were the most favorable ones and were determined in the course of previously carried out studies. The variable parameter during sludge UD was the amount of specific energy (E$_S$), which depends on TS content. The values characterizing the ultrasound field and the amount of energy supplied to the process were calculated by the following Equations (1)–(4):

$$E_V = \frac{P \times t}{V} \tag{1}$$

$$E_S = \frac{P \times t}{V \times TS} \tag{2}$$

$$U_D = \frac{P}{V} \tag{3}$$

$$U_I = \frac{P}{A_E} \tag{4}$$

where: E$_V$—energy supplied per sludge volume (kWh·m^{-3}) [16]; E$_S$—specific energy (kJ·kg$_{TS}$$^{-1}$) [17]; U$_D$—ultrasonic density (W·cm^{-3}) [36]; U$_I$—ultrasonic intensity (W·cm^{-2}) [16]; P—power of the ultrasonic generator (W; kW); V—volume of a sludge sample (m^3·cm^3); t—sonication time (s); TS—total solids; (g·L^{-1}); A$_E$—surface area of the disintegration (cm^2).

2.2.2. Anaerobic Digestion

The anaerobic digestion was conducted in the installation consisting of the anaerobic glass digesters with a working volume of 0.5 L, water bath—in order to maintain a constant temperature, columns to collect and measure the volume of evolved biogas and MULTITEC 540 gauge (Sewerin, Warszawa, Poland)—applied for composition analysis. AD was conducted under mesophilic conditions (37 °C ± 0.5 °C) over 20 days, each time. The examined sample was a mixture of: digested sludge (30%) collected from the fermentation chamber of the Central WWTP in Gliwice and ultrasonically pretreated one (70%) from the same WWTP. Control sample constituted mixture of untreated and digested sludge. During the AD process each digester was shaken manually three times per day to prevent the sludge from settling. The process was conducted once a month (for a period of 7 months). The research position was made by Wytwórnia Przyrządów Laboratoryjnych (WPL) Gliwice, according to DIN 38414-8:1985 [37].

2.3. Direct and Technological Effects

In order to determine the direct and technological effects of sludge UD process, selected indicators were used (defined in Section 2.3.1). The assessment of the direct effects included changes of: pH value, concentration of SCOD, P$_{TOT}$, N$_{TOT}$, proteins and carbohydrates, as well as CST measurement and sludge flocs optical microscopy analysis. Whereas, the technological effects observed after completion of sludge AD included the same parameters as for the direct once (except SCOD, extracellular polymeric substances (EPS) and microscopic analysis). Moreover, process of AD was also controlled by the volume and composition of the evolved biogas, as well as by measuring the rate of TS and VS reduction.

2.3.1. Indicators of the Direct and Technological Effects

In order to evaluate the direct and technological effects of WAS ultrasonic disintegration, a comprehensive analysis of the obtained results was conducted. The disintegration degree (DD$_{COD}$) was determined using indicator proposed by Müller [14] (Equation (4). Whereas, the magnitude of the obtained effects was assessed using author's indicators (ID$_i$, IT$_i$, IT$_d$) based on the ratio of the concentration or value of specific compound resulting from changes in the sludge characteristics, as a result of its pretreatment [13] (Equations (5) and (7)):

$$DD_{COD} = \frac{SCOD_{UD} - SCOD_0}{SCOD_{NaOH} - SCOD_0} \times 100 \tag{5}$$

where: DD_{COD}—disintegration degree of (%), $SCOD_0$ and $SCOD_{UD}$—supernatant COD of the original and disintegrated sample (mg·L^{-1}), $SCOD_{NaOH}$—the maximum COD_{NaOH} obtained by alkaline hydrolysis (0.5 M NaOH, ratio of 1:1 for 22 hours at 20 °C) (mg·L^{-1}).

$$IDi; ITi = \frac{C_{UD}}{C_{NUD}} \tag{6}$$

$$IDd; ITd = \frac{C_{NUD}}{C_{UD}} \tag{7}$$

where: IDi—direct effects indicator relating to the increase of concentration or value of specific compound in sludge/supernatant, in the process of UD; ITi—technological effects indicator relating to the increase of concentration or value of specific compound in sludge/supernatant, in the process of UD, observed after AD; IDd—direct effects indicator relating to the decrease of concentration or value of specific compound in sludge/supernatant, in the process of UD; ITd—technological effects indicator relating to the decrease of concentration or value of specific compound in sludge/supernatant, in the process of UD, observed after AD; C_{NUD} and C_{UD}—concentration or value of a specific compound in the sludge/supernatant of non—disintegrated and disintegrated sludge, respectively (mg·L^{-1}), (s).

2.4. Statistical Analysis

In order to evaluate the obtained results, a comprehensive statistical analysis was conducted. The calculations were performed with Statistica 12.0 (StatSoft) and Excel 2013 (Microsoft Office Standard). To check the differences in the mean concentrations of specific compounds between related groups of variables (sludge characteristics before and after UD and AD processes), *T*-Test was used. The occurrence of a linear correlation between analyzed variables was evaluated by Pearson's correlation coefficient (r). To determine whether any of the differences between the means are statistically significant, one-way analysis of variance (ANOVA) was used. Tests were carried out with a confidence level of 95%.

3. Results and Discussion

3.1. Direct Effects

The characteristics of WAS before and after ultrasonic pretreatment are shown in Table 4. The values of parameters analyzed in WAS (collected periodically for 7 months), were in the range of: 6.9–7.2 (pH); 24.8–53.2 g·L^{-1} (TS); 24.8–53.2 g·L^{-1} (VS); 46.6–82.7 mg·L^{-1} (SCOD); 8.2–12.4 mg·L^{-1} (N_{TOT}), 16.2–37.7 mg·L^{-1} (P_{TOT}); 624.6–1228.2 mg·L^{-1} (proteins); 241.7–821.7 mg·L^{-1} (carbohydrates) and 7–13 s (CST). Among all examined parameters, the pH was characterized by the lowest variability (CV = 2.0%), whilst the concentration of carbohydrates by the highest one (CV = 35.6%).

As a result of WAS pretreatment, the pH value slightly decreased and was in the range from 6.7 to 7.2 and 6.5 to 7.1, for sludge disintegrated in the WK-2010 (WAS_A) and ultrasonic washer (WAS_B), respectively. The highest variability was observed for sludge pretreated in the ultrasonic washer (CV = 2.7%). The decrease of pH was also confirmed by IDd_{pH}, which values were lower than 1.0 (for this reason, it was not included in Figure A1). The above observations were in good agreement with author's previous research [13]. The influence of ultrasonic pretreatment on decreasing of pH value, was also confirmed by other scientist [38], who claimed that it is probably posed by formation of acidic compound resulted from flocs disintegration. Whereas, other researches indicated that ultrasonic pretreatment did not change the pH of sludge [39]. It was also suggested that for effective sludge ultrasonic disintegration, the value of pH must be adjusted to a suitable level [15,40].

Table 4. The characteristics of the WAS before and after ultrasonic pretreatment—direct effects (n = 7).

Parameter	Unit	WAS_NUD	WAS_A	WAS_B
			Mean; CV	
pH	-	7.0 (2.0%)	6.9 (2.2%)	6.8 (2.7%)
TS	$g \cdot L^{-1}$	41.1 (26.9%)	-	-
VS	$g \cdot L^{-1}$	29.8 (29.2%)	-	-
$SCOD_0$ /$SCOD_{UD}$	$mg \cdot L^{-1}$	66.8 (22.8%)	1061.7 (34.8%)	4573.4 (39.2%)
N_{TOT}	$mg \cdot L^{-1}$	10.9 (16.3%)	139.9 (23.8%)	470.4 (37.5%)
P_{TOT}^{1}	$mg \cdot L^{-1}$	25.1 (31.4%)	135.6 (22.5%)	177.0 (18.7%)
Proteins	$mg \cdot L^{-1}$	888.9 (24.9%)	784.0 (35.1%)	927.5 (29.2%)
Carbohydrates	$mg \cdot L^{-1}$	598.4 (35.6%)	480.7 (47.5%)	507.0 (46.9%)
CST	s	9.0 (24.8%)	343.1 (70.0%)	1175.1 (10.2%)
ΔT	°C	-	14.0 (39.7%)	34.0 (29.5%)

WAS_NUD—WAS before pretreatment; WAS_A—WAS after pretreatment in the WK-2010; WAS_B—WAS after pretreatment in the ultrasonic washer; CV—coefficient of variation; TS—total solids; VS—volatile solids; $SCOD_0$—SCOD of the supernatant of the original sludge; $SCOD_{UD}$—SCOD value of the supernatant of the disintegrated sludge; TN—total nitrogen; TP—total phosphorus; CST—capillary suction time; ΔT—temperature increase.

The concentrations of SCOD in the supernatant after sludge pretreatment, were in the range of 741.0–1827.0 $mg \cdot L^{-1}$ (WAS_A) and 1842.0–6935.0 $mg \cdot L^{-1}$ (WAS_B). Higher variability of examined parameter was observed for sludge undergone disintegration in the ultrasonic washer (CV = 39.2%). The biggest increase of SCOD concentration in the sludge supernatant was noted in July and September, when the TS content was in the range of 24.8–35.5 $g \cdot L^{-1}$. The above observations were confirmed by DD_{COD} and IDi_{SCOD}, which maximum values amounted: 37.5 and 91.7% (Figure 1), as well as 24.1; 113.0 (Figure A1) for WAS pretreated in WK-2010 and ultrasonic washer, respectively. The influence of TS content on SCOD release was also confirmed by the other researchers, who claimed that its optimal value should be in the range of 2.0–3.2% [11,41]. However, there have been many studies reporting an increase of SCOD concentration in the supernatant after sludge ultrasonic pretreatment. For example, it was indicated that after 40 min of sludge pretreatment conducted at E_S = 9690 $kJ \cdot kg^{-1}_{TS}$, DD_{COD} reached 57.9% [18]. Moreover, other researchers indicated that the SCOD solubilisation and disintegration degree may depend on the amount of energy input, sonication time or frequency of ultrasonic wave [3,10,16]. However, there is one more important factor, which affects the release of SCOD into supernatant, i.e., the increase of sludge temperature. The conducted experiment showed, that depending on the type of device used for sludge sonication, the increase of sludge temperature ranged from 8 to 22 °C (WK-2010) and 32 to 37 °C (ultrasonic washer). Furthermore, the highest values of ΔT were observed at the lowest TS content and amounted: 22 and 37 °C, respectively (Figure 2). The above observations were confirmed by other researcher, who indicated that at a constant energy input (E_V = 100 $kWh \cdot m^{-3}$), higher increase of sludge temperature was obtained at lower TS content, i.e., from 27.2 to 38.2 °C at TS = 2.0% and 18.0 to 29.0 °C at TS = 4.2% [23]. Moreover, the results of another research revealed, that the increase of sludge temperature may increase linearly with increasing value of specific energy input to the process, i.e., temperature of sludge increased from 22 °C in original sample to 72 °C for disintegrated one, at a maximum E_S = 15,880 $kJ \cdot kg^{-1}$ TS [18]. It is important to note that, the increase of WAS temperature during ultrasonic pretreatment generally has a positive influence on sludge solubilisation, but to avoid of any undesirable effects relating, e.g., with recycling back of leachate (generated after sludge pretreatment) to the biological process, it must be undergo testing and further controlled.

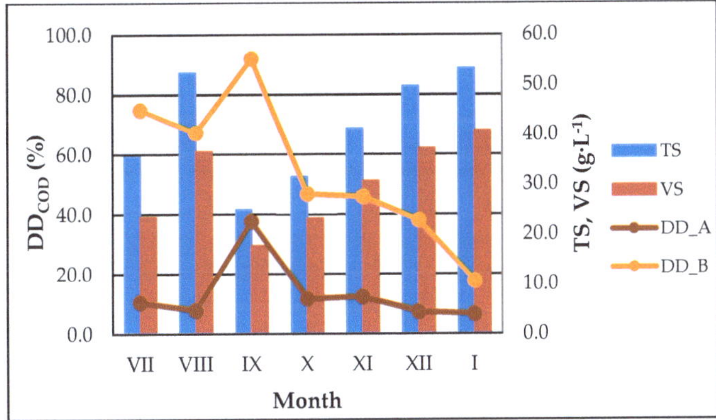

Figure 1. Evolution of disintegration degree (DD$_{COD}$) vs total solids (TS) and volatile solids (VS) content of waste activated sludge (WAS).

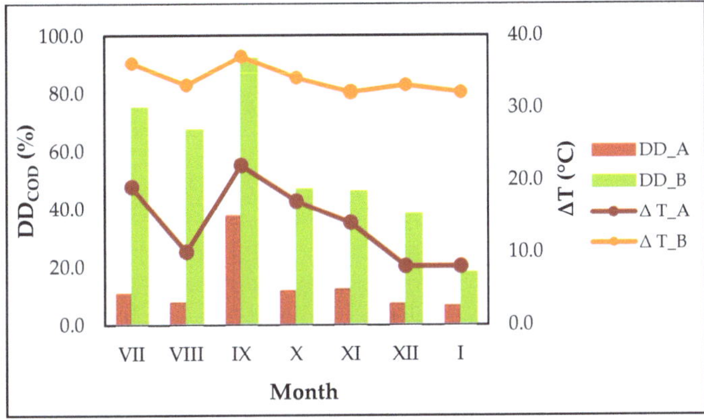

Figure 2. Evolution of disintegration degree (DD$_{COD}$) vs temperature increase (ΔT) of waste activated sludge (WAS).

This study revealed that ultrasonic pretreatment of WAS increased the concentration of biogenic substances in the sludge supernatant. The concentration of total nitrogen ranged from 99.5 to 196.0 mg·L^{-1} (WAS_A) and 210.0 to 719.0 mg·L^{-1} (WAS_B), while for the total phosphorus from 91.5 to 187.0 mg·L^{-1} and 126.0 to 223.0 mg·L^{-1}, respectively. The concentration of N$_{TOT}$ was characterized by higher variability during the experiment duration. The value of CV for N$_{TOT}$ and P$_{TOT}$ were in the range of 23.8–37.5% and 18.7–22.5%, respectively. Almost for all examined sludge samples, the highest values of IDi$_{NTOT}$ and IDI$_{PTOT}$ were obtained in October, i.e., 20.0 (WAS_A), 63.6 (WAS_B) and 8.7 (WAS_A); 11.4 (WAS_B) respectively (Figure A1), while the lowest in January. The above effects were probably additionally strengthened by the increase of temperature during sludge ultrasonic pretreatment. The obtained results are in good agreement with author's earlier research [13]. Moreover, in the scientific papers little attention is given to the release of biogenic substances in the supernatant during the ultrasonic pretreatment of sludge. However, some of the conducted research indicated that increase of specific energy increased the concentration of total nitrogen and phosphorus by 716% and 207.5%, respectively (at E$_S$ = 9690 kJ·kg^{-1} TS) [18]. It is important to note that most part of nitrogen and phosphorus in the sludge supernatant existed in the form of organic products. Thus, the increase of biogenic substances concentration in the supernatant after sludge pretreatment may be associated with the concentration of SCOD or extracellular polymeric substances [17,18]. The above finding was confirmed in this work.

As a result of ultrasonic disintegration of WAS, for most of the analyzed sludge samples the decrease in proteins concentration were observed. The only exception was shown for the sludge samples pretreated in ultrasonic washer, where the increase of proteins content was observed (except of September and November, when TS content was the lowest and ΔT almost the highest). The proteins concentration in the supernatant fluctuated in the range from: 465.3 to 1213.5 mg$\cdot$L^{-1} (WAS_A) and 618.5 to 1310.9 mg$\cdot$L^{-1} (WAS_B). The above results were confirmed by the values of IDi$_{PROT.}$ and IDd$_{PROT.}$, which ranged from: -1.0 to -1.4 (WAS_A) and -1.3 to 1.3 (WAS_B) (Figure A1). Moreover, similar observations were made for carbohydrates, where their concentration in the supernatant was in the range of: 108.3–754.2 mg$\cdot$L^{-1} (WAS_A) and 115.0–790.0 mg$\cdot$L^{-1} (WAS_B). The values of IDd$_{CARBS}$ amounted: -1.1 to -2.2 (WAS_A) and -1.0 to -2.1 (WAS_B) (Figure A1). The highest variability of above extracellular polymeric substances was observed for sludge samples disintegrated in the WK-2010, i.e., CV = 35.1% and CV = 47.5%, for proteins and carbohydrates, respectively. Moreover, sum of EPS, during whole time of experiments duration were decreased. The obtained results were in the good agreement with those obtained by other researchers. For example, it was claimed that the increase of EPS concentration in the supernatant after sludge ultrasonic pretreatment (ES in the range of 0.1–50 kJ$\cdot$kg^{-1} TS) increased the concentration of above substances initially, but with increasing energy supplied to the process their content decreased [19]. Whereas, other researchers indicated that ultrasonic pretreatment of WAS, conducted at high level of specific energy (up to 26,000 kJ$\cdot$kg^{-1} TS) increased the concentration of EPS in the sludge supernatant [17]. However, it must be emphasized, that in the above mentioned study, sludge temperature raised only by 4 °C. It is important to note that proteins and carbohydrates constitute one of the most important components of EPS (they are part of sludge flocs). Thus, its presence in sludge play a significant role in regulating sludge dehydration ability [42], which was confirmed in this work. For example, some researchers indicated that sludge dewaterability may initially increase with the increase of EPS and then decreased when EPS content exceeded a certain threshold [43].

Ultrasonic pretreatment of WAS caused deterioration of the sludge dewaterability expressed by the increase of CST value, which ranged from 147 to 627 (A) and 1042 to 1396 (B). The highest values of CST were noted in September, while the lowest in January, when the TS content was 24.8 g$\cdot$L^{-1} and 53.2 g$\cdot$L^{-1}, respectively. The values of IDi$_{CST}$ were in the range of 6.2–102.6 (WAS_A) and 78.2–199.4 (WAS_B) (Figure A1). Similar observations were made by other researcher, who claimed, that at constant energy input to the process of sludge ultrasonic pretreatment, higher values of CST were obtained at lower TS content [11]. Other scientists indicated that CST may increase with an increase of specific energy input to the process [7]. Moreover, generally it was revealed that CST = 20 s is regarded as representative for a good dewatered sludge [44]. In this study, deterioration of sludge dewaterability was probably caused by the fragmentation of solid fraction and weakens the internal structure, which resulted in the increase of sludge surface characterized with very small particles. Deterioration of the sludge dehydration could also be associated with the release of EPS during sludge pretreatment [42].

In order to compare the changes in flocs structure, occurring as a result of WAS pretreatment, three samples were selected and subjected to microscopic examination. They differed in TS content. The first one (S1) was characterized by a lowest TS (24.8 g$\cdot$L^{-1}); second one (S2) by mean (35.5 g$\cdot$L^{-1}) and third one by a highest TS content (53.2 g$\cdot$L^{-1}), among of all samples analyzed during seven months of the experiment conducting. The collection of above samples was conducted in: September, July and January, respectively. The results were presented depending on the type of disintegrator used for sludge pretreatment, i.e., WK-2010 (A1–A3) and ultrasonic washer (B1–B3) (Figure 3). The optical microscopy analysis indicated significant differences in the flocs structure before and after sludge pretreatment. The most visible changes were observed for WAS disintegrated in the ultrasonic washer. A strong breakdown and dispersion of the flocs were observed. The experiment showed, that both sludge characteristics and technical conditions of process conducting, have a considerable impact on the sludge structure. The above observations were confirmed in author's previous studies [13,24]. Moreover, the usefulness of microscopic analysis in assessing the direct effects of sludge ultrasonic disintegration, has also been demonstrated by other researchers [5,17].

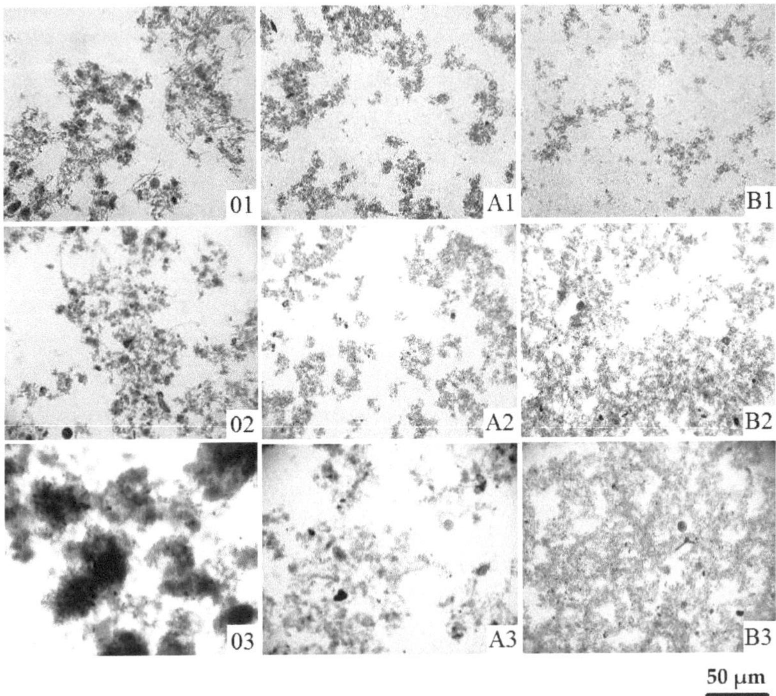

Figure 3. Photomicrographs of the: non-disintegrated (**01–03**) and disintegrated WAS (**A**–WK-2010; **B**–ultrasonic washer) at different total solids (TS) content: 24.8; 35.5 and 53.2 g·L^{-1}, respectively (100× magnification).

3.2. Technological Effects

The characteristics of inoculum, as well as samples containing untreated or disintegrated WAS (mixed with inoculum), before (S_0; S_A; S_B, respectively) and after the process of AD (S_0*; S_A*; S_B*, respectively) are shown in Table 5.

It was indicated that mixtures containing WAS after ultrasonic pretreatment (S_A; S_B), were characterized by higher concentrations and values of examined parameters, compared to the control sample (S_0). The highest variability in the mixtures before AD was expressed by P_{TOT} concentration and CST values. Similar relations were observed in with reference to mixtures after AD process. For both samples, before and after AD, the lowest variability was observed for pH value.

The conducted experiment indicated, that after AD process, following changes in the examined mixtures were observed: the increase of pH, N_{TOT} and P_{TOT} concentration, as well as the reduction of TS and VS content and increase of biogas production. While, CST values were generally decrease. The above observations were confirmed by ITi and ITd (Figure A2). Moreover, due to the low values of Iti_{pH} (<1.0), this parameter was not included in Figure A2.

Table 5. The characteristics of the sludge sample before and after anaerobic digestion—technological effects.

Parameter	Unit	Inoculum	S_0	S_0*	S_A	S_A*	S_B	S_B*
					Mean; CV (%)			
pH	-	7.4 $\pm$ (1.7%)	7.5 (1.4%)	7.5 (1.1%)	7.3 (1.7%)	7.6 (1.2%)	7.3 (1.6%)	7.6 (1.3%)
TS	$g \cdot L^{-1}$	26.2 (15.1%)	36.8 (19.8)	30.2 (21.3%)	37.5 (19.0%)	29.2 (20.6%)	38.7 (17.9%)	28.6 (19.4%)
VS	$g \cdot L^{-1}$	15.9 (14.3%)	25.2 (20.2%)	18.7.1 (21.6%)	26.7 (20.5%)	18.2 (21.8%)	27.4 (18.8%)	17.7 (20.3%)
N_{TOT}	$mg \cdot L^{-1}$	728.1 (11.3%)	210.6 (7.1%)	925.6 (18.5%)	277.0 (7.5%)	951.8 (19.4)	436.6 (21.9%)	1064.6 (16.8%)
P_{TOT}	$mg \cdot L^{-1}$	116 (44.5%)	68.7 (52.6)	288.6 (54.7%)	130.4 (37%)	270.1 (60.3%)	145.8 (35.7%)	258.8 (64.1%)
CST	s	186.4 (26.6%)	42.6 (30.6%)	78.6 (63.8%)	352.4 (55.9%)	215.3 (43.8%)	1068.3 (10.7%)	342.7 (42.2%)
Total biogas production	cm^3	-	-	2014.0 (32.4%)	-	2276.0 (34.0%)	-	2411.0 (29.0%)
CH_4	%vol.	-	-	65.6 (5.7%)	-	66.0 (5.1%)	-	67.9 (4.6%)
CO_2	%vol.	-	-	25.9 (16.6%)	-	25.6 (15.8%)	-	25.9 (12.1%)
H_2S	ppm	-	-	<1.0	-	<1.0	-	<1.0

S_0—sample before AD containing inoculum and original sludge; S_0*—sample after AD containing inoculum and original sludge; S_A—sample before AD containing inoculum and WAS_A; S_A*—sample after AD containing inoculum and WAS_A; S_B—sample before AD containing inoculum and WAS_B; S_B*—sample after AD containing inoculum and WAS_B; TS—total solids; VS—volatile solids; N_{TOT}—total nitrogen; P_{TOT}—total phosphorus; CST—capillary suction time; methane (CH_4), carbon dioxide (CO_2) and hydrogen sulfide (H_2S) content in evolved biogas; vol.—volume.

As a result of ultrasonic disintegration of WAS, in the mixture derived from anaerobic digestion process a high increase of N_{TOT} and P_{TOT} concentration, were observed. The values of ITi_{NTOT} and ITi_{PTOT} ranged: 3.2–5.5; 2.5–4.9; 1.6–4.0 and 3.1–4.9; 1.4–2.9; 1.2–2.5, for S_0*; S_A* and S_B*, respectively (Figure A2). The lowest increase of above mentioned parameters was observed for the mixtures containing sludge after pretreatment in the ultrasonic washer, for which the temperature increase was the highest. It was also indicated, that the increase of biogenic substances in sludge supernatant after AD process was higher in the control sample. The above observations were confirmed by author's previous work, in which ITi_{NTOT} values for the reference samples were in the range of 3.2–5.7 and for mixtures containing disintegrated sludge equaled 2.4–5.5. While in respect to $ITiP_{TOT}$, those values were from 3.5 to 4.9 and 1.5 to 2.8, respectively [13]. Moreover, the increase of biogenic substances in the supernatant after anaerobic digestion of WAS, was associated with the increase of TS and VS content, as well as the decrease of CST value in the samples before the process, which resulted from characteristics of sludge undergone ultrasonic disintegration. Furthermore, the changes in the concentration of total nitrogen in the supernatant after and before sludge AD, were correlated with the increase of biogas production. Whereas, in the case of the total phosphorus, changes in the concentration of this compound, were related with the amount of evolved biogas. The increase in concentration of biogenic substances in the supernatant after sludge AD, was also observed by other researchers [45]. Moreover, these findings can be positive with respect to necessity of recycling the leachate generated after dehydration of digested sludge into the technological line of WWTP.

The reduction of TS and VS content in the samples containing sludge after UD was in the ranged of: 17.8–26.8%; 22.6–29.6% and 29.5–35.6%; 33.1–38.0%, for S_A*; S_B*, respectively. It was showed that the obtained TS and VS reduction was higher in comparison to the control sample by: 4.1; 8.0% and 5.8; 9.5%, for S_A*; S_B*; respectively (Figure 4). The highest decrease of TS and VS content was observed for samples containing WAS after pretreatment in the ultrasonic washer, compared to the control sample. Moreover, the reduction of TS and VS content increased the biogas production. The above results were in good agreement with other researchers, who stated, that ultrasonic pretreatment of sludge, increased VS reduction and biogas production by: 19% and 26%, respectively [46]. Whereas, the results of other investigation, indicated that ultrasonic pretreatment of sludge prior anaerobic digestion, ensured: 12% increase of TS content, compared to the control sample [21].

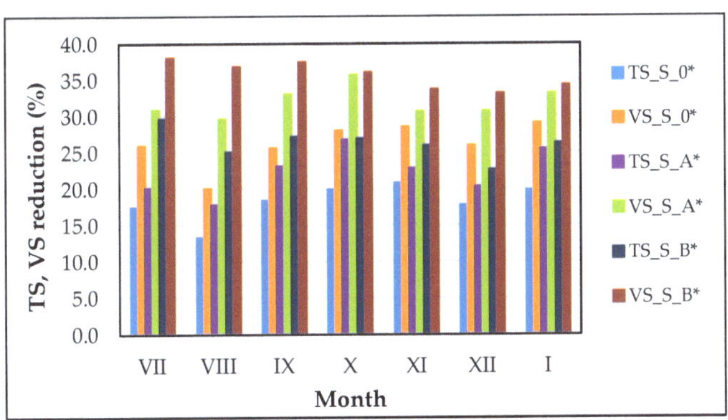

Figure 4. Total solids (TS) and volatile solids (VS) reduction after anaerobic digestion (AD) process.

The changes of CST values in the sludge mixtures after anaerobic digestion revealed, that the best effects of sludge dehydration were obtained for samples containing WAS, after pretreatment in the ultrasonic washer. The values of ITi_{CST} and ITd_{CST} were in the range of: −1.3 to 2.8; −3.9 to 2.4; −1.8 to −7.6, for S_0*; S_A* and S_B*, respectively (Figure A2). The best effects were obtained in the first, third and fourth month of experiments procurement. The obtained results are probably dependent on the characteristic of sludge samples after ultrasonic pretreatment, i.e., the worse susceptibility of sludge to dewatering (larger surface area of the flocks), the greater degree of sludge defragmentation, which favors the course of the anaerobic digestion process [8]. The above observations were in a good agreement with those obtained by other researchers, who achieved 49% reduction of CST after AD process of disintegrated sludge, compared to control sample [47]. Whereas, other scientists reported a 7-fold decrease of CST value (from 2000 s to 267 s) in the examined sludge, in compare to reference sample [48].

The amount of total biogas production observed after the process of sludge anaerobic digestion is presented in Figure 5. Obtained results confirmed the positive effect of WAS ultrasonic pretreatment on the increase of biogas production. The higher volume of evolved biogas was obtained for a mixture containing disintegrated sludge, i.e., 13.0% (S_A*) and 19.7% (S_B*), in compare to control sample. The amount of biogas production did not undergoing a large variability during the experiment conducting, regardless of the technical conditions of WAS pretreatment, before AD process. The highest volume of evolved biogas was obtained from the mixture containing sludge after pretreatment in the ultrasonic washer (2411 cm^3). The above observations were in a good agreement with the results obtained by other researchers. For example, it was stated that ultrasonic pretreatment of WAS at 20 kHz and P = 200 W, increased biogas production by 6.3% [40]. Whereas, other scientists achieved of 8.6% to 31.4% improvement of biogas production, conducting ultrasonic disintegration, at specific energy, in the range of: 15,000 to 35,000 kJ·kg^{-1} TS [10].

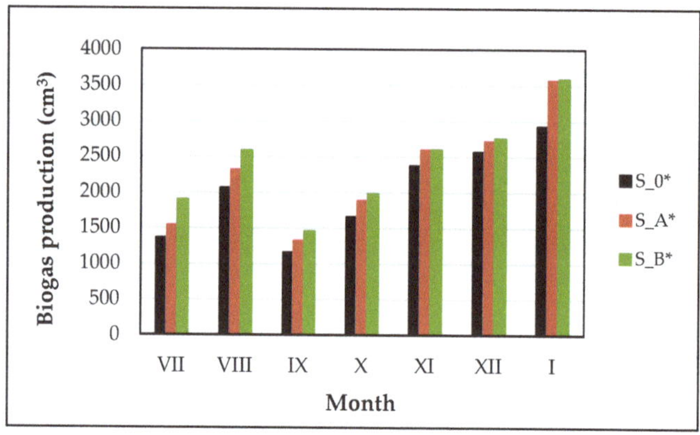

Figure 5. Biogas production after anaerobic digestion (AD) process.

This study revealed, that there were no significant differences between the qualities of biogas originating from the mixtures containing untreated and disintegrated WAS. The concentrations of CH_4, CO_2 and H_2S ranged from: 65.3 to 67.9% and 25.6 to 25.9%, as well as <1.0 ppm, respectively (Table 5). Similar composition of evolved biogas was indicated by other researchers, i.e., CH_4 from 50 to 70%; CO_2 from 25 to 30% and H_2S, H_2, N_2 < 1% [49,50].

3.3. Statistical Analysis

In order to present the effects of ultrasonic disintegration as a function of WAS characteristics and technical conditions of the conducted process, a comprehensive analysis was carried out. In order to determine if there are any significant differences between examined variables, expressed as the changes in the characteristics of WAS before and after ultrasonic disintegration the *T*-Test was used. The strength of the relationship between periodical changes in the WAS characteristics (parameters) and obtained effects, was expressed as Pearson's correlation coefficient (r). Moreover, to indicate the differences between effects obtained in various technical conditions (disintegrators) of process conducting, the one-way analysis of variance (ANOVA) was carried out.

3.3.1. Analysis of Direct Effects

The results of *T*-Test confirmed existence of significant differences in the characteristics of WAS before and after ultrasonic pretreatment ($p < 0.05$). The only exceptions were: pH for samples disintegrated in WK-2010, as well as proteins for ultrasonic washer ($p > 0.05$) (Table A1).

The results of Pearson's correlation are shown in Table A2 (WK-2010) and Table A3 (ultrasonic washer). A detailed analysis of correlation matrix revealed the existence of different relationships between the characteristics of sludge supernatant before and after WAS pretreatment. Moreover, apart from the statistically significant correlations, the ones for which "r" was higher than 0.65, were also taken into consideration. Taking into account the number of correlations between analyzed variables, it can be stated that the most important characteristics of WAS, which affects the magnitude of the direct effects, are: TS, VS, ΔT > EPS > SCOD > CST > N_{TOT}, P_{TOT} > pH. Moreover, data analysis also indicated that, the increase of sludge temperature during ultrasonic pretreatment of WAS exerted a significant impact on its final characteristics.

3.3.2. Analysis of Technological Effects

The *T*-Test results showed a significant differences in the characteristics of mixtures consisting of inoculum and pretreated WAS, prior and after anaerobic digestion ($p < 0.05$). The only exception was: CST for samples containing WAS after pretreatment in the ultrasonic washer ($p > 0.05$) (Table A4).

Furthermore, the results of Pearson's correlation revealed statistically significant differences between the characteristics of mixtures prior and after completion of anaerobic digestion (Tables A5 and A6). The above observations occurred regardless of the technical conditions of ultrasonic disintegration. In conclusion, it was found that the most important sludge parameters, which affect the magnitude of the technological effects, are: TS, VS > CST > N_{TOT}, P_{TOT} > pH. The sequence of above parameters was determined based on the number of correlations between analyzed variables. Moreover, taking into account VS reduction and increase of biogas production, it can be said that increase of sludge temperature during ultrasonic disintegration most likely causes the difference in its biodegradation.

3.3.3. Analysis of Technological Conditions of Process Conducting

The analysis of variance (ANOVA) showed that in the case of direct effects, there is a basis for rejecting the null hypothesis (H0) on the about the absence of statistically significant differences between the variables in considered groups (except pH, proteins and carbohydrates) (Table A7). It means, that above mentioned effects differ depending on the type of experimental device used in the process ($p < 0.05$). Whereas, the results of statistical analysis in accordance to technological effects indicated existence of significant differences between analyzed variables, but only in the case of changes in the CST values (Table A8). It means that technological effects are dependent on the characteristics of sludge subjected to ultrasonic disintegration, but only indirectly. However, in contradiction to results of the analysis of variance, the values of disintegration indicators indicated that there were a differences in the magnitude of technological effects depending on the type of experimental device used in the process of WAS ultrasonic disintegration.

In conclusion, according to the results of statistical analysis and values of applied indicators, it could be inferred that the most promising conditions for ultrasonic pretreatment of WAS (conducting at constant energy input), are: low power, long sonication time, large surface area of the emitter.

4. Conclusions

This article presents the results of a comprehensive analysis of direct and technological effects of disintegration as a function of periodical changes in the characteristics of WAS, as well as technical conditions of the conducted process. In this purpose selected parameters were considered, i.e., pH value, SCOD, N_{TOT}, P_{TOT}, proteins and carbohydrates concentrations, CST value, floc disruption,

TS and VS content, before and after sludge ultrasonic disintegration or anaerobic digestion. To evaluation of direct and technological effects the commonly applied (DD_{COD}) and author's (IDi, IDd) indicators were used.

As a result of ultrasonic pretreatment, in the sludge or supernatant the increase of all examined parameters, except proteins and carbohydrates, were observed. Moreover, it was also showed that after completion of sludge anaerobic digestion, the increase of pH, N_{TOT} and P_{TOT} concentration, sludge dewaterability, as well as the reduction of TS and VS content and increases of biogas production were observed. Furthermore, the conducted experiment revealed that ultrasonic disintegration reduces the content of biogenic substances in the sludge supernatant after its anaerobic digestion. This information is positive, especially with respect to necessity of recycling the leachate, generated after dehydration of sludge undergone anaerobic digestion into the technological line.

The result of *T*-Test showed that there were significant differences between the characteristics of untreated and disintegrated WAS, regardless of the technological conditions of process conducting. The above observations were related both to direct and technological effects. Whereas, Pearson's correlation confirmed, that changes in the characteristics of untreated WAS influencing effects of ultrasonic pretreatment. It was indicated that, the most important parameters, which affected the magnitude of the direct and technological effects, were: TS, VS, ΔT > EPS > SCOD > CST > N_{TOT}, P_{TOT} > pH and TS, VS > CST > N_{TOT}, P_{TOT} > pH, respectively. Furthermore, the analysis of variance showed, that the most significant differences between the effects obtained in various experimental devices were observed for: SCOD, N_{TOT}, P_{TOT}, CST, ΔT (direct effects) and CST (technological effects). It was found, that the most favorable effects of sludge ultrasonic pretreatment can be obtained conducting the process in the device characterized with low power, long sonication time and large surface area of the emitter. The results obtained in this study also confirmed the significant impact of the increase of sludge temperature during ultrasonic disintegration on the obtained effects.

Funding: This work was supported by the "DoktoRis"—Program for innovative Silesia 2007–2013, co-financed by the European Union under the European Social Found [No. 100127].

Acknowledgments: The author thank Ewa Zielewicz of Silesian University of Technology, Poland for her support throughout this study and lending the devices (disintegrators) used for experiments.

Conflicts of Interest: The author declares no conflict of interest.

Appendix A

Table A1. The results of *T*-Test—direct effects.

Variables	WAS_A	WAS_B
pH	0.149546	**0.021015** *
SCOD	**0.000456** *	**0.000558** *
N_{TOT}	**0.000060** *	**0.000463** *
P_{TOT}	**0.000037** *	**0.000010** *
Proteins	**0.018871** *	0.478267
Carbohydrates	**0.000543** *	**0.000457** *
CST	**0.010617** *	**0.000000** *
T	**0.000540** *	**0.000000** *

* bold—significant, $p < 0.05$; SCOD—soluble chemical oxygen demand; N_{TOT}—total nitrogen; P_{TOT}—total phosphorus; CST—capillary suction time; T—temperature.

Table A2. Pearson's correlation coefficients for the direct effects of WAS ultrasonic disintegration WK-2010).

Variables	pH	SCOD	N_{TOT}	P_{TOT}	Proteins	Carbohydrates	CST
pH	0.00	−0.37	−0.17	−0.35	−0.30	−0.05	−0.01
SCOD	−0.32	−0.75 **	−0.56	0.19	0.42	0.68 **	−0.71 **
N_{TOT}	0.29	−0.72 **	−0.61	−0.10	0.37	**0.85 ***	**−0.87 ***
P_{TOT}	−0.51	−0.20	−0.74	0.54	0.70	0.46	−0.39
Proteins	−0.68 **	−0.35	−0.27	**0.92 ***	**0.96 ***	0.57	−0.48
Carbohydrates	−0.21	**−0.88 ***	−0.43	0.45	**0.79 ***	**0.98 ***	**−0.97 ***
CST	−0.44	−0.65 **	−0.60	0.47	0.66 **	0.75 **	−0.72 **
TS	−0.31	**−0.85 ***	−0.54	0.37	0.70 **	**0.90 ***	**−0.94 ***
VS	−0.43	**−0.83 ***	−0.52	0.51	**0.80 ***	**0.90 ***	**−0.92 ***
ΔT	0.47	**0.84 ***	0.53	−0.52	**−0.80 ***	**−0.87 ***	**0.88 ***

* bold—significant correlations, $p < 0.05$; ** non-significant correlations taken into account in statistical analysis ($r \geq 0.65$); SCOD—soluble chemical oxygen demand; N_{TOT}—total nitrogen; P_{TOT}—total phosphorus; CST—capillary suction time; ΔT—temperature increase.

Table A3. Pearson's correlation coefficients for the direct effects of WAS ultrasonic disintegration (ultrasonic washer).

Variables	pH	SCOD	N_{TOT}	P_{TOT}	Proteins	Carbohydrates	CST
pH	0.21	0.44	0.60	−0.16	−0.36	−0.07	−0.24
SCOD	−0.28	0.04	0.06	0.25	0.33	0.65 **	**−0.79 ***
N_{TOT}	0.29	0.22	0.02	−0.01	0.53	**0.84 ***	−0.58
P_{TOT}	−0.42	−0.55	−0.51	0.55	0.51	0.44	−0.34
Proteins	−0.66 **	**−0.92 ***	**−0.95 ***	**0.95 ***	**0.87 ***	0.58	−0.58
Carbohydrates	−0.21	−0.31	−0.47	0.56	**0.88 ***	**0.98 ***	**−0.91 ***
CST	−0.48	−0.28	−0.31	0.44	0.57	0.73 **	−0.74 **
TS	−0.28	−0.17	−0.26	0.46	0.67	**0.88 ***	**−0.91 ***
VS	−0.40	−0.32	−0.40	0.60	0.75 **	**0.89 ***	**−0.94 ***
ΔT	0.59	0.43	0.50	−0.71 **	**−0.77 ***	**−0.76 ***	**0.94 ***

* bold—significant correlations, $p < 0.05$; ** non-significant correlations taken into account in statistical analysis ($r \geq 0.65$); SCOD—soluble chemical oxygen demand; N_{TOT}—total nitrogen; P_{TOT}—total phosphorus; CST—capillary suction time; ΔT—temperature increase.

Table A4. The results of *T*-Test—technological effects.

Variables	WAS_A	WAS_B
pH	**0.019198 ***	**0.000932 ***
TS	**0.000013 ***	**0.000003 ***
VS	**0.000009 ***	**0.000005 ***
N_{TOT}	**0.000113 ***	**0.000463 ***
P_{TOT}	**0.023667 ***	**0.047379 ***
CST	0.191863	**0.000351 ***

* bold—significant, $p < 0.05$; TS—total solids; VS—volatile solids; N_{TOT}—total nitrogen; P_{TOT}—total phosphorus; CST—capillary suction time.

Table A5. Pearson's correlation coefficients for the technological effects of WAS ultrasonic disintegration (inoculum + WAS_A).

Variables	pH	TS	VS	N_{TOT}	P_{TOT}	CST	^{3} Biogas
pH	−0.44	0.03	−0.19	−0.26	−0.65 **	−0.18	−0.42
TS	−0.53	**0.98 ***	**0.99 ***	**0.92 ***	0.57	0.74 **	**0.85 ***
VS	−0.34	**0.92 ***	**0.99 ***	**0.97 ***	0.73 **	**0.81 ***	**0.94 ***
N_{TOT}	0.20	−0.67 **	−0.75 **	−0.73 **	−0.74 **	−0.64	**−0.79 ***
P_{TOT}	0.47	0.22	0.48	0.63	**0.88 ***	0.59	0.74 **
CST	0.61	**−0.91 ***	**−0.88 ***	−0.75 **	−0.37	−0.66 **	**−0.77 ***

* bold—significant correlations, $p < 0.05$; ** non−significant correlations taken into account in statistical analysis ($r \geq 0.65$); Biogas—total biogas production; TS—total solids; VS—volatile solids; N_{TOT}—total nitrogen; P_{TOT}—total phosphorus; CST—capillary suction time.

Table A6. Pearson's correlation coefficients for the technological effects of WAS ultrasonic disintegration (inoculum + WAS_B).

Variables	pH	TS	VS	N_{TOT}	P_{TOT}	CST	Biogas
pH	−0.02	−0.11	−0.32	−0.19	−0.75 **	−0.31	−0.43
TS	−0.64	**0.99 ***	**0.93 ***	**0.88 ***	0.29	**0.78 ***	**0.76 ***
VS	−0.63	**0.97 ***	**0.99 ***	**0.92 ***	0.50	**0.87 ***	**0.89 ***
N_{TOT}	−0.16	−0.26	−0.51	−0.49	**−0.94 ***	−0.62	**−0.76 ***
P_{TOT}	0.17	0.15	0.41	0.31	**0.91 ***	0.57	0.63
CST	0.28	**−0.81 ***	**−0.92 ***	**−0.94 ***	−0.72 **	**−0.88 ***	**−0.93 ***

* bold—significant correlations, $p < 0.05$; ** non−significant correlations taken into account in statistical analysis ($r \geq 0.65$); Biogas—total biogas production; TS—total solids; VS—volatile solids; N_{TOT}—total nitrogen; P_{TOT}—total phosphorus; CST—capillary suction time.

Table A7. The results of ANOVA Test—direct effects.

Variable	Unit	*p* Value
pH	-	0.233545
SCOD	$mg \cdot L^{-1}$	**0.000273 ***
N_{TOT}	$mg \cdot L^{-1}$	**0.000387 ***
P_{TOT}	$mg \cdot L^{-1}$	**0.031615 ***
Proteins	$mg \cdot L^{-1}$	0.344718
Carbohydrates	$mg \cdot L^{-1}$	0.729555
CST	s	0.000003
ΔT	°C	**0.000001 ***

* bold—significant correlations, $p < 0.05$; SCOD—soluble chemical oxygen demand; N_{TOT}—total nitrogen; P_{TOT}—total phosphorus; CST—capillary suction time; ΔT—temperature increase.

Table A8. The results of ANOVA Test—technological effects.

Variable	Unit	*p* Value
pH	-	0.082577
TS	$g \cdot L^{-1}$	0.884421
VS	$g \cdot L^{-1}$	0.792592
N_{TOT}	$mg \cdot L^{-1}$	0.324162
P_{TOT}	$mg \cdot L^{-1}$	0.942013
CST	s	**0.000801 ***
Biogas production	cm^3	0.577505

* bold—significant correlations, $p < 0.05$, TS—total solids; VS—volatile solids; N_{TOT}—total nitrogen; P_{TOT}—total phosphorus; CST—capillary suction time.

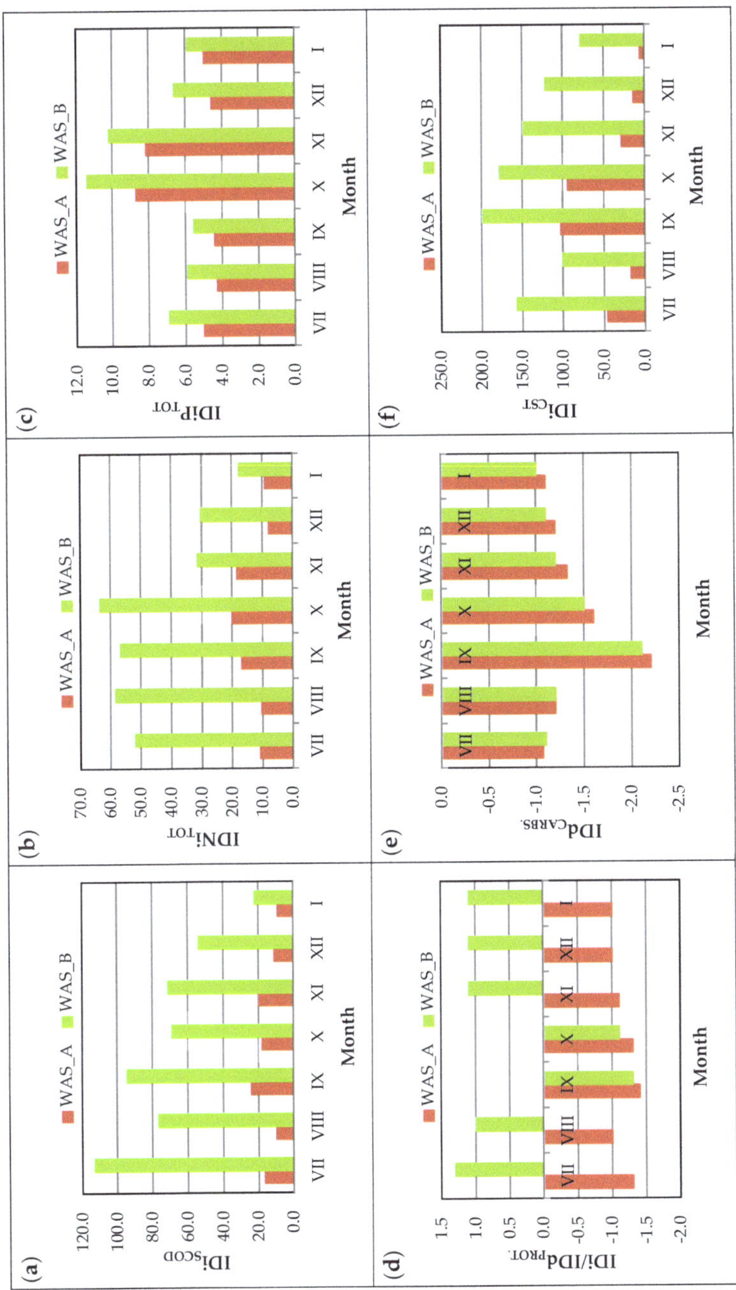

Figure A1. Values of indicators of the direct effects of WAS ultrasonic disintegration: (**a**) IDi_{SCOD}; (**b**) IDi_{NTOT}; (**c**) $IDiP_{TOT}$; (**d**) Idi/IDd_{PROT}; (**e**) IDd_{CARBS}; (**f**) IDi_{CST}. WAS_A—waste activated sludge after pretreatment in the WK-2010; WAS_B—waste activated sludge after pretreatment in the ultrasonic washer.

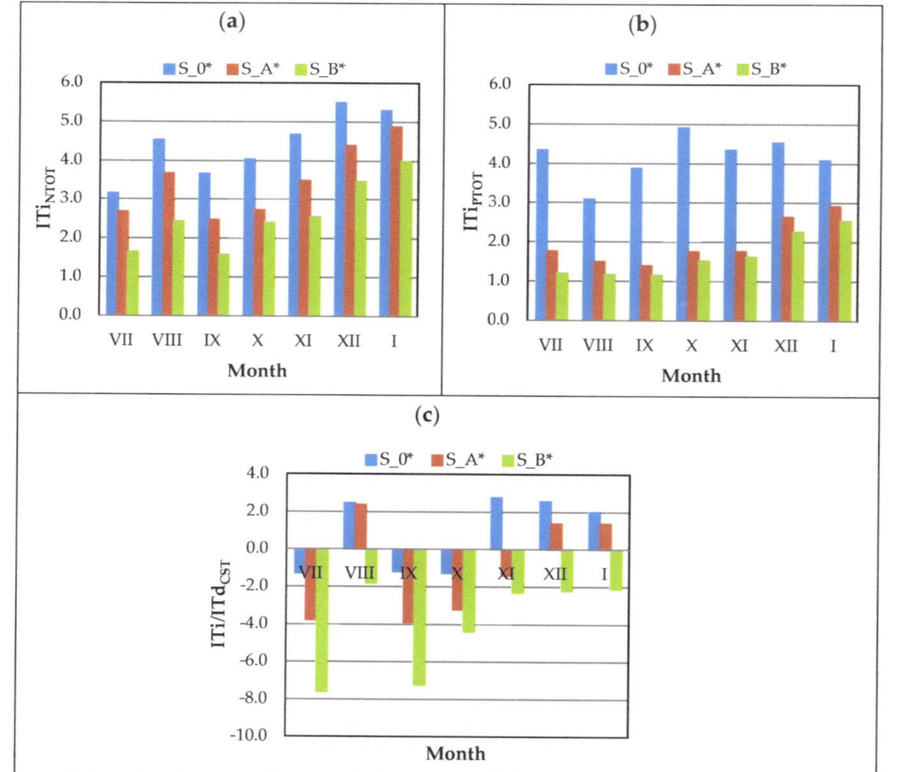

Figure A2. Values of indicators of the technological effects of WAS ultrasonic disintegration: (**a**) ITi$_{NTOT}$; (**b**) ITi$_{PTOT}$; (**c**) ITi/ITd$_{CST}$. S_0*—sample after AD containing inoculum and original sludge; S_A*—sample after AD containing inoculum and WAS_A; S_B*—sample after AD containing inoculum and WAS_B.

References

1. Council Directive 91/271/EEC of 21 May 1991 Concerning Urban Waste—Water Treatment. Available online: https://eur$-$lex.europa.eu/legal$-$content/EN/TXT/PDF/?uri=CELEX:31991L0271&from=EN (accessed on 30 May 1991).
2. Pilli, S.; Bhunia, P.; Yan, S.; LeBlanc, R.J.; Tyagi, R.D.; Surampalli, R.Y. Ultrasonic pretreatment of sludge: A review. *Ultrason. Sonochem.* **2011**, *18*, 1–18. [CrossRef] [PubMed]
3. Lippert, T.; Bandelin, J.; Musch, A.; Drewes, J.E.; Koch, K. Energy-positive sewage sludge pre-treatment with a novel ultrasonic flatbed reactor at low energy input. *Bioresour. Technol.* **2018**, *264*, 298–305. [CrossRef] [PubMed]
4. Milieu Ltd. Final report for the European Commission, Milieu Ltd, WRc, RPA, DG Environment 2008. Environmental, Economic and Social Impacts of the Use of Sewage Sludge on Land. Final Report, Part II: Project Interim (DGENV.G.4/ETU/2008/0076r.). Available online: http://ec.europa.eu/environment/archives/waste/sludge/pdf/part_ii_report.pdf (accessed on 29 April 2015).
5. Simonetti, M.; Rossi, G.; Cabbai, V.; Goi, D. Tests on the effect of ultrasonic treatment on two different activated sludge waste. *Environ. Prot. Eng.* **2014**, *40*, 23–34.
6. Le, N.T.; Julcour-Lebigue, C.; Delmas, H. An executive review of sludge pretreatment by sonication. *J. Environ. Sci.* **2015**, *37*, 139–153. [CrossRef] [PubMed]
7. Skórkowski, Ł.; Zielewicz, E.; Kawczyński, A.; Gil, B. Assessment of excess sludge ultrasonic, mechanical and hybrid pretreatment in relation to the energy parameters. *Water* **2018**, *10*, 551. [CrossRef]
8. Koch, K.; Lippert, T.; Hauck Sabadini, N.; Drewes, J.E. Tube reactors as a novel ultrasonication system for trouble—free treatment of sludges. *Ultrason. Sonochem.* **2017**, *37*, 464–470. [CrossRef] [PubMed]
9. Gonzalez, A.; Hendriks, A.T.W.M.; van Lier, J.B.; de Kreuk, M. Pre—treatments to enhance the biodegradability of waste activated sludge: Elucidating the rate limiting step. *Biotechnol. Adv.* **2018**, *36*, 1431–1469. [CrossRef] [PubMed]
10. Lizama, A.C.; Figueiras, C.C.; Herrera, R.R.; Pedreguera, A.Z.; Espinoza, J.E.R. Effects of ultrasonic pretreatment on the solubilization and kinetic study of biogas production from anaerobic digestion of waste activated sludge. *Int. Biodeterior. Biodegrad.* **2017**, *123*, 1–9. [CrossRef]
11. Zielewicz, E. Effects of ultrasonic disintegration of excess sewage sludge. *Top Curr. Chem.* **2016**, *374*, 149–174.
12. Cimochowicz–Rybicka, M. Minimization of Sewage Sludge Production—European Trends and Selected Technologies. Available online: https://www.kth.se/polopoly_fs/1.651120!/JPSU18P12.pdf (accessed on 10 October 2018).
13. Tytła, M.; Zielewicz, E. The impact of temporal variability of excess sludge characteristics on the effects obtained in the process of its ultrasonic disintegration. *Environ. Technol.* **2018**, *39*, 1–13. [CrossRef] [PubMed]
14. Müller, J. Mechanischer Klärschlammaufschluß, Schriftenereihe "Berichte aus der Verfahrenstechnik". Ph.D. Thesis, der Fakultät für Maschinenbau und Elektrotechnik der, Universität Braunschweig. Shaker Verlag, Aachen, Germany, 1996.
15. Wang, F.; Wang, Y.; Ji, M. Mechanisms and kinetics models for ultrasonic waste activated sludge disintegration. *J. Hazard. Mater.* **2005**, *123*, 145–150. [CrossRef] [PubMed]
16. Tiehm, A.; Nickel, K.; Zellhorn, M.; Neis, U. Ultrasound waste activated sludge disintegration for improving anaerobic stabilization. *Water Res.* **2001**, *35*, 2003–2009. [CrossRef]
17. Feng, X.; Lei, H.Y.; Deng, J.C.; Yu, Q.; Li, H. Physical and chemical characteristics of waste activated sludge treated ultrasonically. *Chem. Eng. Process.* **2009**, *48*, 187–194. [CrossRef]
18. Erden, G.; Filibeli, A. Ultrasonic pre—treatment of biological sludge: Consequences for disintegration, anaerobic biodegradability, and filterability. *J. Chem. Technol. Biotechnol.* **2010**, *85*, 145–150. [CrossRef]
19. Cho, S.K.; Shin, H.S.; Kim, D.H. Waste activated sludge hydrolysis during ultrasonication: Two-step disintegration. *Bioresour. Technol.* **2012**, *121*, 480–483. [CrossRef] [PubMed]
20. Ruiz-Hernando, M.; Martinez—Elorza, G.; Labanda, J.; Llorens, J. Dewaterability of sewage sludge by ultrasonic, thermal and chemical treatments. *Chem. Eng. J.* **2013**, *230*, 102–110. [CrossRef]
21. Tomczak-Wandzel, R.; Ofverstrom, S.; Dauknys, R.; Medrzycka, K. Effect of disintegration pretreatment of sewage sludge for enhanced anaerobic digestion. In *The 8th International Conference "Environmental Engineering": Selected Papers, Vilnius, Lithuania, 19—20 May 2011*; Čygas, D., Froehner, K.D., Eds.; Vilnius Gediminas Technical University: Vilnius, Lithuania, 2011; pp. 679–683.

22. Zielewicz, E.; Tytła, M. Effects of ultrasonic disintegration of excess sludge obtained in disintegrators of different constructions. *Environ. Technol.* **2015**, *36*, 2210–2216.

23. Zielewicz, E. Effects of ultrasonic disintegration of excess sewage sludge. *Appl. Acoust.* **2016**, *103*, 182–189. [CrossRef]

24. Tytła, M.; Zielewicz, E. The effect of ultrasonic disintegration process conditions on the physicochemical characteristics of excess sludge. *Arch. Environ. Prot.* **2016**, *42*, 19–26. [CrossRef]

25. Karapanagiotis, N.K.; Rudd, T.; Sterritt, R.M.; Lester, J.N. Extraction and characterisation of extracellular polymers in digested sewage sludge. *J. Chem. Technol. Biotechnol.* **1989**, *44*, 107–120. [CrossRef]

26. Raszka, A.; Surmacz–Górska, J.; Żabczyński, S. Extracellular polymeric substances in the nitrifying activated sludge. *ACEE* **2010**, *3*, 115–119.

27. Polish Committee for Standardization. *Characterization of Sludge–Determination of pH–Value*; PN–EN 12176:2004; Polish Committee for Standardization: Warszawa, Poland, 2004.

28. Polish Committee for Standardization. *Characteristics of Sewage Sludge, Determination of Dry Residue and Water Content*; PN–EN 12880:2004; Polish Committee for Standardization: Warszawa, Poland, 2004.

29. Polish Committee for Standardization. *Characteristics of Sewage Sludge, Determination of Loss on Ignition of Dry Matter*; PN–EN 12879:2004; Polish Committee for Standardization: Warszawa, Poland, 2004.

30. International Organization for Standardization. *Water Quality–Determination of the Chemical Oxygen Demand Index (ST–COD)–Small–Scale Sealed Tube Method*; ISO 15705:2002; International Organization for Standardization: Geneva, Switzerland, 2002.

31. International Organization for Standardization. *Water Quality–Determination of Nitrogen—Part 1: Method using Oxidative Digestion with Peroxodisulfate*; ISO 11905–1:1997; International Organization for Standardization: Geneva, Switzerland, 1997.

32. International Organization for Standardization. *Water Quality–Determination of Phosphorus–Ammonium Molybdate Spectrometric Method*; ISO 6878:2004; International Organization for Standardization: Geneva, Switzerland, 2004.

33. Lowry, O.; Rosebrough, N.; Farr, A.L.; Randall, R. Protein measurement with the Folin phenol reagent. *J. Biol. Chem.* **1951**, *193*, 265–275. [PubMed]

34. Dubois, M.; Gilles, K.A.; Hamilton, J.K.; Reber, P.A.; Smith, F. Colorimetric method for determination of sugars and related substances. *Anal. Chem.* **1956**, *28*, 350–356. [CrossRef]

35. Polish Committee for Standardization. *Characterization of Sludges–Filtration Properties—Part 3: Capillary Suction Time (CST)*; PN–EN 14701–1:2007; Polish Committee for Standardization: Warszawa, Poland, 2004.

36. Neis, U.; Nickel, K.; Tiehm, A. Enhancement of anaerobic sludge digestion by ultrasonic disintegration. *Water Sci. Technol.* **2000**, *42*, 73–80. [CrossRef]

37. German Institute for Standardisation. *German Standard Methods for the Examination of Water, Waste Water and Sludge; Sludge and Sediments (Group S); Determination of the Amenability to Anaerobic Digestion (S 8)*; DIN 38414–8:1985–06; German Institute for Standardization: Berlin, Germany, 1985.

38. Gündüz, Ç. Ultrasonic Disintegration of Sewage Sludge. Master's Thesis, Dokuz Eylül University, Izmir, Turkey, 2009.

39. Zhang, P.; Zhang, G.; Wang, W. Ultrasonic treatment of biological sludge: Floc disintegration, cell lysis and inactivation. *Biores. Technol.* **2007**, *98*, 207–210. [CrossRef] [PubMed]

40. Sahinkaya, S. Disintegration of municipal waste activated sludge by simultaneous combination of acid and ultrasonic pretreatment. *Process Saf. Environ. Prot.* **2015**, *93*, 201–205. [CrossRef]

41. Show, K.Y.; Mao, T.; Lee, D.J. Optimization of sludge disruption by sonication. *Water Res.* **2007**, *41*, 4741–4747. [CrossRef] [PubMed]

42. Zhou, J.; Zheng, G.; Zhang, X.; Zhou, L. Influences of extracellular polymeric substances on the dewaterability of sewage sludge during bioleaching. *PLoS ONE* **2014**, *9*, e102688. [CrossRef] [PubMed]

43. Houghton, J.I.; Quarmby, J.; Stephenson, T. Municipal wastewater sludge dewaterability and the presence of microbial extracellular polymer. *Water Sci. Technol.* **2011**, *44*, 373–379. [CrossRef]

44. Vesilind, P.A.; Davis, H.A. Using the CST device for characterizing sludge dewaterability. *Water Sci. Technol.* **1988**, *20*, 203–205. [CrossRef]

45. Tomczak-Wandzel, R.; Mędrzycka, K.; Cimochowicz-Rybicka, M. Wpływ dezintegracji ultradźwiękowej na przebieg fermentacji metanowej. Available online: https://mostwiedzy.pl/pl/publication/wplyw-dezintegracji-ultradzwiekowej-na-przebieg-fermentacji-metanowej-the-effect-of-ultrasonic-disin,110542-1 (accessed on 18 October 2018).

46. Braguglia, C.M.; Gianico, A.; Mininni, G. Comparison between ozone and ultrasound disintegration on sludge anaerobic digestion. *J. Environ. Manag.* **2012**, *95*, S139–S143. [CrossRef] [PubMed]

47. Xu, H.; He, P.; Yu, G.; Shao, L. Effect of ultrasonic pretreatment on anaerobic digestion and its sludge dewaterability. *J. Environ. Sci.* **2011**, *23*, 1472–1478. [CrossRef]

48. Martínez, E.; Rosas, J.; Morán, A.; Gómez, X. Effect of ultrasound pretreatment on sludge digestion and dewatering characteristics: Application of particle size analysis. *Water* **2015**, *7*, 6483–6495. [CrossRef]

49. Cimochowicz-Rybicka, M.; Rybicki, S. Application of respirometric tests for assessment of methanogenic bacteria activity in wastewater sludge processing. *J. Ecol. Eng.* **2013**, *14*, 44–52. [CrossRef]

50. Salihu, A.; Alam, M.D. Pretreatment methods of organic wastes for biogas production. *J. Appl. Sci.* **2016**, *16*, 124–137. [CrossRef]

International Journal of
*Environmental Research
and Public Health*

Article

Poultry Slaughterhouse Wastewater Treatment Using Submerged Fibers in an Attached Growth Sequential Batch Reactor

Hamidi Abdul Aziz [1,2,*], Nur Nasuha Ahmad Puat [1], Motasem Y. D. Alazaiza [1] and Yung-Tse Hung [3]

[1] School of Civil Engineering, Engineering Campus, Universiti Sains Malaysia, Nibong Tebal 14300, Penang, Malaysia; nurnasuhaahmadpuat@gmail.com (N.N.A.P.); my.azaiza@gmail.com (M.Y.D.A.)
[2] Solid Waste Management Cluster, Science and Technology Research Centre, Engineering Campus, Universiti Sains Malaysia, Nibong Tebal 14300, Penang, Malaysia
[3] Department of Civil and Environmental Engineering, Cleveland State University, Cleveland, OH 44115, USA; yungtsehung@yahoo.com
* Correspondence: cehamidi@usm.my; Tel.: +60-4599-6215

Abstract: In this study, a sequential batch reactor (SBR) with different types of fibers was employed for the treatment of poultry slaughterhouse wastewater. Three types of fibers, namely, juite fiber (JF), bio-fringe fiber (BF), and siliconised conjugated polyester fiber (SCPF), were used. Four SBR experiments were conducted, using the fibers in different reactors, while the fourth reactor used a combination of these fibers. The treatment efficiency of the different reactors with and without fibers on biochemical oxygen demand (BOD), chemical oxygen demand (COD), ammonia-nitrogen (NH_3-N), phosphorus (P), nitrite (NO_2), nitrate (NO_3), total suspended solids (TSS), and oil-grease were evaluated. The removal efficiency for the reactors with fibers was higher than that of the reactor without fibers for all pollutants. The treated effluent had 40 mg/L BOD_5 and 45 mg/L COD with an average removal efficiency of 96% and 93%, respectively, which meet the discharge limits stated in the Environmental Quality Act in Malaysia.

Keywords: poultry slaughterhouse wastewater; sequential batch reactor; fiber; BOD; COD

1. Introduction

Poultry slaughterhouses discharge a significant volume of highly polluted wastewater, principally during the slaughtering process and the periodic washing of residual particles, which cause a significant variation in the biodegradable organic matter concentration. Organic matter is considered the primary pollutant in the effluents of slaughterhouses [1]. The contribution of organic load to these effluents usually comes from different materials such as undigested food, blood, fat and lard, loose meat, paunch, colloidal particles, soluble proteins, and suspended materials [2,3]. Due to the mentioned components in the slaughterhouses wastewater, these wastewaters have a high concentration of organics such as chemical oxygen demand (COD), biochemical oxygen demand (BOD), phosphorous, and nitrogen [4]. Therefore, before discharging these wastewaters into receiving water bodies, an efficient treatment process should be carried out to prevent severe environmental pollution. In the last few decades, several treatment methods for the slaughterhouse wastewater have been reported. Biological (aerobic and anaerobic) treatment methods have been traditionally used for slaughterhouse wastewater treatment. However, both biological techniques have some limitations. For example, aerobic treatment processes require high energy consumption for aeration and generate a high amount of sludge [1]. The anaerobic treatment process of the poultry slaughterhouse wastewater is often impaired or slowed down because of the accumulation of suspended solids and floating fats in the reactor, which in turn leads to reduction in methanogenic activity and biomass washout [2]. Moreover, the anaerobic treatment process is more suitable in treating high organic loading wastewater [5,6].

Sequential Batch Reactors (SBR) are one of the biological processes applied to remove several types of pollutants. The SBR process is different from conventionally activated sludge techniques, because SBR merges all treatment units and operations into a single basin or tank, whereas traditional systems rely on various tanks. SBR has been successfully used for the treatment of domestic, municipal, industrial, dairy, synthetic, toxic and slaughterhouse wastewaters, swine manure, and landfill leachates [7–12].

Recently, the application of biomass carriers in the SBR process has been investigated by various researchers [13–15]. Fiber-based biomass carriers exhibit a good performance in removing pollutants, especially nitrogenous substances [16,17]. Previous studies that applied the swim bed technologies in SBR using bio-fringe (acryl fiber) revealed high treatment efficiency in removing pollutants, especially nitrogenous substances [17]. Several types of fibers have been used previously in wastewater treatments, such as plastic fibers [18,19], geotextiles [20], bio fringe acryl fiber [17], fibrous packing [21], and polyester fiber [22]. However, the application of fibers as attachment materials in SBR for poultry slaughterhouses wastewater treatment has not been well investigated.

The aim of this paper is to examine the potential use of various types of fibers as biomass carriers for slaughterhouses wastewater treatment by evaluating the removal efficiency of the pollutants with and without fiber in the reactor. The fibers involved are natural white Jute fiber (JF), synthetic siliconised conjugated polyester fiber (SCPF), bio-fringe (acrylic fiber) (BF), and the combination of three fibers in the reactor, called composite fiber (CF). The treatment efficiency of the different reactors with and without fibers on BOD, COD, ammonia-nitrogen (NH_3-N), phosphorus (P), nitrite (NO_2-N), nitrate (NO_3), TSS, and oil-grease were evaluated. Parameters, such as BOD, COD, and NH_3-N, were monitored every day during the experiments. However, the other parameters were evaluated based on the optimum value obtained.

2. Materials and Methods

2.1. Wastewater Source and Characteristics

The wastewater used in this study was collected from a local poultry slaughterhouse plant with a 13,000 birds per day capacity, located in the city of Nibong Tebal, Penang state, Malaysia, generating approximately 140 tons of wastewater daily. This wastewater, which is produced from different operations such as chickens cutting, chilling, scalding, packing and plant cleanup, was collected from the final collection tank after the screening of internal organs and feathers (partially treated using physical treatment). Wastewater samples of 150 to 200 L were collected twice per week, during the period from 23 May 2012 to 11 March 2013. Following the sampling procedure, the wastewater samples obtained were characterized based on pollutant concentration. Samples were preserved by storing in a cold room at 4 °C and were only taken out to room temperature 2 h before the experiment began. Characteristics of the raw wastewater are shown in Table 1.

2.2. Activated Sludge and Characteristics

The activated sludge (AS) used in this study was collected from the sludge dewatering system at the Jelutong Sewerage Treatment Plant (JSTP), Penang State, Malaysia. The AS in this study acts as microorganisms that are responsible for transforming the pollutants into acceptable end products. The AS also followed the poultry slaughterhouses wastewater storing procedures. Characteristics of the AS are shown in Table 1.

Table 1. Characteristics of raw wastewater and activated sludge (No. of samples = 7).

Parameter	Min	Max	Average	Std. Dev.
Raw wastewater				
BOD (mg/L)	573	1177	875	427.09
COD (mg/L)	777	1825	1301	741.04
NH_3-N (mg/L)	56.7	104	80.35	33.44
Nitrite (mg/L)	45.3	80	62.65	24.53
Nitrate(mg/L)	52.6	178.4	115.5	88.95
Oil-grease (mg/L)	2361.5	3616	2988.75	887.06
TSS (mg/L)	395	783	589	274.35
pH	6.3	6.9	6.6	0.4242
Activated sludge				
BOD (mg/L)	1246	1548	1397	213.54
COD (mg/L)	51,248	59,345	55,296.5	5725.44
DO (mg/L)	0.65	0.68	0.665	0.0212
MLSS (mg/L)	47,000	59,000	53,000	8485.28
pH	6.75	6.85	6.8	0.0707

BOD: biochemical oxygen demand; COD: chemical oxygen demand; TSS: total suspended solids; DO: Dissolved oxygen; MLSS: Mixed liquor suspended solids.

2.3. Fiber Preparation

Three types of fibers were used in this study as mentioned earlier. The first type was bio-fringe (BF) fiber made of acrylic fiber and imported from Japan. The other two types were Jute fiber (JF) and siliconised conjugated polyester fiber (SCPF). Composite fibers (CF) are a combination of these three fibers where all types were put together in the reactor. Both JF and SCPF were prepared similar to the size of the ready-made BF. The fibers were sewed neatly into pieces of yarns. Table 2 shows the physical properties of the fibers.

Table 2. Fiber characteristics.

Characteristic	Types of Fibers			
	BF	JF	SCPF	CF
Support filament length (cm)	50	50	50	50
Yarn length (cm)	10 ± 0.5	10 ± 0.5	10 ± 0.5	10 ± 0.5
Yarn diameter (cm)	0.1–0.2	0.5–1.0	1.0–1.5	0.1–1.0
Yarn per string	65	20	20	20
Total weight (g)	19.3	52.65	42.15	114.1

BF: bio-fringe; JF: Jute fiber; SCPF: siliconised conjugated polyester fiber; CF: composite fibers.

2.4. Reactor Setup

Two identical, laboratory scale Plexiglas reactors were used as SBR reactors for this study. Each reactor has the following dimensions: 80 cm × 40 cm × 25 cm with a total volume of 80 L. However, the experimental volume of the liquid for each reactor was 60 L. The first reactor was only operated with activated sludge without adding the fibers, while the other reactor was operated with activated sludge in the presence of fibers. Figure 1 shows the schematic diagram of the SBR reactor. The first cycle started with seeding of the AS collected from the JSTP. Following this, the reactor was fed with the collected raw poultry slaughterhouse wastewater during the filling phase and was aerated and mixed for a certain period of time during the aerating phase. The pH was adjusted approximately to 7.0 ± 0.5 and the mixed liquor suspended solids (MLSS) were maintained at a minimum range of 1500 mg/L to 4000 mg/L during the whole experiment. The adjustments were conducted before the aeration phase. The pH value was adjusted by adding either acid (0.5 M of H_2SO_4) or base solutions (0.5 M of NaOH). A 24-h cycle was selected, and the wastewater was operated for 20 h with the aeration rate of 60 L/min to make sure the wastewater and AS were mixed homogeneously. The final MLSS was

3782 mg/L. An air pump was used for the aeration and water circulation in the reactors. The aerated phase was stopped at the end of the aeration phase (after 20 h) and before the start of the settling phase (3 h). The decanting and discharging phase was the last process in the cycle, which meant that a cycle had been completed. After the first cycle was completed, the SBR reactor was filled with raw poultry slaughterhouse wastewater, aerated, settled, and decanted to repeat the second day treatment. Table 3 summarizes the operation design parameters of SBR reactor.

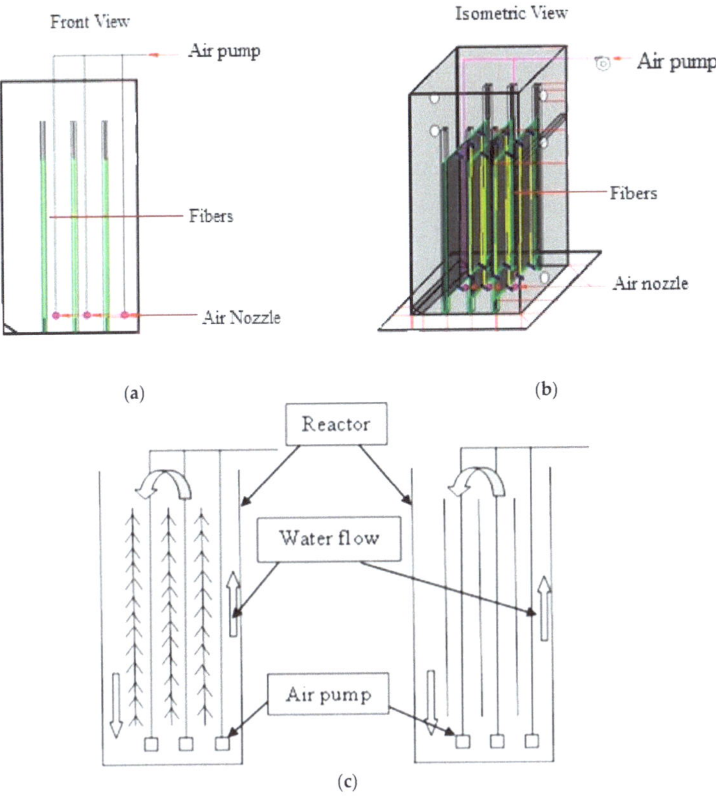

Figure 1. Schematic drawing of SBR reactor (**a**) Front view; (**b**) Isometric view; (**c**) Full details.

Table 3. Summary of operating design parameters.

Parameters	Value
Volume (L)	80
Operating liquid volume (L)	60
Dimension (m)	$0.4 \times 0.4 \times 0.25$
Hours per cycle (h)	20
F/M ratio (day^{-1})	0.2
MLSS (mg/L)	1500
Feeding rate (L/day)	21
Hydraulic Retention Time (HRT) (days)	2.9
Sludge Retention Time (SRT) (days)	7.5
Temperature (°C)	25
Sludge Volume Index (SVI) (mg/L)	50

MLSS: mixed liquor suspended solids.

2.5. Operating Conditions

To maintain 1500 mg/L of MLSS, the poultry slaughterhouse wastewater feed was set at 21 L/day. The concentration of MLSS was checked during every aeration phase (1.00 p.m.). The ratio of food to microorganism (F/M) was set at 0.2, where F refers to BOD (mg) applied per day to the reactor and M refers to TSS (mg) in the reactor. The F/M ratio is an important parameter that represents the amount of substrate available for the microorganisms in activated sludge. A typical F/M ratio for SBR ranges from 0.04 to 0.1 [23], whereas for SBR nutrient removal it ranges from 0.2 to 0.6 [24]. Either a too low or too high value of F/M may cause filamentous bulking or foaming, which leads to poor settlebility. AS was added if the concentration of MLSS was below 1500 mg/L, which meant that the F/M ratio during the reactors operation was maintained below 0.25. Filamentous bulking might occur if the MLSS exceeded 4000 mg/L. In the case where the MLSS exceeded 4000 mg/L, some sludge would be wasted until the MLSS dropped to the desired level (1500–4000 mg/L). Meanwhile, the hydraulic retention time (HRT) and sludge retention time (SRT) have been calculated and kept at 72 h and 176 h, respectively. The reactors were operated at room temperature (25 °C) without any temperature controlling system. Both Filling and decanting processes were also conducted manually without any pumping system. Several experiment runs with the different types of fibers were carried out in order to achieve the objectives of the study. The filling period was 30 min, the aeration period was 20 h, and the settling period was 3 h.

2.6. Analytical Methods

In this study, the performance of the reactors was evaluated based on the values of BOD, COD, NH_3-N, NO_2, NO_3, Phosphate, Oil-grease, TSS, and color. All of the mentioned parameters were determined and carried out as described in the Standard Methods for Water and Wastewater Examination [25]. The COD concentration was measured using a DR 2800 Spectrophotometer while NH_3-N and PO_4^{3-} concentration were calculated using a Hach DR 2500 Spectrophotometer. The removal efficiency of BOD, COD, and color was calculated using the following formula:

$$\text{Removal Efficiency (\%)} = (C_i - C_f)/C_i \times 100 \tag{1}$$

where C_i and C_f are the initial and final concentrations of parameters, respectively. Results from this study were analyzed using the Statistical Package for Social Sciences (SPSS) Version 17, via a one-way analysis of variance (One-Way ANOVA).

3. Results and Discussion

3.1. BOD and COD Removal

During the experiment, the pH of the activated sludge reactor was maintained at around 7 ± 0.5, the optimum range of pH for microbial growth. The MLSS was maintained in the system in the range between 1500 to 4000 mg/L.

The initial concentration of COD was 950 mg/L before treatment. Maximum removal efficiency of COD was achieved in day 12 with 69% and 293 mg/L COD. For the BOD value, the initial concentration was 350 mg/L where the maximum removal was achieved in day 13 with 88% removal efficiency with 105 mg/L BOD after treatment in the reactor without fibers. The growth of microbes started to become slow and stable (between days 11 to day 16) because the microbes did not have any shelter to regenerate before the cycle completes. As compared to the reactor with fibers, the reactor without fibers does not have shelter for microbes to attach.

On the other hand, after using the different types of fibers in the reactor, the removal efficiency of BOD and COD increased over the time in all reactors. As an overall result, the performance of SBR using fibers as an attachment material has demonstrated better results compared to SBR without fibers in the reactor. In general, the BOD removal for each type of fiber was efficient, with an average removal efficiency of higher than 90%. The observations showed that the CF reactor gives the higher removal efficiency of BOD with 96% (40 mg/L). For the JF and BF reactors, the optimum BOD removal was achieved on day 14 and day 12 with 62 mg/L (93%) and 45 mg/L (95%), respectively. Apart from that, the SCPF reactor showed optimum removal on day 13 with 50 mg/L BOD (94%).

For COD, the BF reactor showed a higher performance in the removal efficiency of COD with 93%. The COD removal achieved maximum removal on day 12 at 93% with 45 mg/L COD as shown in Figure 2.

3.2. Ammonia-Nitrogen Removal

The daily observations of the NH_3-N showed that there was no removal for the first two days of treatment as shown in Figure 2. This observation was due to the increase in the concentration of ammonia-nitrogen due to the occurrence of nitrification [26]. According to the Fontenot et al. [27], in SBR, the nitrogen (protein or lipid) removal process was designed for the aerobic carbon removal and nitrification followed by an anoxic de-nitrification with the addition of an external carbon source.

Before using the fiber in reactors, the maximum removal was achieved on day 13, with the removal of 77% and concentration pollutant of 20 mg/L. It was observed that when the pH of the wastewater in the system was not basic, an oxidation of ammonia took place.

When the oxidation occurred, the pH of the wastewater quickly dropped, and this simultaneously produced nitrite. Ammonia can exist as molecular ammonia or ammonium gas. These two forms in water are strongly dependent on pH and temperature. Nitrification and de-nitrification occurred in good condition due to the aerobic and anaerobic zones inside the same system. The higher concentrations of ammonia were shown to have inhibitory effects of pH level during the anaerobic process [28].

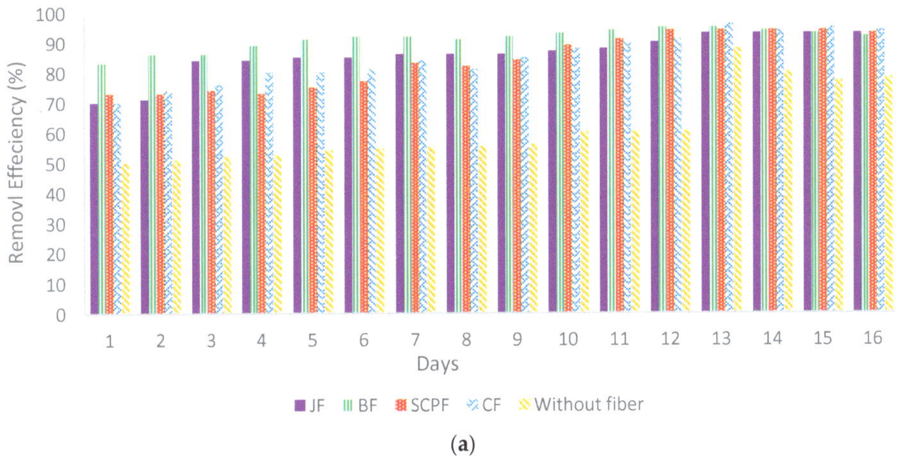

(**a**)

Figure 2. *Cont.*

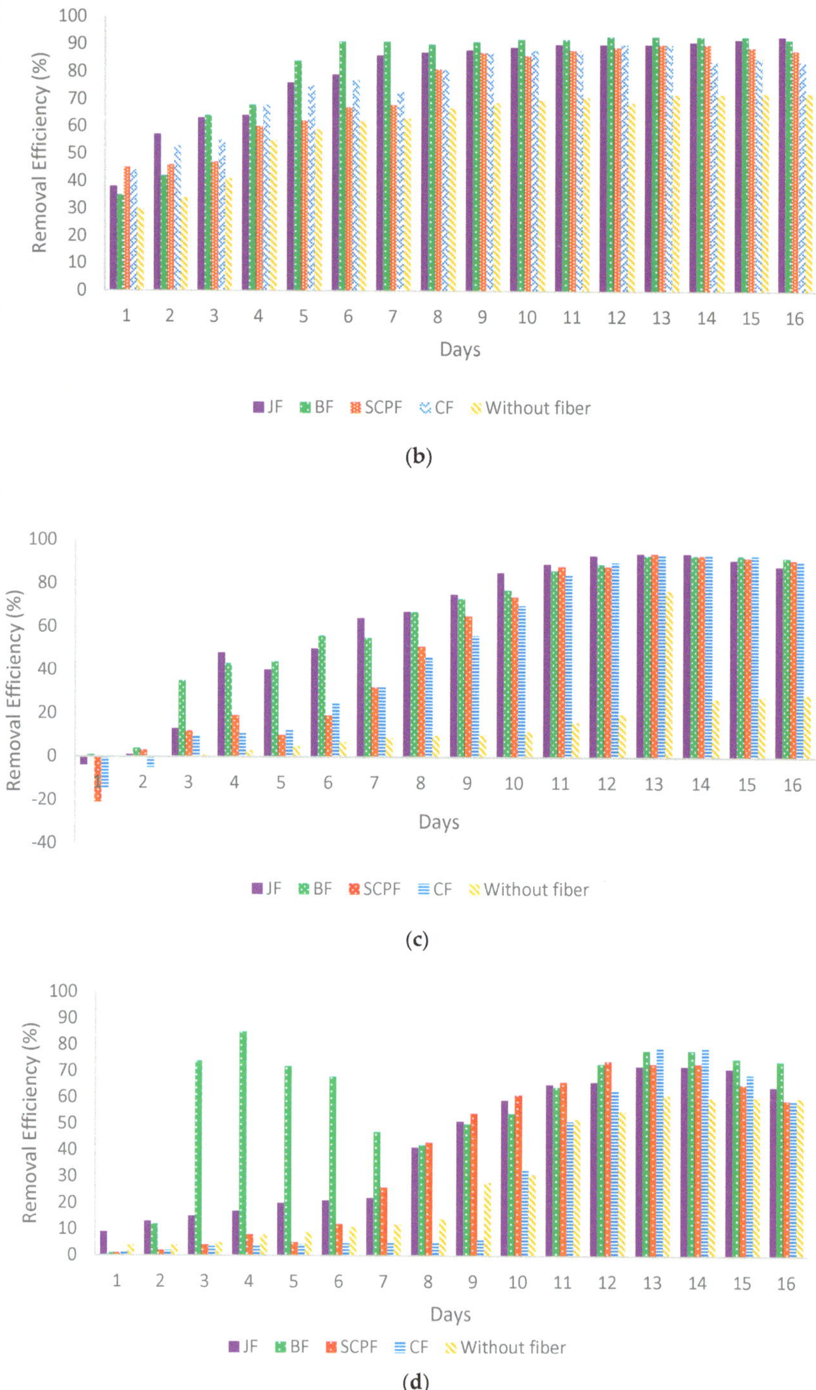

Figure 2. *Cont.*

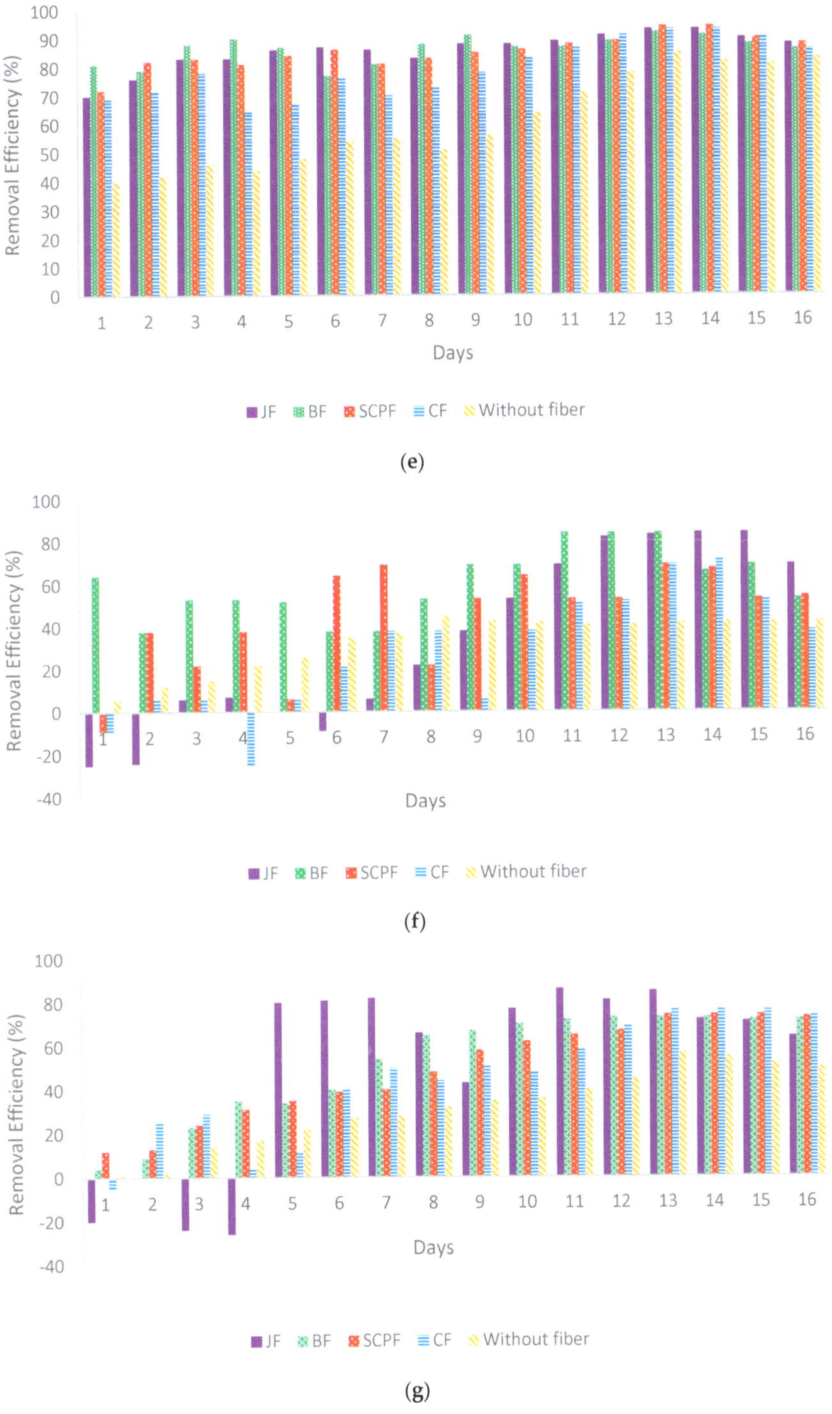

(e)

(f)

(g)

Figure 2. *Cont.*

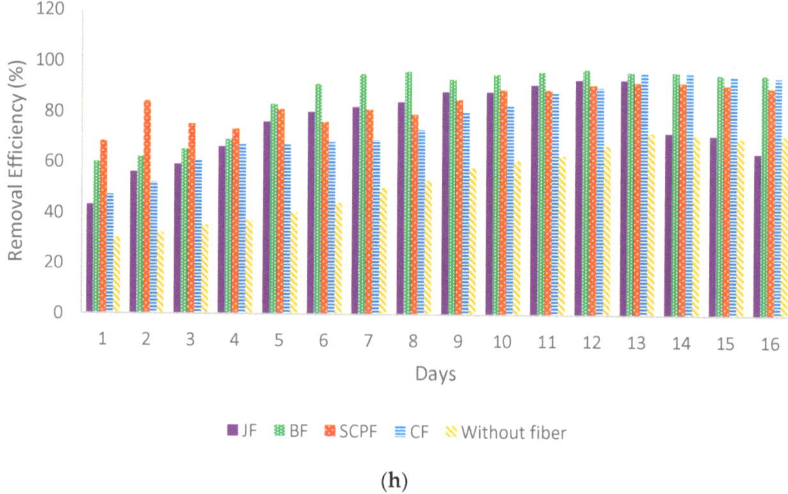

(**h**)

Figure 2. Daily monitoring for the removal efficiency of (**a**) BOD, (**b**) COD, (**c**) ammonia-nitrogen, (**d**) phosphate, (**e**) NO$_2$, (**f**) NO$_3$, (**g**) Oil-grease, (**h**) TSS, in all reactors.

On the other hand, and after using the fiber with SBR reactors, the NH$_3$-N removal was considered to be high at 94%, 94%, 93% and 94% for CF, JF, BF and SCPF reactors, respectively. In the case of CF reactor, during the first and second day of treatment, no removal due to the increase in ammonia-nitrogen concentration from 85 mg/L to 98 mg/L (data not shown). However, in day 3 and 4 of treatment process, a fluctuation in removal efficiency was observed. The maximum removal efficiency was obtained at day 13 at 94%. The initial concentration of ammonia-nitrogen before treatment was 106 mg/L and diminished 5 mg/L after treatment. In general, biological nitrogen removal can be categorized into two separate steps: nitrification and denitrification. In the nitrification process, ammonium is usually converted to nitrate under aerobic condition, whereas the de-nitrification process converted the nitrate into nitrogen gas (N$_2$) [29]. Therefore, when a higher aeration was used, a better removal efficiency of NH$_3$-N was observed due to the fact that more DO was provided to the nitrifying bacteria in order to convert ammonium to nitrate.

3.3. Nitrite (NO$_2$) and Nitrate (NO$_3$) Removal

As can be seen in Figure 2, the maximum removal efficiency for NO$_2$ was achieved on day 8 with the removal of 45% in the reactor without fiber. According to Erses et al. [28], when oxidation ammonia takes place, nitrite is produced simultaneously.

$$NH_4^+ + 1.5O_2 \rightarrow NO_2^- + 2H^+ + H_2O \qquad (2)$$

However, the maximum removal efficiency for NO$_3$ was achieved on day 13 with the removal of 85% when the remaining concentration in the reactor was 12 mg/L. After using the fibers in the reactors, the maximum removal efficiency for NO$_2$ obtained in the JF reactor was 84%, which is approximately twice the removal efficiency of the reactor without fibers. In addition, the maximum removal efficiency for NO$_3$ was 94%, which was achieved using SCPF reactor.

3.4. Phosphate Removal

The maximum removal efficiency for phosphate was achieved on day 13 with a removal efficiency of 61%, where the remaining concentration of 15 mg/L in the case of reactor without fiber, as shown in

Figure 2. The removal showed a fluctuated trend on day 6, due to the increased concentration of nitrate in the anaerobic zone and phosphate in the aerobic zone (data not shown). The phosphorus content in the sample may also be affected by the apparatus used during the experiment if the apparatus was contaminated with detergent [30]. However, after using the fibers in reactors, the maximum removal efficiency of phosphate was 85% for the BF reactor during the experiment. No removal was obtained during the first day due to the increase of phosphate concentration from 39 mg/L to 65 mg/L. This was owing to the occurrence of the polyphosphate bacteria which started to accumulate large quantities of phosphate within their cells [31].

3.5. Oil-Grease and TSS Removal

In the reactor without fiber, the maximum removal efficiency for oil-grease was achieved on day 13 with 57% (117 mg/L). In addition, the same trend was observed for the TSS values. The maximum removal of TSS was recorded on day 13 with the remaining concentration at 72 mg/L (84%) from 532 mg/L. After applying the fibers in reactors, the maximum removal efficiency of the oil-grease and TSS was obtained using a JF reactor and BF reactor with 86% and 97% removal efficiency, respectively, as shown in Figure 2. The pattern of removal showed the fluctuated trend until day 4 of the treatment. Starting from day 5, the treatment slowly showed an increase in the removal over time. Maximum removal efficiency for oil-grease was obtained on day 11. The TSS removal efficiency showed an increase trend over that time. A maximum removal efficiency of TSS was achieved on day 13. The fluctuated trend that occurred in the early part of the treatment may be due to the fact that a pump was not used when the sample was taken. This led to the additional concentration of oil and grease and TSS. From the observation and the properties of the raw wastewater, the value of oil-grease increased because of not using the pump during the sample collection. The oil-grease may have been partially filtered using the pump.

3.6. Statistical Analysis

Table 4 summarized the maximum removal efficiency for all SBR. A One-way (ANOVA) test was used for multiple comparisons between the different types of the reactors. The test shows a significant difference between mean concentration BF and AS, CF, JF, and SCPF for all parameters tested. For example, for the BOD concentration after treatment, the ANOVA test shows that the concentration of BOD in the CF reactor was small as compared to the other reactors, which emphasize the experimental results. In addition, the BF reactor has a lower mean concentration for COD as compared to other reactors, which means that the BF reactor achieved the higher removal efficiency, proving the experimental results. For other parameters, similar results with experiments were obtained. The characteristics of final effluents for all reactors are summarized in Table 5.

Table 4. Summary of maximum removal efficiencies for all reactors.

	Max. Removal Efficiency (%)				
Parameters	**Without Fiber**	**BF**	**JF**	**SCPF**	**CF**
BOD	88	95	93	94	96
COD	69	98	93	91	90
NH_3-N	77	93	94	94	96
Phosphate	61	85	72	74	79
NO_2	45	81	84	69	71
NO_3	86	92	93	94	93
TSS	84	97	93	92	96
Oil-grease	57	73	86	74	77

BF: bio-fringe; JF: Jute fiber; SCPF: siliconised conjugated polyester fiber; CF: composite fibers.

Table 5. Summary of final effluents of all parameters for all reactors.

Parameters	Without Fiber	BF	JF	SCPF	CF
BOD (mg/L)	120	70	65	65	65
COD (mg/L)	309	74	75	112	150
NH_3-N (mg/L)	23	7	10	7	7.9
P (mg/L)	17	10	14	16	16
NO_2 (mg/L)	51	30	20	37	40
NO_3 (mg/L)	14	12	10	10	12
TSS (mg/L)	75	25	39	55	30
Oil-grease	855	801	431	751	723

BF: bio-fringe; JF: Jute fiber; SCPF: siliconised conjugated polyester fiber; CF: composite fibers.

4. Conclusions

This paper investigates the potential application of different types of fibers as biomass carriers in SBR reactors for slaughterhouses wastewater treatment, by evaluating the removal efficiency of the pollutants with and without fibers in a SBR reactor. The study showed that the removal efficiency of the SBR with fibers achieved a higher performance for all tested parameters as compared with the SBR without fibers. The SBR reactor with fibers as attachment materials enabled the attachment of suspended solids to the fibers, which increased the biomass concentration in the reactor and provided a better treatment efficiency. The treated effluent had 40 mg/L BOD and 45 mg/L COD with an average removal efficiency of 96% and 98%, respectively, which meet the discharge limits stated in the Environmental Quality Act in Malaysia. Moreover, all other parameters also satisfied the limits of the Environmental Quality Act in Malaysia.

Author Contributions: H.A.A. conceived and designed the experiments; N.N.A.P. performed the experiments and data analysis; M.Y.D.A. wrote and revised the paper, Y.-T.H. revised and proofread the paper.

Funding: This research was funded by [University Sains Malaysia] grant No. [Grant No. 1001/CKT/870023].

Acknowledgments: This work was funded by Universiti Sains Malaysia under Iconic grant scheme (Grant No. 1001/CKT/870023) for research associated with the Solid Waste Management Cluster, Engineering Campus, Universiti Sains Malaysia.

Conflicts of Interest: The authors declare no conflict of interest.

References

1. Awang, Z.; Bashir, M.; Kutty, S.; Isa, M. Post-treatment of slaughterhouse wastewater using electrochemical oxidation. *Res. J. Chem. Environ.* **2011**, *15*, 229–237.
2. Kobya, M.; Senturk, E.; Bayramoglu, M. Treatment of poultry slaughterhouse wastewaters by electrocoagulation. *J. Hazard. Mater.* **2006**, *133*, 172–176. [CrossRef] [PubMed]
3. Ün, Ü.T.; Koparal, A.S.; Öğütveren, Ü.B. Hybrid processes for the treatment of cattle-slaughterhouse wastewater using aluminum and iron electrodes. *J. Hazard. Mater.* **2009**, *164*, 580–586.
4. Matsumura, E.; Mierzwa, J. Water conservation and reuse in poultry processing plant—A case study. *Resour. Conserv. Recycl.* **2008**, *52*, 835–842. [CrossRef]
5. Massé, D.I.; Masse, L. The effect of temperature on slaughterhouse wastewater treatment in anaerobic sequencing batch reactors. *Bioresour. Technol.* **2001**, *76*, 91–98. [CrossRef]
6. Nunez, L.; Martinez, B. Anaerobic treatment of slaughterhouse wastewater in an expanded granular sludge bed (EGSB) reactor. *Water Sci. Technol.* **1999**, *40*, 99–106. [CrossRef]
7. Uygur, A.; Kargı, F. Biological nutrient removal from pre-treated landfill leachate in a sequencing batch reactor. *J. Environ. Manag.* **2004**, *71*, 9–14. [CrossRef] [PubMed]
8. Tsilogeorgis, J.; Zouboulis, A.; Samaras, P.; Zamboulis, D. Application of a membrane sequencing batch reactor for landfill leachate treatment. *Desalination* **2008**, *221*, 483–493. [CrossRef]
9. Al-Rekabi, W.S.; Qiang, H.; Qiang, W.W. Review on sequencing batch reactors. *Pak. J. Nutr.* **2007**, *6*, 11–19.

10. Mahvi, A. Sequencing Batch Reactor: A Promising Technology in Wastewater Treatment. *Iran J. Environ. Health Sci. Eng.* **2008**, *5*, 79–90.

11. Neczaj, E.; Kacprzak, M.; Kamizela, T.; Lach, J.; Okoniewska, E. Sequencing batch reactor system for the co-treatment of landfill leachate and dairy wastewater. *Desalination* **2008**, *222*, 404–409. [CrossRef]

12. Li, J.; Healy, M.G.; Zhan, X.; Norton, D.; Rodgers, M. Effect of aeration rate on nutrient removal from slaughterhouse wastewater in intermittently aerated sequencing batch reactors. *Water Air Soil Pollut.* **2008**, *192*, 251–261. [CrossRef]

13. Sivic, A.; Atanasova, N.; Puig, S.; Bulc, T.G. Ammonium removal in landfill leachate using SBR technology: Dispersed versus attached biomass. *Water Sci. Technol.* **2018**, *77*, 27–38. [CrossRef] [PubMed]

14. Daverey, A.; Chen, Y.-C.; Dutta, K.; Huang, Y.-T.; Lin, J.-G. Start-up of simultaneous partial nitrification, anammox and denitrification (SNAD) process in sequencing batch biofilm reactor using novel biomass carriers. *Bioresour. Technol.* **2015**, *190*, 480–486. [CrossRef] [PubMed]

15. Masłoń, A.; Tomaszek, J.A. A study on the use of the BioBall® as a biofilm carrier in a sequencing batch reactor. *Bioresour. Technol.* **2015**, *196*, 577–585. [CrossRef] [PubMed]

16. Wang, Q.; Feng, C.; Zhao, Y.; Hao, C. Denitrification of nitrate contaminated groundwater with a fiber-based biofilm reactor. *Bioresour. Technol.* **2009**, *100*, 2223–2227. [CrossRef] [PubMed]

17. Qiao, S.; Kawakubo, Y.; Koyama, T.; Furukawa, K. Partial Nitritation of Raw Anaerobic Sludge Digester Liquor by Swim-Bed and Swim-Bed Activated Sludge Processes and Comparison of Their Sludge Characteristics. *J. Biosci. Bioeng.* **2008**, *106*, 433–441. [CrossRef] [PubMed]

18. Rodríguez, D.C.; Ramírez, O.; Mesa, G.P. Behavior of nitrifying and denitrifying bacteria in a sequencing batch reactor for the removal of ammoniacal nitrogen and organic matter. *Desalination* **2011**, *273*, 447–452. [CrossRef]

19. Lydmark, P. Effects of environmental conditions on the nitrifying population dynamics in a pilot wastewater treatment plant. *Environ. Microbiol.* **2007**, *9*, 2220–2233. [CrossRef] [PubMed]

20. Alimahmoodi, M.; Yerushalmi, L.; Mulligan, C.N. Development of biofilm on geotextile in a new multi-zone wastewater treatment system for simultaneous removal of COD, nitrogen and phosphorus. *Bioresour. Technol.* **2012**, *107*, 78–86. [CrossRef] [PubMed]

21. Li, J.; Xing, X.-H.; Wang, B.-Z. Characteristics of phosphorus removal from wastewater by biofilm sequencing batch reactor (SBR). *Biochem. Eng. J.* **2003**, *16*, 279–285. [CrossRef]

22. Xia, S.; Li, J.; Wang, R. Nitrogen removal performance and microbial community structure dynamics response to carbon nitrogen ratio in a compact suspended carrier biofilm reactor. *Ecol. Eng.* **2008**, *32*, 256–262. [CrossRef]

23. Tchobanoglous, G.; Eddy, M. *Wastewater Engineering Treatment*; McGraw-Hill: New York, NY, USA, 2004.

24. Mishoe, G.L. F/M Ratio and the Operation of an Activated Sludge Process. *Fla. Water Resour. J.* **1999**, *20*, 20–27.

25. American Public Health Association (APHA). *Standard Methods for Examination of Water and Wastewater*; American Water Works Association: Washington, DC, USA, 2002.

26. Manahan, S.E. *Environmental Chemistry*; CRC Press LLC: Boca Raton, FL, USA, 2005.

27. Fontenot, Q.; Bonvillain, C.; Kilgen, M.; Boopathy, R. Effects of temperature, salinity, and carbon: Nitrogen ratio on sequencing batch reactor treating shrimp aquaculture wastewater. *Bioresour. Technol.* **2007**, *98*, 1700–1703. [CrossRef] [PubMed]

28. Erses, A.S.; Onay, T.T.; Yenigun, O. Comparison of aerobic and anaerobic degradation of municipal solid waste in bioreactor landfills. *Bioresour. Technol.* **2008**, *99*, 5418–5426. [CrossRef] [PubMed]

29. Wei, Y.; Ji, M.; Li, R.; Qin, F. Organic and nitrogen removal from landfill leachate in aerobic granular sludge sequencing batch reactors. *Waste Manag.* **2012**, *32*, 448–455. [CrossRef] [PubMed]

30. Kumar, M.; Galil, M.S.A.; Suresha, M.; Sathish, M.; Nagendrappa, G. A simple spectrophotometric determination of phosphate in sugarcane juices, water and detergent samples. *J. Chem.* **2007**, *4*, 467–473.

31. Alcántara, C.; Blasco, A.; Zúñiga, M.; Monedero, V. Accumulation of polyphosphate in *Lactobacillus* spp. and its involvement in stress resistance. *Appl. Environ. Microbiol.* **2014**, *80*, 1650–1659. [CrossRef] [PubMed]

International Journal of
*Environmental Research
and Public Health*

Article

Effect of Bioaugmentation on Biogas Yields and Kinetics in Anaerobic Digestion of Sewage Sludge

Magdalena Lebiocka [1,*], Agnieszka Montusiewicz [1] and Agnieszka Cydzik-Kwiatkowska [2]

[1] Faculty of Environmental Engineering, Lublin University of Technology, 40B Nadbystrzycka, 20-618 Lublin, Poland; a.montusiewicz@pollub.pl
[2] Faculty of Environmental Sciences, University of Warmia and Mazury in Olsztyn, 45g Słoneczna, 10-709 Olsztyn, Poland; agnieszka.cydzik@uwm.edu.pl
* Correspondence: m.lebiocka@pollub.pl; Tel.: +48-81-538-4325; Fax: +48-81-538-1997

Abstract: Bioaugmentation with a mixture of microorganisms (Bacteria and Archaea) was applied to improve the anaerobic digestion of sewage sludge. The study was performed in reactors operating at a temperature of 35 °C in semi-flow mode. Three runs with different doses of bioaugmenting mixture were conducted. Bioaugmentation of sewage sludge improved fermentation and allowed satisfactory biogas/methane yields and a biodegradation efficiency of more than 46%, despite the decrease in hydraulic retention time (HRT) from 20 d to 16.7 d. Moreover, in terms of biogas production, the rate constant k increased from 0.071 h^{-1} to 0.087 h^{-1} as doses of the bioaugmenting mixture were increased, as compared to values of 0.066 h^{-1} and 0.069 h^{-1} obtained with sewage sludge alone. Next-generation sequencing revealed that *Cytophaga* sp. predominated among Bacteria in digesters and that the hydrogenotrophic methanogen *Methanoculleus* sp. was the most abundant genus among Archaea.

Keywords: anaerobic digestion; wastewater sludge; bioaugmentation; metagenome; NGS

---

1. Introduction

The development of wastewater treatment technology, together with the implementation of environmental legislation, has successfully protected the aquatic environment from pollution. However, at the same time, sewage sludge is generated as the by-product of the wastewater treatment. Sewage sludge is becoming a worldwide environmental problem because of its increasing production and its high contents of organic matter, pathogens, and heavy metals.

Anaerobic digestion is a biological process that can degrade organic material by the concerted action of a wide range of microorganisms in the absence of oxygen. However, the advantages of the anaerobic digestion process in the treatment of sewage are still far from being optimized. Regardless of the temperature conditions, only around 50% to 60% of the organic matter can be degraded, leaving a large potential of increasing the biogas production [1]. A better understanding of the basic mechanisms occurring in the digester, conducting the process at high temperatures, application of different kinds of pre-treatment methods (freezing/thawing; cavitation), phase separation, and, recently, bioaugmentation has been applied to improve the anaerobic digestion.

Bioaugmentation is the addition (augmentation) of specialized microbial cultures which are typically grown separately under well-defined conditions to perform a specific task in a given environment (in situ or in a bioreactor) [2,3]. This approach has been used for hazardous waste remediation, as well as for the biodegradation of wastewater in wastewater treatment plants and many others biological treatment processes [4]. In aerobic wastewater treatment, bioaugmentation has resulted in more reliable nitrification, biological phosphorous removal, improved sludge settling, enhanced grease degradation, and accelerated transformation of xenobiotic organic contaminants, such as pentachlorophenol [5–8]. Bioaugmentation has also been studied at laboratory scale to increase the methane production during the anaerobic digestion of animal manure [9], cellulose-rich [10,11], and lipid-rich waste [12] as well as seed biomass [13]. Schmidt et al. [14] used bioaugmentation during the anaerobic digestion of sewage sludge to improve the polycyclic aromatic hydrocarbons removal. Moreover, the bioaugmentation of anaerobic digestion communities by the adapted hydrolytic consortia increased the biogas yield by 10–29% in the anaerobic digestion of maize silage [15].

Interestingly, bioaugmentation was investigated as a method of decreasing the recovery period of anaerobic digesters exposed to a transient toxic event [16,17]. The comprehensive review by Tale et al. [18] described the beneficial effects of bioaugmentation on the efficiency of biochemical processes under temporary organic overloading of reactors. Bioaugmentation has been considered as a useful strategy, playing an excellent role for performance enhancement and recovery in biosystems under various stresses due to the improvement of detrimental conditions, retention, and adaptation of bioaugmentation consortium. However, bioaugmentation faces the potential risk of functional failure, even negative effects on biosystems [19]. Failure of bioaugmentation has been reported to be associated with numerous factors that include the growth rate being lower than the rate of washout, insufficient inoculum size, and substrate availability. Limitations of bioaugmentation can be overcome through the techniques that include increased inoculum dosing, a longer period of inoculum acclimatization in reactors, the addition of nutrients and surfactants, and application of sufficient acclimatization periods. Surveys of the literature show that a key area for the further research should be towards acquiring a better understanding of the degradation pathways where bioaugmentation is applied [20–22]. The present study examines the influence of bioaugmentation on the efficiency of anaerobic digestion of sewage sludge from a municipal wastewater treatment plant. The novel aspect of the study involved using a mixture of wild-living Bacteria and Archaea from Yellowstone National Park, USA, for bioaugmentation. The effect of this bioaugmentative mixture on the kinetics of biogas production and the microbial structure in semi-flow anaerobic reactors was determined.

2. Materials and Methods

2.1. Material Characteristics

Sewage sludge that included two-source residues (originated from primary and secondary clarifiers) were obtained from the Puławy municipal wastewater treatment plant (WWTP), Poland. Sludge was sampled once a week in the WWTP. Under laboratory conditions, the sludge was mixed at the volume ratio of 60:40 (primary: waste sludge), then homogenized, manually screened through a 3 mm screen and partitioned. The sludge samples were stored at 4 °C in a laboratory fridge for no longer than one week. Sludge prepared in this manner was fed to the digester as mixed sewage sludge (SS). The main characteristics of SS during the entire cycle of experiments was as follows (the average value and standard deviation are given): chemical oxygen demand (COD)—43.00 ± 5.49 g dm^{-3}, volatile fatty acids (VFA)—1179 ± 733 mg dm^{-3}, total solids (TS)—37.0 ± 2.9 g kg^{-1}, volatile solids (VS)—28.6 ± 2.11 g kg^{-1}, pH—6.75 ± 0.33, and alkalinity—846 ± 268 mg CaCO$_3$ dm^{-3}.

A liquid solution containing a mixture of Bacteria and Archaea was prepared in a continuous mode throughout the experiment. A nylon pouch filled with a powdery substrate (ArcheaSolutions Inc., Evansville, IN, USA) was placed inside a generator. The microbial composition of the powdery substrate after 1 day of cultivation at 37 °C under constant mixing conditions in distilled water showed that about 36% of microorganisms belonged to Archaea with *Methanosaeta* as s predominant genus and that about 59% of microorganisms belonged to Bacteria with *Exiguobacterium*, *Janthinobacterium*, *Acinetobacter*, and *Stenotrophomonas* as the most abundant genera (Figure 1).

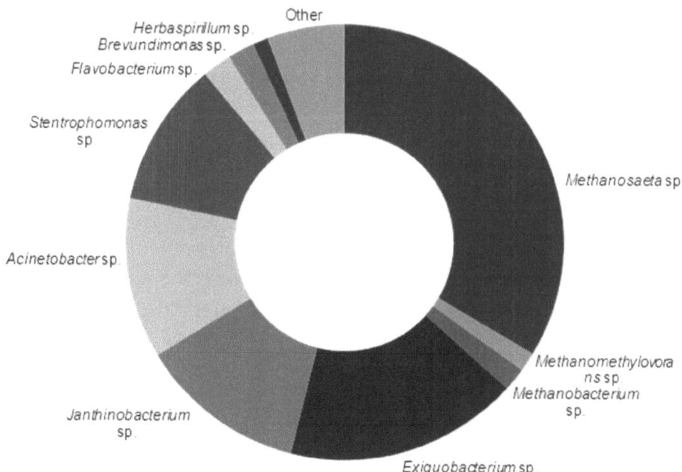

Figure 1. Microbial genera present in the bioaugmenting substrate; "other" includes unclassified bacteria (1%) and Archaea (1%), and taxa with abundance below 1% (3%); sequencing results are present in the Sequence Read Archive SRA (National Center for Biotechnology Information NCBI, BioProject PRJNA431048).

The analysis of the substrate was performed as described in *Molecular analysis*, sequences were placed in NCBI (BioProject PRJNA431048). The substrate was packed in a vinyl alcohol coating, which dissolved upon contact with dechlorinated pipe water flowing through the device. The release of an appropriate microbial content of 1.08 g L^{-1} h^{-1} required a continuous flow of water at a level of about 0.5 L min^{-1}. After 30 days, the pouch containing a mixture of microorganisms was replaced by a new one. The generator was linked to two serially-connected storage tanks with a total active volume of 320 L. The liquor prepared in this manner was stored at room temperature. The average total solids (TS) of the liquor differed slightly during the experiments and reached 0.48 and 0.45 g kg^{-1} in phase 1 and phase 2, respectively, while the average volatile solids (VS) oscillated around 0.04 g kg^{-1}. The COD values were 22 g·m^{-3} in phase 1 and 26 g·m^{-3} in phase 2. Similarly, the VFA concentrations were 21 and 24 g·m^{-3}, respectively. The alkalinity of 330 g $CaCO_3$·m^{-3} and pH value of 7.16 were obtained for both phases.

2.2. Laboratory Installation and Operational Set-Up

The study was performed in anaerobic reactors operating at a temperature of 35 °C in semi-flow mode. The laboratory installation consisted of three completely mixed digesters (with an active volume of 40 dm^3) working in parallel, equipped with a gaseous installation (consisting of pipelines, pressure equalization unit and a mass flow meter), an influent peristaltic pump and storage vessels (Figure 2).

An inoculum for the laboratory reactors was taken from Puławy wastewater treatment plant as a collected digest from a mesophilic anaerobic digester with a volume of 2500 m^3 that was operated at 35–37 °C and a hydraulic retention time (HRT) of about 25 days. The adaptation of the digester biomass was achieved after 30 days.

Two phases of the experiment were conducted, differing in the organic loading rates (OLRs). Each phase lasted 90 days (30 days for acclimatization and 60 days for measurements). In the first phase, three reactors were operated. The first reactor (R1—control one) was fed daily with 2 dm^3 of mixed sludge. The HRT reached 20 days, and the OLR was time-dependent with an average value of 1.55 kg VS m^{-3} day^{-1}.

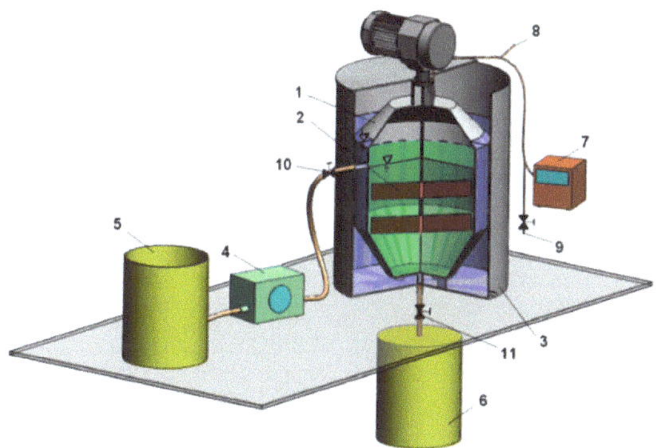

Figure 2. Laboratory installation. 1—anaerobic reactor, 2—mechanical stirrer, 3—heating jacket, 4—influent peristaltic pump, 5—influent storage vessel, 6—effluent storage vessel, 7—drum gas meter, 8—gaseous installation and gas sampler with a rubber septum, 9—dewatering valve, 10—inlet valve, 11—outlet valve.

The second reactor (R2) was operated following the same schedule. However, the reactor was fed with sludge bioaugmented with microbial mixture at a volumetric ratio of 91:9 (the influent consisted of a mixture of 2 dm^3 sludge and 0.2 dm^3 of bioaugmenting mixture). The HRT was shortened to 18.2 days, and the average value of the OLR was 1.54 kg VS m^{-3} day^{-1}.

The third reactor (R3) arrangement was the same as in R2, but this time the volumetric ratio was set at 87:13 (influent consisted of a mixture of 2 dm^3 sludge and 0.3 dm^3 of bioaugmenting mixture). The HRT shortened to 17.4 days, and the average value of the OLR was 1.53 kg VS m^{-3} day^{-1}.

In the second phase, two reactors were operated. The fourth reactor (R4) was operated analogously to the R1 (as a control), although at OLR of only 1.30 kg VS m^{-3} day^{-1}.

Reactor R5 was operated also at low OLR of 1.33 kg VS m^{-3} day, and the sludge to bioaugmenting mixture volumetric ratio was set at 83:17 (influent consisted of a mixture of 2 dm^3 sludge and 0.4 dm^3 of bioaugmenting mixture). The HRT was shortened to 16.7 days.

The reactors were bioaugmented in continuous mode because of the long-term wide microbial growth (from 20 h to 20 days) as well as their possible wash out from the system.

2.3. Analytical Methods

In the mixed sludge (SS), total chemical oxygen demand (COD), total solids (TS) and volatile solids (VS), were analyzed once a week. The same schedule was used for determining the values of the parameters that characterized the supernatant (sludge liquid phase) before digestion. These parameters included soluble chemical oxygen demand (SCOD), volatile fatty acids (VFA), alkalinity and pH level. The supernatant samples were obtained by centrifuging the sludge at 4000 rpm for 30 min. All analyses were carried out in accordance with the procedures listed in the Standard Methods for the Examination of Water and Wastewater [23].

The bioaugmenting mixture was examined twice a week as an averaged and a collected sample taken from its storage tank. The following parameters were analyzed: COD, TS, VS, VFA, alkalinity, and pH.

In the digested sludge, the specified parameters were determined two times a week, in accordance with the timetable adopted. Similarly, the supernatant of the digested sludge was examined using the same schedule.

Anaerobic digestion efficiency was controlled by the daily evaluation of the biogas yield and its composition (CH$_4$, CO$_2$, and other gases). Moreover, the volatile solids removal efficiency was evaluated according to the American Public Health Association (APHA) [23].

The biogas production was determined using Aalborg (USA) digital mass flow meter. Its composition was measured using Trace GC-Ultra gas chromatograph coupled with a thermal conductivity detector (TCD) fitted with divinylbenzene (DVB) packed columns. The Rt-Q-Bond column was used to determine the CH_4 and CO_2 concentrations. The parameters used for the analysis were as follows—injector 50 °C and detector 100 °C. The carrier gas was helium with a flux rate of 1.5 $cm^3 \cdot min^{-1}$. Peak areas were determined by the computer integration program (CHROM-CARD).

The biogas production curves were made on the basis of the averaged experimental data acquired from an XFM Control Terminal. Both the rate constant k and maximum biogas production V_{max} were obtained by means of nonlinear regression method with Statsoft Statistica software (version 10, TIBCO Software Inc., Palo Alto, CA, USA). The strength of the relationships between groups of the results was determined using Pearson's correlation coefficient (R) and determination coefficient R^2.

2.4. Molecular Analysis

Metagenome of the digesters in the first phase was analyzed after 90 days of each reactor operation. The collected samples of biomass were stored at −20 °C. After thawing, DNA was isolated from the biomass using GenElute™ Bacterial Genomic DNA Kit (Sigma-Aldrich Chemie Gmbh, Munich, Germany) according to the producer's protocol, including lysozyme digestion. The presence of DNA was confirmed by agarose electrophoresis. Purity and concentration of the isolated DNA were measured spectrophotometrically using a Biophotometer (Eppendorf, Hamburg, Germany). A universal 515F (GTGCCAGCMGCCGCGGTAA) and 806R (GGACTACHVGGGTWTCTAAT) primer set was used to amplify the archaeal and bacterial 16S rDNA gene. Next-generation sequencing (NGS) using the MiSeq Illumina platform was applied to sequence the amplicons. The sequencing was performed in the Research and Testing Laboratory (USA). Over 126 thousands of full sequences were obtained.

The sequences were analyzed bioinformatically as described in Świątczak et al. [24]. In short, to detect and remove chimeras from the raw reads, UCHIME [25] was applied. The readouts were condensed into FASTA format and clustered into operational taxonomic units (OTUs) using USEARCH global alignment [26]. A .NET algorithm utilizing BLASTN+ was used to query FASTA formatted file with seed sequences for each cluster against a database of NCBI derived sequences. Sequences were aligned by Infernal [27] and clustered by Complete Linkage Clustering (a module of the RDPipeline, https://rdp.cme.msu.edu/). The Shannon-Wiener index of diversity (H′) [28] was calculated (at a level of species). Samples were from the same run and have a similar number of reads; therefore, normalization was not performed to avoid data loss. Rarefaction analysis was performed using a module of the RDPipeline. The sequences have been deposited in the Sequence Read Archive (SRA) NCBI within BioProjectPRJNA380917 as an experiment 'Metagenome of bioaugmented anaerobic digesters.'

3. Results and Discussion

3.1. Impact of Bioaugmentation on the Efficiency of Anaerobic Digestion of Sewage Sludge

The characteristics of both the mixtures feeding the reactors and the digests are presented in Figure 3. As indicated in the figure, the characteristics of SS varied in the control reactors (R1 and R4), which probably resulted from the changes in the chemical composition of sewage (Figure 1, Table 1).

Bioaugmentation by a mixture of Bacteria and Archaea decreased all parameters in the feedstock (due to dilution) as compared to sewage sludge. Decreases in COD concentration from 45.8 g dm^{-3} to 39.8 g dm^{-3} and 38.6 g dm^{-3}, VS from 31.0 g kg^{-1} to 28.1 and 26.4 g kg^{-1}, and TS from 40.1 g kg^{-1} to 36.4 g kg^{-1} and 34.2 g kg^{-1} were observed for R1, R2, and R3, respectively. The concentration of SCOD in the bioaugmented reactors was 1.25 and 1.20 g dm^{-3} and was lower than that in the control (1.40 g dm^{-3}). Similarly, in R4 and R5, the concentration of VS and TS decreased by 15% (from 26.1 to 22.1 g kg^{-1} and from 33.9 to 28.8 g kg^{-1}, respectively), COD by 13.5% (from 38.3 to 33.1 g dm^{-3}), and SCOD by 9% (from 2.5 to 2.3 g dm^{-3}). The higher the dose of the bioaugmenting mixture, the greater the decrease in the aforementioned characteristics as compared to the control reactors.

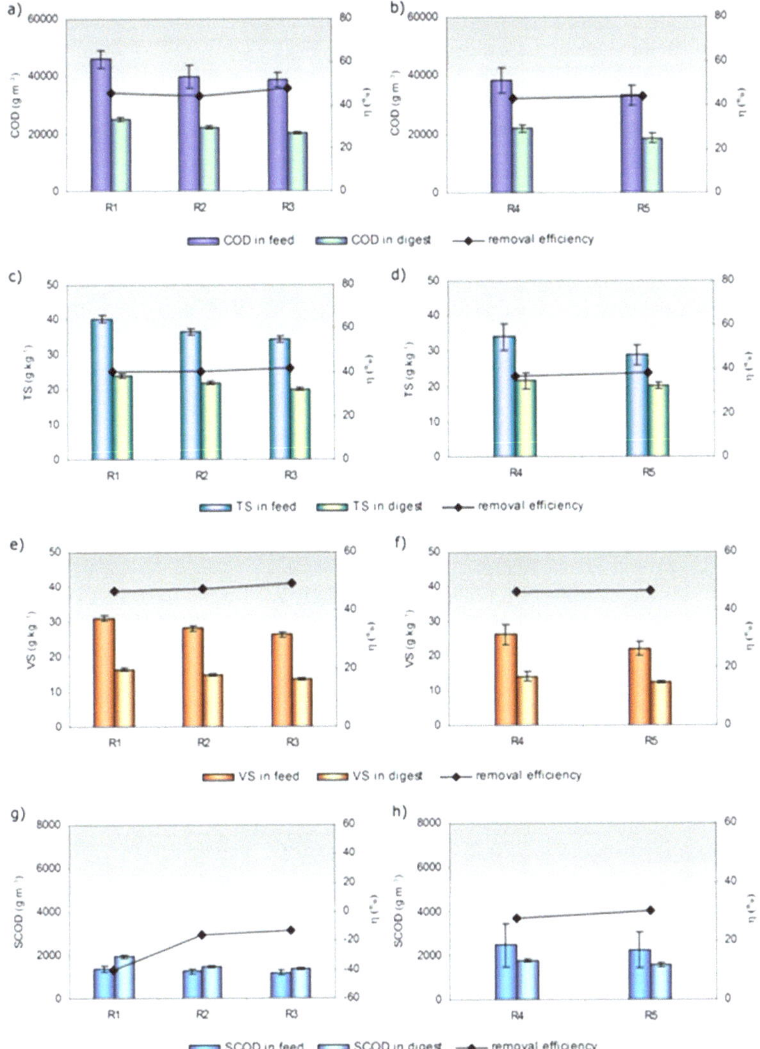

Figure 3. Concentration of organic compounds (expressed as chemical oxygen demand (COD), total solids (TS), volatile solids (VS), and soluble chemical oxygen demand (SCOD)) in feed and digest, as well as related removal efficiencies (average values are given, error bars represent 95% confidence limits for means): (**a**) COD changes in phase 1, (**b**) COD changes in phase 2, (**c**) TS changes in phase 1, (**d**) TS changes in phase 2, (**e**) VS changes in phase 1, (**f**) VS changes in phase 2, (**g**) SCOD changes in phase 1, (**h**) SCOD changes in phase 2.

The degrees of removal pertaining to VS and other parameters were used to evaluate the process efficiency. On average, the removal efficiencies of VS in R2, R3, and R5 (the bioaugmented reactors) amounted to 47.3%, 49.2%, and 46.4%, respectively. These values were greater than the ones obtained for sewage sludge (Figure 3); however, the difference was not statistically significant. The highest removal of VS was observed in the presence of 13% v/v dose of the bioaugmenting mixture (R3), despite the shortening of HRT from 20 to 17.4 d. On the other hand, the removal efficiency of TS reached 40.5%, 41.7%, and 37.9% in R2, R3, and R5, respectively; for the control reactors, it was slightly lower, equaling 40.0% and 35.6% for R1 and R4 control reactors, respectively. As far as the removal efficiency of COD is concerned, comparable values were observed for all reactors. However, the maximum COD removal efficiency, reaching 47.6%, was obtained with 13% v/v dose of the bioaugmenting mixture (R3). This observation is in line with the findings of Yu et al. [29], who reported that the application of bioaugmentation during the mature landfill leachate treatment slightly improved the COD removal efficiency. Interestingly, visible differences were observed for the efficiency of SCOD removal (Figure 3g,h). In comparison with the non-bioaugmented reactors, the SCOD removal efficiency increased by 24% and 27%, in the reactors (R2, R3, R5) supplemented with 9% and 13% of bioaugmenting mixture. In contrast, in the presence of 17% v/v of bioaugmenting mixture, the increase was only by 2.5%. To sum up, the removal efficiency of organic compounds increased, although the hydraulic retention time was shortened from 20 to 16.7 d. Hailei et al. [30] also reported that the addition of mixed microorganisms could shorten the sludge acclimation time, as well as improve the treatment efficiency.

3.2. Process Stability

To evaluate the stability of anaerobic digestion, pH, alkalinity, concentration of VFA and the VFA/alkalinity ratio in digest supernatant should be analyzed. The average values of these parameters in the feedstock and digest are given in Table 1.

Table 1. Volatile fatty acids (VFA) concentration, alkalinity and pH values in feed and digest for specified reactors.

Reactor	Value	VFA (g m^{-3})		Alkalinity (g CaCO$_3$ m^{-3})		pH	
		Feed	Digest	Feed	Digest	Feed	Digest
R1	Average	838	235	812	3722	6.81	7.99
	lower/upper 95% mean	719/957	66/404	764/860	3614/3830	6.68/6.94	7.95/8.03
R2	Average	771	241	769	3391	6.83	7.86
	lower/upper 95% mean	678/864	112/370	732/806	3268/3514	6.70/6.96	7.80/7.92
R3	Average	660	291	749	3336	6.82	7.72
	lower/upper 95% mean	542/778	203/379	712/786	3229/3443	6.68/6.96	7.66/7.78
R4	Average	1520	149	880	3670	6.68	7.71
	lower/upper 95% mean	147/2893	115/183	610/1150	3524/3816	6.44/6.92	7.44/7.98
R5	Average	1062	253	754	3195	6.75	7.59
	lower/upper 95% mean	163/1961	167/339	578/930	3078/3312	6.53/6.97	7.37/7.81

The results indicate that the pH of feedstock was at a comparable level in the bioaugmented and non-bioaugmented reactors. Conversely, alkalinity decreased in the bioaugmented reactors (R2, R3, and R5) by 5.5%, 8.4%, and 16.7%, respectively, but the differences were not statistically significant. An increase in pH was observed in the digest. The values for all runs were in a range favorable for methanogens, i.e., from pH 7.59 to pH 7.99. Similarly, the digest alkalinity increased more than 4-fold. In the bioaugmented reactors, pH and alkalinity were lower compared with the sewage sludge, and these differences were statistically significant. Bioaugmentation caused a decrease in the VFA of the mixture fed to the reactor, but the difference was not statistically significant. Higher VFA removal was reported for anaerobic digestion of sewage sludge: 71.9%—in R1, and 90.2%—in R4. For bioaugmented reactors, these values were 68.8% (R2), 55.6% (R3), and 76.2% (R5). The digest was characterized by low concentrations of VFA—they did not exceed the value of 290 g m^{-3} in any reactor. In the presence of the bioaugmenting mixture, the VFA/alkalinity ratio increased to 0.071 and 0.087 (for 9% and 13 *v/v* dose, respectively) and 0.079 (17% *v/v*). For the sludge, the ratio equaled 0.063 in R1 and 0.041 in R4. The process was carried out under stable conditions in both phases.

3.3. Biogas Production

The results of the investigations are shown in Table 2 (average data are reported). It should be noted that the feed conditions varied through phases (R1–R3 and R4–R5), which was attributed both to the changes in sludge characteristics and the addition of microorganisms from the bioaugmenting mixture. In terms of organic matter removed via anaerobic digestion, regardless of the feedstock, the lowest biogas yield was observed for the removed COD, while the highest for the removed VS. Considering organic substances fed to the reactor, a higher biogas yield was noted in terms of VS and lower regarding TS (Table 2).

The addition of microorganisms from the bioaugmenting mixture had no significant influence on the biogas production (the differences of means were not statistically significant). However, the biogas yields, as well as daily biogas production, were enhanced compared to the control (particularly for R2 and R3). These occurred despite the progressive decrease in the hydraulic retention time (HRT) from 20 d to 16.7 d that led to the faster washing out of the microorganisms from the semi-flow system. Moreover, similar methane content was noted independently of the mixture dose. It was also shown that bioaugmentation enabled to obtain higher biodegradation efficiency regarding the VS removal within shorter HRT. The above suggests the beneficial bioaugmentation effect. The results achieved could be attributed to the increase in rates of both biogas production and organics decomposition. It was due to the differences between the microbial communities of the bioaugmented digesters and the enhanced activity of microorganisms involved in bioaugumentation of anaerobic digestion. This explanation is consistent with the research by Duran et al. [31] regarding bioaugmentation with selected strains belonging to *Baccillus* sp., *Pseudomonas* sp., and *Actinomycetes* sp. The slight increase in biogas yields throughout bioaugmentation could result both from the HRT shortening and the TS/VS feedstock dilution by bioaugmenting mixture. The second one followed by the procedure of mixture preparation recommended by ArcheaSolutions Inc. To significantly improve the biogas yields, the higher concentration of the bioaugmenting mixture would be profitable. This could be achieved in the future research using a microfiltration module. The study of Poszytek et al. [32] using a novel bacterial strain *Ochrobactrum* sp. POC9 for bioaugmentation of sewage sludge anaerobic digestion revealed much better results than the ones presented here. In that case, the cumulative biogas production increased by 22.06% compared to the control, although the study was conducted in batch mode.

Table 2. Biogas production and corresponding yields as well as methane content for bioaugmented and non-bioaugmented reactors.

Parameter	Unit	R1 (Control)	R2	R3	R4 (Control)	R5
Daily biogas production [a]	$dm^3\ d^{-1}$	23.23 ± 3.7 [b]	24.26 ± 3.8	24.53 ± 3.8	19.59 ± 2.2	19.86 ± 2.1
Biogas yield	$m^3\ kg^{-1}$ VS added	0.38 ± 0.07	0.40 ± 0.07	0.40 ± 0.07	0.38 ± 0.05	0.37 ± 0.05
	$m^3\ kg^{-1}$ TS added	0.29 ± 0.05	0.31 ± 0.05	0.31 ± 0.05	0.29 ± 0.04	0.29 ± 0.04
	$m^3\ kg^{-1}$ VS removed	0.83 ± 0.20	0.86 ± 0.21	0.84 ± 0.21	0.82 ± 0.22	0.83 ± 0.22
	$m^3\ kg^{-1}$ TS removed	0.75 ± 0.18	0.78 ± 0.21	0.77 ± 0.19	0.79 ± 0.27	0.78 ± 0.29
	$m^3\ kg^{-1}$ COD removed	0.52 ± 0.09	0.53 ± 0.11	0.53 ± 0.11	0.55 ± 0.09	0.55 ± 0.13
Methane content	%	56.25 ± 1.93	56.56 ± 1.58	56.16 ± 2.06	55.22 ± 1.98	55.57 ± 2.50

a—the average value, b—in normal conditions.

3.4. Kinetics

In the quasi-flow system, the reactor was fed once a day with the portion of substrate or substrates and the same volume of digested medium was removed from the reactor. For this reason, the production of biogas between each feeding related to a temporal interval of 0–24 h. Accordingly, the reaction rate constant was expressed in hours, distinct from the typical units used for batch experiments (d^{-1}). The average volume of biogas produced day by day (calculated for the exemplary 30 measurement days) is shown in Figure 4.

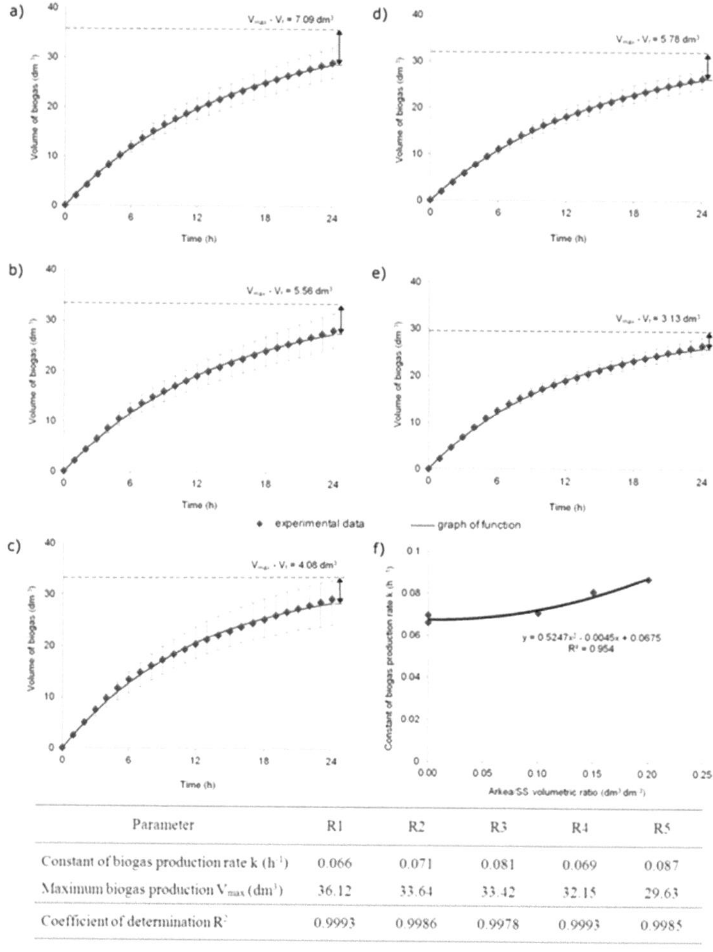

Parameter	R1	R2	R3	R4	R5
Constant of biogas production rate k (h^{-1})	0.066	0.071	0.081	0.069	0.087
Maximum biogas production V_{max} (dm^3)	36.12	33.64	33.42	32.15	29.63
Coefficient of determination R^2	0.9993	0.9986	0.9978	0.9993	0.9985

Figure 4. Biogas production in time (the average values from 30 measurement days and standard deviations are given), the values of kinetic constants and coefficients of determination for specified reactors: (**a**) R1, (**b**) R2, (**c**) R3, (**d**) R4, (**e**) R5, and (**f**) k constant as a function of Arkea/SS volumetric ratio.

For a mathematical description of the changes in the biogas volume (V) produced in time (t), the most appropriate was the equation of first-order reaction [33] as $V = V_{max} (1 - \exp(-k \cdot t))$. This was confirmed by the determination coefficients (R^2). The experimental data allowed to determine the reaction rate constant (k) and the maximum gas volume (V_{max}), which theoretically can be derived from the feed portion feeding the reactor once a day.

The results indicate that the biogas production rate constants for sewage sludge were comparable and equaled 0.066 h^{-1} and 0.069 h^{-1} for R1 and R4, respectively. In the bioaugmented reactors, the k values were 0.071 h^{-1}, 0.081 h^{-1}, and 0.087 h^{-1} in R2, R3, and R5, respectively. It was probably caused by the enhanced microorganisms' activity in the bioaugmented systems. This was confirmed by the increased biogas production rate constant k. The k value increased with the increasing the dose of bioaugmenting mixture, according to the equation $y = 0.5247x^2 - 0.0045x + 0.0675$ (Figure 4f).

In each phase, the differences between the maximum biogas production from the reactor feed (V_{max}) and the actual value of biogas production obtained after the period of 24 h (V_f) were determined (Figure 4). The difference $V_{max} - V_f$ corresponds to the value of biogas potential in the digest and varied from 3.13 to 7.09 dm^3 in the reactors. The best results were obtained in the bioaugmented reactors, with increasing doses of bioaugmenting mixture in the feedstock. In such cases, the untapped biogas potential decreased and amounted to 16.5% (R2), 12.0% (R3), and 10.6% (R5). In the control runs, the untapped biogas potential was higher and was up to 19.6% and 17.8% (R1 and R4, respectively). Bioaugmentation of digested sewage sludge resulted in an enhancement of the metabolism transformation rate, which was associated with the increase of process efficiency.

Importantly, the influence of the feed changes on the kinetics results determined on the basis of continuous data acquisition throughout experiment time was shown. This was revealed in terms of standard deviation growing within 24 h for the analyzed 30 measurement days.

3.5. Changes in Digest Morphology

The scanning electron microscopy was used to observe the microbial aggregates in the digest structure, the space relationship between microorganisms, as well as the presence of extracellular polymeric substances (EPS) and other materials.

While analyzing the structure of the digested sludge, it was noted that despite the shortened HRT, larger agglomerates of microorganisms were formed in the bioaugmented reactors compared to the non-bioaugmented sludge (Figure 5). This was beneficial for the subsequent digest dewatering because of the extended sludge sediments capability (unpublished data). On the basis of the operation strategy ensuring comparable OLR value for both the bioaugmented and non-bioaugmented reactors, the observed effect could be largely attributed to extended EPS secretion by microorganisms [34,35] through bioaugmentation, and thus lower shear sensitivity and lower degree of dispersion [36]. According to the study of Yu et al. [37], the microbial community composition and its activity affected both the EPS production and composition. Interestingly, in the structure of the bioaugmented digest, the changes in the morphology of EPS were clearly visible. EPS began to collapse and condense into fiber-like structures [38]. Similarly, the differences referring to methanogens seem to indicate a response in their cell shapes to bioaugmentation that was reported by Zhang et al. [39].

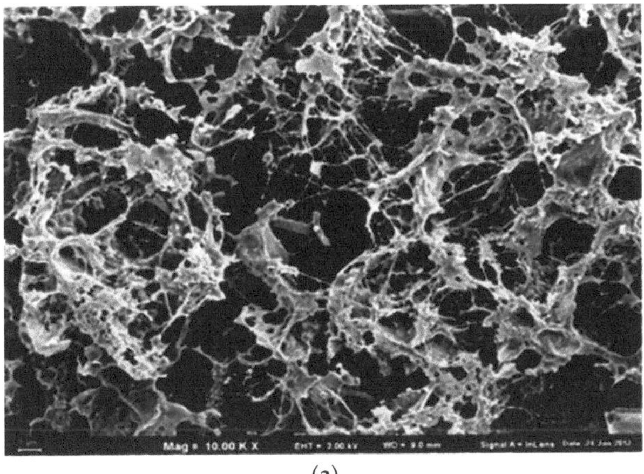

(a)

Figure 5. *Cont.*

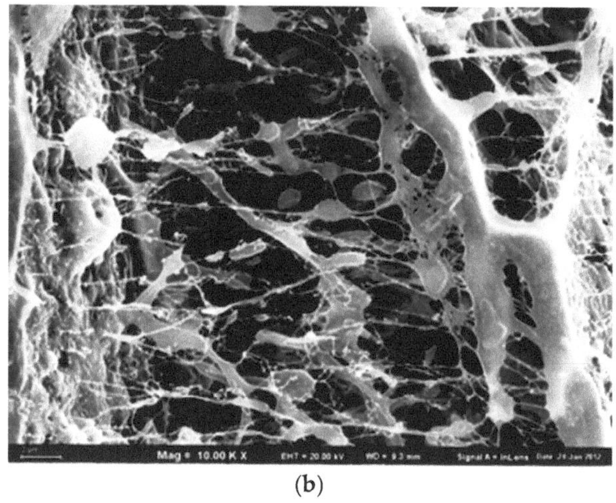

(**b**)

Figure 5. SEM micrograph of non-bioaugmented (**a**) and bioaugmented (**b**) digested sewage sludge (magnification 10×).

3.6. Microbial Structure of Digesters

Sludge fermentation starts with hydrolysis followed by acidogenesis. In both these processes, the predominant microorganisms are obligatory anaerobes and facultative bacteria. Microorganisms conducting acetogenesis produce hydrogen and are in symbiosis with methanogenic archaea, which consume hydrogen (syntrophy or interspecies hydrogen transfer—IHT). The balance between methanogens and microorganisms involved in acido- and acetogenesis is crucial because if the activity of the latter is too high, anaerobic digestion will fail due to the acidification of the reactor.

The microbial structure of the biomass from the reactors operated in the first phase of the experiment was analyzed using next-generation sequencing. Rarefaction analysis was used to characterize the richness of taxa in the experimental digesters. At the genus level, the curves leveled off, indicating acceptable sampling and coverage of the richness in the samples (data not shown). The microbial diversity of the biomass was the highest in the control reactor (H′ = 2.86) and decreased gradually to H′ = 2.52 and H′ = 2.16 in R2 and R3, respectively, as the dose of bioaugmenting mixture was increased. In the reactors, Bacteria predominated and Archaea constituted between 1.3% and 1.4% of identified sequences (Table 3).

Table 3. Percentage of bacterial taxa in biomass from the experimental reactors.

Kingdom; Phylum; Class; Order; Family; Genus	R1	R2	R3
Bacteria; Verrucomicrobia; Verrucomicrobiae; Verrucomicrobiales; Unclassified	3.1	1.0	0.4
Bacteria; Unclassified	31.0	38.8	52.7
Bacteria; Thermotogae; Thermotogae; Thermotogales; Unclassified	4.4	2.8	4.2
Bacteria; Synergistetes; Synergistia; Synergistales; Synergistaceae; Synergistes	0.5	0.3	0.1
Bacteria; Spirochaetes; Unclassified	2.8	3.9	3.9
Bacteria; Spirochaetes; Spirochaetia; Unclassified	0.8	0.3	0.3
Bacteria; Proteobacteria; Unclassified	0.4	0.5	0.3
Bacteria; Proteobacteria; Gammaproteobacteria; Xanthomonadales; Xanthomonadaceae; *Thermomonas*	0.5	0.4	0.3

Table 3. *Cont.*

Kingdom; Phylum; Class; Order; Family; Genus	R1	R2	R3
Bacteria; Proteobacteria; Gammaproteobacteria; Pseudomonadales; Pseudomonadaceae; *Pseudomonas*	0.8	3.9	3.6
Bacteria; Proteobacteria; Deltaproteobacteria; Syntrophobacterales; Unclassified; Unclassified	0.8	1.2	0.8
Bacteria; Proteobacteria; Deltaproteobacteria; Syntrophobacterales; Syntrophaceae; *Syntrophus*	1.4	1.8	1.7
Bacteria; Proteobacteria; Deltaproteobacteria; Syntrophobacterales; Syntrophaceae; *Smithella*	0.9	0.8	0.7
Bacteria; Proteobacteria; Deltaproteobacteria; Desulfobacterales; Desulfobacteraceae; *Desulfofaba*	0.7	1.3	1.0
Bacteria; Proteobacteria; Betaproteobacteria; Rhodocyclales; Rhodocyclaceae; *Dechloromonas*	0.5	0.4	0.3
Bacteria; Proteobacteria; Betaproteobacteria; Burkholderiales; Comamonadaceae; *Rhodoferax*	1.3	1.0	0.7
Bacteria; Proteobacteria; Betaproteobacteria; Burkholderiales; Comamonadaceae; *Diaphorobacter*	0.8	0.8	0.5
Bacteria; Proteobacteria; Betaproteobacteria; Burkholderiales; Comamonadaceae; *Acidovorax*	0.5	0.4	0.4
Bacteria; Firmicutes; Unclassified	0.8	0.7	0.3
Bacteria; Firmicutes; Clostridia; Clostridiales; Clostridiaceae; *Clostridium*	0.5	0.5	0.3
Bacteria; Cloacimonetes; Unclassified	7.1	6.5	3.6
Bacteria; Chloroflexi; Unclassified	1.2	0.7	0.5
Bacteria; Chloroflexi; Anaerolineae; Anaerolineales; Unclassified	1.4	0.8	0.4
Bacteria; Bacteroidetes; Sphingobacteriia; Sphingobacteriales; Unclassified; Unclassified	0.7	0.4	0.3
Bacteria; Bacteroidetes; Cytophagia; Cytophagales; Cytophagaceae; *Cytophaga*	18.9	16.8	12.3
Bacteria; Bacteroidetes; Bacteroidia; Bacteroidales; Bacteroidaceae; *Bacteroides*	1.9	0.7	0.6
Bacteria; Actinobacteria; Actinobacteria; Micrococcales; Intrasporangiaceae; *Tetrasphaera*	0.5	0.4	0.4
Bacteria; Actinobacteria; Actinobacteria; Micrococcales; Dermatophilaceae; *Dermatophilus*	1.1	0.8	0.7
Archaea; Unclassified	0.5	0.4	0.3
Archaea; Euryarchaeota; Methanomicrobia; Methanomicrobiales; Methanomicrobiaceae; *Methanoculleus*	0.8	1.0	1.0
Low abundance *	9.2	7.5	5.4
No Hit	4.5	3.5	2.1

* In the table only bacterial taxa with abundance over 0.5% were presented.

This value was one order of magnitude lower than for example the one noted in mesophilic full-scale digesters with sewage sludge [24]; however, it was similar to the values noted in mesophilic reactors for co-digesting of fish waste and cow manure (about 1%) [7] or solid-state digesters fed with kitchen waste, pig manure and excess sludge (about 0.5%) [40]. In this study, despite being only a small fraction of the entire microbial community, Archaea ensured efficient production of methane-rich biogas.

The structure of the biomass in the reactors differed from that of the mixture for bioaugmentation. This indicates that most of the microbes in the mixture were not able to survive in the reactors; however, the ones that survived improved the efficiency of methane fermentation as concluded based on the rate constant for biogas production and efficiency of SCOD removal. Within the biomass, the percentage of unclassified bacteria increased with increasing dose of bioaugmenting mixture showing that high diversity of yet unknown microorganisms was present in the reactor as a result of bioaugmentation.

Among Bacteria, the core communities belonged to the phyla Thermotogae, Spirochaetes, Cloacimonetes, Actinobacteria, Bacteroidetes, Chloroflexi, Firmicutes, and Proteobacteria. Bacteroidetes, Chloroflexi, Firmicutes, and Proteobacteria contain most of the identified species of acidogenic bacteria that support the hydrolysis stage [41]. Microorganisms belonging to class Cytophagia predominated in the biomass (12.3–18.9%). However, as doses of the bioaugmenting mixture were increased, the percentage of *Cytophaga* sp. decreased; this was a strong association, with an R^2 value of 0.91, indicating that 91% of the variation in *Cytophaga* sp. abundance was associated with the dose of bioaugmenting mixture. Similarly, an increased dose of this mixture was associated with decreases in the percent abundance of the order Verrucomicrobiales ($R^2 = 1.00$), the phyla Cloacimonetes ($R^2 = 0.83$), and Choroflexi ($R^2 = 1.00$), including order Anaerolineales ($R^2 = 0.99$).

Members of the genus *Cytophaga* are important for anaerobic decomposition of biopolymers, such as xylan or cellulose. *Cytophaga xylanolytica* is a mesophilic anaerobe that grows by fermentation of mono-, di-, and polysaccharides (but not cellulose) to acetate, propionate, succinate, CO_2, and H_2; xylan-grown cells of this species have xylanase and various glycosidase activities. *Cytophaga hutchinsonii* can rapidly digest crystalline cellulose without free cellulases or cellulosomes [42]. As extracellular polymeric substances may comprise up to 30% of activated sludge, these activities of *Cytophaga* sp. may be crucial for efficient degradation of organics in fermented sludge.

The abundance of *Pseudomonas* sp. increased from 0.4% in R1 to nearly 4% in bioaugmented R2 and R3 and it was the only identified genus whose abundance increased by such a large amount. Such an increase can be advantageous for anaerobic digestion because Duran et al. [31] observed that the presence of selected strains of genera *Pseudomonas*, *Bacillus*, and *Actinomycetes* improved the anaerobic digestion of biosolids, increasing net CH_4 production by 29% and diminished odor formation. In addition, Xia et al. [43] reported that an increase in the proportion of some functional organisms, including *Pseudomonas* sp., led to an increase in the efficiency of anaerobic digestion when the proportion of more biodegradable, low molecular weight fractions (<20 kDa) was increased 10 times because of solubilization of some of the proteins, polysaccharides, nucleic acids, and humic-like substances. In our study, the concentration of low molecular weight SCOD in the digest was negatively correlated ($R^2 = 0.97$) with the abundance of *Pseudomonas* sp. in the biomass in the bioaugmented reactors, indicating that this genus played an important role in SCOD degradation.

Syntrophic bacteria comprised a significant part of the biomass. From 3.6 to 7.1% of the identified sequences belonged to syntrophic prokaryotes from phylum Cloacimonetes. The analysis of the proteome of *Candidatus Cloacimonas acidaminovorans* indicated that this bacterium derives carbon and energy from the fermentation of amino acids and that it is a syntroph producing H_2 and CO_2 from formate [44]. On the other hand, the abundance of phylum Cloacimonetes was linked with lowered methane production in reactors fed with protein-rich substrates [7]. In this study, the abundance of other syntrophic microorganisms belonging to *Syntrophus* sp. and *Smithella* sp. was relatively stable in the reactors (1.4–1.8% and 0.7–0.9%, respectively). *Smithella* sp. are syntrophic acetogens involved in propionate degradation in anaerobic digesters, while *Syntrophus* sp. oxidates fatty acids and benzoate [41,45].

During the methane fermentation, Methanomicrobiales are usually less numerous than Methanosarcinales [8]. In this study, the predominance of Methanomicrobiales (about 1% of all identified sequences) in the species structure of Archaea indicated hydrogenotrophic methanogenesis as the main pathway of methane generation. Although *Methanosaeta* sp. predominated in the liquor used for bioaugmentation, *Methanoculleus* was the most abundant genus in the reactors. The idea of the study was to bioaugment the reactor with Archaea microorganisms to support methane production because this is the most critical step in the anaerobic digestion conducted by slow-growing microorganisms and prone to changes in environmental conditions. From the microbial analysis of biomass, it can be concluded that the biodiversity of bioaugmenting mixture reflecting the potential of different species to colonize the fermentation reactors was low (only 11 species with abundance over 0.5%). Despite this, bioaugmentation was successful in terms of the most abundant group that is *Methanosaeta* sp. belonging to Methanosarcinales that produce methane via an acetotrophic pathway. This genus was present in the experimental reactors, but its abundance was below 0.5%. *Methanosaeta* sp. is sensitive to OLR. They were abundant in the mesophilic fermentation reactors that were operated at an OLR of 1 kg COD m^{-3} day^{-1} but the increase in OLR to 2 kg COD m^{-3} day^{-1}, at the maintained process temperature, caused their disappearance from the biomass [46]. In the present study, OLR was higher than the optimal for *Methanosaeta* sp.; therefore, although present in the biomass, they were not able to outcompete *Methanoculleus* sp. that predominated in the reactors. *Methanoculleus* sp. commonly occurs in digesters operated in meso- and thermophilic temperatures, including the digesters in which the process is supported by physico-chemical treatment, e.g., microwave radiation [46]. *Methanoculleus* sp. cope well with high OLRs, comprising over 40% of biomass during thermophilic co-digestion of manure and waste whey at an OLR of 60.4 g COD m^{-3} day^{-1} [40]. This fact may explain the predominance in the present study of this genera among Archaea in the experimental reactors operated at the higher OLRs.

4. Conclusions

Bioaugmentation decreased the HRT from 20 d to 16.7 d, but despite this decrease, the observed daily biogas production, methane content in the biogas and the biogas yield per kg of VS were similar in bioaugmented and control reactors. With bioaugmentation, SCOD removal improved, especially in reactors operated at higher OLR, which can be attributed to the increase in the rate of biogas production. Regardless of loading, the value of k was higher in bioaugmented reactors than in control reactors. The structure of the biomass in all reactors was different from that of the mixture for bioaugmentation and bioaugmentation diminished species diversity. In all digesters, bacteria belonging to Thermotogae, Spirochaetes, Cloacimonetes, Bacteroidetes, and Proteobacteria predominated with cellulose-hydrolyzing *Cytophaga* as the most abundant genus. The abundance of *Pseudomonas* sp. increased as the dose of bioaugmentative mixture was increased. The predominance of *Methanoculleus* sp. among Archaea indicated that hydrogenotrophic methanogenesis was the main pathway of methane generation.

Author Contributions: Conceptualization, A.M. and M.L.; Methodology, A.M., M.L. and A.C.-K.; Software, M.L., A.M. and A.C.-K.; Validation, M.L., A.M. and A.C.-K.; Formal Analysis, M.L., A.M. and A.C.-K.; Investigation, A.M., M.L. and A.C.-K.; Resources, A.M., M.L. and A.C.-K.; Data Curation, A.M., M.L. and A.C.-K.; Writing-Original Draft Preparation, M.L., and A.C.-K.; Writing-Review & Editing, A.M., M.L. and A.C.-K.; Visualization, A.M., M.L. and A.C.-K.; Supervision, A.M.; Project Administration, A.M. and A.C.-K.; Funding Acqustion, A.M. and A.C.-K.

Funding: The authors thank for the financial aid from the National Centre of Science (Poland), No.7405/B/T02/2011/40 and financial aid from the Ministry of Science and Higher Education in Poland (Statutory Research, 18.610.006-300).

Conflicts of Interest: The authors declare no conflict of interest.

References

1. Maj Duong, T.H.; Smits, M.; Vestraete, W.; Carballa, M. Enhanced biomethanation of kitchen waste by different pretreatments. *Bioresour. Technol.* **2011**, *102*, 592–599. [CrossRef]
2. Satoh, H.; Okabe, S.; Yamaguchi, Y.; Watanabe, Y. Evaluation of the impact of bioaugmentation and biostimulation by in situ hybridization and microelectrode. *Water Res.* **2003**, *37*, 2206–2216. [CrossRef]
3. Timmis, K.N. *Handbook of Hydrocarbon and Lipid Microbiology*; Springer: Berlin, Germany, 2010; ISBN 978-3-54-077584-3.
4. Cycoń, M.; Mrozik, A.; Piotrowska-Seget, Z. Bioaugmentation as a strategy for remediation of pesticide-polluted soil: A review. *Chemosphere* **2017**, *172*, 52–71. [CrossRef] [PubMed]
5. Chen, Q.; Ni, J.; Ma, T.; Liu, T.; Zheng, M. Bioaugmentation treatment of municipal wastewater with heterotrophic-aerobic nitrogen removal bacteria in a pilot-scale SBR. *Bioresour. Technol.* **2015**, *183*, 25–32. [CrossRef] [PubMed]
6. Neumann, L.; Scherer, P. Impact of bioaugmentation by compost on the performance and ecology of an anaerobic digester fed with energy crops. *Bioresour. Technol.* **2007**, *102*, 2931–2935. [CrossRef] [PubMed]
7. Solli, L.; Håvelsrud, O.E.; Horn, S.J.; Rike, A.G. A metagenomic study of the microbial communities in four parallel biogas reactors. *Biotechnol. Biofuels* **2014**, *7*, 146. [CrossRef] [PubMed]
8. Tabatabaei, M.; Rahim, R.A.; Abdullah, N.; Wright, A.-D.G.; Shirai, Y.; Sakai, K.; Sulaiman, A.; Hassan, M.-A. Importance of the methanogenic archaea populations in anaerobic wastewater treatments. *Process Biochem.* **2010**, *45*, 1214–1225. [CrossRef]
9. Nielsen, H.B.; Mladenovska, Z.; Ahring, B.K. Bioaugmentation of a two-stage thermophilic (68 °C/55 °C) anaerobic digestion concept for improvement of the methane yield from cattle manure. *Biotechnol. Bioeng.* **2007**, *97*, 1638–1643. [CrossRef] [PubMed]

10. Li, P.; Wang, Y.; Wang, Y.; Jiang, Z.; Tong, L. Bioaugmentation of cellulose degradation in swine wastewater treatment with a composite microbial consortium. *Fresen. Environ. Bull.* **2010**, *19*, 3107–3112. [CrossRef]

11. Wang, A.; Ren, N.; Shi, Y.; Lee, D.-J. Bioaugmented hydrogen production from microcrystalline cellulose using co-culture—*Clostridium acetobutylicum* X_9 and *Ethanoigenensharbinense* B_{49}. *Int. J. Hydrogen Energy* **2008**, *33*, 912–917. [CrossRef]

12. Cirne, D.G.; Bjornssom, L.; Alves, M.; Mattiasson, B. Effects of bioaugmentation by an anaerobic lipolytic bacterium on anaerobic digestion of lipid-rich waste. *J. Chem. Technol. Biotechnol.* **2006**, *81*, 1745–1752. [CrossRef]

13. Venkiteshwaran, K.; Milferstedt, K.; Hamelin, J.; Zitomer, D.H. Anaerobic digester bioaugmentation influences quasi steady state performance and microbial community. *Water Res.* **2016**, *104*, 128–136. [CrossRef] [PubMed]

14. Schmidt, J.E.; Larsen, S.B.; Karakashev, D. Ex-situ bioremediation of polycyclic aromatic hydrocarbons in sewage sludge. *WIT Trans. Ecol. Environ.* **2008**, *10*, 189–198. [CrossRef]

15. Poszytek, K.; Pyzik, A.; Sobczak, A.; Lipinski, L.; Sklodowska, A.; Drewniak, L. The effect of the source of microorganisms on adaptation of hydrolytic consortia dedicated to anaerobic digestion of maize silage. *Anaerobe* **2017**, *46*, 46–55. [CrossRef] [PubMed]

16. Li, Y.; Zhang, Y.; Sun, Y.; Wu, S.; Kong, X.; Yuan, Z.; Dong, R. The performance efficiency of bioaugmentation to prevent anaerobic digestion failure from ammonia and propionate inhibition. *Bioresour. Technol.* **2017**, *231*, 94–100. [CrossRef] [PubMed]

17. Schauer-Gimenez, A.E.; Zitomer, D.H.; Maki, J.H.; Struble, C.A. Bioaugmentation for improved recovery of anaerobic digesters after toxicant exposure. *Water Res.* **2010**, *44*, 3555–3564. [CrossRef] [PubMed]

18. Tale, V.P.; Maki, J.S.; Zitomer, D.H. Bioaugmentation of overloaded anaerobic digesters restores function and archaeal community. *Water Res.* **2015**, *70*, 138–147. [CrossRef] [PubMed]

19. Herrero, M.; Stuckey, D.C. Bioaugmentation and its application in wastewater treatment: A review. *Chemosphere* **2015**, *140*, 119–128. [CrossRef] [PubMed]

20. Mehariya, S.; Patel, A.K.; Obulisamy, P.K.; Punniyakotti, E.; Wong, J.W.C. Co-digestion of food waste and sewage sludge for methane production: Current status and perspective. *Bioresour. Technol.* **2018**. [CrossRef] [PubMed]

21. Raper, E.; Stephenson, T.; Anderson, D.R.; Fisher, R.; Soares, A. Industrial wastewater treatment through bioaugmentation. *Process Saf. Environ. Prot.* **2018**, *118*, 178–187. [CrossRef]

22. Zhang, Q.-Q.; Yang, G.-F.; Zhang, L.; Zhang, Z.-Z.; Tia, G.-M.; Jin, R.-C. Bioaugmentation as a useful strategy for performance enhancement in biological wastewater treatment undergoing different stresses: Application and Mechanisms. *Crit. Rev. Environ. Sci. Technol.* **2017**, *47*, 1877–1899. [CrossRef]

23. American Public Health Association. *Standard Methods for the Examination of Water & Wastewater*, Centennial Edition 21 ed.; American Public Health Association: Washington, DC, USA, 2005.

24. Świątczak, P.; Cydzik-Kwiatkowska, A.; Rusanowska, A. Microbiota of anaerobic digesters in full-scale wastewater treatment plant. *Arch. Environ. Prot.* **2017**, *43*, 53–60. [CrossRef]

25. Edgar, R.C.; Haas, B.J.; Clemente, J.C.; Quince, C.; Knight, R. UCHIIME improves sensitivity and speed of chimera detection. *Bioinformatics* **2011**, *27*, 2194–2200. [CrossRef] [PubMed]

26. Edgar, R.C. Search and clustering orders of magnitude faster than BLAST. *Bioinformatics* **2010**, *26*, 2460–2461. [CrossRef] [PubMed]

27. Nawrocki, E.P.; Eddy, S.R. Infernal 1. 1, 100-fold faster RNA homology searches. *Bioinformatics* **2013**, *29*, 2933–2935. [CrossRef] [PubMed]

28. Hill, T.C.J.; Walsh, K.A.; Harris, J.A.; Moffett, B.A. Using ecological diversity measures with bacterial communities. *FEMS Microbiol. Ecol.* **2003**, *43*, 1–11. [CrossRef] [PubMed]

29. Yu, D.; Yang, Y.; Teng, F.; Feng, L.; Fang, X.; Ren, H. Bioaugmentation treatment of mature landfill leachate by new isolated ammonia nitrogen and humic acid resistant microorganisms. *J. Microbiol. Biotechnol.* **2014**, *24*, 987–997. [CrossRef] [PubMed]

30. Hailei, W.; Guosheng, L.; Ping, L.; Feng, P. The effect of bioaugmentation on the performance of sequencing batch reactor and sludge characteristics in the treatment process of papermaking wastewater. *Bioprocess Biosyst. Eng.* **2006**, *29*, 283–289. [CrossRef] [PubMed]

31. Duran, M.; Tepe, N.; Yurtsever, D.; Punzi, V.L.; Bruno, C.; Mehta, R.J. Bioaugmenting anaerobic digestion of biosolids with selected strains of *Bacillus, Pseudomonas,* and *Actinomycetes* species for increased methanogenesis and odor control. *Appl. Microbiol. Biotechnol.* **2006**, *73*, 960–966. [CrossRef] [PubMed]
32. Poszytek, K.; Karczewska-Golec, J.; Jakusz, G.; Krucon, T.; Lomza, P.; Zhenfdong, Y.; Drewniak, L.; Ciok, A.; Decewicz, P.; Dziurzynski, M.; et al. Genome-Guided characterization of *Ochrobactrum* sp. POC9 enhancing sewage sludge utilization—Biotechnological potential and biosafety consideration. *Int. J. Environ. Res. Public Health* **2018**, *15*, 1501. [CrossRef] [PubMed]
33. Gavala, N.H.; Angelidaki, I.; Ahring, B.K. Kinetics and modeling of anaerobic digestion process. *Adv. Biochem. Eng. Biotechnol.* **2002**, *81*, 57–93.
34. Sheng, G.-P.; Yu, H.-Q.; Li, X.-Y. Extracellular polymeric substances (EPS) of microbial aggregates in biological wastewater treatment systems: A review. *Biotechnol. Adv.* **2010**, *28*, 882–894. [CrossRef] [PubMed]
35. Yang, Z.H.; Xu, R.; Zheng, Y.; Chen, T.; Zhao, L.-J.; Li, M. Characterization of extracellular polymeric substances and microbial diversity in anaerobic co-digestion reactor treated sewage sludge with fat, oil, grease. *Bioresour. Technol.* **2016**, *212*, 164–173. [CrossRef] [PubMed]
36. Mikkelsen, L.H.; Keiding, K. Physico-chemical characteristics of full scale sewage sludges with implication to dewatering. *Water Res.* **2002**, *36*, 2451–2462. [CrossRef]
37. Yu, Z.; Wen, X.; Xu, M.; Huang, X. Characteristics of extracellular polymeric substances and bacterial communities in an anaerobic membrane bioreactor coupled with online ultrasound equipment. *Bioresour. Technol.* **2012**, *117*, 333–340. [CrossRef] [PubMed]
38. Dohnalkova, A.C.; Marshall, M.J.; Arey, B.W.; Williams, K.H.; Bukc, E.C.; Fredrickson, J.K. Imaging hydrated microbial extracellular polymers: Comparative analysis by electron microscopy. *Appl. Environ. Microbiol.* **2017**, *77*, 1254–1262. [CrossRef] [PubMed]
39. Zhang, J.; Dong, H.; Zhao, L.; McCarrick, R.; Agrawal, A. Microbial reduction and precipitation of vanadium by mesophilic and thermophilic methanogens. *Chem. Geol.* **2014**, *370*, 29–39. [CrossRef]
40. Li, A.; Chu, Y.; Wang, X.; Ren, R.; Yu, J.; Liu, X.; Yan, J.; Zhang, L.; Wu, S.; Li, S. A pyrosequencing-based metagenomic study of methane-producing microbial community in solid-state biogas reactor. *Biotechnol. Biofuels* **2013**, *6*, 3. [CrossRef] [PubMed]
41. Venkiteshwaran, K.; Bocher, B.; Maki, J.; Zitomer, D.H. Relating Anaerobic Digestion Microbial Community and Process Function. *Microbiol. Insights* **2015**, *8*, 37–44. [CrossRef] [PubMed]
42. Haack, S.H.; Breznak, J.A. *Cytophaga xylanolytica* sp. nov., a xylan-degrading, anaerobic gliding bacterium. *Arch. Microbiol.* **1993**, *159*, 6–15. [CrossRef]
43. Xia, S.; Zhou, Y.; Eustance, E.; Zhang, Z. Enhancement mechanisms of short-time aerobic digestion for waste activated sludge in the presence of cocoamidopropyl betaine. *Sci. Rep.* **2017**, *7*, 13491. [CrossRef] [PubMed]
44. Pelletier, E.; Kreymeyer, A.; Bocs, S.; Rouy, Z.; Gyapay, G.; Chouari, R.; Riviere, D.; Ganesan, P.; Daegelen, A.; Sghir, A.; et al. "*Candidatus Cloacamonas acidaminovorans*": Genome sequence reconstruction provides a first glimpse of a new bacterial division. *J. Bacteriol.* **2008**, *190*, 2572–2579. [CrossRef] [PubMed]
45. Jackson, B.E.; Bhupathiraju, V.K.; Tanner, R.S.; Woese, C.R.; McInerney, M.J. *Syntrophus aciditrophicus* sp. nov., a new anaerobic bacterium that degrades fatty acids and benzoate in syntrophic association with hydrogen-using microorganisms. *Arch. Microbiol.* **1999**, *171*, 107–114. [CrossRef] [PubMed]
46. Zielińska, M.; Cydzik-Kwiatkowska, A.; Zieliński, M.; Dębowski, M. Impact of temperature, microwave radiation and organic loading rate on methanogenic community and biogas production during fermentation of dairy wastewater. *Bioresour. Technol.* **2013**, *129*, 308–314. [CrossRef] [PubMed]

International Journal of
*Environmental Research
and Public Health*

Article

Full-Scale Experimental Study of Groundwater Softening in a Circulating Pellet Fluidized Reactor

Ruizhu Hu [1,2], Tinglin Huang [1,2,*], Aofan Zhi [1,2] and Zhangcheng Tang [1,2]

[1] Key Laboratory of Northwest Water Resource, Environment and Ecology, MOE, Xi'an University of Architecture and Technology, Xi'an 710055, China; www_lonely_com@163.com (R.H.); zhizhiing17@163.com (A.Z.); sdtangzhch@163.com (Z.T.)

[2] Shaanxi Key Laboratory of Environmental Engineering, Xi'an University of Architecture and Technology, Xi'an 710055, China

* Correspondence: huangtinglin@xauat.edu.cn; Tel.: +86-29-8220-1038

Abstract: The softening effect of a new type of circulating pellet fluidized bed (CPFB) reactor on groundwater was studied through a full-scale experiment. The operation of the CPFB reactor in the second water plant in Chang'an District in Xi'an China was monitored for one year, and the results were compared with those for the Amsterdam reactor in The Netherlands. The removal efficiency of Ca^{2+} in the CPFB reactor reached 90%; the removal rate of total hardness was higher than 60%; effluent pH was 9.5–9.8; the turbidity of the effluent and the turbidity after boiling were lower than 1.0 NTU; the unit cost was less than €0.064 per m^3; and the softened effluent was stable. The pellets in the CPFB reactor were circulated, providing higher crystallization efficiency. The diameter of the discharged pellets reached between 3–5 mm, and the fluidized area height of the CPFB reactor was 4 m. The performance parameters of the CFPB reactor were optimized.

Keywords: pellet reactor; circulating fluidization; groundwater softening; full-scale experiment

1. Introduction

Pellet softening in a fluidized bed reactor was developed and introduced in the 1970s in The Netherlands [1]. Almost all of the drinking water in The Netherlands was conditioned in 2016 and approximately 50% was softened by pellet fluidized bed (PFB) reactors [2]. PFB reactors provide evident advantages over the lime or adsorption softening of water [3,4].

The Amsterdam reactor is currently the most widely used PFB reactor. The fluidized bed part is a part of the Amsterdam reactor, which is cylindrical with a height of approximately 5 m [5,6]. Seeds are placed at the bottom of the reactor and fluidized under upward flow. The calcium carbonate ($CaCO_3$) crystallization on the seeds mainly occurs at the bottom of the reactor [6,7]. The pilot and full-scale study of the Amsterdam reactor on the removal of total hardness (TH) and Ca^{2+} and the development of the growth kinetics of pellets were carried out in the Weesperkarspel drinking water treatment and pilot plant of Waternet in Amsterdam, The Netherlands [2,8,9]. For example, Hofman et al. (2006) presented 20 years of experience with PFB reactor softening in The Netherlands. They reported that the pellets discharged from the reactor reached approximately 1.0 mm, and the removal rate of Ca^{2+} was approximately 50% [1]. Van Schagen et al. (2008) adopted mathematical models to show that the pellet size control in a PFB reactor had a significant influence on performance with respect to the water quality parameter. Maintaining the pellet size at the bottom of the reactor at 0.8 mm instead of 1.4 mm reduced the supersaturation of $CaCO_3$ in the water after the reactor by 50%. However, this increased the consumption of the seeding material by 550% [10]. Schetters et al. (2015) studied the reuse of ground pellets as seeding material in the pellet-softening process through a pilot-scale experiment. The effluent TH was in the range of 0.2 mmol/L to 2.0 mmol/L, and the pellet discharge diameter was only between 0.6–1.2 mm [8]. Chen et al. (2016) applied the Amsterdam reactor in recirculating cooling water softening, and analyzed the influence factors, including the pH, height of the fluid bed, particle size, influent flow, and reflux ratio (ratio of the part of the effluent flow refluxed to influent flow and influent flow) on hardness removal. The effluent concentration of Ca^{2+} reached a removal efficiency of 86.6% [11]. Hu et al. (2017) studied the influence of factors including superficial velocity (SV), particle size (L_0), and supersaturation (S) on the pellet growth rate of $CaCO_3$. In addition, they developed two models of pellet growth rate and fixed bed height growth rate in a pilot-scale experiment on the Amsterdam reactor, and reported that the pellet discharge diameter reached 1 mm to 2 mm [12].

It can be seen from the above-mentioned studies that the diameter of discharged pellets can only reach 1 mm to 2 mm, and the size of the pellets was uneven from the bottom to the top of a reactor. However, the size and distribution of the pellets in a PFB reactor can directly determine the crystallization efficiency and the resistance ability against hydraulic impact. Disturbed layers were observed frequently. Bed height depended on flow, and garnets were typically flushed out of the reactor [13]. The removal efficiency of TH and Ca^{2+} can be further improved. This paper introduces a circulating pellet fluidized bed (CPFB) reactor that helps the circulation growth of pellets based on the Amsterdam reactor. The pellet size tends to be uniform in the crystal growth process from the bottom to the top of the reactor. The CPFB reactor can effectively prevent the problem of uneven pellet size, extend the discharge time, improve the crystallization efficiency, reduce the reactor height, increase the discharge size of the pellet, and ensure the high removal efficiency of TH and Ca^{2+}.

2. Materials and Methods

2.1. Material

The water source of the full-scale experiment was the groundwater of the second water plant in Chang'an District, Xi'an, China. Table 1 presents the water quality data, showing that the water is mainly characterized by temporary hardness, which is extremely suitable for PFB reactor softening. The turbidity after boiling should be considered as the standard for softening to ensure drinking water quality.

Table 1. Groundwater quality.

No.	Water Quality	Value
1	pH	7.6–7.7
2	Temperature/°C	18–21
3	Turbidity of raw water/NTU	<1
4	Turbidity after boiling/NTU	90–100
5	Total alkalinity ($CaCO_3$)/(mg/L)	262
6	Bicarbonate alkalinity ($CaCO_3$)/(mg/L)	262
7	Total hardness (TH) ($CaCO_3$)/(mg/L)	286
8	Ca^{2+} (mg/L)	75
9	Mg^{2+} (mg/L)	24

Garnets with a size range of 0.2 mm to 0.4 mm, and a density of 3.93 g/cm^3, were used as seeds in pellet reactors. NaOH was adopted as a softening agent, and its mass concentration was 30% [12]. Hydrochloric acid (HCl) was used to adjust pH, and its mass concentration was 30%.

2.2. Full-Scale Experimental System

Figure 1 shows the diagram of the full-scale pellet softening reactor. The softening system was mainly composed of influent and effluent systems, a NaOH dosing system, an acid-dosing system, a seed-dosing system, and pellet discharge and storage systems. The core of this system was the CPFB reactor, which produces 5000 m^3 of soft water every day.

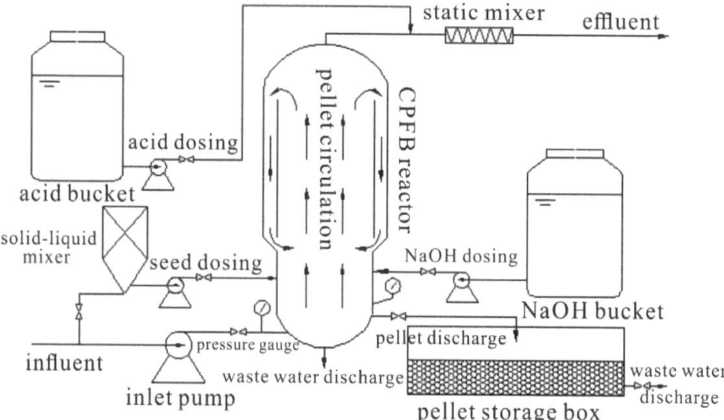

Figure 1. Schematic diagram of the full-scale pellet softening reactor.

The CPFB reactor was different from the single layer cylinder structure of the Amsterdam reactor, which had a double cylinder structure and a different diameter for the upper and lower cylinders. The crystalline pellets can circulate to the bottom in the upper part of the inner cylinder. The structure design improved the growth efficiency of the pellets and stabilized the bed height.

The specific equipment parameters of the system are shown in Table 2. As shown in the table, the superficial velocity of the CPFB reactor was 60 m/h to 100 m/h, and the fluidized area height was 4 m, which was lower than that of the Amsterdam reactor (5–6 m) [6,14].

Table 2. Equipment parameters. CPFB: circulating pellet fluidized bed.

No.	Equipment Name	Parameter	Remarks
1	Pipeline pump	H = 0–25 m, Q = 200 m^3/h, P = 15 kW	Frequency conversion pump
2	CPFB reactor	D = 1.6 m, H = 4.0 m	Stainless steel
3	Acid bucket	V = 12 m^3	Polyethylene, design for 7 days
4	NaOH bucket	V = 12 m^3	Polyethylene, design for 10 days
5	Pellet storage box	V = 15 m^3	Carbon steel

2.3. Experiment Process Description

High hardness groundwater was measured using an electromagnetic flowmeter, pumped into the CPFB reactor, and reacted with NaOH. A certain amount of garnet crystal seeds was pumped into the CPFB reactor after mixing with water through a pellet pump every day in normal operation. Mature pellets were discharged into the pellet storage box every day based on pressure change. Effluent pH was adjusted to 7–8 by HCl after the static mixer.

In the normal water supply process, superficial velocity was controlled by influent flow; then, NaOH dosage and acid dosage were adjusted manually based on influent flow. The hardness removal efficiency, the pressure change process in the CPFB reactor, and the growth kinetics of pellets were studied using water samples and pellet samples that were obtained at different heights of the CPFB reactor and at the outlet every few days. The system operation parameters are shown in Table 3.

Table 3. Operation parameters.

No.	Parameter Name	Value
1	Superficial velocity/m/h	60–100
2	NaOH dosage/mg/L	38–150
3	HCl dosage/mg/L	16–80
4	Pellet discharge/kgCaCO$_3$/day	300–400
5	Garnet dosage/kg/time/day	25–50
6	pH before acidification	9.5–9.9
7	pH after acidification	7.0–8.0

2.4. Analysis Methods

The hardness and Ca^{2+} and Mg^{2+} concentrations of inlet water and outlet water was analyzed through ethylenediaminetetraacetic acid (EDTA) [15]. After drying, pictures of the pellets were taken using the microscope Nikon 50i (Nikon, Tokyo, Japan). The diameters of the pellets were determined by employing the American Society of Testing Materials (ASTM) sieving method [16]. The average pellet diameter was calculated by using Equation (1) [12]. The fluidized bed height after expansion was measured by a meter ruler. Pressure was monitored by an online pressure meter. pH was monitored using an online real-time pH meter and a handheld portable pH meter [10].

$$\overline{d_p} = \frac{1}{\sum \frac{x_i}{d_{pi}}} \tag{1}$$

where $\overline{d_p}$ is the average diameter of the pellets in mm; x_i is the mass fraction of the pellets trapped in the i sieve net layer; and d_{pi} is the average diameter of the pellets trapped in the i sieve net layer in mm.

The information about the influent flow, inlet pressure, fluidized bed pressure, and effluent pH required in the experiment was collected in real time and transmitted to the central control room.

3. Results and Discussion

3.1. Experimental Study on NaOH and HCl Dosage Optimization

Figure 2a shows the variation in the NaOH and HCl dosages and effluent pH with time during operation in 2017. The data can be divided into the experimental data collection stage in months one to eight, and the steady system operation stage in months eight to 12. It can be seen from the figure that effluent pH increased with NaOH dosage, and the pH after adjustment decreased as the HCl dosage increased. In the normal system operation stage, the NaOH dosage was 180 mg/L; the HCl dosage was 50 mg/L, and the pH value can be adjusted to be stable between seven and eight.

Figure 2b depicts how the TH and Ca^{2+} and Mg^{2+} concentrations of the CPFB reactor effluent changed with time. The CPFB reactor evidently removed TH and Ca^{2+}, with removal rates reaching 60% and 90% at 18–21 °C, respectively. According to the dosage and removal rate data, the removal of 1 mM Ca^{2+} needed 2.6 mM NaOH. As expected, the pellet softening process reduced the CaCO$_3$ content in the water, and left the Mg^{2+} concentration unchanged [1], because the removal of Mg^{2+} must ensure that the pH exceeds 11 [17]. The pH of the CPFB reactor was controlled between 9.5–10 (Figure 2a).

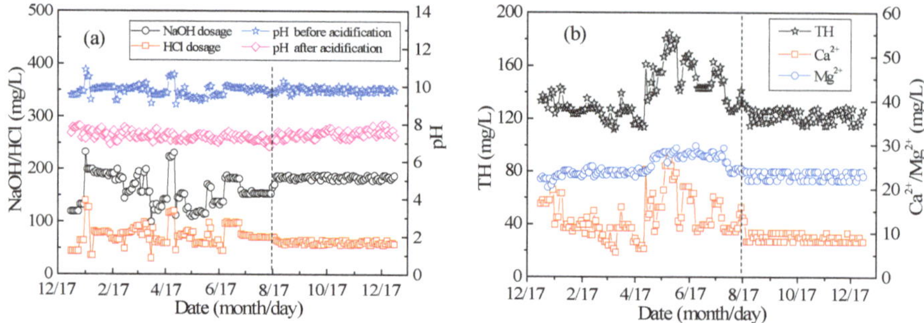

Figure 2. (**a**) Change in NaOH/HCl/pH and (**b**) TH and Ca^{2+}/Mg^{2+} concentrations with time.

Figure 2 shows that the CPFB reactor ran smoothly, and the removal rates of TH and Ca^{2+} slightly fluctuated. Thus, the TH of the effluent, the average concentration of Ca^{2+} in the effluent, the turbidity of the effluent, and the turbidity after boiling under different NaOH dosage conditions can be calculated using the data shown in Figure 2. Figure 3 illustrates the direct relationship between the effluent pH and the ionic concentration and turbidity. Thus, the crystallization and hardness removal effects of the CPFB reactor can be predicted by monitoring the effluent pH, which was useful for automatic control and effluent quality prediction. Figures 2 and 3 depict that the CPFB reactor exhibited a good and stable removal effect for hardness when the NaOH dosage was 180 mg/L (pH reaches 9.7) and HCl dosage was 50 mg/L (pH reaches 7–8). It also ensured that the effluent turbidity and the turbidity after boiling were less than 1.0 NTU.

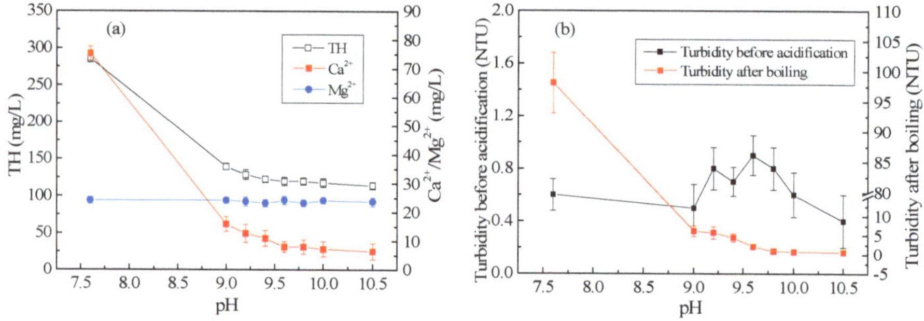

Figure 3. (**a**) Variation in TH and average concentrations of Ca^{2+} and Mg^{2+} with pH and (**b**) Variation in turbidity before acidification and turbidity after boiling with pH.

Various reports presented that the softening effect of the PFB reactor is not exactly the same during operation because of the different drinking water quality standards in different countries and different softening purposes such as drinking and scale inhibition. At the same time, the removal rate of calcium and hardness is affected not only by the crystallization process, but also by the dosage of softeners, the types of softeners, and alkalinity. However, the literature has rarely reported on the CPFB reactor providing removal rates of 90% and 60% for Ca^{2+} and TH, respectively. For example, Hofman et al. (2006) presented the water quality parameters of raw and treated water with 20 years of operation experience in Waternet, Vitens and Brabant Water and found a Ca^{2+} removal rate of less than 50% [1,18]. Hammes et al. (2011) reported that the "Amsterdam-type" pellet softening reactor could reduce Ca^{2+} concentration from 1.65 mM to 0.8 mM; the removal efficiency was only 50% [14].

Therefore, the CPFB reactor has significant advantages in water softening, which can provide data references for engineering with a high Ca^{2+} or TH removal rate.

3.2. Experimental Study on Pellet Distribution and Hardness Removal Characteristics at Different Heights of CPFB Reactor

Figure 4a,b shows the relationship between the average pellet size distribution and the mass percentage of $CaCO_3$ at different heights of the CPFB reactor with an increase in operation time. Pellet size became consistent at different heights of the CPFB reactor as the operation time increased, which is considerably different from the literature, where the pellet diameter in the Amsterdam PFB reactor gradually decreased from the bottom to the top. The pellet size at the bottom can reach 1–2 mm. However, the pellet size at the top was only 0.2–0.3 mm. These pellets were not yet crystallized [10,14]. The pellet size distribution state was affected by the water flow fluctuation. If influent flow increases abruptly, the higher velocity will take small pellets out of the PFB reactor.

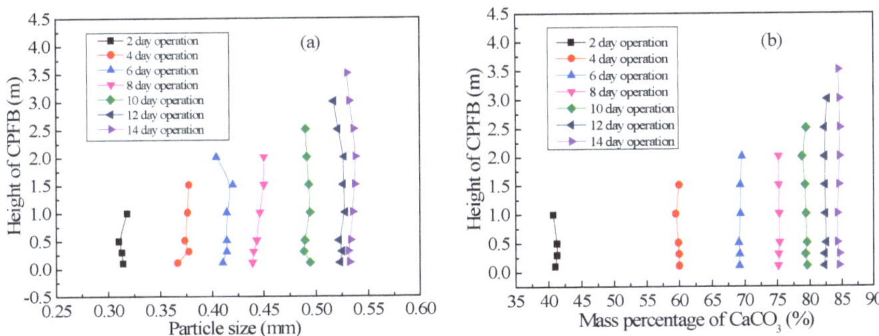

Figure 4. (**a**) Granularity distribution and (**b**) the mass percentage of $CaCO_3$ at different heights of the CPFB reactor.

The pellet size distribution in the CPFB reactor tended to be uniform because of its double cylinder structure design, which can make the circulating pellets grow. The special pellet circulation structure of the CPFB reactor resulted in a different growth kinetics of the pellets compared to the Amsterdam reactor. In the CPFB and Amsterdam reactors, crystallization occurred mainly at the bottom of the reactor [7,19]. As shown in Figure 5, more than 90% of TH and Ca^{2+} were removed below 0.5 m in the CPFB reactor. The water and the chemicals mixed at the bottom of the reactor; thus, the pellets at the bottom of the reactor crystallized first. However, the pellet location in the Amsterdam reactor remained relatively stable under hydraulic screening; thus, the pellet size at the bottom rapidly increased. In the CPFB reactor, the small pellets at the top fell to the bottom under different density flows, then, they were flushed up by the flow. The $CaCO_3$ crystallization was in progress when the pellets were flushed up. The pellet size at the bottom continuously increased, the voidage in the fluidized zone increased, and the pellets were constantly poured into the circulation zone for growth. Finally, the pellet size of the entire fluidized zone tended to be uniform, which was why the average pellet size and the mass percentage of $CaCO_3$ in the CPFB reactor were the same at different heights after operation for two weeks, as shown in Figure 4.

Therefore, the unique structure of the CPFB reactor increased the utilization of the seed material, extended the discharge time of the mature pellets, and importantly, reduced the height of the CPFB reactor. Due to the increase in crystallization efficiency, more than 90% of the hardness was removed below 1.0 m in the CPFB reactor, while the Amsterdam reactor can only remove 75% below 1.5 m [14]. Therefore, the height of the fluidization zone of the CPFB reactor was only 4.0 m, and it was likely to decrease further. The height of the Amsterdam reactor was mostly 5–6 m.

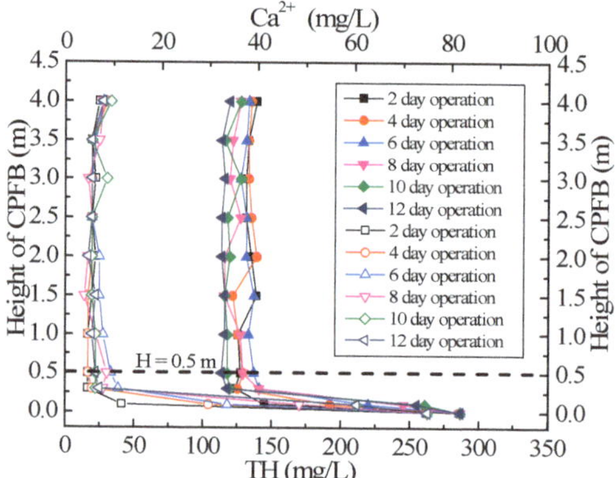

Figure 5. Removal of TH and Ca^{2+} at different running times of the CPFB reactor.

3.3. Experimental Study on Pellet Growth

Figure 6a shows the relationship between the average pellet size and the mass percentage of $CaCO_3$ when the CPFB reactor has just started to run. The average pellet size and the mass percentage of $CaCO_3$ were positively related to running time. However, the growth rate of the average pellet size was the same, while the rate of increase in the mass percentage of $CaCO_3$ presented a slower trend with running time. This is mainly because there is a relationship between the increase in the pellet size and the mass of the crystallized $CaCO_3$. The mass of the crystallized $CaCO_3$ was the same every day under the condition that the influent flow and hardness of the inlet and outlet water were stable. However, the increase in the mass percentage of $CaCO_3$ of every pellet was related to the crystallization efficiency of $CaCO_3$ for each pellet. Accordingly, the crystallization efficiency decreased as the pellet size increased [15]. However, note that a certain function relationship existed between the pellet size and the mass percentage of $CaCO_3$, as shown in Figure 6b [10,20].

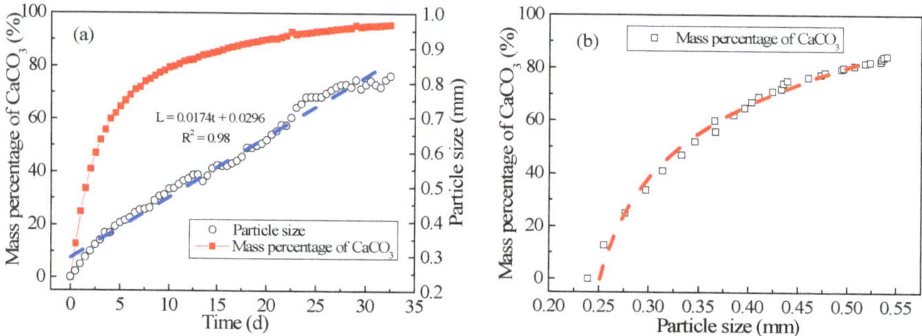

Figure 6. (**a**) Change in the average pellet size and the percentage of $CaCO_3$ with running time at the bottom of CPFB reactor, and (**b**) Relation between the average pellet size and the percentage of $CaCO_3$.

From Figure 6a, it can be seen the average pellet size was only approximately 0.8 mm when the reactor had run for 30 days. The crystallization efficiency of pellets was relatively high, and the mass percentage of $CaCO_3$ increased. The first discharge time was considerably prolonged. As described in Section 2.2, pellets can grow cyclically in the CFPB reactor; therefore, the size of pellets was less at the bottom of the reactor. The pellet growth rate can be calculated by fitting the growth curve of the pellet size, and it was approximately 2.01×10^{-10} m/s, which was not less compared with that in the literature [12,16,21,22]. The functional relationship between the average pellet size and the mass percentage of $CaCO_3$ can be obtained from Figure 6b. Thus, when one of the variables is known, it can be used to calculate another variable [5].

3.4. Experimental Study on the Relationship between Pressure and Bed Height Variation and Pellet Discharge

Figure 7a shows the change in inlet pressure with time. The increase in influent pressure (p) was directly proportional to time (t); the functional relation can be fitted as p = 0.0023 t + 0.0720 (R^2 = 0.99). Similarly, the increase in the bottom pressure (p) was proportional to time (t), and the functional relation can be fitted as p = 0.0016 t + 0.052 (R^2 = 0.99). These two formulas indicate that the influent pressure changed by 2.3 kPa daily, and the bottom pressure changed by 1.6 kPa. The influent pressure and bottom pressure increased with the operation time because of the continuous crystallization of $CaCO_3$ on the crystal seeds every day. Based on the hardness of the influent and effluent, crystallization mass can be calculated to be approximately 430 kg/day, which can produce a pressure of approximately 2.1 kPa at the bottom of the CPFB reactor. The change in the pressure at the bottom of the reactor was the same as that obtained through fitted data.

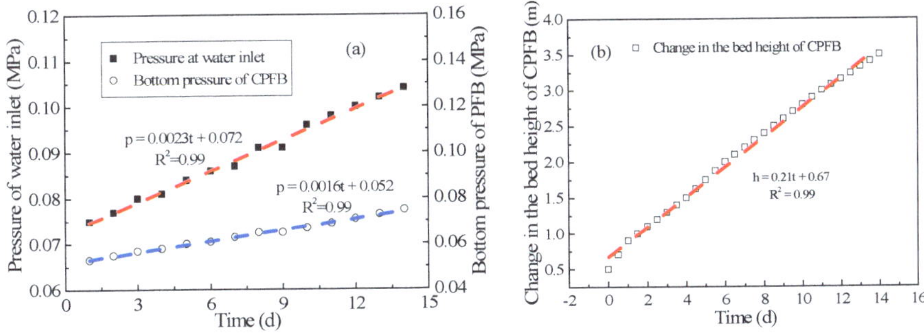

Figure 7. Change in (**a**) pressure and (**b**) bed height with operation time.

The change in the bed height of the PFB reactor was examined in a few studies [9,16,19]. Figure 7b shows the change in bed height with time. The increase in bed height (h) was directly proportional to time (t), and the data can be fitted to h = 0.21 t + 0.67 (R^2 = 0.99). The change in the bed height of the PFB reactor can be calculated to be approximately 0.21 m/day through the formula, which was coincident with that observed in the actual operation.

The actual pressure change and bed height change laws of the CPFB reactor can be obtained through Figure 7. The pellet discharge was closely related to the two change laws during the actual operation. In the early stage of CPFB reactor testing, the balance between the pellet discharge size and the bed height change must be found, and the pressure range of the CPFB reactor can also be determined at this time. The pellet discharge began when the pressure exceeded a certain value; when pressure is below a certain value, the pellets discharge stopped, and crystal seeds began to dose. The dosage of crystal seeds was calculated based on the daily discharge. The control mode of the pellet discharge and the dosage of crystal seeds were similar to those of the Amsterdam reactor [13]. However, owing to the circulatory crystallization of pellets in the CPFB reactor, the size of the discharged pellets was larger, as shown in Figure 8. Most of the pellet sizes in the CPFB reactor during normal operation were approximately 3–5 mm, which was considerably higher than the size of the discharged pellets for the Amsterdam reactor (1–2 mm) [7,14,23]. Importantly, the softening effect was ensured under high crystallization efficiency. The performance was better than that reported in the literature, where small white pellets appeared in the effluent when the pellet discharge diameters >1.1 mm [8]. However, as seen from Figure 8b, the pellets were not particularly uniform, mainly because the size of a few pellets increased until the flow could not suspend them in the process of pellet circulation, and these pellets will always crystallize at the bottom, leading to oversizing (larger than 5 mm). Therefore, it is necessary to discharge oversized pellets as far as possible during the pellet discharge process to prevent the effect of oversized pellets on the water quality.

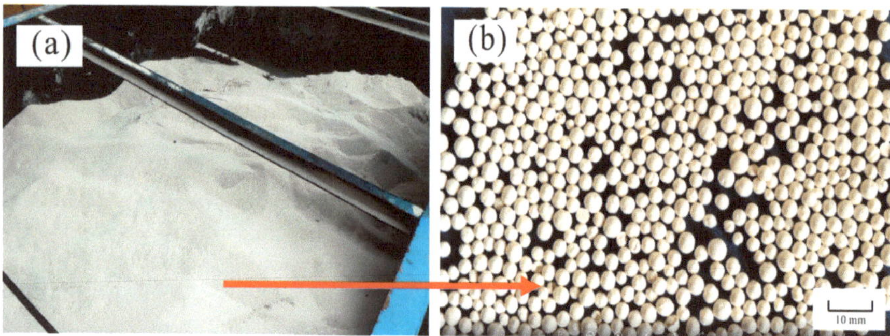

Figure 8. (a) Pellet storage box and (b) mature discharged pellets.

3.5. Costs

Pellet softening on a large scale is relatively inexpensive. This process is more expensive when it is applied at a smaller scale, such as the majority of groundwater treatments in The Netherlands [1]. As described in the literature [1], the average cost for the large-scale treatment is €0.02 per m^3 (101 million m^3/year), but the average cost can be increased to at least €0.25 per m^3 when the Amsterdam reactor is applied at a smaller scale. For the CPFB reactor softening system, the annual cost for the treatment of 5000 m^3/day (installed capacity) mainly includes chemical cost, energy cost, and other costs. Other costs consist of the garnet cost and the labor cost, etc., which can be neglected. The specific cost analysis can be seen in Table 4.

Table 4. Cost analysis.

No.	Cost Composition	Euro	Percentage (%)
1	NaOH	69,228	65
2	HCl	19,230	18
3	Garnet	1154	1
4	Energy	13,461	12
5	Labor	3846	4
6	Total cost	106,919	100
7	Unit cost (€ per m^3)	0.058	/

As shown in the table, the cost of unit water treatment is € 0.058 per m^3, while the chemical cost (NaOH + HCl) is 83% of the total cost, and the labor cost is only 4%. These values are significantly different from the cost composition of the Amsterdam reactor provided in the literature, in which the chemical and labor costs account for 32% and 25% of the total cost, respectively. This difference is mainly caused by the low labor cost in the water treatment plants of China. Even though the variations in cost are primarily caused by the variations in the market prices of NaOH and HCl, the water treatment cost of the CPFB reactor softening system is still lower than that of the Amsterdam reactor [1,8].

4. Conclusions

This study investigated the performance parameters of a CPFB reactor using a full-scale experiment system. The following conclusions were obtained:

1. In the CPFB reactor, the removal rate of Ca^{2+} and TH can reach 90% and 60%, respectively, and the effluent pH can be controlled between 9.5–9.8. The turbidity of the effluent and the turbidity after boiling are stable at less than 1.0 NTU. The unit water treatment cost is less than €0.064 per m^3. The CPFB reactor has advantages in terms of the softening effect and cost.
2. The unique structure of the CPFB reactor improves the crystallization efficiency, increases the utilization of the seed material, and extends the discharge time of mature pellets. The size of the discharged pellets can reach 3–5 mm, and the height of the CPFB reactor is reduced from between 5–6 m to 4 m.

Author Contributions: R.H., T.H., A.Z., and Z.T. worked together. T.H. came up with the idea. R.H. designed the study and interpreted the results. R.H., A.Z., and Z.T. performed the experiments.

Funding: This research is supported by the National Key Research and Development Program of China (2016YFC0400706).

References

1. Hofman, J.; Kramer, O.; Hoek, J.P.V.D.; Nederlof, M.; Groenendijk, M. Twenty years of experience with central softening in The Netherlands: Water quality, environmental benefits, and costs. In Proceedings of the International Symposium on Health Aspects of Calcium and Magnesium in Drinking Water, Baltimore, MD, USA, 27–28 April 2006.
2. Kramer, O.J.I.; Jobse, M.A.; Baars, E.T.; van der Helm, A.W.C.; Colin, M.G.; Kors, L.J.; van Vugt, W.H. Model-based prediction of fluid bed state in full-scale drinking water pellet softening reactors. In Proceedings of the 2nd IWA New Developments in IT & Water Conference, Hague, The Netherlands, 8–10 February 2015.
3. He, C.; Gross, M.; Westerhoff, P.; Fox, P.; Li, K. *Comparing Conventional and Pelletized Lime Softening Concentrate Chemical Stabilization*; Water Research Foundation: Denver, CO, USA; Phoenix Water Services Department: Phoenix, AZ, USA, 2011.
4. Pratomo, U.; Anggraeni, A.; Lubis, R.A.; Pramudya, A.; Farida, I.N. Study of softening hard water using Pistacia vera shell as adsorbent for calcium and magnesium removal. *Procedia Chem.* **2015**, *16*, 400–406. [CrossRef]
5. Van Schagen, K.M.; Babuška, R.; Rietveld, L.C.; Veersma, A.M.J. Model-based dosing control of a pellet softening reactor. *IFAC Proc. Vol.* **2009**, *42*, 267–272. [CrossRef]
6. Graveland, A.; van Dijk, J.C.; de Moel, P.J.; Oomen, J.H.C.M. Developments in water softening by means of pellet reactors. *J.-Am. Water Works Assoc.* **1983**, *75*, 619–625. [CrossRef]
7. Van der Veen, C.; Graveland, A. Central softening by crystallization in a fluidized-bed process. *J.-Am. Water Works Assoc.* **1988**, *80*, 51–58. [CrossRef]
8. Schetters, M.J.A.; van der Hoek, J.P.; Kramer, O.J.; Kors, L.J.; Palmen, L.J.; Hofs, B.; Koppers, H. Circular economy in drinking water treatment: Reuse of ground pellets as seeding material in the pellet softening process. *Water Sci. Technol.* **2015**, *71*, 479–486. [CrossRef] [PubMed]
9. Schetters, M.J.A. Grinded Dutch Calcite as Seeding Material in the Pellet Softening Process. Master's Thesis, TU Delft, Delft, The Netherlands, 2013.
10. Schagen, K.M.V.; Rietveld, L.C.; Babuška, R. Dynamic modelling for optimisation of pellet softening. *J. Water Supply Res. Technol.* **2008**, *57*, 45–56. [CrossRef]
11. Chen, Y.; Fan, R.; An, D.; Cheng, Y.; Tan, H. Water softening by induced crystallization in fluidized bed. *J. Environ. Sci. (Engl. Ed.)* **2016**, *50*, 109–116. [CrossRef] [PubMed]
12. Hu, R.Z.; Huang, T.L.; Wen, G.; Yang, S. Modelling particle growth of calcium carbonate in a pilot-scale pellet fluidized bed reactor. *Water Sci. Technol. Water Supply* **2017**, *17*, 643–651. [CrossRef]
13. Rietveld, L.C.; Schagen, K.M.V.; Kramer, O.J.I. Optimal operation of the pellet softening process. In Proceedings of the AWWA Workshop, Austin, TX, USA, 29–31 January 2006.
14. Hammes, F.; Boon, N.; Vital, M.; Ross, P.; Magic-Knezev, A.; Dignum, M. Bacterial colonization of pellet softening reactors used during drinking water treatment. *Appl. Environ. Microbiol.* **2011**, *77*, 1041–1048. [CrossRef] [PubMed]

15. Jiang, K.; Zhou, K.G.; Yang, Y.C.; Du, H. Growth kinetics of calcium fluoride at high supersaturation in a fluidized bed reactor. *Environ. Technol.* **2014**, *35*, 82–88. [CrossRef] [PubMed]

16. Tai, C.Y. Crystal growth kinetics of two-step growth process in liquid fluidized-bed crystallizers. *J. Cryst. Growth* **1999**, *206*, 109–118. [CrossRef]

17. Su, C.C.; Dulfo, L.D.; Dalida, M.L.P.; Lu, M.C. Magnesium phosphate crystallization in a fluidized-bed reactor: Effects of pH, Mg:P molar ratio and seed. *Sep. Purif. Technol.* **2014**, *125*, 90–96. [CrossRef]

18. Mahvi, A.H.; Shafiee, F.; Naddafi, K. Feasibility study of crystallization process for water softening in a pellet reactor. *Int. J. Environ. Sci. Technol.* **2005**, *1*, 301–304. [CrossRef]

19. Garea, A.; Aldaco, R.; Irabien, A. Improvement of calcium fluoride crystallization by means of the reduction of fines formation. *Chem. Eng. J.* **2009**, *154*, 231–235. [CrossRef]

20. Van Schagen, K.; Rietveld, L.; Babuška, R.; Baars, E. Control of the fluidised bed in the pellet softening process. *Chem. Eng. Sci.* **2008**, *63*, 1390–1400. [CrossRef]

21. Aldaco, R.; Garea, A.; Irabien, A. Particle growth kinetics of calcium fluoride in a fluidized bed reactor. *Chem. Eng. Sci.* **2007**, *62*, 2958–2966. [CrossRef]

22. Tai, C.Y.; Hsu, H.P. Crystal growth kinetics of calcite and its comparison with readily soluble salts. *Powder Technol.* **2001**, *121*, 60–67. [CrossRef]

23. Aldaco, R.; Garea, A.; Irabien, A. Modeling of particle growth: Application to water treatment in a fluidized bed reactor. *Chem. Eng. J.* **2007**, *134*, 66–71. [CrossRef]

International Journal of
Environmental Research and Public Health

Article

Simple Urea Immersion Enhanced Removal of Tetracycline from Water by Polystyrene Microspheres

Junjun Ma [1,†], Bing Li [2,†], Lincheng Zhou [3], Yin Zhu [4], Ji Li [4] and Yong Qiu [1,*]

[1]   State Key Joint Laboratory of Environment Simulation and Pollution Control, School of Environment, Tsinghua University, Beijing 100084, China; mjj16@mails.tsinghua.edu.cn
[2]   School of Energy and Environmental Engineering, University of Science and Technology Beijing, Beijing 100083, China; libing@ustb.edu.cn
[3]   State Key Laboratory of Applied Organic Chemistry, College of Chemistry and Chemical Engineering, Institute of Biochemical Engineering and Environmental Technology, Lanzhou University, Lanzhou 730000, China; zhoulc@lzu.edu.cn
[4]   School of Environmental and Civil Engineering, Jiangnan University, Wuxi 214122, China; zyfreely@vip.jiangnan.edu.cn (Y.Z.); liji@jiangnan.edu.cn (J.L.)
*   Correspondence: qiuyong@tsinghua.edu.cn; Tel.: +86-010-6279-6953; Fax: +86-010-6277-1472
†   These authors contributed equally to this work.

Abstract: Antibiotics pose potential ecological risks in the water environment, necessitating their effective removal by reliable technologies. Adsorption is a conventional process to remove such chemicals from water without byproducts. However, finding cheap adsorbents with satisfactory performance is still a challenge. In this study, polystyrene microspheres (PSM) were enhanced to adsorb tetracycline by surface modification. Simple urea immersion was used to prepare urea-immersed PSM (UPSM), of which surface groups were characterized by instruments to confirm the effect of immersion. Tetracycline hydrochloride (TC) and doxycycline (DC) were used as typical adsorbates. The adsorptive isotherms were interpreted by Langmuir, Freundlich, and Tempkin models. After urea immersion, the maximum adsorption capacity of UPSM at 293 K and pH 6.8 increased about 30% and 60%, achieving 460 mg/g for TC and 430 mg/g for DC. The kinetic data were fitted by first-order and second-order kinetics and Weber–Morris models. The first-order rate constant for TC adsorption on UPSM was 0.41 /h, and for DC was 0.33 /h. The cyclic urea immersion enabled multilayer adsorption, which increased the adsorption capacities of TC on UPSM by two to three times. The adsorption mechanism was possibly determined by the molecular interaction including π–π forces, cation-π bonding, and hydrogen bonding. The simple surface modification was helpful in enhancing the removal of antibiotics from wastewater with similar structures.

Keywords: microsphere resin; urea immersion; adsorption isotherms; surface characterization; kinetics analysis; multilayer adsorption

---

1. Introduction

Antibiotics are a concern as contaminants in the water environment due to their potential ecological risks and ubiquitous distribution in the world [1]. Among the many categories of antibiotics, reagents in the tetracycline group have been used extensively to control disease in human beings and livestock, due to their broad-spectrum antimicrobial activities [2]. Tetracycline (TC) and doxycycline (DC) are typical widely used tetracycline antibiotics. The potential health risks of residual antibiotics in the aqueous environment lie in the possible development of drug resistance in bacteria, challenging the current therapies of known antibiotics [3]. Compared with aerosol and soil, aqueous phase transporting in wastewater, sewers, surface water, and groundwater was thought to be the dominant way to spread antibiotics and antibiotic resistance to remote areas and society [4]. Therefore, antibiotic contamination is a serious environmental issue that needs an effective response.

Many effective techniques have been developed to remove or degrade residual antibiotics in wastewater, such as adsorption [5], advantage oxidation [6,7], activated sludge transformation [8], electrochemical treatment [9], biological composition [10], membrane filtration [11], etc. Adsorption was an old but clean, easy, and efficient process to remove aqueous pollutants [12]. Engineers could adapt the adsorption process rapidly by utilizing existing apparatus such as filter or dosing pumps. Thus large wastewater treatment plants in the pharmaceutical industry could be easily equipped, or on-site small apparatuses could be used in rural areas. During adsorption treatment, very few byproducts are produced and released into the water, resulting in less risk of unknown and uncontrolled products in oxidation processes that could be hazardous to the ecosystem. There were many difficulties in applying adsorption in antibiotics removal, such as unselective adsorption, insufficient capacity, and high cost of adsorbent. Finding cheap adsorbent with satisfactory performance is still a challenge. Thus people have attempted to use carbonaceous materials [13], sludge [14], natural minerals [15], siliceous materials [16,17], and polymer resins [18–20] to remove antibiotics. Therefore, understanding the adsorption properties of antibiotics is important for engineering approaches to improve their performance.

Polymer resin microspheres, with the merits of low cost, high porosity, large surface area, and adsorption capacity, have potential for commercial application [14,19–22]. Surface modification is important to acquire designed functions of adsorbents. Surface modification was a typical way to improve the selectivity and capacity of antibiotics adsorption [22,23], as well as easier phase separation by magnetic forces [14,19,21]. Modification of the functional groups on the surface can increase adsorption capacity by enhancing the chemical bonds to the target chemicals. For example, adsorption of TC has been enhanced by surface functionalization by the amino-ferrous group [24] and amino-copper group [25]. Modification with the anion exchange group improved adsorption capacity to more than 355 mg/g [26].

Tetracycline reagents have multiple functional groups (Supplementary Figure S1), including phenol, amino, alcohol, and ketone groups, which are capable of electronic coupling and various reactions. One idea is to impregnate an amino group ($-NH_2$) on the adsorbent surface as a neutral anion receptor to link an ester group ($-COO-R$) in antibiotic molecules, in order to form chelated complexes with hydrogen bond acceptors like ditopic carboxylates. Another choice is to use a carbonyl group as an electron donor to enhance the π–π interaction and cation-π bonding. As a kind of simple and cheap commercial material, urea ($CO(NH_2)_2$) combined with two amino ($-NH_2$) and one carbonyl ($C=O$) functional group, makes it a good agent to interact with tetracycline molecules.

In this paper, we modified the surface of polystyrene ethylenediaminetetraacetate (EDTA) microsphere (PSM) resin by simply immersing it in urea solution. Then, the urea modified PSM (UPSM) was characterized by Brunauer–Emmett–Teller (BET), x-ray photoelectron spectroscopy (XPS), and infrared (IR) to confirm the effects of impregnation. Later, we conducted adsorption experiments to evaluate their capacity to remove TC and DC from water. Finally, the interaction between antibiotic molecules and UPSM surface is discussed.

2. Materials and Methods

2.1. Microsphere Modification

Polystyrene EDTA microsphere (PSM) resins were synthesized according to a reported thermal-solvent method [26]. The polystyrene microspheres were first synthesized by gentle agitation to control the average diameter to about 900 μm. Then the surface was functionalized by adding EDTA groups by 2 steps of reactions. The formula of PSM is shown in Supplementary Figure S1. The material has been successfully applied to treat aqueous pollutants [27,28].

The urea-immersed PSM (UPSM) was acquired by immersing the PSM into 100 g/L of urea solution for 12 h. The urea solution was prepared by dissolving 5 g urea powder (Alfa Aesar, Johnson Matthey Company, West Chester, PA, USA) in 50 mL of ultrapure water. Then 2 g of PSM was mixed with 50 mL of urea solution in a 100 mL 3-necked glass flask. The mixture was agitated in a thermostatic shaker at 150 rpm and 25 ± 1 °C for 12 h. After filtration, the UPSM were intensively washed by ultrapure water 10 times to remove the excessive urea. Then the UPSM were dried in an oven with desiccants at room temperature overnight.

2.2. Surface Characterization

The surface morphologies of PSM and UPSM were characterized by using a scanning electron microscope (SEM; HT 7700, Hitachi Corp., Tokyo, Japan). The SEM images were similar between PSM and UPSM, as shown in Supplementary Figure S2. Their surface areas in Brunauer–Emmett–Teller (BET) were determined by N_2 adsorption-desorption isotherm at liquid nitrogen temperature (TriStar 3020 II, Micromeritics Instrument Corp., Norcross, GA, USA). The instrumental parameters were default as described in the manual and guidance. The surface area of UPSM was slightly higher than that of PSM, as shown in Supplementary Figure S3. The pore size distributions of PSM and UPSM were close to each other (Supplementary Figure S4).

The surface functional groups of PSM and UPSM were identified by Fourier transform infrared (FT-IR) spectroscopy using the KBr tableting technique on an FT-IR spectrometer (PE Spectrum GX, Perkin-Elmer Corp., Waltham, MA, USA) in transmission mode. The elemental analysis was conducted by x-ray photoelectron spectroscopy (XPS) on an electron spectrometer (PHI Quantera II, Ulvac-Phi Corp., Chigasaki, Japan) using 300 WAl-Ka radiation.

2.3. Chemicals and Analysis

The target antibiotic reagents, tetracycline hydrochloride (TC) and doxycycline hydrochloride (DC), were purchased from a commercial supplier (Inalco spa Milano, Milan, Italy). The molecular structures and other information of urea and tetracycline are shown in Supplementary Table S1. All chemicals were of analytical grade and used without pretreatment. Ultrapure water (Milli-Q, Millipore Co., Burlington, MA, USA) was used to prepare the solutions.

The concentrations of TC and DC in aqueous solution were determined on a UV-Vis spectrophotometer (U-3900, Hitachi Corp., Japan). The specific wavelength for TC was 360 nm and for DC was 325 nm according to the maximum absorption. The concentration was calibrated from linear standard curves (R^2 > 0.999). Control experiments were conducted by mixing 40 mg/L tetracycline with 100 g/L urea with equal volume. After 12 h, the residual concentrations of TC and DC changed by less 1% of the initial values, indicating that tetracycline molecules are inert with urea in the solution.

Hydrochloric acid (HCl) and sodium hydroxide (NaOH) were ordered from a local supplier (Shanghai Chemical Reagents Co., Shanghai, China) to prepare HCl solution in 0.1 mol/L (M) and NaOH in 0.1 M. The bulk pH of the solutions was adjusted before adsorption experiments by dosing drops of above acid and base manually. The instant pH value was read out by a laboratory pH meter 1 min after dosing.

2.4. Adsorption Experiments

Preliminary adsorption experiments were conducted to determine the optimal bulk pH in the gradient range (pH 2.7, 6.8, 8.5 and 10.3). Kinetic curves at the gradient pH were acquired by exposing 60 mg/L of TC or DC to 60 mg of UPSM. The results are shown in Supplementary Figure S5, in which the optimal pH was determined to be 6.8.

For isothermal experiments, tetracycline solutions at 16 levels of initial concentration (100–2000 mg/L) were prepared to expose to 60 mg of PSM or UPSM in a volume of 20 mL. The mixture was agitated in a thermostatic shaker at a speed of 150 rpm and temperature of 25 °C. The experiments lasted for 24 h to ensure that the adsorption reached the equilibrium state.

For kinetic experiments, 60 mg of PSM or UPSM was dosed into tetracycline solutions at 5 levels of initial concentration (100–300 mg/L). The experiments lasted for 100 min and the concentrations of TC and DC were determined to obtain the kinetics.

2.5. Recycling Adsorption by Urea Immersion

Recycling adsorption by UPSM was achieved by using urea immersion after the saturated adsorption of tetracycline. First, 40 mg/L of TC was adsorbed onto 60 mg of UPSM for 12 h. After that, the UPSM were rapidly washed by ultrapure water 10 times to remove urea. Second, the USPM were again immersed in 20 mL of urea solution at a concentration of 100 g/L for 12 h. Finally, the UPSM were cleaned and exposed to 40 mg/L of TC solution for an additional cycle of adsorption. In total there were 10 cycles of adsorption-immersion. The TC concentrations were analyzed for comparison with the control experiments using UPSM.

USPM after 10 cycles were further examined for the desorption property under thermal, acidic, and alkaline treatment. Thermal desorption was conducted at 35 °C by soaking the UPSM in 20 mL of pure water in a water bath for 12 h. The acidic desorption was conducted by immersing the UPSM in 20 mL of 0.1 M HCl for 3 h. The alkaline desorption was achieved under similar conditions. Control experiments were conducted in parallel. The experimental duration was 7 h. The thermal stability of tetracycline was examined in a water bath at 95 °C for 1 h.

2.6. Data Interpretation

We evaluated the solid surface loading (q_t) of tetracycline onto microsphere resin by the following equation, as described in the literature [29]:

$$q_t = (C_0 - C_t) \cdot V / M \tag{1}$$

where q_t is the surface loading of adsorbates on adsorbents (mg/g), M is the mass of adsorbent (g), V is the volume of the solution (L), and C_0 and C_t are bulk tetracycline concentrations initially and at time t of the experiment, respectively (mg/L). In the case of isothermal experiments, subscript e was used instead of t to represent the equilibrium state, as q_e and C_e.

The isothermal models, including Langmuir, Freundlich, and Tempkin, are shown in Table 1. The kinetic models including first-order, second-order, and Weber–Morris models are also shown in the table. The parameters were estimated by using the linear Lineweaver–Burk equation (Excel 2010, Microsoft Corp., Redmond, WA, USA), and the correlation coefficients were used to evaluate the quality of data interpretation.

Table 1. Data interpretation models for adsorption isotherms and kinetics.

Model Name	Equation	Lineweaver–Burk Equation	Coefficients
Langmuir	$q_e = q_m \cdot C_e / (K_L + C_e)$	$C_e / q_e = C_e / q_m + K_L / q_m$	K_L, q_m
Freundlich	$q_e = K_f \cdot C_e^{1/n}$	$\ln q_e = \ln K_f + 1/n \cdot \ln C_e$	K_f, n
Tempkin	$q_e = RT/b \cdot \ln(a \cdot C_e)$	$q_e = RT/b \cdot \ln a + RT/b \cdot \ln C_e$	$a, RT/b$
First-order	$dq_t/dt = K_1 \cdot (q_e - q_t)$	$\ln(q_e - q_t) = \ln q_e - K_1 \cdot t$	K_1
Second-order	$dq_t/dt = K_2 \cdot (q_e - q_t)^2$	$1/t = (K_2 \cdot q_e^2)(1/q_t - 1/q_e)$	K_2
Weber–Morris	$q_t = q_e \cdot K_w \cdot t^{1/2}$	$q_t/q_e = K_w \cdot t^{1/2}$	K_w

Note: C_e (mg/L) is the concentration of TC or DC at equilibrium in the solution, q_e (mg/g) is the amount of TC and DC adsorbed per unit weight of the adsorbents. The constant q_m (mg/g) is the maximal adsorption capacity in the Langmuir equation, K_L is related to the energy of adsorption (L/mg). K_f and n are constants of the Freundlich equation, which relate to adsorption capacity and intensity, respectively. T is the temperature of the solution (°C), R is the molar gas constant equal to 8.314 J/(K·mol), a and b are parameters of the Tempkin model. Parameters q_e and q_t (mg/g) are the amounts of TC adsorbed at equilibrium time and at time t (h) in the adsorption process, respectively. K_1 (/h), K_2 (g/(mg h)), and K_w (/h$^{1/2}$) are the rate constants for first-order, second-order, and Weber–Morris kinetics, respectively.

3. Results and Discussion

3.1. Solid Surface Characterization

3.1.1. Surface Characterization

Similar SEM images of PSM and UPSM (Supplementary Figure S2) suggest that urea immersion did not physically modify the surface of the microspheres. The pore size distribution curves of UPSM and PSM were also similar to each other (Supplementary Figure S4), but the pore structures were slightly different (Supplementary Table S2). The average pore size of UPSM (20.4 nm) was 10% lower than that of PSM (22.6 nm) and pore volume (0.34 cm^3/g) was 13% higher than PSM's. Consequently, the BET surface area of UPSM (112.42 m^2/g) was about 60% higher than that of PSM (71.69 m^2/g), according to Supplementary Figure S3.

3.1.2. FT-IR Analysis

The FT-IR spectra (Figure 1) suggest more amino and imino groups on UPSM than on PSM. Although both PSM and UPSM showed vibration of imino bond (N–H) in the amino/imino (–NH$_2$/–NH) group in the range of 3300–3500 cm^{-1} [15], UPSM had a broader curve at about 3400 cm^{-1}. Carbonyl groups in UPSM and PSM were slightly different, in the range of 2500–3400 cm^{-1}. The broader band of UPSM at 1640–1650 cm^{-1} refers to the more singular hydrogen bonds between amino and carbonyl groups.

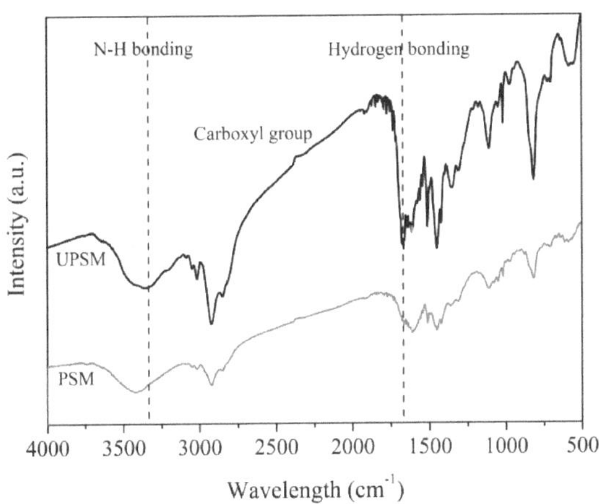

Figure 1. FT-IR spectra of fresh polystyrene microspheres (PSM) and urea-immersed PSM (UPSM).

3.1.3. XPS Analysis

XPS analysis (Figure 2) was used to confirm the variation of surface functional groups. PSM and UPSM had similar full XPS spectra (Figure 2a,c), as well as C 1s and N 1s deconvolution spectra (Supplementary Figure S6) because of their similar chemical components. PSM and UPSM had slightly different O 1s deconvolution spectra (Figure 2b,d). PSM showed subpeaks of C=O at 531.5 eV (O1), O–C–O at 532.5 eV (O2), and C–O at 533.1 eV (O3), while UPSM showed two small peaks at 532.7 eV (O2a) and 531.3 eV (O2a). The highest subpeak of PSM was 531.3 eV (O1) but that of UPSM shifted to 533.1 eV (O3). This shift might be related to the hydrogen bonding formed on the surface, such as H–N...H–C=O.

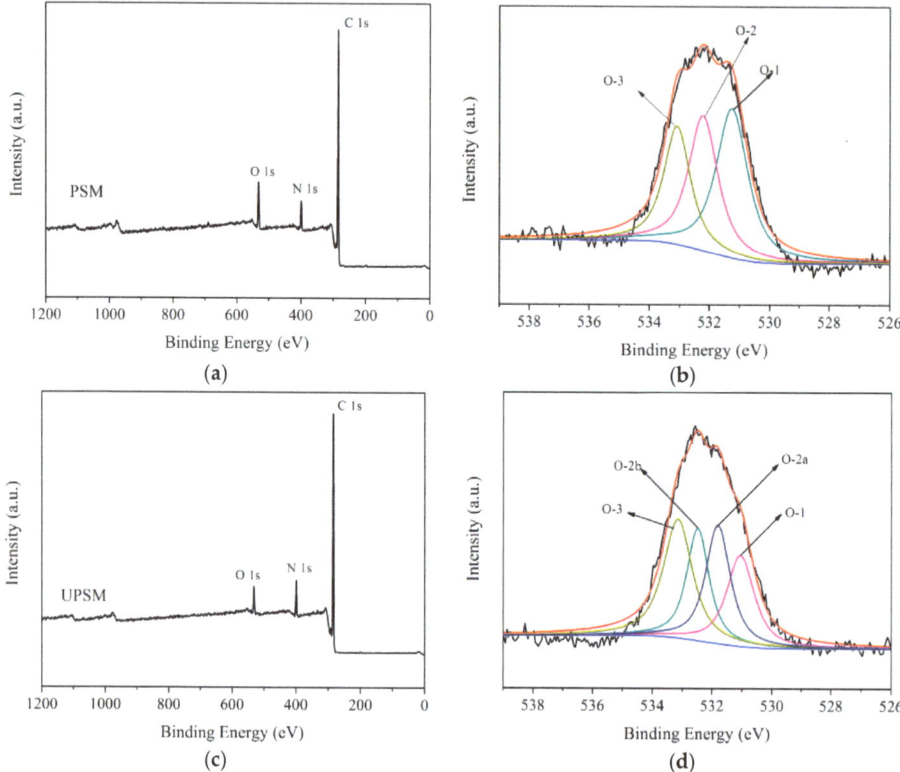

Figure 2. X-ray photoelectron spectroscopy (XPS) spectra of fresh PSM and UPSM. (**a**) Full spectrum of PSM; (**b**) O1s deconvolution of PSM; (**c**) full spectrum of PSM; (**d**) O1s deconvolution of UPSM.

3.2. Adsorption Isotherms and Kinetics

3.2.1. Isothermal Modeling

The parameters of isothermal models were close for TC and DC, as shown in Table 2. According to the correlation coefficient (R^2), the Langmuir and Freundlich models were not perfect to fit the data but better than the Tempkin model. The data-fitting results by the Langmuir model are shown in Figure 3 as an example. The maximum loading rate (q_m) of TC on UPSM was 460 mg/g at pH 6.8 and 25 °C. This capacity was about 60% higher than that on PSM (290 mg/g). The adsorption capacity of DC on UPSM (430 mg/g) was 30% higher than that on PSM (330 mg/g). The higher adsorption capacity of UPSM than PSM was accordant to the larger BET surface area than PSM.

Table 2. Isothermal fitting for tetracycline hydrochloride (TC) and doxycycline (DC) on PSM and UPSM at pH 6.8 and T = 298 K.

Adsorbent	Adsorbate	Langmuir			Freundlich			Tempkin		
		q_m	K_L	R^2	K_f	n	R^2	a	RT/b	R^2
PSM	TC	290	180	0.960	15	2.4	0.996	0.06	59	0.952
	DC	330	290	0.923	9.4	2.1	0.983	0.14	83	0.919
UPSM	TC	460	120	0.966	25	2.3	0.955	0.06	57	0.871
	DC	430	56	0.985	35	2.5	0.940	0.47	67	0.830

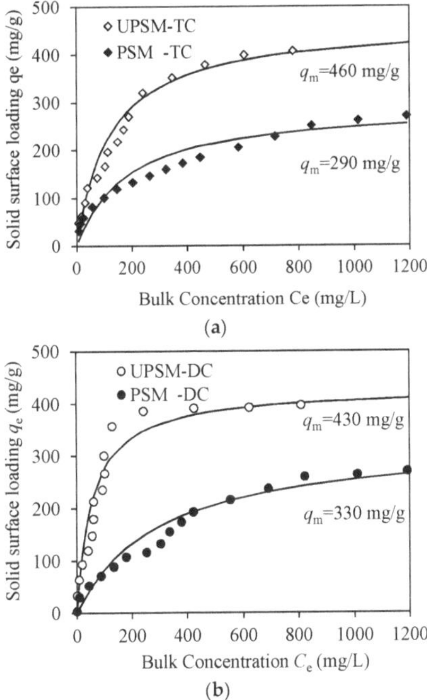

Figure 3. Adsorption isotherms of tetracycline antibiotics on microspheres. (**a**) TC on UPSM and PSM; (**b**) DC on UPSM and PSM. Observing data are fitted by the Langmuir model and the maximum adsorption capacity (q_m) is shown near the curves.

3.2.2. Kinetics Modeling

The model parameters of the three kinetic models for tetracycline on UPSM are shown in Table 3. Similar to previous studies of microsphere adsorption [29,30], all three models were capable of explaining the kinetic data satisfactorily. As an example, the data curving fitting results obtained by the second-order kinetic model are shown in Figure 4. The performance of the global fitting was not as good as expected according to the correlation coefficient, partly because the initial concentrations showed small effects on the shape of the kinetic curves. The rate constants for each kinetic experiment showed a correlation to the initial concentrations as shown in Supplementary Table S3.

Kinetics is useful to determine the optimal volume of reactors to ensure certain removal efficiency. The duration of half reduction can be calculated as $t_{1/2} = \ln2/K_1$, where K_1 is the rate constant of first-order kinetics. According to Table 3, the rate constant K_1 for TC and DC adsorption is 0.33 and 0.41 /h, respectively, indicating that 50% removal of initial concentration occurred in 1.7 and 2.1 h. This duration covers the practical hydraulic retention time in filters.

Table 3. Global estimation of kinetics for tetracycline adsorption on UPSM.

Target	First Rate Constant K_1	R^2	Second Rate Constant $K_2q_e^2$	R^2	Weber–Morris Constant K_w	R^2
	/h		mg/g/h		/h$^{1/2}$	
TC	0.41	0.942	1.8	0.939	0.64	0.963
DC	0.33	0.929	2.1	0.933	0.63	0.962

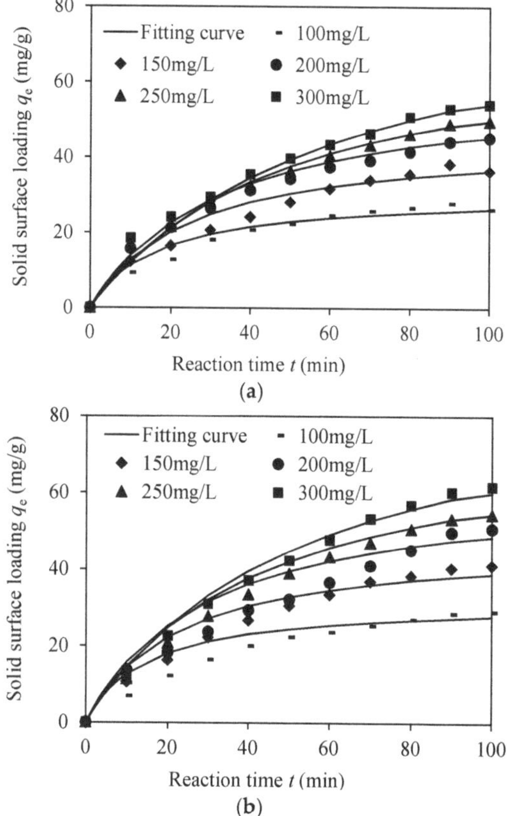

Figure 4. Adsorption kinetics of (**a**) TC and (**b**) DC on UPSM. The data fitting curves were printed in the same line type for each experiment.

3.2.3. Adsorption Performance

Table 4 shows a comparison of adsorption capacities of UPSM in this study with other materials in the literature. Polymer resins showed a promising capacity to remove TC, especially after surface modification by graphene oxide, which reached 198 mg/g [31]. Using a fabricated nanosheet can increase the capacity to 315 mg/g [32]. The silicon-based material enhanced the capacity to 303 mg/g [33]. Activated carbon achieved satisfactory performance in removing TC [34]. In this study, the adsorption capacity of TC by UPSM (460 mg/g) was slightly higher than that of the above materials.

The magnetic particles on graphene oxide nanosheets showed impressively high adsorption capacity for tetracycline at 714 mg/g [35]. By comparison with UPSM, the material of the graphene oxide nanosheet is a little bit expensive and lacks commercialization. Additionally, structures of such material are not easy to integrate with existing filters in practice. UPSM in this study are cheap. Moreover, they are recyclable by heating to destroy the urea. What is most important is that UPSM can be easily used in conventional filters due to their physical size and hardness.

Table 4. Comparison of TC adsorption capacity in the literature.

New Adsorbent	q_m, mg/g	Reference
Magnetic multiamine resins	117	[35]
Magnetic polystyrene resins	166	[29]
Magnetic polydopamine resins	152	[36]
Polystyrene microsphere/graphene oxide	198	[31]
Polymer resins/anion exchange group	355	[37]
Magnetic microsphere/graphene oxide nanosheet	714	[38]
Nanosheet-layered double hydroxide	98	[39]
TiO_2 nanosheets	213	[40]
Magnetic polyacrylonitrile nanofiber mat	315	[32]
Amino-ferrous functionalized silica	188	[24]
La-impregnated silicates	303	[33]
Activated carbons from hazelnut shell	303	[34]
Urea functionalized polystyrene resins	460	This study

It is better to estimate adsorption performance in similar concentrations of adsorbates in the actual matrix. Tetracycline in a surface water environment, e.g., rivers and lakes, is generally at trace levels of several μg/L [1]. Its concentration in municipal wastewater might be as high as 150 μg/L [41]. The concentration levels in this study were selected to prove the concept and feasibility. Evaluating UPSM for tetracycline removal at trace levels is beyond the scope of this paper. Nevertheless, in the case of industrial wastewater treatment, when tetracycline concentration may exceed the mg/L level, the results in this study can be used as a reference.

3.3. Recycling Adsorption and Desorption

3.3.1. Adsorption by Cycle Urea Immersion

The isotherms of TC on UPSM by single and cycle urea immersion are compared in Figure 5. The Freundlich equation was used to interpret the data. The values of parameter n for both modes (n = 0.93 and n = 1.26) were close to 1, indicating linear isotherm in low concentrations. In single immersion mode, the bulk concentration was saturated at about 37 mg/L after 5 cycles, while it took 10 cycles to saturate at about 35 mg/L in cycle immersion mode. This result suggests that multilayer adsorption increased the capacity but such effect reduced gradually with the increasing number of layers. At C_e = 20 mg/L, the adsorption capacity was twice that in single immersion mode. The improvement was three times at C_e = 30 mg/L. The cycle urea immersion obviously increased the adsorption capacity of UPSM by two to three times.

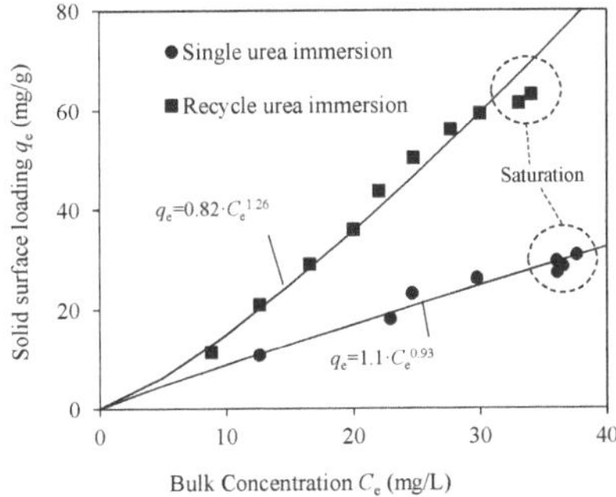

Figure 5. Comparison of isotherms of TC on UPSM with urea immersion in single and recycling modes. Experimental conditions include pH 6.8 and 25 °C for 12 h. Data is fitted by the Freundlich equation. The dashed circles highlight the saturation level of bulk concentration after repeated exposure of UPSM to TC solution.

3.3.2. Desorption Performance

A desorption curve of TC from loaded UPSM by thermal treatment is shown in Figure 6a. The TC molecule was thermally stable at 35 °C according to the control curve. UPSM released 2 mg/L of TC in 7 h in a linear trend. According to Figure 6b, 12.4 mg/L of TC was released by 0.01 M HCl, which is higher than that by 0.01 M NaOH (6.7 mg/L). The mass of TC on UPSM was equal to 180 mg/L of TC in bulk solution, thus desorption rates of TC by heating, NaOH, and HCl in 2 h were 0.4%, 3.7%, and 6.9%, respectively.

The insufficient desorption by HCl indicated chemical association during the adsorption, such as π–π electron donor–acceptor (EDA) interaction and hydrogen bonding. Hydrogen bonding can be formed between the amino groups of tetracycline and the carbonyl groups of EDTA. The ketone group in tetracycline molecules is a π-electron acceptor and the ester group (–COO–R) in EDTA is a strong electron donor. Adsorption might be enhanced by different couples of π–π EDA interactions, e.g., the interaction between the conjugated π-electron moiety, the cation-π bonding between the amino groups, and the π-electron rich structures on the solid surface.

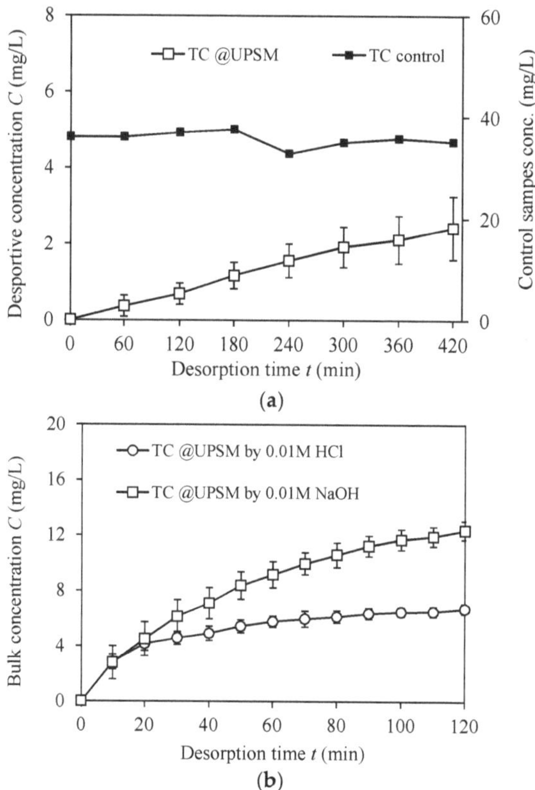

Figure 6. Desorption of recycling adsorbed antibiotics from loaded UPSM. (**a**) TC desorption in pure water at 35 °C for 7 h and pure TC solution used as control; (**b**) TC desorption by 0.01 M NaOH and 0.01 M HCl for 2 h. Error bars represent data deviation of duplicate experiments.

3.4. Adsorption Mechanisms

3.4.1. FT-IR Analysis after Adsorption

After adsorption, specific peaks of pure tetracycline (Supplementary Figure S7) in the range of 3000–3500 cm^{-1} appeared in the FT-IR spectra of loaded UPSM (Figure 7a). The peaks at 1354 and 1305 cm^{-1} disappeared and the peak at 1650 cm^{-1} was blue shifted to about 1600 cm^{-1}, which might be related to strong intermolecular hydrogen interactions between urea and tetracycline (Figure 7b).

Figure 7. FT-IR spectra analysis of fresh, TC-loaded, and DC-loaded UPSM. (**a**) Full-spectrum curves; (**b**) zoomed spectra for carboxyl and hydrogen bonding.

3.4.2. Role of Urea for Multilayer Adsorption

Figure 8 shows the concepts of the bridging mechanism by urea immersion. During urea immersion, the ketone and amino groups of urea interact with the amino group and oxygen atoms of EDTA on the surface of PSM (Figure 8a). During the adsorption process, ketone groups of both urea and EDTA act as hydrogen acceptors to form hydrogen bonding (C–H … O) with carbonyl groups in tetracycline molecules (Figure 8b). During cycle urea immersion, two or more TC molecules can be linked due to their interaction with the urea molecules (Figure 8c). It is difficult for the tetracycline molecules to interact with each other directly due to their structural shape. However, the small-sized urea molecules acted as a lubricant to reduce such structural incompatibility, making the multilayer adsorption possible and feasible.

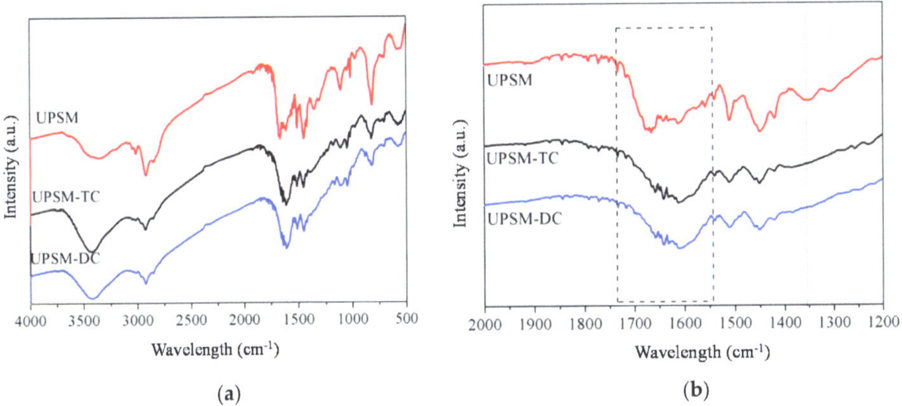

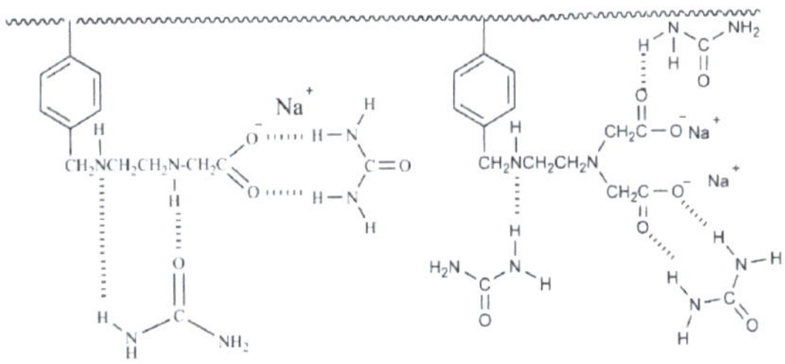

(**a**)

Figure 8. *Cont.*

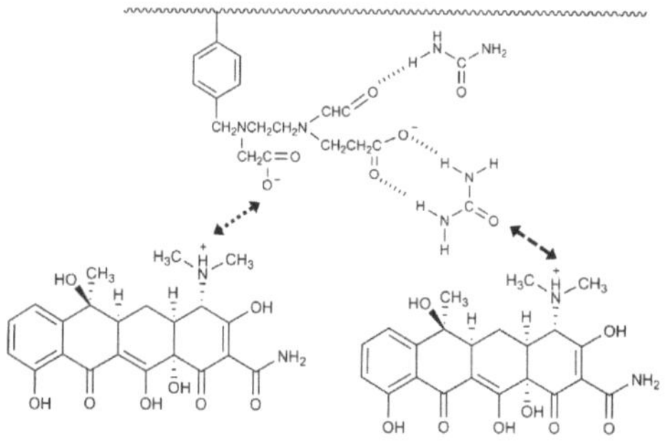

(b)

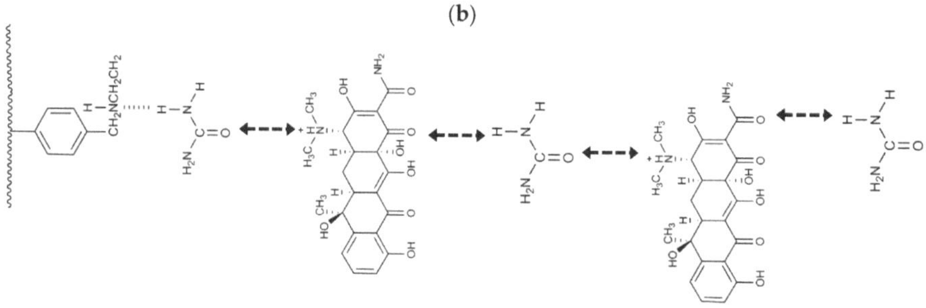

(c)

Figure 8. Proposed intermolecular hydrogen bond net corresponding to urea immersion and TC attraction. (**a**) Hydrogen bonds between urea and EDTA on PSM; (**b**) hydrogen bonds between tetracycline and urea or EDTA; (**c**) hydrogen bonds between urea and tetracycline to support multilayer adsorption.

4. Conclusions

Urea immersion was used to modify microsphere resin to enhance its capacity of removing tetracycline reagents from water. The surface characteristics by XPS and FT-IR analysis confirmed successful urea immersion and tetracycline adsorption. The adsorption isotherms were explained by the Langmuir, Freundlich, and Tempkin models. The adsorption capacity of UPSM for TC was 460 mg/g and for DC was 430 mg/g, which were increased by 30% and 60%, respectively, by urea immersion. Adsorptive kinetic data were interpreted by first-order, second-order, and Weber–Morris models. The rate constant for TC adsorption on UPSM was 0.41/h and for DC was 0.33/h, indicating that the durations of 2.1 and 1.7 h were necessary for 50% removal of TC and DC, respectively.

Desorption experiments revealed pure dissociation of TC from UPSM. The releasing rates were 0.4%, 3.7%, and 6.9% in 2 h by heating, NaOH, and HCl, respectively. Possible chemical bonding such as hydrogen bonding and π–π interaction may contribute to the adsorption enhancement. Using urea molecules as bridges, multilayer adsorption of tetracycline was possible, which was confirmed by repeated adsorption experiments on UPSM with cycle urea immersion. The adsorption capacity was two to three times higher than the UPSM with single urea immersion. In summary, urea immersion is satisfactory to modify the surface of microsphere resin to enhance the removal of tetracycline antibiotics from water.

Supplementary Materials: The following are available online at http://www.mdpi.com/1660-4601/15/7/1524/s1, Figure S1: Structural formula of Tetracycline HCl (TC), Doxycycline HCl (DC) and original PSM, Figure S2: SEM image of original PSM and urea-immersed UPSM. (a-c) microsphere PSM in different scales, (d) urea modified microsphere UPSM before adsorption, (e) UPSM after adsorption of tetracycline, (f) UPSM after adsorption of tetracycline. Scale bar represents 500 μm in (a,d), 50 μm in (b,e) and 5μm in (c,f), Figure S3: The BET surface of original PSM and urea-immersed UPSM, Figure S4: The pore structures of original PSM and urea-immersed UPSM, Figure S5: Optimization of the initial pH value for tetracycline adsorption by comparing their kinetic curves. (a) Kinetic curves of TC on UPSM; (b) Kinetic curves of DC on UPSM; C_0 is 60 mg/L, mass of UPSM is 60 mg, Figure S6: The XPS spectra analysis of fresh PSM and UPSM. (a) C 1s at PSM, (b) N 1s at PSM, (c) C 1s at UPSM and (d) N 1s at UPSM, , Figure S7: FT-IR spectra of TC and DC, Table S1: The molecular information of chemicals used in this study, Table S2: Porous structure information of the microspheres, Table S3: The kinetic parameters of tetracycline adsorption at different initial concentrations.

Author Contributions: J.M. and B.L. contributed equally to the paper. Y.Q., J.M. and L.Z. conceived the idea; J.M., B.L. and Y.Z. conducted the experiments and gathered the data; B.L. and Y.Q. analyzed the data; J.M. and Y.Q. composed the draft; Y.Q. and B.L. revised the paper; Y.Q. and J.L. managed the project to give financial support; all authors contributed to proofreading.

Funding: National Natural Science Foundation of China (51778325, 51708023), Science and Technology Project in Jiangsu Province (BE2015622), Special Fund of State Key Joint Laboratory of Environment Simulation and Pollution Control (17K03ESPCT), and Fundamental Research Funds for the Central Universities (FRF-TP-16-058A1).

Acknowledgments: The authors would like to express their gratitude to the referees for the comments to improve the quality of the manuscript, and to Peng Liang, Xianghua Wen, and Xia Huang at School of Environment in Tsinghua University for proofreading and valuable comments.

Conflicts of Interest: The authors declare no conflict of interests.

References

1. Kulkarni, P.; Olson, N.; Raspanti, G.; Rosenberg Goldstein, R.; Gibbs, S.; Sapkota, A.; Sapkota, A. Antibiotic Concentrations Decrease during Wastewater Treatment but Persist at Low Levels in Reclaimed Water. *Int. J. Environ. Res. Public Health* **2017**, *14*, 668. [CrossRef] [PubMed]

2. Gao, P.; Mao, D.; Luo, Y.; Wang, L.; Xu, B.; Xu, L. Occurrence of sulfonamide and tetracycline-resistant bacteria and resistance genes in aquaculture environment. *Water Res.* **2012**, *46*, 2355–2364. [CrossRef] [PubMed]

3. Hladicz, A.; Kittinger, C.; Zarfel, G. Tigecycline Resistant *Klebsiella pneumoniae* Isolated from Austrian River Water. *Int. J. Environ. Res. Public Health* **2017**, *14*, 1169. [CrossRef] [PubMed]

4. Zhang, H.; Li, X.; Yang, Q.; Sun, L.; Yang, X.; Zhou, M.; Deng, R.; Bi, L. Plant Growth, Antibiotic Uptake, and Prevalence of Antibiotic Resistance in an Endophytic System of Pakchoi under Antibiotic Exposure. *Int. J. Environ. Res. Public Health* **2017**, *14*, 1336. [CrossRef] [PubMed]

5. Ji, L.; Wan, Y.; Zheng, S.; Zhu, D. Adsorption of tetracycline and sulfamethoxazole on crop residue-derived ashes: Implication for the relative importance of black carbon to soil sorption. *Environ. Sci. Technol.* **2011**, *45*, 5580–5586. [CrossRef] [PubMed]

6. Pereira, J.H.; Queirós, D.B.; Reis, A.C.; Nunes, O.C.; Borges, M.T.; Boaventura, R.A.; Vilar, V.J. Process enhancement at near neutral pH of a homogeneous photo-Fenton reaction using ferricarboxylate complexes: Application to oxytetracycline degradation. *Chem. Eng. J.* **2014**, *253*, 217–228. [CrossRef]

7. López Peñalver, J.J.; Sánchez Polo, M.; Gómez Pacheco, C.V.; Rivera Utrilla, J. Photodegradation of tetracyclines in aqueous solution by using UV and UV/H_2O_2 oxidation processes. *J. Technol. Biotechnol.* **2010**, *85*, 1325–1333. [CrossRef]

8. Shi, Y.; Wang, X.; Qi, Z.; Diao, M.; Gao, M.; Xing, S.; Wang, S.; Zhao, X. Sorption and biodegradation of tetracycline by nitrifying granules and the toxicity of tetracycline on granules. *J. Hazard. Mater.* **2011**, *191*, 103–109. [CrossRef] [PubMed]

9. Dirany, A.; Sirés, I.; Oturan, N.; Özcan, A.; Oturan, M.A. Electrochemical treatment of the antibiotic sulfachloropyridazine: Kinetics, reaction pathways, and toxicity evolution. *Environ. Sci. Technol.* **2012**, *46*, 4074–4082. [CrossRef] [PubMed]

10. Chai, R.; Huang, L.; Li, L.; Gielen, G.; Wang, H.; Zhang, Y. Degradation of Tetracyclines in Pig Manure by Composting with Rice Straw. *Int. J. Environ. Res. Public Health* **2016**, *13*, 254. [CrossRef] [PubMed]

11. Kovalova, L.; Siegrist, H.; Singer, H.; Wittmer, A.; McArdell, C.S. Hospital wastewater treatment by membrane bioreactor: Performance and efficiency for organic micropollutant elimination. *Environ. Sci. Technol.* **2012**, *46*, 1536–1545. [CrossRef] [PubMed]

12. Khamparia, S.; Jaspal, D.K. Adsorption in combination with ozonation for the treatment of textile waste water: A critical review. *Front. Environ. Sci. Eng.* **2017**, *11*, 8. [CrossRef]

13. Acosta, R.; Fierro, V.; de Yuso, A.M.; Nabarlatz, D.; Celzard, A. Tetracycline adsorption onto activated carbons produced by KOH activation of tyre pyrolysis char. *Chemosphere* **2016**, *149*, 168–176. [CrossRef] [PubMed]

14. Shan, D.; Deng, S.; Zhao, T.; Wang, B.; Wang, Y.; Huang, J.; Yu, G.; Winglee, J.; Wiesner, M.R. Preparation of ultrafine magnetic biochar and activated carbon for pharmaceutical adsorption and subsequent degradation by ball milling. *J. Hazard. Mater.* **2016**, *305*, 156–163. [CrossRef] [PubMed]

15. Parolo, M.E.; Savini, M.C.; Vallés, J.M.; Baschini, M.T.; Avena, M.J. Tetracycline adsorption on montmorillonite: pH and ionic strength effects. *Appl. Clay Sci.* **2008**, *40*, 179–186. [CrossRef]

16. Zhang, Z.; Liu, H.; Wu, L.; Lan, H.; Qu, J. Preparation of amino-Fe (III) functionalized mesoporous silica for synergistic adsorption of tetracycline and copper. *Chemosphere* **2015**, *138*, 625–632. [CrossRef] [PubMed]

17. Turku, I.; Sainio, T.; Paatero, E. Thermodynamics of tetracycline adsorption on silica. *Environ. Chem. Lett.* **2007**, *5*, 225–228. [CrossRef]

18. Yang, W.; Zheng, F.; Lu, Y.; Xue, X.; Li, N. Adsorption interaction of tetracyclines with porous synthetic resins. *Ind. Eng. Chem. Res.* **2011**, *50*, 13892–13898. [CrossRef]

19. Zhou, Q.; Zhang, M.C.; Shuang, C.D.; Li, Z.Q.; Li, A.M. Preparation of a novel magnetic powder resin for the rapid removal of tetracycline in the aquatic environment. *Chin. Chem. Lett.* **2012**, *23*, 745–748. [CrossRef]

20. Chao, Y.; Zhu, W.; Ye, Z.; Wu, P.; Wei, N.; Wu, X.; Li, H. Preparation of metal ions impregnated polystyrene resins for adsorption of antibiotics contaminants in aquatic environment. *J. Appl. Polym. Sci.* **2015**, *132*, 41803. [CrossRef]

21. Ma, Y.; Zhou, Q.; Li, A.; Shuang, C.; Shi, Q.; Zhang, M. Preparation of a novel magnetic microporous adsorbent and its adsorption behavior of *p*-nitrophenol and chlorotetracycline. *Hazard. Mater.* **2014**, *266*, 84–93. [CrossRef] [PubMed]

22. Chao, Y.; Zhu, W.; Yan, B.; Lin, Y.; Xun, S.; Ji, H.; Wu, X.; Li, H.; Han, C. Macroporous polystyrene resins as adsorbents for the removal of tetracycline antibiotics from an aquatic environment. *J. Appl. Polym. Sci.* **2014**, *131*, 40561. [CrossRef]

23. Hao, R.; Xiao, X.; Zuo, X.; Nan, J.; Zhang, W. Efficient adsorption and visible-light photocatalytic degradation of tetracycline hydrochloride using mesoporous BiOI microspheres. *J. Hazard. Mater.* **2012**, *209*, 137–145. [CrossRef] [PubMed]

24. Zhang, Z.; Li, H.; Liu, H. Insight into the adsorption of tetracycline onto amino and amino-Fe3+ gunctionalized mesoporous silica: Effect of functionalized groups. *J. Environ. Sci.* **2018**, *65*, 171–178. [CrossRef] [PubMed]

25. Lv, J.; Ma, Y.; Chang, X.; Fan, S. Removal and removing mechanism of tetracycline residue from aqueous solution by using Cu-13X. *Chem. Eng. J.* **2015**, *273*, 247–253. [CrossRef]

26. Yang, L.; Li, Y.; Wang, L.; Zhang, Y.; Ma, X.; Ye, Z. Preparation and adsorption performance of a novel bipolar PS-EDTA resin in aqueous phase. *J. Hazard. Mater.* **2010**, *180*, 98–105. [CrossRef] [PubMed]

27. Li, X.; Yang, L.; Li, Y.; Ye, Z.; He, A. Efficient Removal of Cd^{2+} from Aqueous Solutions by Adsorption on PS-EDTA Resins: Equilibrium, Isotherms, and Kinetic Studies. *J. Environ. Eng. ASCE* **2012**, *138*, 940–948. [CrossRef]

28. Wang, L.; Yang, L.; Li, Y.; Zhang, Y.; Ma, X.; Ye, Z. Study on adsorption mechanism of Pb(II) and Cu(II) in aqueous solution using PS-EDTA resin. *Chem. Eng. J.* **2010**, *163*, 364–372. [CrossRef]

29. Li, B.; Ma, J.; Zhou, L.; Qiu, Y. Magnetic microsphere to remove tetracycline from water: Adsorption, H_2O_2 oxidation and regeneration. *Chem. Eng. J.* **2017**, *330*, 191–201. [CrossRef]

30. He, J.; Dai, J.; Xie, A.; Tian, S.; Chang, Z.; Yan, Y.; Huo, P. Preparation of macroscopic spherical porous carbons@carboxymethylcellulose sodium gel beads and application for removal of tetracycline. *RSC Adv.* **2016**, *6*, 84536–84546. [CrossRef]

31. Chen, L.; Lei, S.; Wang, M.; Yang, J.; Ge, X. Fabrication of macroporous polystyrene/graphene oxide composite monolith and its adsorption property for tetracycline. *Chin. Chem. Lett.* **2016**, *27*, 511–517. [CrossRef]

32. Liu, Q.; Zheng, Y.; Zhong, L.; Cheng, X. Removal of tetracycline from aqueous solution by a Fe_3O_4 incorporated PAN electrospun nanofiber mat. *J. Environ. Sci.* **2015**, *28*, 29–36. [CrossRef] [PubMed]

33. Vu, B.K.; Snisarenko, O.; Lee, H.S.; Shin, E.W. Adsorption of tetracycline on La-impregnated MCM-41 materials. *Environ. Technol.* **2010**, *31*, 233–241. [CrossRef] [PubMed]

34. Fan, H.; Shi, L.; Shen, H.; Chen, X.; Xie, K. Equilibrium, isotherm, kinetic and thermodynamic studies for removal of tetracycline antibiotics by adsorption onto hazelnut shell derived activated carbons from aqueous media. *RSC Adv.* **2016**, *6*, 109983–109991. [CrossRef]

35. Zhu, Z.; Zhang, M.; Wang, W.; Zhou, Q.; Liu, F. Efficient and synergistic removal of tetracycline and Cu(II) using novel magnetic multi-amine resins. *Sci. Rep.* **2018**, *8*, 4762. [CrossRef] [PubMed]

36. Mao, B.; An, Q.; Xiao, Z.; Zhai, S. Hydrophilic, hollow Fe_3O_4@PDA spheres with a storage cavity for efficient removal of polycyclic structured tetracycline. *New J. Chem.* **2017**, *41*, 1235–1244. [CrossRef]

37. Zhou, Q.; Wang, M.; Li, A.; Shuang, C.; Zhang, M.; Liu, X.; Wu, L. Preparation of a novel anion exchange group modified hyper-crosslinked resin for the effective adsorption of both tetracycline and humic acid. *Front. Environ. Sci. Eng.* **2013**, *7*, 412–419. [CrossRef]

38. Hu, X.; Zhao, Y.; Wang, H.; Tan, X.; Yang, Y.; Liu, Y. Efficient Removal of Tetracycline from Aqueous Media with a Fe_3O_4 Nanoparticles@graphene Oxide Nanosheets Assembly. *Int. J. Environ. Res. Public Health* **2017**, *14*, 1495. [CrossRef] [PubMed]

39. Soori, M.M.; Ghahramani, E.; Kazemian, H.; Al-Musawi, T.J.; Zarrabi, M. Intercalation of tetracycline in nano sheet layered double hydroxide: An insight into UV/VIS spectra analysis. *J. Taiwan Inst. Chem. Eng.* **2016**, *63*, 271–285. [CrossRef]

40. Fu, D.; Huang, Y.; Zhang, X.; Kurniawan, T.A.; Ouyang, T. Uncovering potentials of integrated $TiO_2(B)$ nanosheets and H_2O_2 for removal of tetracycline from aqueous solution. *J. Mol. Liq.* **2017**, *248*, 112–120. [CrossRef]

41. Borghi, A.A.; Palma, M.S.A. Tetracycline: Production, waste treatment and environmental impact assessment. *Braz. J. Pharm. Sci.* **2014**, *50*, 25–40. [CrossRef]

International Journal of
*Environmental Research
and Public Health*

Article

Genome-Guided Characterization of *Ochrobactrum* sp. POC9 Enhancing Sewage Sludge Utilization—Biotechnological Potential and Biosafety Considerations

Krzysztof Poszytek [1], Joanna Karczewska-Golec [1], Anna Ciok [2], Przemyslaw Decewicz [2], Mikolaj Dziurzynski [2], Adrian Gorecki [2], Grazyna Jakusz [1], Tomasz Krucon [1], Pola Lomza [1], Krzysztof Romaniuk [2], Michal Styczynski [2], Zhendong Yang [1], Lukasz Drewniak [1] and Lukasz Dziewit [2,*]

[1]   Laboratory of Environmental Pollution Analysis, Faculty of Biology, University of Warsaw, Miecznikowa 1, 02-096 Warsaw, Poland; kposzytek@biol.uw.edu.pl (K.P.); karczewska@biol.uw.edu.pl (J.K.-G.); grazyna.jakusz@biol.uw.edu.pl (G.J.); tkrucon@biol.uw.edu.pl (T.K.); pola.lomza@biol.uw.edu.pl (P.L.); zyang@biol.uw.edu.pl (Z.Y.); ldrewniak@biol.uw.edu.pl (L.D.)

[2]   Department of Bacterial Genetics, Institute of Microbiology, Faculty of Biology, University of Warsaw, Miecznikowa 1, 02-096 Warsaw, Poland; aciok@biol.uw.edu.pl (A.C.); decewicz@biol.uw.edu.pl (P.D.); mikolaj.dziurzynski@biol.uw.edu.pl (M.D.); agorecki@biol.uw.edu.pl (A.G.); romaniuk@biol.uw.edu.pl (K.R.); mstyczynski@biol.uw.edu.pl (M.S.)

*   Correspondence: ldziewit@biol.uw.edu.pl; Tel.: +48-225-541-406

Abstract: Sewage sludge is an abundant source of microorganisms that are metabolically active against numerous contaminants, and thus possibly useful in environmental biotechnologies. However, amongst the sewage sludge isolates, pathogenic bacteria can potentially be found, and such isolates should therefore be carefully tested before their application. A novel bacterial strain, *Ochrobactrum* sp. POC9, was isolated from a sewage sludge sample collected from a wastewater treatment plant. The strain exhibited lipolytic, proteolytic, cellulolytic, and amylolytic activities, which supports its application in biodegradation of complex organic compounds. We demonstrated that bioaugmentation with this strain substantially improved the overall biogas production and methane content during anaerobic digestion of sewage sludge. The POC9 genome content analysis provided a deeper insight into the biotechnological potential of this bacterium and revealed that it is a metalotolerant and a biofilm-producing strain capable of utilizing various toxic compounds. The strain is resistant to rifampicin, chloramphenicol and β-lactams. The corresponding antibiotic resistance genes (including bla_{OCH} and *cmlA/floR*) were identified in the POC9 genome. Nevertheless, as only few genes in the POC9 genome might be linked to pathogenicity, and none of those genes is a critical virulence factor found in severe pathogens, the strain appears safe for application in environmental biotechnologies.

Keywords: antibiotic resistance; biosafety; biogas production; *Ochrobactrum* sp. POC9; methane; sewage sludge utilization

---

1. Introduction

Natural and anthropogenically-shaped environments are frequently screened for the presence of bacterial strains potentially useful in environmental biotechnologies [1–4]. Municipal wastewaters and the products of their transformation (including sewage sludge)—as the environments rich in various toxic compounds—are potentially a good source of microorganisms metabolically active against numerous contaminants. Such microorganisms are well adapted to a variety of life-limiting factors, including toxic organic compounds and heavy metals [5–7], and thus are desired in technologies dedicated to sewage sludge transformation and utilization. However, pathogenic and antibiotic-resistant strains can also be found amongst bacteria isolated from wastewaters [8–10]. Therefore, for biosafety reasons, each isolate should be thoroughly analysed and, ideally, its genome should be explored to track, e.g., antibiotic resistance and virulence genes.

Ochrobactrum spp. are aerobic, non-fermenting, Gram-negative bacteria thriving in various environments, including soil, water, plants, and animals [11–15]. Members of the *Ochrobactrum* genus have frequently been isolated from environments persisting under strong anthropogenic pressure and some representatives were defined as animal and human pathogens. For example, *O. anthropi* and *O. intermedium* were recognized as relatively benign opportunistic human pathogens, infecting mostly immunocompromised patients [11,15–19]. However, cases of life-threatening infections, e.g., endocarditis caused by *O. anthropi*, were also reported [20]. *Ochrobactrum* spp. are metabolically versatile and exhibit some unique features, e.g., *O. intermedium* MZV101 is able to produce lipase and biosurfactants at pH 10 and temperature of 60 °C [21]. What is more, *Ochrobactrum* spp. seem to be well adapted for living in contaminated environments, which are rich in xenobiotics, organic pollutants and heavy metals. *Ochrobactrum* strains produce a variety of hydrolyzing enzymes that enable them to utilize even hardly-degradable compounds, e.g., phenols, organophosphorus pesticides and petroleum hydrocarbons [22–26]. Therefore, bacteria of this genus gained attention as promising candidates for use in diverse fields of biotechnology. They have been frequently employed in bioremediation technologies, e.g., in continuously stirred tank bioreactors to treat hydrocarbon-rich industrial wastewaters, or in biopiles and land farming to remove petroleum pollutions [27–30].

In this study, a novel *Ochrobactrum* strain, *Ochrobactrum* sp. POC9, was isolated from sewage sludge originated from the wastewater treatment plant (WWTP) "Czajka" in Warsaw, Poland. The strain exhibited various enzymatic activities, and significantly enhanced biogas production by improving sewage sludge utilization. The draft genomic sequence of *Ochrobactrum* sp. POC9 was obtained and thoroughly analyzed, which—together with functional analyses—provided insight into the biotechnological potential and the biosafety of the strain.

2. Materials and Methods

2.1. Isolation of Ochrobactrum sp. POC9, Culture Conditions and Screening of Enzymatic Activities

A raw sewage sludge sample collected from the wastewater treatment plant "Czajka" (Warsaw, Poland) was serially diluted using 0.8% (*w*/*v*) saline solution. Then, 100 µL of each dilution was spread on lysogeny broth (LB) solidified by the addition of 1.5% (*w*/*v*) agar [31]. The plates were incubated at 37 °C for 72 h. A random selection of 100 isolates differing in their colony morphology was subjected to enzymatic activity examination. For testing of the proteolytic activity, Frazier agar (BTL, Lodz, Poland) and nutrient agar medium [31] with skim milk (10% *v*/*v*) were used. For screening of the lipolytic activity, tributyrin agar (Sigma-Aldrich, St. Louis, MO, USA) was used. For screening of the cellulolytic and amylolytic activities, CMC-Red Congo agar [32] and nutrient agar medium [31] supplemented with a specific soluble chromogenic substrate (Megazyme, Bray, Ireland), respectively, were used. In each case, the material from a single colony was transferred onto the specific medium and the plates were incubated at 37 °C for 72 h, during which the visual observation was performed. Formation of clearing zones around the bacterial colonies indicated the hydrolysis of a particular substrate. Each experiment was carried out in triplicate. Based on the results of the performed analyses, the fastest growing strain (*Ochrobactrium* sp. POC9) exhibiting proteolytic, lipolytic, cellulolytic and amylolytic activities was selected for further studies.

2.2. Amplification and Sequencing of the 16S rRNA Gene

Genomic DNA was extracted from the bacterial cells using Genomic Mini purification kit (A&A Biotechnology, Gdynia, Poland). The 16S rRNA gene fragment was amplified by PCR with universal primers 27f and 1492r [33]. The amplified 16S rDNA fragment was used as a template for DNA sequencing with ABI3730xl DNA Analyzer (Applied Biosystems, Thermo Fisher Scientific, Foster City, CA, USA).

2.3. Draft Genome Sequencing

Genomic DNA of the POC9 strain was isolated using the CTAB/Lysozyme method [31]. An Illumina TruSeq library was constructed following the manufacturer's instructions. The genomic libraries were sequenced on Illumina MiSeq instrument (using the v3 chemistry kit) (Illumina, San Diego, CA, USA) in the DNA Sequencing and Oligonucleotide Synthesis Laboratory (oligo.pl) at the Institute of Biochemistry and Biophysics, Polish Academy of Sciences, Warsaw. The reads trimmed with CutAdapt v 1.9.1 [34] were further assembled using Newbler De Novo Assembler v3.0 (Roche, Basel, Switzerland).

2.4. Bioinformatic Analyses

The POC9 genome was automatically annotated using RAST [35] on PATRIC 3.5.11 [36] web service. Similarity searches were performed using BLAST programs [37] and Pfam database [38]. Metabolic features were identified applying KEGG database [39]. The COG numbers were assigned to each gene by local RPS-BLAST search against the COG database (last modified 22 January 2015) with 1e-5 e-value threshold by considering only the best BLAST hits [40]. Putative rRNA and tRNA sequences were identified using the Rfam [41], tRNAScan-SE [42] and ARAGORN programs [43]. To identify genetic determinants responsible for the heavy metal resistance phenotype, the genome was screened using the BacMet: antibacterial biocide and metal resistance genes database [44]. To identify antibiotic resistance genes, the Resistance Gene Identifier (RGI; Comprehensive Antibiotic Resistance Database) software was used [45]. Potential virulence factors were identified using VFDB–Virulence Factors Database [46].

2.5. Analytical Methods

Dry organic matter (VS) analysis was performed according to standard methods described by the American Public Health Association (APHA, 1998) [47]. The volatile fatty acids (VFAs) concentration and soluble chemical oxygen demand (sCOD) were determined using Nanocolor® kits (Machery-Nagel GmbH, Düren, Germany). The volume of the produced biogas was monitored using Milligascounter MGC-1 (Ritter, Bochum, Germany). Methane content was analyzed with a gas analyzer GA5000 (Geotech, Leamington Spa, UK).

2.6. Simulation of the Anaerobic Digestion Process

The effect of bioaugmentation of sewage sludge anaerobic digestion with *Ochrobactrum* sp. POC9 was investigated in laboratory-scale anaerobic batch experiments, which were performed in 1-L glass bottles GL 45 (SCHOTT Poland, Warsaw, Poland) connected with Dreschel-type scrubbers. To each reactor, a 1-L Tedlar gas bag (Sigma-Aldrich, St. Louis, MO, USA) was attached to collect biogas. Each bioreactor was filled with (i) the liquid phase from a separated fermentation chamber from the wastewater treatment plant "Krym" (Wolomin, Poland) [11 g dry organic matter per l (11 gvs·L^{-1})], containing a methanogenic consortium inoculum and (ii) sewage sludge from the same WWTP [11 g dry organic matter per l (11 gvs L^{-1})]. Then, bioreactors were supplemented with 4 mL of the bacterial (*Ochrobactrum* sp. POC9) suspension (approx. 10^7 cells·mL^{-1}). The negative controls for the experiment were cultures containing only methanogenic consortium and sewage sludge from the WWTP "Krym" without the addition of *Ochrobactrum* sp. POC9. At the beginning of the experiment, the total dry organic matter content in all bioreactors was 22 gvs·L^{-1}, soluble chemical oxygen demand (sCOD) was 1.97 g L^{-1} and VFAs concentration was 5.03 g·L^{-1}. Anaerobic batch assays were run at 37 °C for 30 days without refeeding. Analyses were carried out at the beginning of the experiment and after 3, 7, 14, 21, and 30 days. The experiment was performed in triplicate.

2.7. Antibiotic Susceptibility Testing

To determine the antimicrobial susceptibility patterns of *Ochrobactrum* sp. POC9, MICs of 10 antimicrobial agents were assessed using Etest™ (Liofilchem, Roseto degli Abruzzi, Italy). The analysis was conducted according to European Committee on Antimicrobial Susceptibility Testing (EUCAST) recommendations [48]. The following antibiotics (selected based on bioinformatic analyses that identified putative antibiotic resistance genes) were used: aminoglycosides–gentamicin (CN; concentration of antibiotic: 0.064–1024 µg mL^{-1}), β-lactams (penicillin derivatives)–ampicillin (AMP; 0.016–256 µg·mL^{-1}), β-lactams (cephalosporins)–cefixime (CFM; 0.016–256 µg·mL^{-1}), β-lactams (cephalosporins)–cefotaxime (CTX; 0.016–256 µg·mL^{-1}), β-lactams (cephalosporins)–ceftriaxone (CRO; 0.016–256 µg mL^{-1}), fluroquinolones– ciprofloxacin (CIP; 0.002–32 µg·mL^{-1}), fluroquinolones–moxifloxacin (MXF; 0.002–32 µg·mL^{-1}), phenicols–chloramphenicol (C; 0.016–256 µg·mL^{-1}), ryfamicins–rifampicin (RD; 0.016–256 µg·mL^{-1}), tetracyclines–tetracycline (TE; 0.016–256 µg·mL^{-1}). The susceptibility testing was performed at 37 °C for 20 h. After incubation, plates were photographed and MICs were defined. Antimicrobial susceptibility data were interpreted according to the EUCAST breakpoint table (version 8.0) [48].

2.8. Heavy Metal Resistance Testing

Analytical grade salts ($NaAsO_2$, $Na_2HAsO_4 \times 7H_2O$, $3CdSO_4 \times 8H_2O$, $CoSO_4 \times 7H_2O$, K_2CrO_4, $CuSO_4$, $NiSO_4 \times 7H_2O$, $ZnSO_4 \times 7\ H_2O$) (Sigma-Aldrich) were used in a resistance assay performed in 96-well plates, as described previously [49]. Triplicate cultures of the tested strain were challenged with a range of concentrations of those heavy metal salts. If the strain grew in the presence of the following heavy metal ion concentrations, it was considered resistant: (i) 10 mM As (V), (ii) 1 mM Cd^{2+}, Co^{2+}, $CrO4^{2-}$, Cu^{2+}, Ni^{2+}, Zn^{2+}, (iii) 2.5 mM As (III) [50–53].

2.9. Adherence of Bacteria to Artificial Surface (Biofilm Formation) Testing

The modified crystal violet staining method was used [54]. Bacteria were cultivated overnight in LB medium at 37 °C and then diluted to obtain OD (optical density) = 0.1 at 600 nm (OD_{600}; CFU $\approx 1.3 \times 10^8$) in the same medium. In the next step, 200 µL of three biological replicates were transferred into 24 sterile 96-well plates. Bacteria were incubated at 37 °C for 24 h, 48 h, and 72 h. The OD_{600} of cultures was then measured using Sunrise™ plate reader (Tecan, Männedorf, Switzerland) with Magellan software (Tecan). In the next step, the medium with pelagic cells (i.e., cells that did not adhere to artificial surface, and thus were not a part of a formed biofilm or alternatively were released from a formed biofilm) from each well was removed, and the wells were rinsed with saline solution and dried at 37 °C for about 15 min. The adhered (biofilm-forming) bacteria were stained with crystal violet (200 µL/well) for 10 min at room temperature. Then, an excess dye was removed, wells were rinsed with saline solution, and dried again at 37 °C. Dried and stained biofilm was then dissolved with 98% ethanol (200 µL/well). The OD_{570} measurement of the obtained suspension was carried out using a Sunrise™ instrument equipped with Magellan software. The significance of statistical results was determined by the Student *t*-test [55].

2.10. Nucleotide Sequence Accession Number

The whole-genome shotgun project of *Ochrobactrum* sp. POC9 has been deposited in the NCBI GenBank (https://www.ncbi.nlm.nih.gov/genbank) database under the accession number SAMN09237519.

3. Results and Discussion

3.1. Isolation and Identification of Ochrobactrum sp. POC9

Ochrobactrum sp. POC9 was isolated from raw sewage sludge collected from the "Czajka" wastewater treatment plant (Warsaw, Poland) in 2014. An initial screening for culturable bacteria was aimed at isolating strains capable of utilizing a broad spectrum of organic compounds. In such preliminary tests, the POC9 strain exhibited unequivocal lipolytic, proteolytic, cellulolytic, and amylolytic activities. For identification of this bacterium, its 16S rRNA gene was amplified and sequenced. The obtained sequence was compared with the 16S rDNA sequences gathered in the NCBI database, and the POC9 strain was classified into the *Ochrobactrum* genus.

A phylogenetic analysis based on 16S rDNA sequences of the POC9 strain and 20 other reference *Ochrobactrum* species was then performed. Analysis of the phylogenetic tree topology revealed the presence of two separate clusters. The first one gathered 12 species with the *O. endophyticum* EGI 60010 as an outlier. The second cluster grouped nine representatives of *Ochrobactrum* spp. with *Ochrobactrum* sp. POC9 as an outlier. Based on 16S rDNA sequence analysis results we may speculate that *Ochrobactrum* sp. POC9 is most closely related to *O. intermedium* and *O. ciceri* (Figure 1).

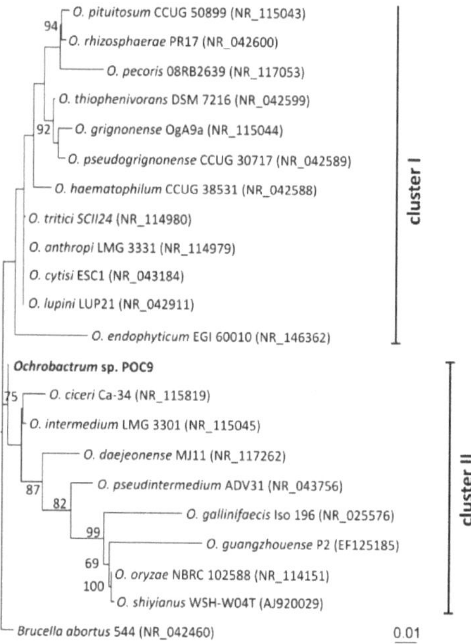

Figure 1. Phylogenetic tree for 16S rDNA sequences of *Ochrobactrum* spp. The tree was constructed by applying the Maximum Likelihood method based on the Tamura-Nei model. Statistical support for the internal nodes was determined by 1000 bootstrap replicates and values of $\geq$50% are shown. Initial tree(s) for the heuristic search were obtained automatically by applying Neighbor-Join and BioNJ algorithms to a matrix of pairwise distances estimated using the Maximum Composite Likelihood (MCL) approach. A discrete Gamma distribution was used to model evolutionary rate differences among sites (5 categories (+G, parameter = 0.6173)). The tree is drawn to scale, with branch lengths measured in the number of substitutions per site. The analysis involved 22 nucleotide sequences with 16S rDNA sequence of *Brucella abortus* 544 used as an outlier. All positions containing gaps and missing data were eliminated. There were a total of 1353 positions in the final dataset. GenBank accession numbers of the 16S rDNA sequences used for the phylogenetic analysis are given in parentheses. The 16S rDNA of the POC9 strain, analyzed in this study, is in bold text.

3.2. Bioaugmentation of Sewage Sludge Anaerobic Digestion

Bioaugmenation is a method that relies on adding specific microorganisms or microbial consortia to biological systems to enhance a desired activity [56]. This technology can be also applied in anaerobic digestions, e.g., to increase biogas (methane) production. Examples of various bioaugmentation procedures applied in anaerobic digestion of wastes are presented in Table 1.

Table 1. Examples of various bioaugmentation procedures applied in anaerobic digestion of wastes.

Strain/ Microbial Consortium	Scale	Substrate for Anaerobic Digestion	Effect	Reference
Caldicellulosiruptor saccharolyticus, *Enterobacter cloacae*	Laboratory	Waste water sludge, pig manure slurry and dried plant biomass from Jerusalem artichoke	Increased biogas production of up to 160–170%	[57]
Caldicellulosiruptor lactoaceticus 6A, *Dictyoglomus* sp. B4a	Laboratory (batch experiments and CSTR bioreactor)	Cattle manure	Increased methane yield of up to 93%	[58]
Hemicellulolytic consortium immobilized on activated zeolite	Laboratory (batch experiments and CSTR bioreactor)	Xylan from birch wood	Increased methane yield of up to 5%	[59]
Clostridium thermocellum, *Melioribacter roseus*	Laboratory scale (batch experiments and CSTR bioreactor)	Wheat straw	Increased methane yield of up to 34%	[60]
Ruminococcus flavefaciens 007C, *Pseudobutyrivibrio xylanivorans* Mz5T, *Fibrobacter succinogenes* S85 and *Clostridium cellulovorans*	Laboratory	Brewery spent grain	Increased biogas production of up to 5–18%	[61]
Methanoculleus bourgensis MS2	Laboratory	Ammonia-rich substrates (mixed pig and chicken manure, slaughterhouse residues, and food industry waste)	Increased methane yield of up to 34%	[62]
Microbial consortium with high cellulolytic activity (MCHCA)	Laboratory (two-stage anaerobic digestion)	Maize silage (lignocellulose biomass)	Increased biogas production of up to 38%, increased methane yield of up to 64%	[63]

Lipids, proteins and polysaccharides in sewage sludge are a potent source of energy and substrates for biogas production [64]. However, such use of sewage sludge is limited due to its heterogeneity, complex structure, and the presence of toxic compounds, e.g., heavy metals [65]. The initial breakdown of sewage sludge components can be substantially improved by the addition of bacteria capable of hydrolyzing complex compounds [66].

The POC9 strain exhibited diverse enzymatic (lipolytic, proteolytic, cellulolytic and amylolytic) activities under laboratory conditions. To determine whether the strain is able to break down complex residues present in sewage sludge, and thus whether it could be used for bioaugmentation of anaerobic digestion of sewage sludge, batch experiments were performed. The bioaugmentation with *Ochrobactrum* sp. POC9 was carried out once, at the beginning of the experiment. During a 30-day simulation of anaerobic digestion of sewage sludge, the daily biogas production yield and methane content in biogas were measured.

The cumulative biogas production in cultures supplemented with the POC9 strain increased by 22.06% compared to the control (non-bioaugmented) cultures. The cumulative biogas production for anaerobic digestion of sewage sludge with the addition of *Ochrobactrum* sp. POC9 was 294.58 ± 44.98 dm^3/kg of VS, while in the control experiment it was 229.58 ± 13.92 dm^3/kg (Table 2).

Interestingly, the analysis of the daily biogas production revealed that the increased biogas production occurred mostly during the first eight days of culturing (Figure S1).

Table 2. Cumulative biogas production and methane content in biogas of control and POC9-supplemented variants of the experiment simulating anaerobic digestion of sewage sludge.

Parameter (unit)	Control			Culture with the POC9 Strain		
	3 days	7 days	30 days	3 days	7 days	30 days
Cumulative biogas production (L/kgvs)	229.58 ± 13.92			294.58 ± 44.98		
CH$_4$ content (%)	43.41	61.34	49.18	46.00	66.48	58.87

The methane content during anaerobic digestion of sewage sludge after bioaugmentation with *Ochrobactrum* sp. POC9 increased from 46.00% in the third day of culturing to 58.87% after 30 days of culturing, and was highest on the seventh day—66.48% (Table 2). At the same time, methane content in biogas in the control variant increased from 43.41% to 49.18%, and was highest also on the seventh day, when it reached 61.34% (Table 2).

The results showed that the bioaugmentation with the POC9 strain improved the overall biogas production and methane content during anaerobic utilization of sewage sludge. This might be a consequence of improved hydrolysis occurring at the first stage of an anaerobic decomposition process. During this step, the combined lipolytic, proteolytic, cellulolytic and amylolytic activities of the POC9 strain probably enabled specific "pretreatment" of the raw sewage sludge, leading to increased production of easily utilizable, simple compounds (e.g., simple sugars, fatty acids, amino acids, etc.) that may be readily used by indigenous microorganisms (i.e., acetogenic bacteria and metahnogenic arachaea) for biogas production.

3.3. Genome-Based Insight into the Metabolic Potential of the POC9 Strain

Sequencing of the *Ochrobactrum* sp. POC9 genome on the Illumina MiSeq platform generated 1,275,451 paired-reads and 76,6385,632 nucleotides. As the result of the assembly, 298 contigs of a total length of 4,976,112 bp were obtained. The genome sequence was automatically annotated using RAST on PATRIC 3.5.11 web service and its general features are presented in Table 3. Based on the results of the manual inspection of the genome assembly and annotation, three potential plasmid contigs: contig00029 (35,460 bp; QGST01000029.1), contig00037 (31,160 bp; QGST01000037.1), and contig00057 (5064 bp; QGST01000057.1) were identified. In these contigs, complete *repABC*-type replication-partitioning modules, typical of large replicons of *Alphaproteobacteria*, were found [67].

Table 3. General features of the *Ochrobactrum* sp. POC9 draft genome.

Genomic feature	Calculation
Number of contigs	298
Estimated genome size (bp)	4,976,112
GC content (%)	55.68%
Coding density (%)	89.07%
Number of genes	5217
Number of tRNA genes	66
Number of 16S-23S-5S rRNA clusters	3

Genes identified within the POC9 genome were blasted against the COG database [40]. COG numbers were assigned to 3894 genes, considering only the best RPSBLAST hit with the e-value threshold of 1e-5. The genes were classified into appropriate COG categories (Figure 2).

This analysis revealed that 2005 (51.5%) genes with assigned COG numbers were associated with the cellular metabolism, and the most numerous fractions of the classified genes were: E (grouping proteins associated with amino acid transport and metabolism)—13.2%, G (carbohydrate transport and metabolism)—9.6%, and P (inorganic ion transport and metabolism)—7.7%. These observations suggested that the POC9 strain possesses complex metabolic networks and is able to utilize or transform a broad spectrum of organic and inorganic compounds.

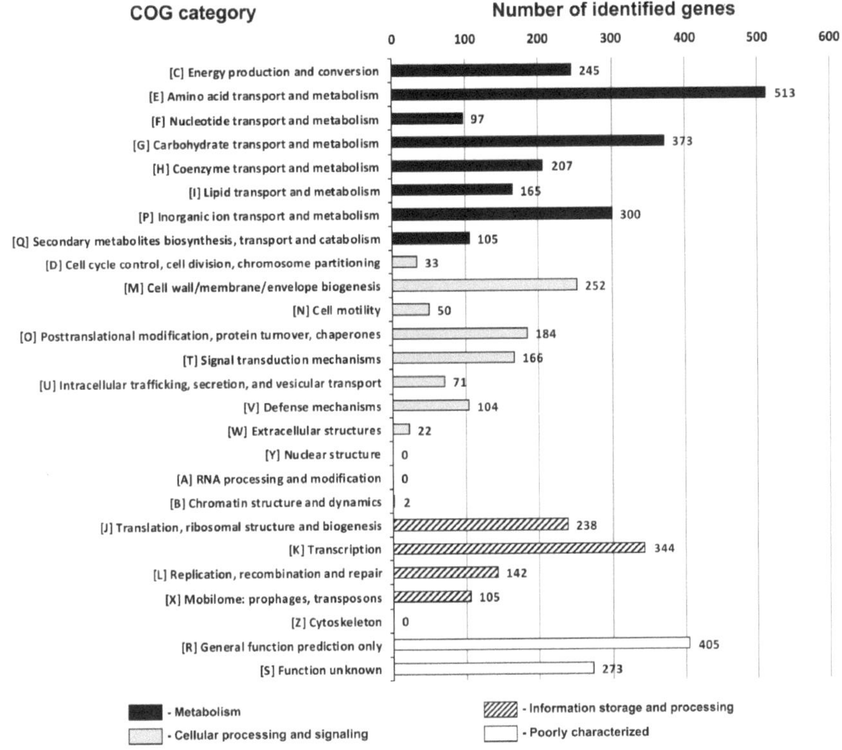

Figure 2. Number of genes classified into particular COG functional categories.

To gain a deeper insight into the POC9 metabolic potential, its draft genome was subjected to an analysis with KEGG Automatic Annotation System (KAAS), which allowed for reconstruction of its metabolic pathways. Apart from the basic metabolic pathways, such as glycolysis and the Krebs cycle, the KEGG-mapping revealed the presence of genes encoding both the enzymes enabling utilization of complex and toxic compounds and other features important in biotechnological applications, and especially advantageous in sewage sludge utilization. *Ochrobactrum* sp. POC9 possesses complete gene repertoires for the following pathways: (i) phenol degradation through benzoate utilization [23], (ii) propanoate utilization [68], (iii) denitrification [69], and (iv) assimilatory sulfate reduction [70] (Figure S2). Given the potential application of *Ochrobactrum* sp. POC9 for enhanced biogas production, it is also important to emphasize its possible contribution to CO_2 availability for methanogenesis. The POC9 strain can potentially produce CO_2 in the citric acid cycle and also through decarboxylation of formate, formaldehyde, and *N*-formyl derivatives [71,72] (Figure S2).

3.4. Heavy Metal Metabolism

We demonstrated that *Ochrobactrum* sp. POC9 enhances sewage sludge utilization. However, sewage sludge usually contains various toxic contaminants, including heavy metals and metalloids (e.g., As, Cd, Co, Cr, Cu, Fe, Hg, Ni, Pb and Zn) [73–75]. The presence of such toxic compounds may decrease the overall efficiency of the utilization treatment due to their bactericidal effect on the exogenous microorganisms used in the treatment. Hence, bacterial strains used in bioaugmentation during sewage sludge utilization should tolerate heavy metals at the highest possible concentrations [76].

Therefore, in the next step, the POC9 genome was screened for the presence of genes putatively involved in heavy metal ions transformation and resistance. We found 10 such genetic modules, encoding: (i–vi) six heavy metal translocating P-type ATPases ZntA (COG2217), which may potentially confer resistance to various divalent ions, i.e., Cd(II), Co(II), Cu(II), Ni(II), Pb(II), and Zn(II) [77–79]; (vii) chromate transport protein ChrA (COG2059), which actively transports chromate ions across cell membrane [80]; (viii–ix) two divalent metal ion transporters of the cation diffusion facilitator (CDF) family, i.e., FieF (COG0053), which usually function as Fe(II) efflux system but may also confer resistance to Cd(II), Co(II), Ni(II), and Zn(II), as well as CzcD (COG1230), which protects the cell from Cd(II), Co(II), and Zn(II) [81,82]; and (x) arsenic resistance module (ARS), which may be involved in resistance to arsenate and arsenite (Table 4). The ARS module encodes (i) arsenite efflux pump ArsB (COG0798) transporting As(III) ions out of the cell, (ii) arsenate reductase ArsC (COG1393), which reduces As(V) to As(III), and (iii) ArsH, the function of which is not completely understood, but it was hypothesized that the protein may play a regulatory role [83].

Table 4. Heavy metal resistance genes identified within the *Ochrobactrum* sp. POC9 draft genome.

Protein Name	Localization within the POC9 Draft Genome Sequence (GenBank acc. no.)	Predicted Protein Function	Homologous Protein Based on Best BLASTP Hit (GenBank acc. no.)
arsB	contig00008 (QGST01000008.1) coordinates: 148,637-147,576	Export of As(III) ions	arsenic transporter of *O. anthropi* FRAF13 (KXO76567)
arsC	contig00008 (QGST01000008.1) coordinates: 147,579-147,169	Reduction of As(V) to As(III)	arsenate reductase of *O. intermedium* LMG 3301 (EEQ95705)
arsH	contig00008 (QGST01000008.1) coordinates: 147,172-146,459	Unknown function, probably regulatory protein	arsenical resistance protein ArsH of *Ochrobactrum* sp. 30A/1000/2015 (PJT26941)
chrA	contig00005 (QGST01000005.1) coordinates: 67,497-66,199	Export of Cr(VI)	chromate transporter of *Ochrobactrum* sp. EGD-AQ16 (ERI13917)
czcD	contig00031 (QGST01000031.1) coordinates: 6786-7745	Export of Cd(II), Co(II), and Zn(II)	cation transporter of *Ochrobactrum* sp. MYb71 (PQZ25943)
fieF	contig00005 (QGST01000005.1) coordinates: 103,790-102,813	Export of Cd(II), Co(II), Fe(II), Ni(II), and Zn(II)	cadmium transporter of *O. anthropi* FRAF13 (KXO76051)
zntA	contig00025 (QGST01000025.1) coordinates: 28,104-26,260	Export of Cd(II), Co(II), Cu(II), Ni(II), Pb(II), and Zn(II)	haloacid dehalogenase of *O. anthropi* FRAF13 (KXO73917)
zntA	contig00032 (QGST01000032.1) coordinates: 21,766-24,129	Export of Cd(II), Co(II), Cu(II), Ni(II), Pb(II), and Zn(II)	lead, cadmium, zinc and mercury transporting ATPase; copper-translocating P-type ATPase of *O. haematophilum* FI11154 (SPL62610)
zntA	contig00034 (QGST01000034.1) coordinates: 5822-7666	Export of Cd(II), Co(II), Cu(II), Ni(II), Pb(II), and Zn(II)	cadmium-translocating P-type ATPase of *O. rhizosphaerae* PR17 (OYR19288)
zntA	contig00065 (QGST01000065.1) coordinates: 984-2825	Export of Cd(II), Co(II), Cu(II), Ni(II), Pb(II), and Zn(II)	heavy metal translocating P-type ATPase of *O. anthropi* ATCC 49188 (ABS17306)
zntA	contig00001 (QGST01000001.1) coordinates: 97,090-99,573	Export of Cd(II), Co(II), Cu(II), Ni(II), Pb(II), and Zn(II)	ATPase of *O. anthropi* FRAF13 (KXO79927)
zntA	contig00004 (QGST01000004.1) coordinates: 22,193-19,692	Export of Cd(II), Co(II), Cu(II), Ni(II), Pb(II), and Zn(II)	copper-translocating P-type ATPase of *O. lupini* LUP21 (OYR29555)

We tested whether the predicted genes and gene clusters identified in the POC9 genome are able to confer resistance to heavy metals. The effect of As(III), As(V), Cd(II), Co(II), Cr(VI), Cu(II), Ni(II), and Zn(II) ions on *Ochrobactrum* sp. POC9 growth was examined. These metals were selected based on the predicted metal specificity of the putative resistance genes identified within the POC9 genome (Table 4). The MIC values were determined to establish the levels of metal resistance of the strain. The obtained results are as follows: As(III)—MIC value of 3 mM, As(V)—1200 mM, Cd(II)—1 mM, Co(II)—1 mM, Cr(VI)—3 mM, Cu(II)—5 mM, Ni(II)—2 mM, and Zn(II)—3 mM.

Ochrobactrum sp. POC9 exhibited an extremely high-level of As(V) resistance and a moderate level of As(III) resistance. The observed resistance to arsenic compounds is probably mediated by the presence of the ARS module enabling reduction of arsenate to arsenite and its further extrusion from the cell. The POC9 strain showed also a moderate level of resistance to Cr(VI), Cu(II), Ni(II), and Zn(II) ions, which may be linked to the presence of the above mentioned transporters of broad-range specificities, i.e., ChrA, CzcD, FieF, and ZntA.

The results indicated that the POC9 strain shows at least moderate resistance to six heavy metal ions that are frequently present in sewage sludge. This makes it a robust candidate for application in biotreatment of metal-contaminated waste, as the strain is well equipped to withstand the toxic effect of heavy metals in these environments.

3.5. Antibiotic Resistance Genes and Virulence Factors

Wastewater treatment plants are considered a point source of emerging pollutants, including antibiotics, antibiotic resistant and/or pathogenic bacteria as well as antibiotic resistance and virulence genes [84]. Therefore, with an aim to apply *Ochrobactrum* sp. POC9—a wastewater isolate–in biotechnology, and taking into account biosafety considerations, genome of this strain was analyzed for the presence of antibiotic resistance and virulence genes.

To determine the antimicrobial resistance profile of *Ochrobactrum* sp. POC9, bioinformatic analyses and MIC characterization using Etests were performed. It total, five antibiotic resistance genes (bla_{OCH}, *qacH*, *cmlA/floR*, *acc (6')*, and *tetG*) and two clusters of genes encoding a multidrug efflux system (*acrAB-TolC*) potentially conferring resistance to various types of β-lactams, aminoglycosides, fluoroquinolones, tetracyclines, phenicols and ryfamicins were identified within the POC9 genome (Table 5).

Results of the analysis with Etests indicated that *Ochrobactrum* sp. POC9 is resistant to various β-lactams (including ampicillin, cefexime, cefotaxime, and ceftriaxone), rifampicin and chloramphenicol (Table 5). Broad spectrum resistance to various classes of β-lactams was previously observed in other (also non-pathogenic) *Ochrobactrum* strains [85]. Resistance phenotype in those strains was linked to the presence of a class C β-lactamase gene (bla_{OCH}) and to an upstream gene encoding a LysR family regulator [85]. Within the POC9 genome, the bla_{OCH} gene and an upstream regulatory *gcvA* gene were also found. Interestingly, protein products of these genes in POC9 exhibited 98% sequence identity with class C beta-lactamase of *Ochrobactrum tritci* (GenBank acc. no. SME85995) and transcriptional regulator GcvA of *Ochrobactrum* anthropi (GeneBank acc. no. WP_061347328). It seems that the presence of highly conserved bla_{OCH} gene is common in *Ochrobactrum* genomes. Furthermore, the POC9 strain was also resistant to chloramphenicol, which well reflects the presence of the *cmlA/floR* gene in its genome [86].

The Etest results revealed that the POC9 strain was susceptible to ciprofloxacin, moxifloxacin, gentamicin and tetracycline, which suggests that *qacH*, *acc (6')* and *tetG* were inactive under the tested laboratory conditions (Table 5). Further analyses with a broader range of antibiotics belonging to fluoroquinolones, aminoglycosides and tetracyclines should be carried out to test whether these genes are specific to other antibiotics of these groups.

The results of the susceptibility testing did not enable us to unequivocally determine whether the above mentioned *acrAB-TolC* modules encoding resistance-nodulation-cell division (RND) type multidrug efflux systems are active or not, as their predicted activities partially overlap with the specificities of proteins encoded by bla_{OCH} and *cmlA/floR* genes [85–87] (Table 5).

Table 5. Antibiotic resistance genes identified within the *Ochrobactrum* sp. POC9 draft genome and the resistance profile of the strain.

Gene/Gene Cluster Name	Localization within the POC9 Draft Genome Sequence (GenBank acc. no.)	Protein	Best BLAST Hits: [% Identity] Organism (GeneBank acc. no.)	Predicted Antimicrobial Resistance Profile	Tested Antibiotics	Profile
gcvA- *bla*$_{OCH}$	contig00009 (QGST01000009.1) coordinates: 164,062–165,431	class C beta-lactamase / transcriptional regulator GcvA	[98%] *Ochrobactrum tritici* C8846-N36 (SME85995) / [98%] *Ochrobactrum anthropi* (WP_061347328)	penams, penems, cephalosporins, cephamycins, monobactams	AMP / CFM / CTX / CRO	R / R / R / R
acrAB-TolC	contig00001 (QGST01000001.1) coordinates: 191,438–196,605	transcriptional regulator TetR / efflux RND transporter periplasmic adaptor subunit / efflux RND transporter permease subunit	[93%] *Ochrobactrum oryzae* (WP_104756164) / [98%] *Ochrobactrum anthropi* (WP_061344971) / [99%] *Ochrobactrum anthropi* (WP_061344972)	tetracyclines, cephalosporins penams, phenicols, ryfamycins, fluoroquinolones	AMP / CIP / CFM / CTX / CRO / TE / MXF / RIF	R / S / R / R / R / S / S / R
acrAB-TolC	contig00010 (QGST01000010.1) coordinates: 49,666–54,163	efflux RND transporter periplasmic adaptor unit / efflux RND transporter permease subunit	[98%] *Ochrobactrum* sp. (WP_024900215) / [99%] *Ochrobactrum oryzae* (WP_104755654)			
qacH	contig00007 (QGST01000007.1) coordinates: 192,104–191,772	efflux SMR transporter	[100%] *Ochrobactrum* sp. (WP_010661279)	fluoroquinolones	CIP / MXF	S / S
cmlA/floR	contig00016 (QGST01000016.1) coordinates: 56,907–55,714	CmlA/floR chloramphenicol efflux MFS transporter	[94%] *Ochrobactrum anthropi* (WP_061345584)	chloramphenicol	C	R
acc (6')	contig00016 (QGST01000016.1) coordinates: 92,931–92,485	aminoglycoside 6'-acetyl-transferase	[93%] *Ochrobactrum anthropi* FRAF13(KXO77791)	aminoglycosides	CN	S
tetG-tetR	contig00014 (QGST01000014.1) coordinates: 7148–9063	Tet(A/B/C) family MFS transporter / transcriptional regulator tetR	[87%] *Ochrobactrum oryzae* (WP_104755825) / [89%] *Ochrobactrum oryzae* (WP_104755986)	tetracyclines	TE	S

Abbreviations: AMP—ampicillin; C—chloramphenicol; CN—gentamicin; CFM—cefixime; CTX—cefotaxime, CRO—ceftriaxone; CIP—ciprofloxacin; TE—tetracycline; MXF—moxifloxacin; RIF—rifampicin; R—resistant; S—susceptible.

The POC9 genome was also screened for the presence of putative virulence genes. Only a few genetic modules that may potentially be linked to pathogenicity were found. These were: (i–iii) Sec-SRP (the general secretion route), Tat (Twin-arginine translocation pathway), Vir-like type IV secretion systems (T4SS) [88], and (iv–v) genes responsible for the synthesis of flagella and sigma-fimbriae subsystem (sF-Chap, sF-UshP, sF-Adh), which may be responsible for biofilm formation and adherence of cells to various surfaces [89]. Within the genome of *Ochrobactrum* sp. POC9, we also identified several genes encoding proteins responsible for the synthesis of lipopolysaccharides (LpxA, LpxB, LpxC, LpxD, LpxH, LpxK, LpxL and KdtA) as well as ABC-type systems (*rfb* and *lpt*) responsible for the transport of lipopolysaccharides [90–92]. Whereas these genes may be somehow linked to the POC9 strain pathogenicity, they are common among *Alphaproteobacteria*, including non-pathogenic strains [93].

The presence of genes involved in the synthesis of lipopolysaccharides constituting biofilm matrix as well as several genes that may indirectly influence biofilm formation (e.g., sigma-fimbriae subsystem) inspired us to analyze biofilm formation ability in the POC9 strain. Whereas bacterial biofilms are usually linked to the virulence of clinical strains, they also play a major ecological role in maintaining complex multispecies ecosystems in diverse environments [94]. Moreover, biofilms are also beneficial in bioremediation, e.g., they are crucial for the proper functioning of the attached growth systems in WWTPs [95]. The crystal violet staining method was applied to determine the adherence potential of *Ochrobactrum* sp. POC9. The results showed that the POC9 strain is able to adhere to surface after 24 h of incubation without shaking and then maintains the biofilm up to 48 h of incubation. Interestingly, after 72 h, the biofilm began to disperse, resulting in a significant ($p < 0.005$) reduction of the adhered biomass (Figure S3). This observation suggested that the POC9 strain is able to form biofilm or to at least initiate its formation by adhesion to artificial surfaces. However, stable maintenance of the biofilm may require specific (and not yet recognized) environmental conditions or cooperation with other microorganisms.

4. Conclusions

In this study, a novel bacterial strain, *Ochrobactrum* sp. POC9, was isolated from sewage sludge. The strain exhibited diverse enzymatic (lipolytic, proteolytic, cellulolytic and amylolytic) activities, which suggested that it might be beneficial in the biodegradation of complex organic compounds. Therefore, the strain was tested for its ability to enhance the process of sewage sludge utilization, and was found to substantially improve the overall biogas yield and methane content during anaerobic digestion of sewage sludge. This may be a consequence of enhanced hydrolysis occurring at the first stage of the process. The obtained results suggest that *Ochrobactrum* sp. POC9 carries out specific "pretreatment" of the raw sewage sludge, which leads to increased production of easily utilizable simple organic compounds for other microorganisms (i.e., acetogenic bacteria and methanogenic arachaea) involved in the biogas production.

The analysis of the POC9 genome content offered a deeper insight into the biotechnological potential of this bacterium and revealed its denitrifying, biofilm forming, and toxic compound (e.g., phenol) utilization abilities. Genomic investigations combined with physiological analyses indicated also that *Ochrobactrum* sp. POC9 is a metalotolerant bacterium, and carries several heavy metal resistance genes in its genome. This is a beneficial feature, as heavy metals are common in wastes, and may generate bactericidal effect. We also demonstrated that the POC9 strain is resistant to various β-lactams (including ampicillin, cefexime, cefotaxime, and ceftriaxone), rifampicin and chloramphenicol, which well correlates with the presence of several antibiotic resistance genes, including bla_{OCH} and *cmlA/floR*, in the *Ochrobactrum* POC9 genome. Nevertheless, as only few genes in the POC9 genome were recognized as potentially linked to pathogenicity, and none of these genes is a critical virulence factor found in severe pathogens, the strain appears safe for environmental biotechnology applications.

5. Patents

Polish patent No. PL413998. Drewniak L., Poszytek K., Dziewit L., Sklodowska A. 2018. Consortium of microorganisms capable of hydrolysis of the proteins and lipids in the sewage sludge and/or contaminated soil, the formulation comprising them, the application of the consortium and method of hydrolysis of proteins, lipids and hardly degradable compounds in sewage sludge and/or organic compounds in soils.

Supplementary Materials: The following are available online at http://www.mdpi.com/1660-4601/15/7/1501/s1, Figure S1: Daily biogas production during anaerobic digestion of sewage sludge. Figure S2: Metabolic pathways of a putative biotechnological value recognized within the POC9 draft genome. Figure S3: Adherence of *Ochrobactrum* sp. POC9 to an artificial (plastic) surface after three time intervals (24 h, 48 h and 72 h).

Author Contributions: Conceptualization, K.P., L.D. (Lukasz Drewniak) and L.D. (Lukasz Dziewit); Methodology, K.P., L.D. (Lukasz Drewniak) and L.D. (Lukasz Dziewit); Software, P.D. and M.D.; Validation, K.P., J.K.-G., L.D. (Lukasz Drewniak) and L.D. (Lukasz Dziewit); Formal Analysis, K.P., J.K.-G., L.D. (Lukasz Drewniak) and L.D. (Lukasz Dziewit); Investigation, K.P., A.C., P.D., M.D., A.G., G.J., T.K., K.R., M.S. and Z.Y.; Resources, L.D. (Lukasz Drewniak) and L.D. (Lukasz Dziewit); Data Curation, K.P., L.D. (Lukasz Drewniak) and L.D. (Lukasz Dziewit); Writing-Original Draft Preparation, K.P., A.C., P.D., M.D., A.G., G.J., T.K., P.L., K.R., M.S., Z.Y. and L.D. (Lukasz Dziewit); Writing-Review & Editing, K.P., J.K.-G., L.D. (Lukasz Drewniak) and L.D. (Lukasz Dziewit); Visualization, K.P., P.D., M.D. and M.S.; Supervision, L.D. (Lukasz Drewniak) and L.D. (Lukasz Dziewit); Project Administration, L.D. (Lukasz Drewniak) and L.D. (Lukasz Dziewit); Funding Acquisition, L.D. (Lukasz Drewniak) and L.D. (Lukasz Dziewit).

Funding: This research was funded by the Gekon 2 (Generator of Ecological CONcepts) programme of the National Centre for Research and Development (Poland) and the National Fund for Environmental Protection and Water Management (Poland) [project number GEKON2/O2/266405/7/2015].

Acknowledgments: Library construction and genome assembly were carried out at the DNA Sequencing and Oligonucleotide Synthesis Laboratory of the IBB Polish Academy of Science using the CePT infrastructure financed by the European Union—the European Regional Development Fund [Innovative economy 2007–13, Agreement POIG.02.02.00-14-024/08-00]. We thank Jan Gawor for his technical assistance.

Conflicts of Interest: The authors declare no conflict of interest. The funders had no role in the design of the study; in the collection, analyses, or interpretation of data; in the writing of the manuscript, and in the decision to publish the results.

References

1. Malla, M.A.; Dubey, A.; Yadav, S.; Kumar, A.; Hashem, A.; Abd_Allah, E.F. Understanding and designing the strategies for the microbe-mediated remediation of environmental contaminants using omics approaches. *Front. Microbiol.* **2018**, *9*, 1132. [CrossRef] [PubMed]

2. Danilovich, M.E.; Sanchez, L.A.; Acosta, F.; Delgado, O.D. Antarctic bioprospecting: In pursuit of microorganisms producing new antimicrobials and enzymes. *Polar Biol.* **2018**, 1–17. [CrossRef]

3. Head, I.M.; Jones, D.M.; Roling, W.F. Marine microorganisms make a meal of oil. *Nat. Rev. Microbiol.* **2006**, *4*, 173–182. [CrossRef] [PubMed]

4. Harayama, S.; Kasai, Y.; Hara, A. Microbial communities in oil-contaminated seawater. *Curr. Opin. Biotechnol.* **2004**, *15*, 205–214. [CrossRef] [PubMed]

5. Sarkar, J.; Kazy, S.K.; Gupta, A.; Dutta, A.; Mohapatra, B.; Roy, A.; Bera, P.; Mitra, A.; Sar, P. Biostimulation of indigenous microbial community for bioremediation of petroleum refinery sludge. *Front. Microbiol.* **2016**, *7*, 1407. [CrossRef] [PubMed]

6. Aislabie, J.; Saul, D.J.; Foght, J.M. Bioremediation of hydrocarbon-contaminated polar soils. *Extremophiles* **2006**, *10*, 171–179. [CrossRef] [PubMed]

7. Vogt, C.; Richnow, H.H. Bioremediation via in situ microbial degradation of organic pollutants. *Adv. Biochem. Eng. Biotechnol.* **2014**, *142*, 123–146. [CrossRef] [PubMed]

8. Piotrowska, M.; Przygodzinska, D.; Matyjewicz, K.; Popowska, M. Occurrence and variety of β-lactamase genes among *Aeromonas* spp. isolated from urban wastewater treatment plant. *Front. Microbiol.* **2017**, *8*, 863. [CrossRef] [PubMed]

9. Ye, L.; Zhang, T. Pathogenic bacteria in sewage treatment plants as revealed by 454 pyrosequencing. *Environ. Sci. Technol.* **2011**, *45*, 7173–7179. [CrossRef] [PubMed]

10. Lood, R.; Erturk, G.; Mattiasson, B. Revisiting antibiotic resistance spreading in wastewater treatment plants—Bacteriophages as a much neglected potential transmission vehicle. *Front. Microbiol.* **2017**, *8*, 2298. [CrossRef] [PubMed]

11. Aujoulat, F.; Romano-Bertrand, S.; Masnou, A.; Marchandin, H.; Jumas-Bilak, E. Niches, population structure and genome reduction in *Ochrobactrum intermedium*: Clues to technology-driven emergence of pathogens. *PLoS ONE* **2014**, *9*, e83376. [CrossRef] [PubMed]

12. Kampfer, P.; Huber, B.; Busse, H.J.; Scholz, H.C.; Tomaso, H.; Hotzel, H.; Melzer, F. *Ochrobactrum pecoris* sp. nov. isolated from farm animals. *Int. J. Syst. Evol. Microbiol.* **2011**, *61*, 2278–2283. [CrossRef] [PubMed]

13. Jackel, C.; Hertwig, S.; Scholz, H.C.; Nockler, K.; Reetz, J.; Hammerl, J.A. Prevalence, host range, and comparative genomic analysis of temperate *Ochrobactrum* phages. *Front. Microbiol.* **2017**, *8*, 1207. [CrossRef] [PubMed]

14. Bathe, S.; Achouak, W.; Hartmann, A.; Heulin, T.; Schloter, M.; Lebuhn, M. Genetic and phenotypic microdiversity of *Ochrobactrum* spp. *FEMS Microbiol. Ecol.* **2006**, *56*, 272–280. [CrossRef] [PubMed]

15. Chain, P.S.; Lang, D.M.; Comerci, D.J.; Malfatti, S.A.; Vergez, L.M.; Shin, M.; Ugalde, R.A.; Garcia, E.; Tolmasky, M.E. Genome of *Ochrobactrum anthropi* ATCC 49188 T, a versatile opportunistic pathogen and symbiont of several eukaryotic hosts. *J. Bacteriol.* **2011**, *193*, 4274–4275. [CrossRef] [PubMed]

16. Menezes, F.G.; Abreu, M.G.; Kawagoe, J.Y.; Warth, A.N.; Deutsch, A.D.; Dornaus, M.F.; Martino, M.D.; Correa, L. *Ochrobactrum anthropi* bacteremia in a preterm infant with cystic fibrosis. *Braz. J. Microbiol.* **2014**, *45*, 559–561. [CrossRef] [PubMed]

17. Moller, L.V.; Arends, J.P.; Harmsen, H.J.; Talens, A.; Terpstra, P.; Slooff, M.J. *Ochrobactrum intermedium* infection after liver transplantation. *J. Clin. Microbiol.* **1999**, *37*, 241–244. [PubMed]

18. Galanakis, E.; Bitsori, M.; Samonis, G.; Christidou, A.; Georgiladakis, A.; Sbyrakis, S.; Tselentis, Y. *Ochrobactrum anthropi* bacteraemia in immunocompetent children. *Scand. J. Infect. Dis.* **2002**, *34*, 800–803. [CrossRef] [PubMed]

19. Aujoulat, F.; Roger, F.; Bourdier, A.; Lotthe, A.; Lamy, B.; Marchandin, H.; Jumas-Bilak, E. From environment to man: Genome evolution and adaptation of human opportunistic bacterial pathogens. *Genes* **2012**, *3*, 191–232. [CrossRef] [PubMed]

20. Mahmood, M.S.; Sarwari, A.R.; Khan, M.A.; Sophie, Z.; Khan, E.; Sami, S. Infective endocarditis and septic embolization with *Ochrobactrum anthropi*: Case report and review of literature. *J. Infect.* **2000**, *40*, 287–290. [CrossRef] [PubMed]

21. Zarinviarsagh, M.; Ebrahimipour, G.; Sadeghi, H. Lipase and biosurfactant from *Ochrobactrum intermedium* strain MZV101 isolated by washing powder for detergent application. *Lipids Health Dis.* **2017**, *16*, 177. [CrossRef] [PubMed]

22. Zu, L.; Xiong, J.; Li, G.; Fang, Y.; An, T. Concurrent degradation of tetrabromobisphenol A by *Ochrobactrum* sp. T under aerobic condition and estrogenic transition during these processes. *Ecotoxicol. Environ. Saf.* **2014**, *104*, 220–225. [CrossRef] [PubMed]

23. El-Sayed, W.S.; Ibrahim, M.K.; Abu-Shady, M.; El-Beih, F.; Ohmura, N.; Saiki, H.; Ando, A. Isolation and identification of a novel strain of the genus *Ochrobactrum* with phenol-degrading activity. *J. Biosci. Bioeng.* **2003**, *96*, 310–312. [CrossRef]

24. Zhang, X.-H.; Zhang, G.-S.; Zhang, Z.-H.; Xu, J.-H.; Li, S.-P. Isolation and characterization of a dichlorvos-degrading strain DDV-1 of *Ochrobactrum* sp. *Pedosphere* **2006**, *16*, 64–71. [CrossRef]

25. Ermakova, I.T.; Shushkova, T.V.; Sviridov, A.V.; Zelenkova, N.F.; Vinokurova, N.G.; Baskunov, B.P.; Leontievsky, A.A. Organophosphonates utilization by soil strains of *Ochrobactrum anthropi* and *Achromobacter* sp. *Arch. Microbiol.* **2017**, *199*, 665–675. [CrossRef] [PubMed]

26. Arulazhagan, P.; Vasudevan, N. Biodegradation of polycyclic aromatic hydrocarbons by a halotolerant bacterial strain *Ochrobactrum* sp. VA1. *Mar. Pollut. Bull.* **2011**, *62*, 388–394. [CrossRef] [PubMed]

27. Gargouri, B.; Karray, F.; Mhiri, N.; Aloui, F.; Sayadi, S. Application of a continuously stirred tank bioreactor (CSTR) for bioremediation of hydrocarbon-rich industrial wastewater effluents. *J. Hazard. Mater.* **2011**, *189*, 427–434. [CrossRef] [PubMed]

28. Katsivela, E.; Moore, E.R.; Maroukli, D.; Strompl, C.; Pieper, D.; Kalogerakis, N. Bacterial community dynamics during in-situ bioremediation of petroleum waste sludge in landfarming sites. *Biodegradation* **2005**, *16*, 169–180. [CrossRef] [PubMed]

29. Katsivela, E.; Moore, E.R.B.; Kalogerakis, N. Biodegradation of aliphatic and aromatic hydrocarbons: Specificity among bacteria isolated from refinery waste sludge. *Water Air Soil Pollut.* **2003**, *3*, 103–115. [CrossRef]

30. Calvo, C.; Silva-Castro, G.A.; Uad, I.; Garcia Fandino, C.; Laguna, J.; Gonzalez-Lopez, J. Efficiency of the EPS emulsifier produced by *Ochrobactrum anthropi* in different hydrocarbon bioremediation assays. *J. Ind. Microbiol. Biotechnol.* **2008**, *35*, 1493–1501. [CrossRef] [PubMed]

31. Sambrook, J.; Russell, D.W. *Molecular Cloning: A Laboratory Manual*; Cold Spring Harbor Laboratory Press: New York, NY, USA, 2001.

32. Hendricks, C.W.; Doyle, J.D.; Hugley, B. A new solid medium for enumerating cellulose-utilizing bacteria in soil. *Appl. Environ. Microbiol.* **1995**, *61*, 2016–2019. [PubMed]

33. Lane, D.J. 16S/23S rRNA sequencing. In *Nucleic Acid Techniques in Bacterial Systematics*; Stackebrandt, E., Goodfellow, M., Eds.; Wiley: New York, NY, USA, 1991; pp. 115–175.

34. Martin, M. Cutadapt removes adapter sequences from high-throughput sequencing reads. *EMBnet J.* **2011**, *17*, 10. [CrossRef]

35. Aziz, R.K.; Bartels, D.; Best, A.A.; DeJongh, M.; Disz, T.; Edwards, R.A.; Formsma, K.; Gerdes, S.; Glass, E.M.; Kubal, M.; et al. The RAST Server: Rapid annotations using subsystems technology. *BMC Genom.* **2008**, *9*, 75. [CrossRef] [PubMed]

36. Wattam, A.R.; Davis, J.J.; Assaf, R.; Boisvert, S.; Brettin, T.; Bun, C.; Conrad, N.; Dietrich, E.M.; Disz, T.; Gabbard, J.L.; et al. Improvements to PATRIC, the all-bacterial Bioinformatics Database and Analysis Resource Center. *Nucleic Acids Res.* **2017**, *45*, D535–D542. [CrossRef] [PubMed]

37. Altschul, S.F.; Madden, T.L.; Schaffer, A.A.; Zhang, J.; Zhang, Z.; Miller, W.; Lipman, D.J. Gapped BLAST and PSI-BLAST: A new generation of protein database search programs. *Nucleic Acids Res.* **1997**, *25*, 3389–3402. [CrossRef] [PubMed]

38. Finn, R.D.; Bateman, A.; Clements, J.; Coggill, P.; Eberhardt, R.Y.; Eddy, S.R.; Heger, A.; Hetherington, K.; Holm, L.; Mistry, J.; et al. Pfam: The protein families database. *Nucleic Acids Res.* **2014**, *42*, D222–D230. [CrossRef] [PubMed]

39. Kanehisa, M.; Furumichi, M.; Tanabe, M.; Sato, Y.; Morishima, K. KEGG: New perspectives on genomes, pathways, diseases and drugs. *Nucleic Acids Res.* **2017**, *45*, D353–D361. [CrossRef] [PubMed]

40. Tatusov, R.L.; Fedorova, N.D.; Jackson, J.D.; Jacobs, A.R.; Kiryutin, B.; Koonin, E.V.; Krylov, D.M.; Mazumder, R.; Mekhedov, S.L.; Nikolskaya, A.N.; et al. The COG database: An updated version includes eukaryotes. *BMC Bioinform.* **2003**, *4*, 41. [CrossRef] [PubMed]

41. Nawrocki, E.P.; Burge, S.W.; Bateman, A.; Daub, J.; Eberhardt, R.Y.; Eddy, S.R.; Floden, E.W.; Gardner, P.P.; Jones, T.A.; Tate, J.; et al. Rfam 12.0: Updates to the RNA families database. *Nucleic Acids Res.* **2014**, *43*, D130–D137. [CrossRef] [PubMed]

42. Lowe, T.M.; Chan, P.P. tRNAscan-SE On-line: Integrating search and context for analysis of transfer RNA genes. *Nucleic Acids Res.* **2016**, *44*, W54–W57. [CrossRef] [PubMed]

43. Laslett, D.; Canback, B. ARAGORN, a program to detect tRNA genes and tmRNA genes in nucleotide sequences. *Nucleic Acids Res.* **2004**, *32*, 11–16. [CrossRef] [PubMed]

44. Pal, C.; Bengtsson-Palme, J.; Rensing, C.; Kristiansson, E.; Larsson, D.G. BacMet: Antibacterial biocide and metal resistance genes database. *Nucleic Acids Res.* **2014**, *42*, D737–D743. [CrossRef] [PubMed]

45. Jia, B.; Raphenya, A.R.; Alcock, B.; Waglechner, N.; Guo, P.; Tsang, K.K.; Lago, B.A.; Dave, B.M.; Pereira, S.; Sharma, A.N.; et al. CARD 2017: Expansion and model-centric curation of the comprehensive antibiotic resistance database. *Nucleic Acids Res.* **2017**, *45*, D566–D573. [CrossRef] [PubMed]

46. Chen, L.; Zheng, D.; Liu, B.; Yang, J.; Jin, Q. VFDB 2016: Hierarchical and refined dataset for big data analysis—10 years on. *Nucleic Acids Res.* **2016**, *44*, D694–D697. [CrossRef] [PubMed]

47. Eaton, A.D.; Clesceri, L.S.; Greenberg, A.E.; Franson, M.A.H. *Standard Methods for the Examination of Water and Wastewater*; American Public Health Association (APHA): Washington, DC, USA, 1998.

48. European Committee on Antimicrobial Susceptibility Testing (EUCAST). Available online: http://www.eucast.org (accessed on 13 May 2018).

49. Dziewit, L.; Pyzik, A.; Matlakowska, R.; Baj, J.; Szuplewska, M.; Bartosik, D. Characterization of *Halomonas* sp. ZM3 isolated from the Zelazny Most post-flotation waste reservoir, with a special focus on its mobile DNA. *BMC Microbiol.* **2013**, *13*, 59. [CrossRef] [PubMed]

50. Nieto, J.J.; Ventosa, A.; Ruiz-Berraquero, F. Susceptibility of halobacteria to heavy metals. *Appl. Environ. Microbiol.* **1987**, *53*, 1199–1202. [PubMed]

51. Nies, D.H. Microbial heavy-metal resistance. *Appl. Microbiol. Biotechnol.* **1999**, *51*, 730–750. [CrossRef] [PubMed]

52. Abou-Shanab, R.A.; van Berkum, P.; Angle, J.S. Heavy metal resistance and genotypic analysis of metal resistance genes in gram-positive and gram-negative bacteria present in Ni-rich serpentine soil and in the rhizosphere of *Alyssum murale*. *Chemosphere* **2007**, *68*, 360–367. [CrossRef] [PubMed]

53. Dib, J.; Motok, J.; Zenoff, V.F.; Ordonez, O.; Farias, M.E. Occurrence of resistance to antibiotics, UV-B, and arsenic in bacteria isolated from extreme environments in high-altitude (above 4400 m) Andean wetlands. *Curr. Microbiol.* **2008**, *56*, 510–517. [CrossRef] [PubMed]

54. O'Toole, G.A. Microtiter dish biofilm formation assay. *J. Vis. Exp.* **2011**. [CrossRef] [PubMed]

55. Andersson, S.; Kuttuva Rajarao, G.; Land, C.J.; Dalhammar, G. Biofilm formation and interactions of bacterial strains found in wastewater treatment systems. *FEMS Microbiol. Lett.* **2008**, *283*, 83–90. [CrossRef] [PubMed]

56. Deflaun, M.F.; Steffan, R.J. Bioaugmentation. In *Encyclopedia of Environmental Microbiology*; Bitton, G., Ed.; Wiley-Interscience: New York, NY, USA, 2002; Volume 1, pp. 434–442.

57. Bagi, Z.; Acs, N.; Balint, B.; Horvath, L.; Dobo, K.; Perei, K.R.; Rakhely, G.; Kovacs, K.L. Biotechnological intensification of biogas production. *Appl. Microbiol. Biotechnol.* **2007**, *76*, 473–482. [CrossRef] [PubMed]

58. Nielsen, H.B.; Mladenovska, Z.; Ahring, B.K. Bioaugmentation of a two-stage thermophilic (68 °C/55 °C) anaerobic digestion concept for improvement of the methane yield from cattle manure. *Biotechnol. Bioeng.* **2007**, *97*, 1638–1643. [CrossRef] [PubMed]

59. Weiss, S.; Tauber, M.; Somitsch, W.; Meincke, R.; Muller, H.; Berg, G.; Guebitz, G.M. Enhancement of biogas production by addition of hemicellulolytic bacteria immobilised on activated zeolite. *Water Res.* **2010**, *44*, 1970–1980. [CrossRef] [PubMed]

60. Tsapekos, P.; Kougias, P.G.; Vasileiou, S.A.; Treu, L.; Campanaro, S.; Lyberatos, G.; Angelidaki, I. Bioaugmentation with hydrolytic microbes to improve the anaerobic biodegradability of lignocellulosic agricultural residues. *Bioresour. Technol.* **2017**, *234*, 350–359. [CrossRef] [PubMed]

61. Cater, M.; Fanedl, L.; Malovrh, S.; Marinsek Logar, R. Biogas production from brewery spent grain enhanced by bioaugmentation with hydrolytic anaerobic bacteria. *Bioresour. Technol.* **2015**, *186*, 261–269. [CrossRef] [PubMed]

62. Fotidis, I.A.; Wang, H.; Fiedel, N.R.; Luo, G.; Karakashev, D.B.; Angelidaki, I. Bioaugmentation as a solution to increase methane production from an ammonia-rich substrate. *Environ. Sci. Technol.* **2014**, *48*, 7669–7676. [CrossRef] [PubMed]

63. Poszytek, K.; Ciezkowska, M.; Sklodowska, A.; Drewniak, L. Microbial Consortium with High Cellulolytic Activity (MCHCA) for enhanced biogas production. *Front. Microbiol.* **2016**, *7*, 324. [CrossRef] [PubMed]

64. Ariunbaatar, J.; Panico, A.; Esposito, G.; Pirozzi, F.; Lens, P.N.L. Pretreatment methods to enhance anaerobic digestion of organic solid waste. *Appl. Energy* **2014**, *123*, 143–156. [CrossRef]

65. Parawira, W.; Tekere, M. Biotechnological strategies to overcome inhibitors in lignocellulose hydrolysates for ethanol production: Review. *Crit. Rev. Biotechnol.* **2011**, *31*, 20–31. [CrossRef] [PubMed]

66. Herrero, M.; Stuckey, D.C. Bioaugmentation and its application in wastewater treatment: A review. *Chemosphere* **2015**, *140*, 119–128. [CrossRef] [PubMed]

67. Cevallos, M.A.; Cervantes-Rivera, R.; Gutierrez-Rios, R.M. The *repABC* plasmid family. *Plasmid* **2008**, *60*, 19–37. [CrossRef] [PubMed]

68. Textor, S.; Wendisch, V.F.; De Graaf, A.A.; Muller, U.; Linder, M.I.; Linder, D.; Buckel, W. Propionate oxidation in *Escherichia coli*: Evidence for operation of a methylcitrate cycle in bacteria. *Arch. Microbiol.* **1997**, *168*, 428–436. [CrossRef] [PubMed]

69. Lv, P.; Luo, J.; Zhuang, X.; Zhang, D.; Huang, Z.; Bai, Z. Diversity of culturable aerobic denitrifying bacteria in the sediment, water and biofilms in Liangshui River of Beijing, China. *Sci. Rep.* **2017**, *7*, 10032. [CrossRef] [PubMed]

70. Tan, T.; Liu, C.; Liu, L.; Zhang, K.; Zou, S.; Hong, J.; Zhang, M. Hydrogen sulfide formation as well as ethanol production in different media by *cysND*- and/or *cysIJ*-inactivated mutant strains of *Zymomonas mobilis* ZM4. *Bioprocess. Biosyst. Eng.* **2013**, *36*, 1363–1373. [CrossRef] [PubMed]

71. Ciriminna, R.; Meneguzzo, F.; Delisi, R.; Pagliaro, M. Citric acid: Emerging applications of key biotechnology industrial product. *Chem. Cent. J.* **2017**, *11*, 22. [CrossRef] [PubMed]

72. Jormakka, M.; Byrne, B.; Iwata, S. Formate dehydrogenase—A versatile enzyme in changing environments. *Curr. Opin. Struct. Biol.* **2003**, *13*, 418–423. [CrossRef]

73. Oleszczuk, P. Phytotoxicity of municipal sewage sludge composts related to physico-chemical properties, PAHs and heavy metals. *Ecotoxicol. Environ. Saf.* **2008**, *69*, 496–505. [CrossRef] [PubMed]

74. Werle, S.; Wilk, R.K. A review of methods for the thermal utilization of sewage sludge: The Polish perspective. *Renew. Energy* **2010**, *35*, 1914–1919. [CrossRef]

75. Cieslik, B.M.; Namiesnik, J.; Konieczka, P. Review of sewage sludge management: Standards, regulations and analytical methods. *J. Clean. Prod.* **2015**, *90*, 1–15. [CrossRef]

76. Dixit, R.; Malaviya, D.; Pandiyan, K.; Singh, B.U.; Sahu, A.; Shukla, R.; Singh, P.B.; Rai, P.J.; Sharma, K.P.; Lade, H.; et al. Bioremediation of heavy metals from soil and aquatic environment: An overview of principles and criteria of fundamental processes. *Sustainability* **2015**, *7*, 2189–2212. [CrossRef]

77. Binet, M.R.; Poole, R.K. Cd(II), Pb(II) and Zn(II) ions regulate expression of the metal-transporting P-type ATPase ZntA in *Escherichia coli*. *FEBS Lett.* **2000**, *473*, 67–70. [CrossRef]

78. Hou, Z.; Mitra, B. The metal specificity and selectivity of ZntA from *Escherichia coli* using the acylphosphate intermediate. *J. Biol. Chem.* **2003**, *278*, 28455–28461. [CrossRef] [PubMed]

79. Hou, Z.J.; Narindrasorasak, S.; Bhushan, B.; Sarkar, B.; Mitra, B. Functional analysis of chimeric proteins of the Wilson Cu(I)-ATPase (ATP7B) and ZntA, a Pb(II)/Zn(II)/Cd(II)-ATPase from *Escherichia coli*. *J. Biol. Chem.* **2001**, *276*, 40858–40863. [CrossRef] [PubMed]

80. Alvarez, A.H.; Moreno-Sanchez, R.; Cervantes, C. Chromate efflux by means of the ChrA chromate resistance protein from *Pseudomonas aeruginosa*. *J. Bacteriol.* **1999**, *181*, 7398–7400. [PubMed]

81. Anton, A.; Grosse, C.; Reissmann, J.; Pribyl, T.; Nies, D.H. CzcD is a heavy metal ion transporter involved in regulation of heavy metal resistance in *Ralstonia* sp. strain CH34. *J. Bacteriol.* **1999**, *181*, 6876–6881. [PubMed]

82. Munkelt, D.; Grass, G.; Nies, D.H. The chromosomally encoded cation diffusion facilitator proteins DmeF and FieF from *Wautersia metallidurans* CH34 are transporters of broad metal specificity. *J. Bacteriol.* **2004**, *186*, 8036–8043. [CrossRef] [PubMed]

83. Ryan, D.; Colleran, E. Arsenical resistance in the IncHI2 plasmids. *Plasmid* **2002**, *47*, 234–240. [CrossRef]

84. Rizzo, L.; Manaia, C.; Merlin, C.; Schwartz, T.; Dagot, C.; Ploy, M.C.; Michael, I.; Fatta-Kassinos, D. Urban wastewater treatment plants as hotspots for antibiotic resistant bacteria and genes spread into the environment: A review. *Sci. Total Environ.* **2013**, *447*, 345–360. [CrossRef] [PubMed]

85. Nadjar, D.; Labia, R.; Cerceau, C.; Bizet, C.; Philippon, A.; Arlet, G. Molecular characterization of chromosomal class C beta-lactamase and its regulatory gene in *Ochrobactrum anthropi*. *Antimicrob. Agents Chemother.* **2001**, *45*, 2324–2330. [CrossRef] [PubMed]

86. Kadlec, K.; Kehrenberg, C.; Schwarz, S. Efflux-mediated resistance to florfenicol and/or chloramphenicol in *Bordetella bronchiseptica*: Identification of a novel chloramphenicol exporter. *J. Antimicrob. Chemother.* **2007**, *59*, 191–196. [CrossRef] [PubMed]

87. Pradel, E.; Pages, J.M. The AcrAB-TolC efflux pump contributes to multidrug resistance in the nosocomial pathogen *Enterobacter aerogenes*. *Antimicrob. Agents Chemother.* **2002**, *46*, 2640–2643. [CrossRef] [PubMed]

88. Natale, P.; Bruser, T.; Driessen, A.J. Sec- and Tat-mediated protein secretion across the bacterial cytoplasmic membrane-distinct translocases and mechanisms. *Biochim. Biophys. Acta* **2008**, *1778*, 1735–1756. [CrossRef] [PubMed]

89. Nuccio, S.P.; Baumler, A.J. Evolution of the chaperone/usher assembly pathway: Fimbrial classification goes Greek. *Microbiol. Mol. Biol. Rev.* **2007**, *71*, 551–575. [CrossRef] [PubMed]

90. Wang, X.; Quinn, P.J.; Yan, A. Kdo2-lipid A: Structural diversity and impact on immunopharmacology. *Biol. Rev. Camb. Philos. Soc.* **2015**, *90*, 408–427. [CrossRef] [PubMed]

91. Polissi, A.; Sperandeo, P. The lipopolysaccharide export pathway in *Escherichia coli*: Structure, organization and regulated assembly of the Lpt machinery. *Mar. Drugs* **2014**, *12*, 1023–1042. [CrossRef] [PubMed]

92. Manning, P.A.; Stroeher, U.H.; Karageorgos, L.E.; Morona, R. Putative O-antigen transport genes within the *rfb* region of *Vibrio cholerae* O1 are homologous to those for capsule transport. *Gene* **1995**, *158*, 1–7. [CrossRef]

93. Marczak, M.; Mazur, A.; Koper, P.; Zebracki, K.; Skorupska, A. Synthesis of rhizobial exopolysaccharides and their importance for symbiosis with legume plants. *Genes* **2017**, *8*, 360. [CrossRef] [PubMed]

94. Davey, M.E.; O'Toole, G.A. Microbial biofilms: From ecology to molecular genetics. *Microbiol. Mol. Biol. Rev.* **2000**, *64*, 847–867. [CrossRef] [PubMed]

95. Loupasaki, E.; Diamadopoulos, E. Attached growth systems for wastewater treatment in small and rural communities: A review. *J. Chem. Technol. Biotechnol.* **2013**, *88*, 190–204. [CrossRef]

International Journal of
*Environmental Research
and Public Health*

Article

Effect of C/N Ratio on the Removal of Nitrogen and Microbial Characteristics in the Water Saturated Denitrifying Section of a Two-Stage Constructed Rapid Infiltration System

Qinglin Fang [1], Wenlai Xu [1,2,3,*], Gonghan Xia [2] and Zhicheng Pan [3]

[1]   State Key Laboratory of Geohazard Prevention and Geoenvironment Protection, Chengdu University of Technology, Chengdu 610059, China; xiaofang20111009@126.com
[2]   State Environmental Protection Key Laboratory of Synergetic Control and Joint Remediation for Soil & Water Pollution, Chengdu University of Technology, Chengdu 610059, China; txgsfy@163.com
[3]   Haitian Water Grp Co Ltd., Chengdu 610059, China; pan12487616@126.com
*   Correspondence: xuwenlai1983@163.com; Tel.: +86-135-510-29646

Abstract: The aim of this study was to improve the removal of nitrogen pollutants from artificial sewage by a modeled two-stage constructed rapid infiltration (CRI) system. The C/N ratio of the second stage influent was elevated by addition of glucose. When the C/N ratio was increased to 5, the mean removal efficiency of total nitrogen (TN) reached up to 75.4%. Under this condition, the number of denitrifying bacteria in the permanently submerged denitrifying section (the second stage) was 22 times higher than that in the control experiment without added glucose. Elevation of the C/N ratio resulted in lower concentrations of nitrate and TN in the second stage effluent, without impairment of chemical oxygen demand removal. The concentration of nitrate and TN in effluent decreased as the abundance of denitrifying bacteria increased. Moreover, the bacterial biofilms that had formed in the sand of the second stage container were analyzed. The secretion of extracellular polymeric substances, a major constituent of biofilms, was enhanced as a result of the elevated C/N ratio, which lead to the improved protection of the bacteria and enhanced the removal of pollutants.

Keywords: two-stage constructed rapid infiltration system; water-saturated denitrifying section; C/N ratio; denitrifying bacteria; extracellular polymeric substances

1. Introduction

The constructed rapid infiltration system (CRI) is a novel ecologically friendly technology for sewage treatment based on a modification of conventional sewage treatment [1]. The core of a CRI is a mixture of river sand and gravel that replaces conventional soil layers as the main infiltration media, which improves hydraulic load [2]. In addition, a dry-wet cycling of water feeding and draining is applied, which alternates the operation mode between an aerobic and (facultative) anaerobic environment, thus employing a more diverse mix of microorganisms. The unique operation mode and the rich diversity of microorganisms colonizing the filling medium of a CRI are favorable for efficient removal of pollutants from sewage [3].

The common practice, with photographs shown in Figure 1, results in satisfying removal performances of organic pollutants and suspended solids (SS); in particular removal rates of chemical oxygen demand (COD) of 85% can be established, while ammonia nitrogen (NH_4^+-N) removal rates can be as high as 90%. In contrast, removal of total nitrogen (TN) is relatively poor, only reaching 10–30% [4]. In a simulation experiment, Wang et al. [5] found that the upper part of the CRI simulation column contained higher amounts of denitrifying bacteria, related to presence of organic carbon sources, better oxygen transmission and aeration, while these were lacking in the anaerobic environment of the lower part of the CRI simulation column. Due to the limitation of denitrifying bacteria and organic carbon sources in that lower part, the overall removal of TN from the effluent was poor [5]. Removal of TN, the sum of NH_4^+-N, NO_3^--N, NO_2^--N and organic nitrogen, is difficult due to fluctuations between these various nitrogen forms. Insufficient removal and excessive discharge of TN into the environment can lead to eutrophication of rivers and lakes, which poses a world-wide environmental problem [6]. Thus, various methods have been employed to improve the nitrogen removal efficiency of a CRI. It was shown that the main factors determining nitrogen removal in a CRI are the concentration of dissolved oxygen, the ratio of C/N in the water, and the residence time of NO_3^--N in the anaerobic section [7]. To improve nitrogen removal performance, a sub-section of the effluent can be fed back into the CRI, which in simulated columns resulted in a 64.8% increase of TN removal efficiency [8]. Matsumoto [9] observed that denitrification by CRI could be improved when the C/N ratio in the influent was increased to 2. Moreover, Chen et al. [10] found that the TN removal rate could reach over 60% by adding 7 mg/L Fe^{3+} to the water-saturated section of CRI. Lastly, Song et al. [11] added corncob waste to the water-saturated section of CRI, which increased the nitrogen removal rate to above 79%. Thus, various methods exist to improve the nitrogen removal efficiency of CRI systems, but few studies determined the optimal C/N ratio in influent of the water-saturated section, or the variation in microbial biomass related to changes in C/N ratio in that section.

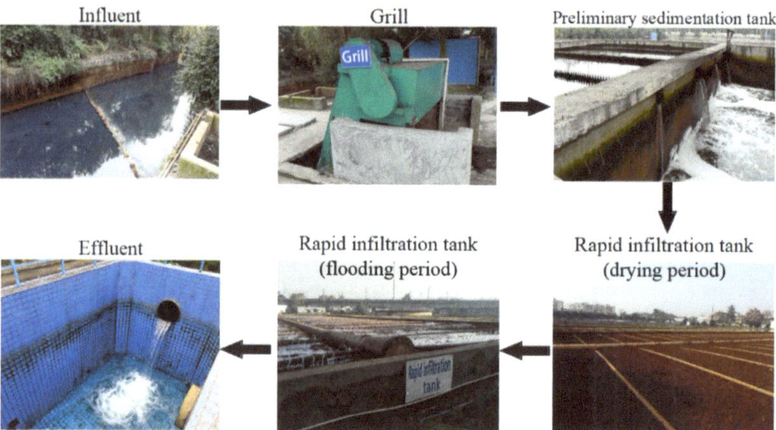

Figure 1. Practical engineering of the Phoniex River CRI system operated successfully for 12 years in Chengdu, China.

The microorganisms active in CRI system mostly form biofilms, and the necessary aerobic, anaerobic and facultative anaerobic processes for efficient water purification depend on large numbers of microbes attached to the surface of the filling medium [12]. As an important component of biofilms, extracellular polymeric substances (EPS) constitute over 80% of biofilm organic matter [13]. These substances form a macromolecular gelatinous matrix with a three-dimensional network structure mainly made up of proteins and carbohydrates [14]. The overall biofilm community has been described as a "microbial city" [15], whereby the EPS forms the "accommodation" for the microbial cells [16]. EPS that can easily be released into solution is described as the soluble EPS (S-EPS) fraction; EPS that remains bound to the bacteria (bound EPS or B-EPS) can be further divided into loosely bound EPS (LB-EPS) and tightly bound EPS (TB-EPS, Figure 2). Due to its typical physico-chemical properties, EPS enables a close packing of microbial cells, which protects them from external adverse effects. Moreover, EPS can serve as an energy and carbon source for the microbes after it is degraded into smaller molecules by extracellular enzymes. Lastly, EPS plays an important role in promoting microbial cell aggregation, as it accelerates biofilm formation and maintains collective structures. Due to these effects, EPS improves CRI performance, and has become the focus of research towards biological sewage treatment [17]. For instance, a positive correlation was found between the EPS content in sludge and its sludge volume index in a sequential batch reactor (SBR) [18]. The concentration of proteins and carbohydrates in LB-EPS of activated sludge correlated with the C/N ratio in influent [19], although in those studies the EPS concentration of the biofilms in the water-saturated section of CRI had not been reported.

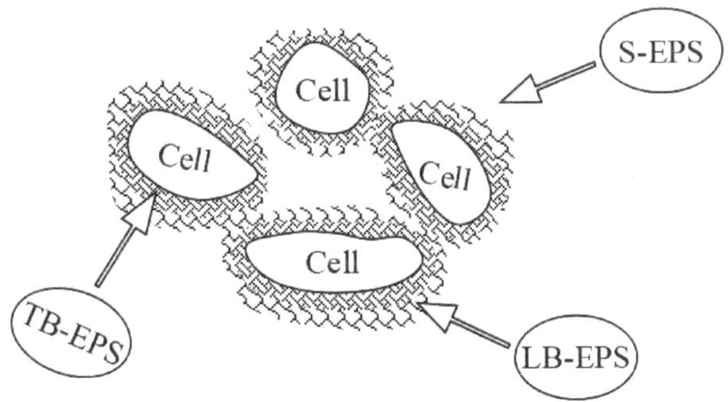

Figure 2. Schematic structure of EPS composition.

The goal of the current study was to use simulation columns of a two-stage CRI system to study the nitrifying simulation column (the first stage) and the water-saturated denitrifying simulation column (the second stage) separately, during treatment simulation of artificial sewage. The effect of variable C/N ratios in the second-stage influent on pollutant removal was determined, and the microbial community and biofilm EPS concentrations were quantified in the water-saturated denitrifying section. This addressed the influence of different C/N ratios in influent on the pollutant removal performance of CRI from a microcosmic perspective, which provides insights to improve nitrogen removal.

2. Materials and Methods

2.1. Experimental Design

A schematic diagram of the simulation columns representing the two stages of the CRI system used in this study is shown in Figure 3. The model is composed of two parts: a nitrifying simulation column (the first stage) and a water-saturated denitrifying simulation column that represents the second stage; these were connected in series. The temperature of the columns was kept constant at 30 ± 1.2 °C by means of a temperature-controlled insulation mantle. The columns were made of polyvinyl chloride (PVC) with a diameter of 8 cm and a height of 30 cm. The filling medium consisted of two layers: a 4 cm high supporting layer of pebbles (5.0–10.0 mm) mixed with gravel (3.0–4.0 mm) was covered by a 21 cm high treatment layer of 90% river sand (0.25–0.30 mm grain size), mixed with 5% marble sand (1.0–2.0 mm) and 5% zeolite sand (1.5–1.7 mm). The sampling outlet of the first stage column was positioned 1 cm above the bottom of the column, while the sampling outlet of the second-stage column was fitted at a height of 26 cm. In this way, the second stage remained continuously submerged. The influent sewage was pumped up so that it entered from the top of the column, moved through the packing medium vertically, and left by the outlet where the water quality was measured. In order to regulate the C/N ratio of the influent of the second stage, a tank with a volume of 1 L was installed between the columns, from which a carbon source could be added.

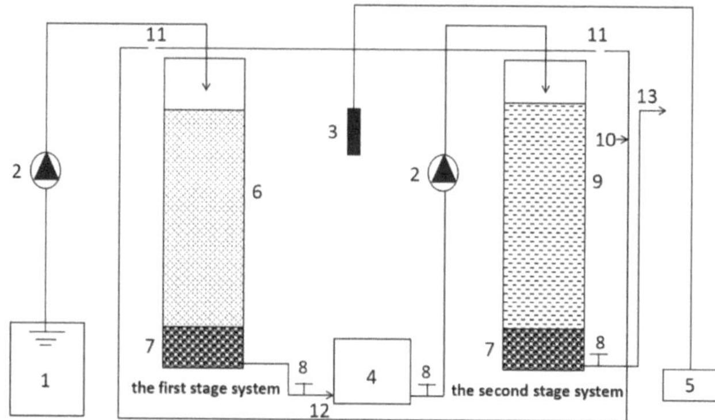

Figure 3. Schematic diagram of the simulated two-stage column CRI. 1: Feeding tank (V = 5 L); 2: Peristaltic pump; 3: Thermometer; 4: Tank to add a carbon source (V = 1 L); 5: Temperature controller; 6: Filling medium; 7: Support layer; 8: Valve; 9: Saturated section; 10: Insulation mantle; 11: Air hole; 12: Effluent outlet of the first stage; 13: Effluent outlet of the second stage.

2.2. Sewage and Operational Conditions

The influent used in this study was synthetic sewage made up of glucose, sodium acetate, ammonium sulfate, ammonium chloride, potassium phosphate, sodium carbonate and peptone, which was refilled every three days. The water quality parameters are shown in Table 1. The sewage influent entered the first stage column via a dry-wet alternating operation mode, with water feeding with a hydraulic load of 0.6 $m^3 \cdot (m^{-3} \cdot d^{-1})$, each feeding time would last for 1.5 h and this was performed twice every 24 h, followed by a drying time lasting 10.5 h, and a water flow of 200 mL/h. The second-stage column received the effluent of the first stage column as influent. This was added by intermittent feeding using the same regime as for the first stage column.

Table 1. Water quality parameters of influent of the first stage.

Water Quality Parameters	Mean Concentration (mg/L)
Chemical Oxygen Demand (COD)	268.3 ± 20.0
NH_4^+-N	50.4 ± 3.5
NO_3^--N	2.6 ± 0.3
NO_2^--N	0.039 ± 0.06
Total nitrogen (TN)	60.8 ± 2.0
pH	8.2

2.3. Batch Experiments

To study the effect of different C/N ratios of the second stage influent, three experimental setups were constructed as described in Section 2.2 and these were started under the same conditions, to give Test 1, Test 2 and Test 3 (T1, T2, T3). After operation for 30 days, the removal percentages of COD and NH_4^+-N from effluent were all stabilized, reaching levels up to 90%, indicating that biofilms had formed successfully in the filling medium [8]. From this time point onwards, T1 served as control treatment not receiving additives; the C/N ratio in the second stage influent of T2 was increased by adding a carbon source at twice the amount of COD per nitrate-nitrogen (C/N 2:1) and that of T3 was increased to a C/N ratio of 5:1. Glucose was used for this, as it is cheap, easy to obtain, non-toxic, and easily biodegraded [20]. Influent and effluent of all three experiments was sampled every two days and used for analysis. At the end of the experiment, after the removal efficiency of TN in the second stage effluent of all experimental tests had stabilized, the filling medium (sand) was sampled from the second stage column for analysis.

2.4. Analytical Methods

2.4.1. Water Quality Analytical Methods

The concentration of COD in the water was determined using the potassium dichromate method, the TN concentration was measured by UV spectrometry and the concentration of NH_4^+-N was determined by the Nessler's reagent colorimetric method. Lastly, NO_3^--N was measured by UV spectrometry, using standard procedures [21].

2.4.2. Microbiological Analysis and EPS Quantization

Bacterial counts of nitrifying and denitrifying bacteria were determined by the most probable number method (MPN) [22]. For this, 10 g of sand was collected from the position 10 cm above the supporting layer of the water-saturated denitrifying column and this was added to 100 mL sterile water in conical flasks that were shaken for 30 min. The suspension was then serially diluted by 10-fold steps and this dilution series was inoculated into culture medium and cultured for 14 days at 28 °C. Finally, the culture medium was titrated by a chromogenic agent, and the number of bacteria was obtained by comparing the number of tubes in the chromogenic medium with the MPN value table.

EPS in biofilms was extracted as previously described [14,23]. Sand samples (10 g) were mixed with 45 mL extraction buffer (2 mmol/L Na_3PO_4, 4 mmol/L NaH_2PO_4, 9 mmol/L NaCl, 1 mmol/L KCl, pH 7.0), treated by ultrasonication for 5 min and centrifuged (20 min 2000× g). The supernatant was filtered through a milipore filter of 0.45 μm to give the extracted S-EPS fraction. The pellet was resuspended into 45 mL extraction buffer and shaken for 1 h after which centrifugation was performed at 5000× g for 20 min. The filtered supernatant from this second centrifugation step resulted in LB-EPS, while the resuspended pellet was heated to 60 °C for 1 h and centrifuged at 10,000× g (20 min). The supernatant of this third centrifugation step was filtered as above to give TB-EPS.

The amount of biofilm in the sand samples was estimated as follows. The samples were briefly rinsed in distilled water and then mixed with 1 M NaOH and incubated at 70 °C for 30 min. Following ultrasonic oscillation for 15 min the biofilm containing extracts were filtered (0.45 μm) using pre-weighed membranes. After filtering, the membranes were dried in an oven at 103 °C for 1 h. The amount of biofilm was estimated by the difference in weight of the membrane filters before and after drying (SS).

The content of protein and carbohydrates was used to characterize EPS content (mg/g SS) as these are the main components of EPS [14]. The carbohydrate content was measured by means of the phenol-sulfuric method [24] with glucose as the standard. The protein content was measured by Coomassie brilliant blue [18] using bovine serum albumin as the standard.

3. Results and Discussion

3.1. Effect of C/N Ratio on Removal Efficiency of Ammonium and Nitrate

In the first stage of the experimental CRI, ammonium is converted to nitrate according to Equation (1):

$$NH_4^+ + 2O_2 = NO_3^- + H_2O + 2H^+ \tag{1}$$

The concentrations and removal efficiencies of ammonium in the effluent of the second stage over time was determined for T1 (control), T2 (glucose added at a ratio of 2:1, C/N), and T3 (C/N increased to 5:1). The results are shown in Figure 4a. The relative removal efficiency was calculated as the difference in concentration between the first-stage influent and the second-stage effluent divided by the concentration in the first stage influent.

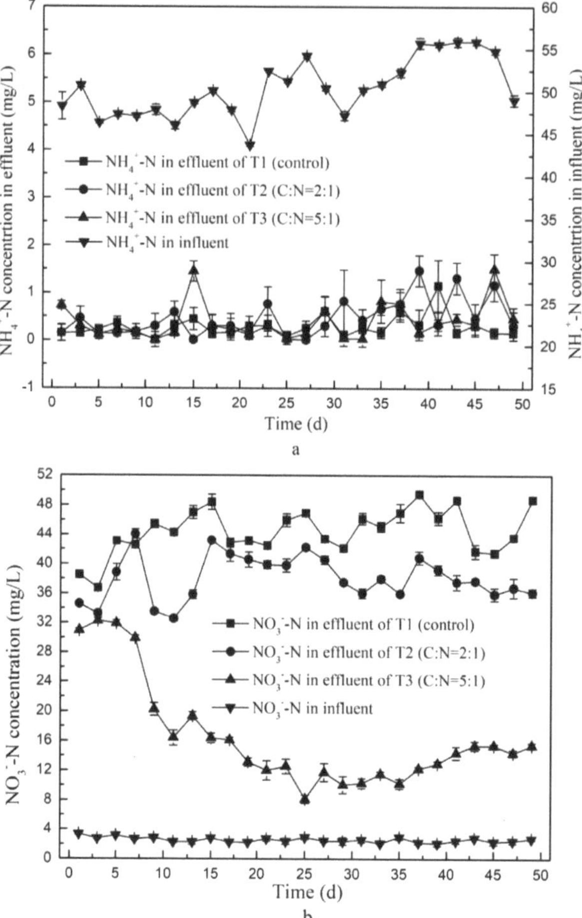

Figure 4. Overall ammonium and nitrate concentrations of a modeled two-stage CRI over time. (a) Absolute concentrations of NH_4^+-N in first-stage influent and in the second-stage effluent of the three experimental setups; (b) The nitrate concentration in influent of the first stage and in the effluent of the second stage in the three experimental setups.

The mean concentration of ammonium in the second stage effluent was lower in the control T1 (0.3 mg/L), than in T2 and T3 (0.4 mg/L and 0.5 mg/L, respectively). The removal percentages of NH_4^+-N reached up to 99.1% with only minor differences between the three experiments, indicating that the modeled CRI resulted in good overall ammonium removal and an increased C/N ratio in the second-stage influent contributed little to improve this. The removal of NH_4^+-N in this CRI mainly occurs through adsorption and nitrification, which take place in the first stage, so that alterations in conditions of the second stage have little effect. Moreover, as a result of the used setup, the denitrifying column provides a constant submerged environment, which restricts the nitrification reaction [5] that would otherwise convert ammonium to nitrate (Equation (1)). As a result, varying the carbon to nitrogen ratio in the influent of the second phase has no effect on the overall removal rate of ammonium.

The nitrate in the effluent is mainly generated by nitrifying bacteria that oxidize ammonium during the dry phase [25]. As shown in Figure 4b, the mean concentration of NO_3^--N in the first-stage influent was only around 2.7 mg/L, but it was much higher in the effluent of the second stage. In the control the mean nitrate concentration reached 45 mg/L in second-stage effluent, accounting for 93% of the TN concentration. Thus, the nitrifying section of the CRI resulted in nearly complete nitrification of the available nitrogen, as a result of the employed dry-wet alternating operation mode. Since negatively charged nitrate is not adsorbed by dielectric particles that are mostly also negatively charged [26], it remained in solution and was efficiently flushed out with the water flow, resulting in high concentration of nitrate and TN in the effluent. Wang et al. [25] showed that an extension of the residence time of the denitrifying section could improve the denitrification capacity of a CRI, as it would allow bacteria more time for nitrogen conversion. We extended the retention time to 6.5 h in the second stage by constructing a permanent water-saturated environment. However, this did not significantly increase the removal efficiency of nitrate in second-stage effluent of T1. Hou et al. [27] reported that a C/N ratio less than 2 in influent provided insufficient amounts of carbon for denitrification, resulting in low TN removal in effluent. We found that an increase in C/N to 2:1 in T2 resulted in average nitrate concentrations in the second-stage effluent of 38.6 mg/L, which represents a decrease of 6.4 mg/L compared to the control T1. With the highest tested C/N ratio of 5:1 (in T3) the nitrate concentration decreased by 71.8% compared to T1, so that only 12.7 mg/L NO_3^--N remained in the second stage effluent. Thus, we confirmed that the denitrifying capacity of a CRI can be improved by adding an external carbon source to the water-saturated denitrifying section, in order to meet the energy demand of denitrifying bacteria.

3.2. Effect of C/N Ratio on Removal Efficiency of Total Nitrogen

CRI has been shown to perform well for removal of COD, NH_4^+-N and SS, but under standard conditions the removal of TN is typically only around 10–30% [4]. This can be improved by an increased C/N ratio in the second stage influent, as shown in Figure 5.

Initially, the TN concentrations in the second-stage effluent of all three experiments were above 35 mg/L during the first 5 days (Figure 5a), which indicates a relatively poor nitrogen removal performance at this early stage of the experiments. Possibly, the denitrifying bacteria in the second phase need time to adapt to the new conditions applying after addition of an external carbon source. From day 9 onwards, the TN concentration in the second stage effluent of T2 and T3 decreased, to 34.2 mg/L and 21.9 mg/L respectively. As the experiments continued, this TN concentration in effluent of T3 kept declining, but in T2 it increased again. This indicates that the C/N ratio of 2:1 enhanced the denitrifying capacity of the second stage section to a certain extent by promoting growth of denitrifying bacteria, but it was insufficient to meet the increasing demand of carbon as the denitrifying bacteria proliferated. This is in line with findings reported by Fan et al. [28], who tested a C/N ratio of 2.5 in influent of a constructed wetland which resulted in a TN removal efficiency of only around 25%.

On day 17 of the experiments, the TN concentration in the second stage effluent of was more or less stable with mean TN concentrations in the second stage effluent of 48.4 mg/L (giving a 19.6% removal rate) in T1, 41.0 mg/L (31.8% removal rate) in T2 and 14.7 mg/L (75.4% removal rate) in T3. The TN removal rate of the second stage effluent had thus nearly doubled in T2 compared to T1, but the resulting concentration was still too high to meet the national sewage discharge B standard (GB18918-2002) (TN $\leq$ 20 mg/L). The second-stage effluent of T3 passed this standard, and even passed the sewage discharge a standard (TN $\leq$ 15 mg/L). The TN removal of T3 had improved by 69.5% compared to T1. This clearly shows that addition of a carbon source in the second stage influent enhanced the nitrogen removal efficiency of the system, whereby a C/N ratio of 5 performed better than a ratio of 2. The best TN performance obtained resulted in 79.8% removal of TN.

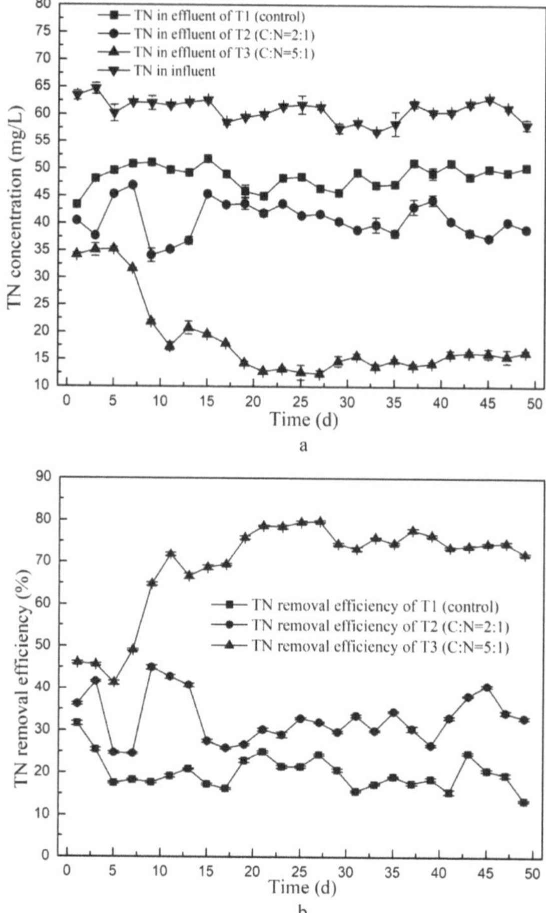

Figure 5. Total nitrogen removal of the second stage of the CRI of the three experimental tests with variable C/N ratios. (**a**) Total nitrogen concentration in influent of the first stage and in effluent of the second stage; (**b**) Total nitrogen removal efficiency.

3.3. Effect of C/N Ratio on Removal Efficiency of Chemical Oxygen Demand

The COD concentration of the first stage influent was 268.3 ± 20.0 mg/L. This reduced during the first stage to 22.8 mg/L, 24.5 mg/L and 20.5 mg/L for effluent of T1, T2 and T3 respectively, so the mean fraction of removed COD reached up to 90% in all three experiments. However, the mean concentration of NO_3^--N in the first stage effluent remained above 48.3 mg/L, resulting in a C/N ratio in the first stage effluent below 1. The effect of increasing this ratio to 2 and 5 in T2 and T3, respectively, on COD removal is shown in Figure 6. Whereas the mean concentration of COD in the second stage effluent of T1 decreased to only 16.2 mg/L, for T2 and T3 it went down to 9.3 mg/L and 8.7 mg/L respectively. Compared with T1, the mean COD concentration in the second stage effluent of T2 and T3 were much lower, indicating that the addition of glucose to the second stage influent had not resulted in an increase of COD in the effluent, but rather promoted the removal of COD during the second stage. Clearly, this external carbon source enhanced the activity and growth of denitrifying bacteria. The amount of glucose required for this was not fully met under experimental condition T2, as the best COD removal was obtained with T3, which received the highest amount of glucose. In another study, Chen et al. [7] found that the removal efficiency of COD could reach up to 85% in the nitrifying section of CRI, while Wang et al. [20] determined an optimal C/N ratio for denitrification of 6–7 when they used glucose as external carbon source. Zhang [29] noticed that nitrogen could be completely removed in SBR by adjusting the C/N ratio to 7.1 with added glucose. Yan et al. [30] reported that if the main purpose of the sewage treatment was to remove nitrogen, the C/N ratio in influent should be slightly lower than the optimum C/N ratio, which could not only guarantee high COD removal rates (resulting in low residual organic matter), but also produce a better nitrogen removal performance.

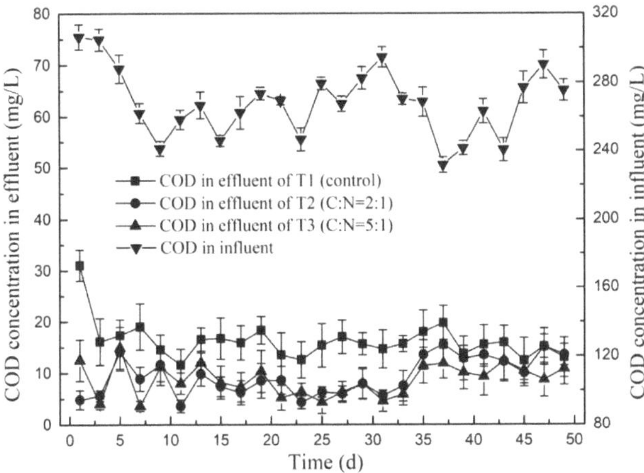

Figure 6. The chemical oxygen demand (COD) concentration in influent of the first stage and in effluent of the second stage of the three experimental setups.

3.4. Analysis of the Microbial Communities and EPS

The bacteria in the water-saturated denitrifying section were quantitatively analyzed by the MPN method (Figure 7). The sand from the second stage of T1 contained 6.5×10^3 CFU (g sand)$^{-1}$ nitrifying bacteria, and approximately half were present in the sand of T2 and T3 (3.3×10^3 and 3.1×10^3 CFU·(g sand)$^{-1}$, respectively). The low amount of dissolved oxygen and NH_4^+-N in the fully submerged second stage most probably limited the growth of these aerobic bacteria [31]. These relatively low counts of nitrifying bacteria support the conclusion that the nitrification reaction was restricted in the fully submerged phase of the CRI. A previous study [32] described that when organic matter was abundant in sewage, heterotrophic bacteria could grow rapidly and these would outcompete nitrifying bacteria for dissolved oxygen and nutrients.

The numbers of denitrifying bacteria were a factor of 1000 higher, in the order of 10^6–10^7 CFU·(g sand)$^{-1}$, In particular in T3 their numbers peaked to 5.1×10^7 CFU·(g sand)$^{-1}$, which was over 20 times higher than in T1, and 10 times higher than in T2. Therefore, under the water-saturated conditions applied, an increase of the C/N ratio in the influent promoted the proliferation of denitrifying bacteria as long as the nitrogen supply was sufficient. When the C/N ratio increased from 2 to 5, the number of denitrifying bacteria increased by more than one order of magnitude. Combining all results, it was found that the concentrations of TN and NO_3^--N in the second stage effluent decreased with an increase in numbers of denitrifying bacteria in the water-saturated denitrifying section. This suggests that the number of denitrifying bacteria presenting this section of the CRI is animportant factor determining the nitrogen removal performance, and this can be influenced by changing the C/N ratio. However, the specific relationship between the bacterial abundance and the nitrogen removal performance needs to be further explored.

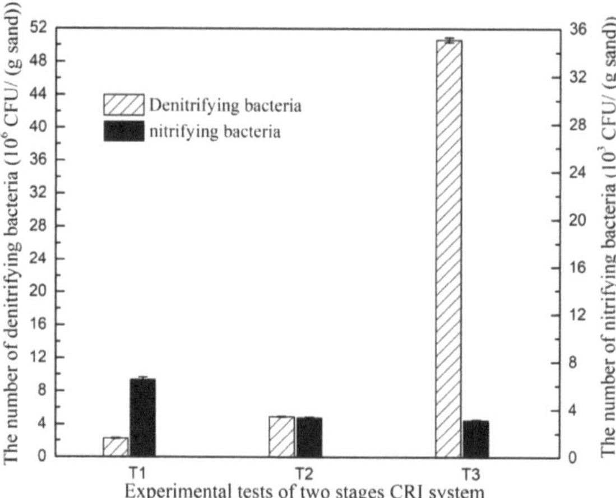

Figure 7. The number of denitrifying bacteria (10^6 CFU·(g sand)$^{-1}$) and nitrifying bacteria (10^3 CFU·(g sand)$^{-1}$) in the sand of the water-saturated denitrifying section.

The amount of S-EPS, LB-EPS and TB-EPS in the formed biofilms was determined for the denitrifying section and the total EPS concentration was calculated as the sum of these fractions. The concentrations of proteins and carbohydrates were also determined, with results shown in Figure 8. Panels 8a and 8b show that the amount of biofilm S-EPS and LB-EPS was higher in T2 than in T1, probably as a result of the increased C/N ratio, though a further increase of this ratio in T3 did not elevate S-EPS and LB-EPS further. Their concentrations all remained below 10 mg/g SS. In contrast, the fraction of TB-EPS (panel 8c) was higher in T2 than in T1, and higher still in T3, with concentrations exceeding 10 mg/g SS. The total EPS reached nearly 30 mg/g SS in T3 (panel 8d). This increase in biofilm EPS as a result of the increase by C/N ratio was mostly attributed to TB-EPS, confirming that TB-EPS is the key factor affecting the concentration of total EPS [19].

The comparisons of protein and carbohydrate fractions of EPS between T1, T2 and T3 produced a less clear picture, with an increase for proteins as a result of increased C/N ratio in S-EPS and TB-EPS but not in LB-EPS. For the carbohydrate fraction, the trends were in the opposite direction (Figure 8). The reason for these observations may be that as long as the nitrogen source was sufficient in the influent, addition of extra glucose could promote the formation of protein in S-EPS, but this limited the production of carbohydrates due to competition. The fractions of both protein and carbohydrate were increased in TB-EPS when the C/N ratio was increased (Figure 8c), which indicated that the increase of carbon source could simultaneously promote the increase of protein and carbohydrate concentrations in biofilm TB-EPS. Moreover, the contents of protein and carbohydrate in total EPS (Figure 8d) showed the same changing trends as those of TB-EPS, which indicates that increasing the C/N ratio in the second stage influent was favorable to enhance the secretion of protein and carbohydrates in total EPS. This would lead to an improved ability of biofilm formation, which would protect the microbial cells and allow more efficient pollutants removal.

It has been reported that the protein fraction of EPS could promote biofilm and granular sludge formation, and maintain the stability of microbial aggregates [33]. The carbohydrate fraction serves as the skeleton of these microbial aggregates, and when a mutation was introduced in a gene that abolishes carbohydrate biosynthesis, the bacteria could no longer form mature biofilms [34]. However, the contents of protein and carbohydrate in biofilm LB-EPS both showed a opposite trend to that of biofilm S-EPS, which may be because the carbon source was primarily used to generate carbohydrate and protein, which was then gradually degraded and utilized by the microorganisms when the carbon source became restricted over time.

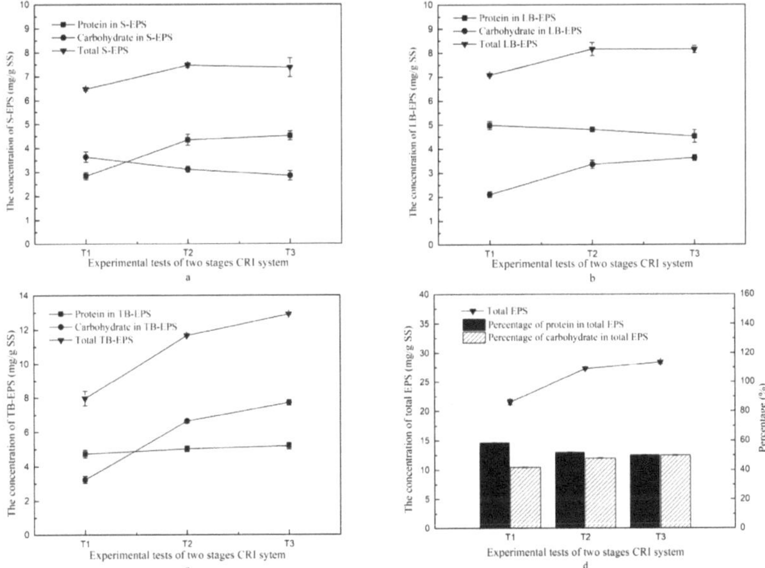

Figure 8. The concentrations of EPS (triangles), and its protein (squares) and carbohydrate (circles) fractions (all expressed as mg/g SS) in the sand of the second stage system. (**a**) S-EPS; (**b**) LB-EPS; (**c**) TB-EPS; (**d**) Total EPS. The percentages of protein and carbohydrate fractions in total EPS are also shown as bars in (**d**).

4. Conclusions

This work shows that the removal efficiency of TN by a two-stage CRI could reach 75.4 ± 2.8%, provided the C/N ratio in the influent of the second stage was elevated to 5 by the addition of glucose. We have shown that elevating the C/N ratio can promote the removal of nitrate and TN in the second-stage effluent, without impairing the COD removal. The quantity of denitrifying bacteria in the sand of the permanently submerged denitrifying section increased as a result of the higher C/N ratio, and the concentrations of NO_3^--N and TN in the subsequent effluent both decreased as the quantity of denitrifying bacteria increased. Moreover, analysis of the bacterial biofilms that had formed in the sand of the denitrifying section showed that the elevated C/N ratio had enhanced secretion of the total biofilm EPS, which leads to improved biofilm formation and enhances the pollutant's removal in the two-stage CRI.

Author Contributions: W.X. conceived and designed the experiments and wrote the paper; Q.F. and Z.P. performed the experiments and contributed reagents and analysis tools; G.X. analyzed the data.

Funding: This research is funded by Chinese National Natural Science Foundations (41502333), Sichuan science and technology support project (2017JY0141, 2018GZ0416), China Postdoctoral Science Foundation (2017M610598, 2018T110963), and the State Key Laboratory of Geohazard Prevention and Geoenvironment Protection Foundation (SKLGP2016Z019). We received above funds for covering the costs to publish in open access.

Conflicts of Interest: The authors declare no conflict of interest.

References

1. He, J.T.; Zhong, Z.S.; Tang, M.G.; Chen, H.H. Experimental research of constructed rapid infiltration wastewater treating system. *China Environ. Sci.* **2002**, *22*, 239–243.

2. He, J.T.; Zhong, Z.S.; Tang, M.G. New method of solving contradiction of rapid infiltration system land using. *Geoscience* **2001**, *15*, 339–345.

3. Xu, W.L.; Zhang, W.; Jian, Y. Analysis of nitrogen removal performance of constructed rapid infiltration system (CRIS). *Appl. Ecol. Environ. Res.* **2017**, *15*, 199–206. [CrossRef]

4. Zhang, J.B. Study on Constructed Rapid Infiltration for Wastewater Treatment. Ph.D. Thesis, University of Geosciences, Beijing, China, 2002.

5. Wang, L.; Yu, Z.P.; Zhao, Z.J. The removal mechanism of ammoniac nitrogen in constructed rapid infiltration system. *China Environ. Sci.* **2006**, *26*, 500–504.

6. Chislock, M.F.; Doster, E.; Zitomer, R.A.; Wilson, A. Eutrophication: Causes, consequences, and controls in aquatic ecosystems. *Nat. Educ. Knowl.* **2013**, *4*, 10.

7. Chen, J.M.; Liu, F.; Fu, Y.S.; Yang, J.F. Nitrogen removal mechanism of the constructed rapid infiltration system. *Technol. Water Treat.* **2009**, *35*, 32–34.

8. Fan, X.J.; Fu, Y.S.; Liu, F.; Xue, D.; Xu, W. Total nitrogen removal efficiency of improved constructed rapid infiltration system. *Technol. Water Treat.* **2009**, *35*, 70–72.

9. Matsumoto, M.R. *Abiotic Nitrogen Removal Mechanisms in Rapid Infiltration Wastewater Treatment Systems*; University of California Water Resources Center, UC Berkeley: Berkeley, CA, USA, 2004.

10. Chen, J.; Zhang, J.Q.; Wen, H.; Zhang, Q.; Yang, X.; Li, J. Effect of Fe^{3+} on nitrogen removal efficiency in constructed rapid infiltration system. *Chin. J. Environ. Eng.* **2016**, *10*, 7058–7062.

11. Song, Z.X.; Zhang, H.Z.; Wang, Z.L.; Ping, Y.H.; Liu, G.Y.; Zhao, Q. Treating sewage by strengthened constructed rapid infiltration system. *Chin. J. Environ. Eng.* **2016**, *10*, 3491–3495.

12. Li, X.N.; Luo, L.X.; Liu, H.; Pei, T.Q. Research on the variation of microorganism quantity and enzyme activity in constructed rapid infiltration system. *Environ. Pollut. Control* **2013**, *35*, 49–52.

13. Nelson, Y.M.; Lion, L.W.; Shuler, M.L. Modeling oligotrophic biofilm formation and lead adsorption to biofilm components. *Environ. Sci. Technol.* **1996**, *30*, 202–207. [CrossRef]

14. Liang, Z.W.; Li, W.H.; Yang, S.Y.; Du, P. Extraction and structural characteristics of extracellular polymeric substances (EPS), pellets in autotrophic nitrifying biofilm and activated sludge. *Chemosphere* **2010**, *81*, 626–632. [CrossRef] [PubMed]

15. Watnick, P.; Kolter, R. Biofilm, city of microbes. *J. Bacteriol.* **2000**, *182*, 2657–2679. [CrossRef]

16. Flemming, H.C.; Neu, T.R.; Wozniak, D.J. The EPS matrix: The "house of biofilm cells". *J. Bacteriol.* **2007**, *189*, 7945–7947. [CrossRef] [PubMed]

17. Zhang, P. Compositions and Surface Characteristic of Microbial Extracellular Polymeric Substances in Wastewater Treatment. Ph.D. Thesis, Chongqing University, Chongqing, China, 2016.

18. Sun, H.W.; Chen, C.Z.; Wu, C.F.; Zhao, H.N.; Yu, X.; Fang, X.H. Influence of operating modes for the alternating anoxic/oxic process on biological nitrogen removal and extracellular polymeric substances of activated sludge. *Environ. Sci.* **2018**, *39*, 256–262.

19. Ye, F.X.; Ye, Y.F.; Li, Y. Effect of C/N ratio on extracellular polymeric substances (EPS) and physicochemical properties of activated sludge flocs. *J. Hazard. Mater.* **2011**, *188*, 37–43. [CrossRef] [PubMed]

20. Wang, L.L.; Zhao, L.; Tan, X.; Yan, B. Influence of different carbon source and ratio of carbon and nitrogen for water denitrification. *Environ. Protec. Sci.* **2004**, *30*, 15–18.

21. Wei, F.S. *The Standard Methods for the Examination of Water and Wastewater*, 4th ed.; China Environmental Science Press: Beijing, China, 2002.

22. Li, Z.G.; Luo, Y.M.; Teng, Y. *The Research Methods of Soil and Environmental Microorganism*; Science Press: Beijing, China, 2008.

23. Gu, C.C.; Guo, Y.H.; Sun, X.C.; Liu, Z.H.; Xue, G.; Jia, H.Z.; Gao, P. Comparative study on extracellular polymeric substance extraction method for biofilms in biological aerated filter. *J. Donghua Univ. (Nat. Sci.)* **2017**, *43*, 720–726.

24. Herbert, D.; Philipps, P.J.; Strange, R.E. Carbonhydrate analysis. *Methods Enzymol.* **1971**, *5B*, 265–277.

25. Wang, F.; Luo, L.X.; Liu, H.; Li, X.N.; Lu, L.B.; Yang, X.M. Study on the contaminant removal efficiency of rapid infiltration pond in constructed rapid infiltration system. *Environ. Pollut. Control* **2013**, *35*, 58–63.

26. Zhang, J.; Huang, X.; Wei, J.; Hu, H.Y.; Shi, H.C. Nitrogen and phosphorus removal mechanism in subsurface wastewater infiltration system. *China Environ. Sci.* **2002**, *22*, 438–441.

27. Hou, L.; Xia, L.; Ma, T.; Zhang, Y.Q.; Zhou, Y.Y.; He, X.G. Achieving short-cut nitrification and denitrification in modified intermittently aerated constructed wetland. *Bioresour. Technol.* **2017**, *232*, 10–17. [CrossRef] [PubMed]

28. Fan, J.; Wang, W.; Zhang, B.; Guo, Y.; Ngo, H.H.; Guo, W.; Zhang, J.; Wu, H. Nitrogen removal in intermittently aerated vertical flow constructed wetlands: Impact of influent COD/N ratios. *Bioresour. Technol.* **2013**, *143*, 461–466. [CrossRef] [PubMed]

29. Zhang, Z.L. Selection of External Carbon Sources for Denitrification. Master's Thesis, Harbin Institute of Technology, Harbin, China, 2009.

30. Yan, N.; Jin, X.B.; Zang, J.Q. A comparison between the processes of denitrification with glucose and methanol as carbon sourse. *J. Shanghai Teach. Univ. Nat. Sci.* **2002**, *31*, 41–44.

31. Xie, Y.X. Analysis of Running State and Distribution of the Microbial Strain in the Three Stages Constructed Rapid Infiltration System. Master's Thesis, University of Geosciences, Beijing, China, 2010.

32. Wang, S.Y.; Qian, F.Y.; Wang, J.F.; Shen, Y.L. Impact of Biodegradable Organic Matter on the Functional Microbe Activities in Partial Nitrification Granules. *Environ. Sci.* **2017**, *38*, 269–275.

33. Lv, J.P.; Wang, Y.Q.; Zhong, C.; Li, Y.C.; Hao, W.; Zhu, J.R. The effect of quorum sensing and extracellular proteins on the microbial attachment of aerobic granular activated sludge. *Bioresour. Technol.* **2014**, *152*, 53–58. [CrossRef] [PubMed]

34. Danese, P.N.; Pratt, L.A.; Kolter, R. Exopolysaccharide production is required for development of Escherichia coli K-12 biofilm architecture. *J. Bacteriol.* **2000**, *182*, 3593–3596. [CrossRef] [PubMed]

International Journal of
*Environmental Research
and Public Health*

Article

Adsorption of Trace Estrogens in Ultrapure and Wastewater Treatment Plant Effluent by Magnetic Graphene Oxide

Xianze Wang [1,2], Zhongmou Liu [3], Zhian Ying [3], Mingxin Huo [1,2,3] and Wu Yang [1,2,3,*]

[1] Science and Technology Innovation Center for Municipal Wastewater Treatment and Water Quality Protection, Jilin Province, Northeast Normal University, Changchun 130117, China; wangxz940@nenu.edu.cn (X.W.); huomx097@gmail.com (M.H.)

[2] Engineering Lab for Water Pollution Control and Resources Recovery, Jilin Province, Northeast Normal University, Changchun 130117, China

[3] School of Environment, Northeast Normal University, Changchun 130117, China; liuzm338@nenu.edu.cn (Z.L.); yingza451@nenu.edu.cn (Z.Y.)

* Correspondence: yangw104@gmail.com; Tel.: +86-137-5647-8864

Abstract: In the current study, graphene oxide, Fe^{3+}, and Fe^{2+} were used for the synthesis of magnetic graphene oxide (MGO) by an in situ chemical coprecipitation method. Scanning electron microscopy, transmission electron microscopy, Fourier transform infrared spectroscopy, and X-ray diffraction were used to characterize the well-prepared MGO. The prepared MGO was used as an adsorbent to remove five typical estrogens (estrone (E1), 17β-estradiol (E2), 17α-ethinylestradiol (17α-E2), estriol (E3), and synthetic estrogen (EE2)) at the ppb level from spiked ultrapure water and wastewater treatment plant effluent. The results indicated that the MGO can efficiently remove estrogens from both spiked ultrapure water and wastewater treatment plant effluent in 30 min at wide pH ranges from 3 to 11. The temperature could significantly affect removal performance. A removal efficiency of more than 90% was obtained at 35 °C in just 5 min, but at least 60 min was needed to get the same removal efficiency at 5 °C. In addition, an average of almost 80% of the estrogens can still be removed after 5 cycles of MGO regeneration but less than 40% can be reached after 10 cycles. These results indicate that MGO has potential for practical applications to remove lower levels of estrogens from real water matrixes and merits further evaluation.

Keywords: estrogens; magnetic graphene oxide; adsorption; wastewater treatment plant effluent

---

1. Introduction

China has the largest population in the world and has been facing a serious water crisis and water pollution problem in the past decades. According to the annual statistical yearbook for urban construction, 4.8×10^{10} m^3 of wastewater was discharged in 2016 and more than 93% of the wastewater was treated by wastewater treatment plants. 77.6% of the treated wastewater from wastewater treatment plants (WWTPs) was directly discharged into rivers and the other 22.4% was used to yield reclaimed water. The rate of reclaimed water use was 44.9% (4.5×10^9 m^3) [1]. These figures are increasing as a consequence of government policies [2,3]. The safety of WWTP effluent has attracted scrutiny, even after it has met the criteria for discharge or reuse [4–6]. However, there are still persistent residual chemicals, such as endocrine-disrupting compounds (EDCs), pesticides, and pharmaceutical and personal care products [5,7,8], which are characterized by low concentrations (ng/L), high variety, and complicated physicochemical properties [9].

Endocrine-disrupting activity caused by natural steroid estrogens, including estrone (E1), 17β-estradiol (E2), 17α-ethinylestradiol (17α-E2), estriol (E3), and synthetic estrogen (EE2), has been widely detected in the effluent of WWTPs and in their receiving water bodies, as well as in reclaimed water using the effluent as a water source [4,10–12]. The levels varied from concentrations of pg/L to μg/L. Due to their potential health risk, adsorption, biodegradation, advanced oxidation process, and photodegradation are frequently used for removing estrogens. Adsorption is considered as the most efficient and economical method.

Abundant oxygen-containing functional groups (such as hydroxyl, carboxyl, and epoxy groups) on the surface of graphene oxide (GO) make it extremely hydrophilic and gives it the capability to be used in aqueous environments as a superior sorbent for removing various pollutants, such as metal ions [13,14], tetracycline antibiotics [15], microcystin [16], and polycyclic aromatic hydrocarbons [17]. However, due to its high dispersibility, it is difficult to separate from the aqueous solution, which may lead to secondary pollution [18,19]. In recent years, magnetic materials have been widely used in the water and wastewater treatment for removing of various pollutants since they can be efficiently separated by magnetic separation technology [20–25]. Magnetic graphene oxide (MGO) combines the easy separation of magnetic particles and the high adsorption capacity of GO. In recent years, MGO was reported as a superior sorbent for the removal of antibiotics [26], dyes [27], and metal ions [28,29] with concentrations up to hundreds of mg/L. However, to our knowledge, there is an absence of information focused on the removal of a low level of estrogens with MGO.

The objective of this study was to investigate the adsorption efficiency of MGO for five typical estrogens in both ultrapure and reclaimed water at a ppb level. MGO was synthesized using a chemical coprecipitation method. Estrogens with different concentrations were mixed as samples for investigating MGO adsorption efficiency. Various factors influence the adsorption process, such as pH and temperature, and these were studied. The current study will help to address some of the knowledge gaps about the removal of estrogenic hormones in reclaimed water.

2. Materials and Methods

2.1. Materials

Analytical standards of E1, E2, 17α-E2, EE2, E3, and acetonitrile were purchased from Aladdin (Shanghai, China). Flake graphite (99.95%, 325 meshes) was provided by Jinrilai Co., Ltd. (Qingdao, China). $NH_4Fe(SO_4)_2 \cdot 12H_2O$ and $FeCl_2 \cdot 4H_2O$ were purchased from Sinopharm Chemical Reagent Co., Ltd. (Shanghai, China). CH_2Cl_2, C_6H_{14}, Sulfuric acid (H_2SO_4, 98%), $KMnO_4$, H_2O_2 (30%), ammonia solution (25%), and hydrochloric acid were produced by Beijing Chemicals Corporation (Beijing, China).

2.2. Preparation and Characterization of MGO

The synthesis of MGO was fulfilled by an in situ chemical coprecipitation of Fe^{3+}, Fe^{2+}, and GO. GO was synthesized by a pressurized oxidation method described by Bao et al. [30]. Firstly, 100 mL of GO (5 mg/mL) was sonicated for 30 min to form a stable suspension. Then, 8.33 g $NH_4Fe(SO_4)_2 \cdot 12H_2O$ and 1.7 g $FeCl_2 \cdot 4H_2O$ were dissolved in 100 mL of ultrapure water under nitrogen protection, followed by a rapid addition of 10 mL of 25% ammonia. Then, the GO suspension was injected dropwise into the solution while being strongly stirred and the solution was keep at 85 °C for 1 h. The product was collected with a magnet and washed with ethanol and ultrapure water three times, then dried at 65 °C for 12 h.

The prepared MGO was characterized by scanning electron microscopy (SEM) (XL30-ESEM, FEI, Hillsboro, OR, USA), transmission electron microscopy (TEM) (TECNAI F20, FEI, Hillsboro, OR, USA), Fourier transform infrared spectroscopy (FTIR) (Nicolet 6700, Thermo Fisher Scientific, Waltham, MA, USA), and X-ray diffraction (XRD) (D8 ADVANCE, Bruker, Karlsruhe, Germany). Additionally, the zeta potential of MGO was also measured (Nano ZS 90, Malvern, UK).

2.3. Adsorption Experiments in Estrogen-Spiked Ultrapure Water

The initial concentrations of five estrogens in the 100 mL solutions were 200 µg/L. The experiment was carried out in a 25 °C thermostatic room. Duplicate 1 mL water samples were taken at regular time intervals of 1, 2, 5, 10, 15, and 30 min for HPLC-MS/MS analysis. All the flasks were kept in the dark and agitated on a shaker at 200 rpm, a speed at which adsorption onto the glassware and stripping can be neglected. The estrogen reduction in the blank tests spiked with 200 µg/L ranged from $1.2 \pm 1.1\%$ to $3.3 \pm 1.8\%$ in 2 h. Therefore, estrogen reduction due to glassware adsorption, soluble organic matter adsorption, and photodegradation were considered minimal. The MGO concentrations were 0.1 g/L. The effect of initial pH was testing by mixing 0.1 g/L MGO with 200 µg/L estrogen mixture solutions at various pH values (pH of 3–11) for 30 min. The pH solution was adjusted with 0.1 mol/L HCl or NaOH solutions.

2.4. Adsorption Experiments in WWTPs Effluent

WWTP effluent samples were taken from three local WWTPs and filtered through a 0.45 µm filter. The filtrate was collected for further use. For adsorption experiments, samples were adjusted to a pH of 5. MGO dosage was at 0.1 g/L and the contact time set as 30 min.

Solid phase extraction (SPE)-HPLC-MS/MS was used to analyze estrogen concentration before and after adsorption. SPE was carried out using the Aqua Trace 899 (GL Science, Kyoto, Japan) automated solid-phase extraction instrument. A 1000 mL sample was consecutively extracted by a C_{18} column (6 mL, 500 mg). The C_{18} column was preconditioned consecutively with 2 mL Milli-Q water, 2 mL acetonitrile, and 2 mL methylene dichloride. The filtered sample was loaded onto the SPE column at a flow rate of 10 mL/min. Then, the column was rinsed with 5 mL Milli-Q water and 5 mL hexyl hydride, followed by column elution with 4 mL hexyl hydride at a flow rate of 3 mL/min and desiccation with nitrogen gas for 1 h. Finally, 3 mL acetonitrile was used to redissolve estrogens using a vortex oscillation system for 5 s. The organic eluent was eventually concentrated down to 0.5 mL under a high purity nitrogen stream in a 40 °C water bath and, within a week, Milli-Q water was added to make 1 mL for HPLC-MS/MS analysis.

2.5. HPLC-MS/MS Analysis for Estrogens

In this study, a Shimadzu LC-20AD HPLC system (Shimazu, Kyoto, Japan) consisting of an Eclipse Plus C_{18} column (50×2.1 mm, 3.5 µm particle size) (Agilent, Santa Clara, CA, USA) was used for estrogen separation. The mobile phase, with a flow rate of 0.3 mL/min, was composed of acetonitrile-water (45:55, v/v). The sample injection volume was 20 µL.

Analyses were performed using Qtrap 5500 mass spectrometry (Applied Biosystems Sciex, Toronto, ON, Canada) with a Turbo Ion Spray source. Data acquisition was performed in the negative ion mode, and the optimized parameters were as follows: a source temperature of 120 °C, a desolvation temperature of 380 °C, a capillary voltage of 3.2 kV, a desolvation gas flow of 700 L/h, and a cone gas flow of 80 L/h. Argon (99.999%) was used as the collision gas. Quantitative analysis was performed in the multiple reaction monitoring (MRM) mode. The optimal conditions for MS/MS analysis are listed in the Table 1.

The overall method recoveries for the target analytes were between 82.6% and 113.2%, with a relative standard deviation (RSD) less than 13.4%. The limits of quantification (LOQ) of the target analytes were between 2 and 8 ng/L in the pure water and WWTP effluent.

Table 1. Main mass fragments of the target compounds.

Compound	Precursor Ion	Product Ion	Declustering Potentials (V)	Collision Energy (eV)
E1	269.5	145.1	−70	−52
E2	270.8	145.1	−70	−58
17α-E2	270.8	145.1	−70	−58
EE2	294.9	145.1	−60	−53
E3	286.6	145.1	−70	−58

3. Results and Discussion

3.1. Characterization of Adsorbent

3.1.1. SEM

The morphological structure of MGO was observed using SEM and TEM (Figure 1). Compared with the smooth surface and wrinkles of GO (Figure 1a), it can be seen that Fe_3O_4 nanoparticles were successfully coated on the surface of GO to form MGO (Figure 1b). Figure 1c,d show the TEM images of MGO, which suggests that Fe_3O_4 nanoparticles with a diameter of about 10–20 nm were well-dispersed on GO sheets. The composition of MGO was verified by energy dispersive X-ray spectroscopy (EDX), as shown in Figure 2. The spectrum showed peaks corresponding to C, O, and Fe. The mass and atom ratio of Fe in MGO was 79.18% and 51.07%, respectively, which also suggested that Fe_3O_4 nanoparticles were well-dispersed on GO sheets.

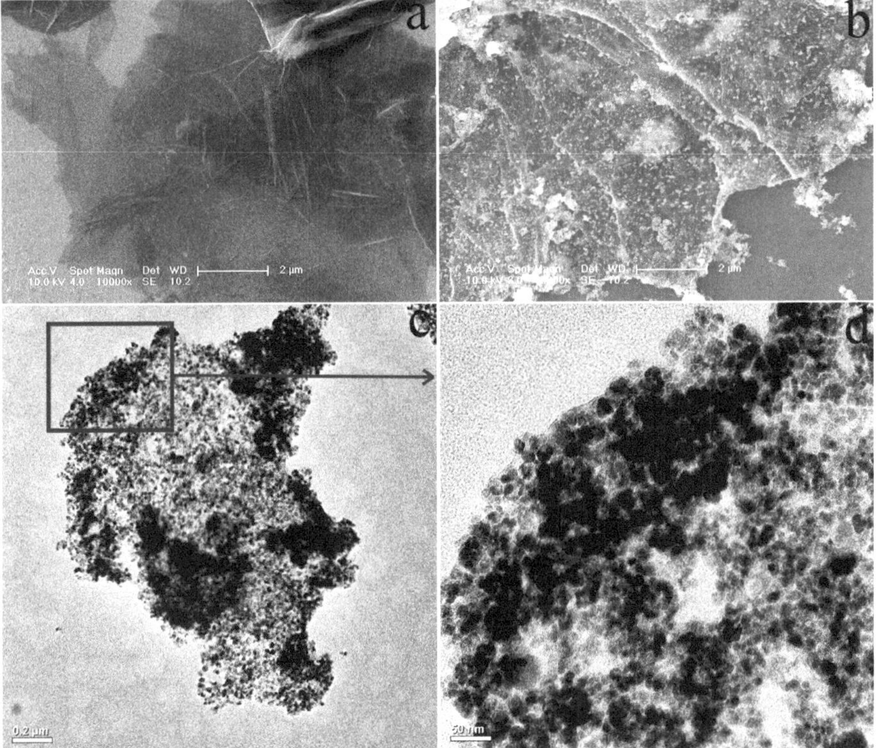

Figure 1. SEM images of (**a**) graphene oxide (GO) and (**b**) magnetic GO (MGO); (**c**,**d**) TEM images of MGO at different resolutions.

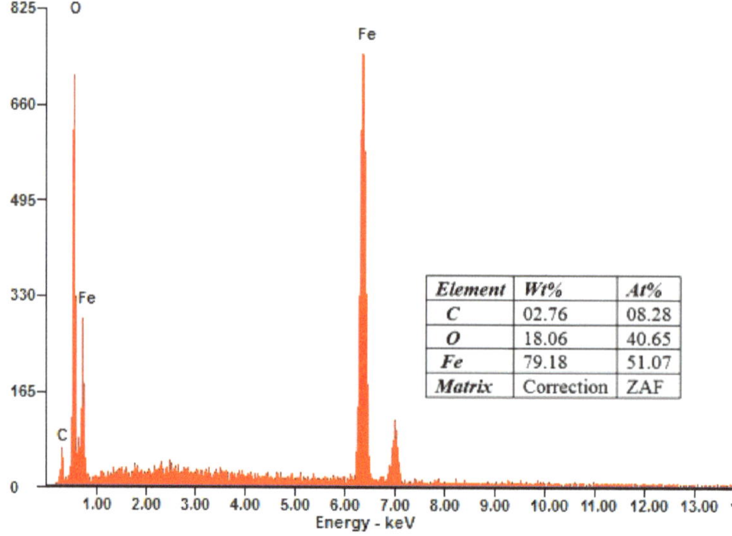

Element	Wt%	At%
C	02.76	08.28
O	18.06	40.65
Fe	79.18	51.07
Matrix	Correction	ZAF

Figure 2. EDX spectrum of MGO.

3.1.2. FTIR Spectroscopy

FTIR is a valuable technique for understanding the mechanism of adsorption. The FTIR spectra for GO and MGO are shown in Figure 3. For GO, the peak at 1722 cm^{-1} corresponds to the stretching band of C=O in carboxylic acid or carbonyl moieties. The intense peaks at 3431 cm^{-1} are attributed to the stretching of the O–H band. The peak at 1613 cm^{-1} (aromatic C=C) can be assigned to the skeletal vibrations of unoxidized graphitic domains. For the FTIR spectrum of MGO, two new vibrational peaks appear at around 1122 and 1182 cm^{-1}. These can be assigned to the formation of either a monodentate complex or a bidentate complex between the carboxyl group and Fe. The appearance of new peaks suggests that Fe_3O_4 are covalently bonded to the surface of GO nanosheets. Moreover, the peaks at 561 cm^{-1} can be ascribed to the lattice absorption of Fe_3O_4, indicating that Fe_3O_4 nanoparticles were loaded onto the surface of GO successfully and the MGO was synthesized successfully.

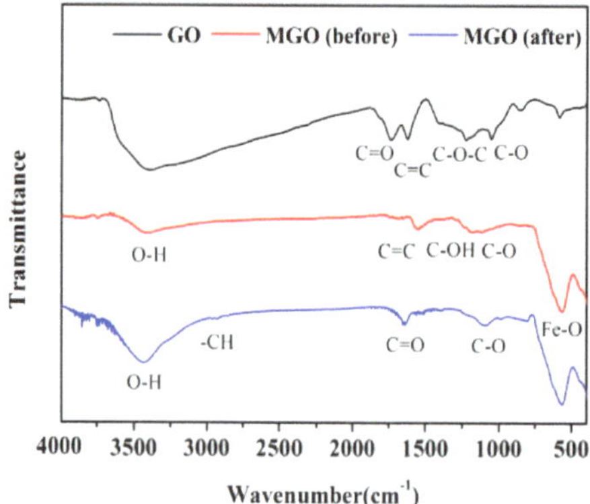

Figure 3. FTIR spectra of GO and MGO before and after the adsorption of estrogens.

3.1.3. XRD

The XRD patterns of the GO and MGO composite are presented in Figure 4. The strongest peaks at $2\theta = 11.4°$ (001) can be appointed to the reflection of the GO, and the peaks at $2\theta = 30.3°$ (220), 35.7° (311), 43.5° (400), 53.9° (422), 57.5° (511), and 63.0° (440) are consistent with the standard XRD data of Fe_3O_4. After modification with Fe_3O_4, the iron oxides cover up the weak carbon peaks when there is a disappearance of GO at the diffraction peak ($2\theta = 11.4°$). In addition, the presence of magnetite reduces the aggregation of graphene sheets, which results in more monolayer graphene. This, in turn, leads to weaker peaks from carbon being observed [31].

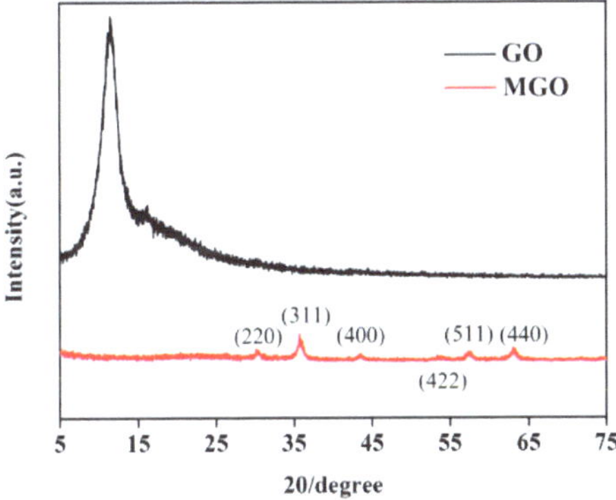

Figure 4. XRD pattern of GO and MGO.

3.1.4. Magnetization

Figure 5 represents the "S"-type hysteresis loops of MGO and their magnetization saturation (Ms) at 278 and 300 K, respectively. The Ms of MGO is about 1.93 and 1.97 emu/g at 278 and 300 K, respectively. This evidence demonstrates that the MGO made by coprecipitation synthesis was given stronger magnetization. Though the Ms is not very high, it is enough for magnetic separation, as can be seen from Figure 4. The figure also showed that the effect of temperature on the magnetic of MGO is not significant since the hysteresis loops of MGO at 278 and 300 K tend to coincide. This phenomenon suggests that the variation of temperature does not influence the magnetization of MGO in the subsequent adsorption experiment.

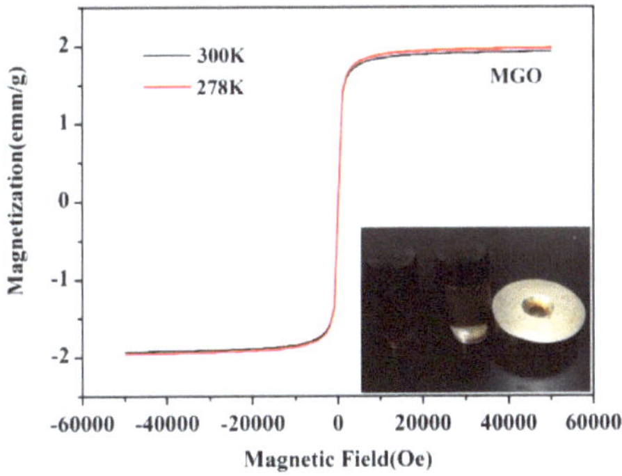

Figure 5. Magnetization curve of MGO. The inset shows magnetic separation of adsorbents from the solution.

3.2. Effect of Initial Solution pH

pH is an essential environmental factor and the most important external factor in influencing the surface charge on the adsorbents and the potential ionization of chemicals. The effect is determined by conducting experiments at initial pH values ranging from 3 to 11. MGO and estrogen are set at 0.1 and 200 µg/L, respectively, and the solutions are unbuffered.

Figure 6 shows the effect of pH on the MGO adsorption of estrogens. Overall, the sorption decreased with the increasing pH values. E3 is the most sensitive to pH conditions compared to the other estrogens. The highest sorption capacity (86.7%) of E3 was observed at pH 3, while only 7.3% was observed at pH 11. However, aqueous phase pH showed a negligible influence on E1 and EE2 sorption onto MGO. More than 90% of the sorption capacity was obtained at pH 11.

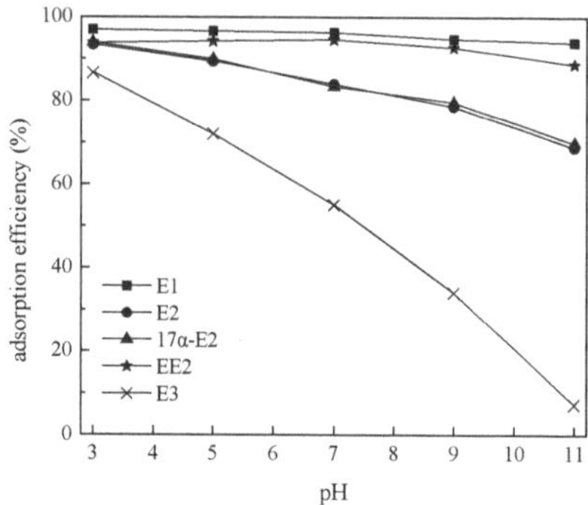

Figure 6. Effect of pH on MGO's adsorption of estrogens (MGO = 0.05 g, T = 308 K, *t* = 30 min, and estrogens' concentrations = 200 µg/L).

Lower estrogen sorption efficiency was observed at higher pH levels, especially basic pH conditions. This might be attributed to the increase in hydroxyl ions, leading to the formation of aqua complexes which retard the sorption phenomena [32]. On the other hand, the benzene ring and phenol hydroxyl of estrogen's molecular structure were more reactive at acidic conditions and more easily accepting of electrons. This results in the benzene ring rupture further oxidizing to form carboxylic acid functional groups [33], which have a greater affinity to MGO. As shown in Figure 7, the zeta potentials of MGO were negative when pH > 5.4, and the negative charge was enhanced with increasing pH. The repulsive electrostatic interaction established between the negative surface charge of MGO and the estrogens might lead to the lower adsorptive capacity the higher pH ranges. On the other hand, when pH < 5.4, the electrostatic attraction will play a major role in the adsorption of estrogens to positively charged surfaces of MGO.

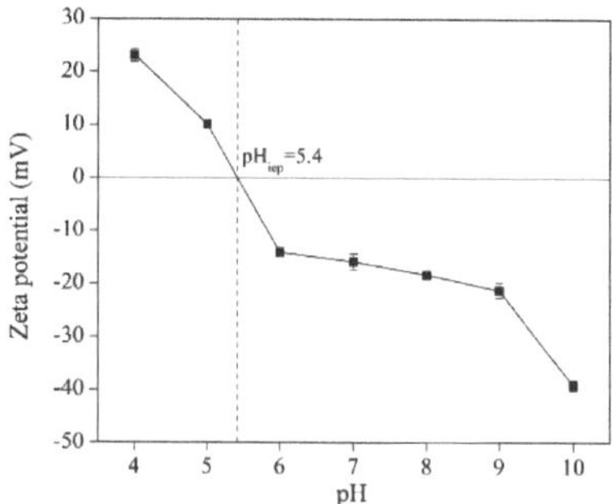

Figure 7. Zeta potential of MGO at different pH values.

The sorption efficiency of estrogens onto MGO followed the order of E1 $\approx$ EE2 > E2 $\approx$ 17α-E2 > E3 at all the pH ranges, especially at alkaline conditions. This may be attributed to the quantity of the hydroxyl group contacted in estrogen's molecular structure. There are three hydroxyl groups in the E3 molecule, two in E2 and 17α-E2, and only one in E1 and EE2. More hydroxyl groups lead to a higher negative surface charge of the estrogen molecule, especially in alkaline conditions. Thus, stronger repulsive electrostatic interactions occurred.

3.3. Adsorption Kinetics

Pseudo-first-order and pseudo-second-order models (expressed as Equations (1) and (2), respectively) were employed to describe the kinetics of adsorption:

$$\ln(q_e - q_t) = \ln q_e - k_1 t \tag{1}$$

$$t/q_t = \frac{1}{k_2 q_e{}^2} + t/q_2 \tag{2}$$

where q_e and q_t are adsorption capacity (mg/g) at equilibrium and at time t (min), respectively, and k_1 and k_2 are the pseudo-first-order constant (min^{-1}) and the pseudo-second-order rate constant (g/(μg·min)), respectively.

The kinetic parameters for the two models were determined and listed in Table 2. The results show that the correlation coefficient (r) for the pseudo-first-order model is relatively low, and there is a large difference between the calculated adsorption capacity (q_e (cal)) and the experimental value (q_{exp}), especially for E1, EE2, and E3. The plots of pseudo-first-order and pseudo-second-order kinetic models are shown in Figure 8. As shown in the figure, experimental data had a much better fit with the pseudo-second-order kinetic model. The coefficient factor for this model is very high ($r > 0.97$), and the calculated adsorption capacity agrees well with experimental values.

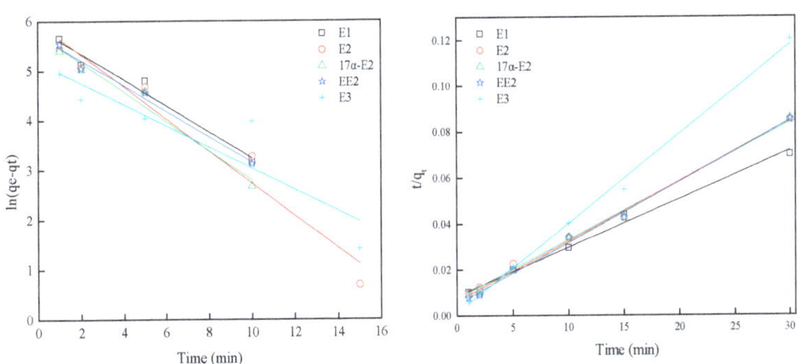

Figure 8. Plots of pseudo-first-order and pseudo-second-order kinetic models.

Table 2. Kinetic parameters for pseudo-first-order and pseudo-second-order models.

Kinetic Model	Estrogen	k_1 (min^{-1})	k_2 g/(μg·min)	q_e (cal) (μg/g)	q_{exp} (μg/g)	r
	E1	0.436		633.9	387.6	0.94
	E2	0.324		394.6	333.6	0.97
pseudo-first-order	17α-E2	0.295		317.4	333.6	0.98
	EE2	0.470		625.7	378.4	0.93
	E3	0.213		176.4	294.8	0.88
	E1		0.553	400.0	387.6	0.97
	E2		1.001	386.1	333.6	0.99
pseudo-second-order	17α-E2		1.112	375.9	333.6	0.99
	EE2		0.902	442.5	378.4	0.99
	E3		3.877	289.8	294.8	0.97

3.4. Effect of Temperature

Temperature is another essential and important external factor which can significantly influence adsorption ability. The effect is determined by conducting experiments at 5, 15, 25, and 35 °C. The solution's initial pH set at 5. MGO and estrogen are 0.1 g/L and 200 µg/L, respectively.

Figure 9 shows the effect of temperature on the MGO adsorption of estrogens. Overall, temperature significantly influences adsorption efficiency. The adsorption efficiency varied from 41.3% to 95.2% for E1, 51.4% to 79.4% for E2, 52.4% to 98.2% for 17α-E2, 60.1% to 93.2% for EE2, and 88.4% to 73.3% for E3. Except for E3, the sorption efficiency was increased with the increasing temperature, which was consistent with previous works [34–36]. However, E3 expresses a different trend related to temperature. Lower temperatures seem favorable for the sorption of E3, which is inconsistent with previous work where activated carbon was used as an adsorbent for removing E3 from aqueous solutions [32].

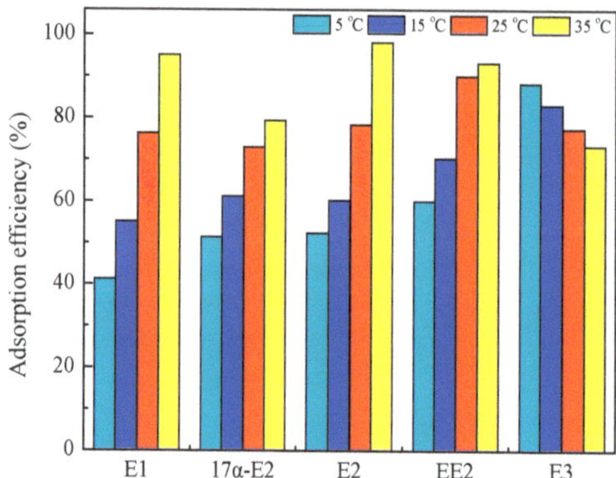

Figure 9. Effect of temperature on the MGO adsorption of estrogens (MGO = 0.05 g, pH = 5, *t* = 30 min, and estrogens' concentrations = 200 µg/L).

3.5. Regeneration and Reusability

To study the reusability of MGO, the particles were separated after the adsorption process by using a magnet. Estrogens were desorbed by stirring in 10 mL ethanol for 30 min, and the MGO were dried naturally at room temperature. The recycled adsorbents were used for the next adsorption runs. The results of recycling the experiment for 10 cycles are shown in Figure 10. It is observed that the removal efficiency of all the five estrogens decreased slowly with the increasing regeneration cycles in the first five cycles and dramatically decreased after five cycles. With E1, for example, it is observed that about 96% of the E1 was removed after the first cycle. More than 85% removal efficiency was observed after the 5th cycle, but only 42% was observed after the 10th cycle. These results demonstrated that MGO could be regenerated effectively by ethanol and has the potential for reusability in about five cycles.

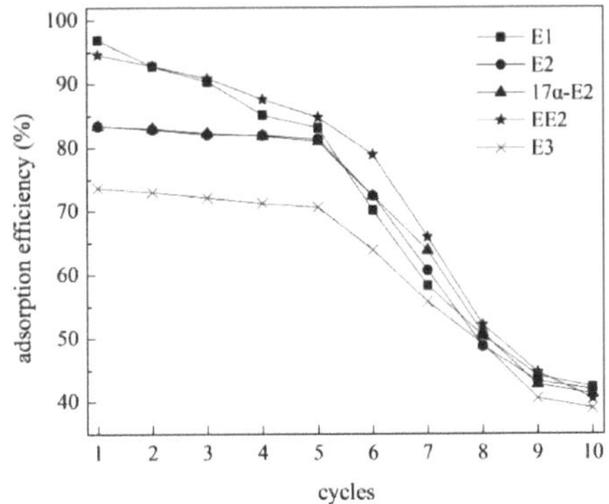

Figure 10. The efficiency of regenerated and reused MGO to adsorption of estrogens.

3.6. Application to WWTP Effluent

WWTPs are considered a significant source of estrogen into the receiving environment [11]. Thus, three WWTP effluent samples were collected from three local WWTPs to investigate the removal of estrogens by MGO. The samples were adjusted to a pH of 5. MGO dosage was 0.1 g/L and the contact time set as 30 min. The concentration of estrogens before and after adsorption are shown in Table 2.

As shown in Table 3, though wastewater was treated by traditional bioprocess, dozens of ng/L estrogens were detected in local WWTPs effluent. Overall, MGO exhibits an effective ability for adsorption removal of estrogens from WWTP effluent. After adsorption, E2 was completely removed from all three samples. More than 90% of E1, 17α-E2, and EE2 were also removed. Just like in pure water, E3 was found to be relatively recalcitrant to MGO in WWTP effluent. More than 4.8 ng/L of E3 was detected in all the three samples after adsorption, which means more than 30% of the E3 remained in samples. The above experiments demonstrated that though the WWTP effluent matrix is much more complex than that of pure water and the estrogen concentration is of low magnitude, MGO can effectively sorb the trace estrogens. This indicates that the MGO has potential for practical applications to remove lower levels of estrogens from real water matrixes.

Table 3. Removal of estrogens in wastewater treatment plant (WWTP) effluent by using MGO.

Estrogens	Samples	Concentration (ng/L)	
		Before Adsorption	After Adsorption
E1	A	31	1.5
	B	56	n.d.
	C	38	n.d.
17α-E2	A	27	n.d.
	B	16	0.8
	C	21	n.d.
E2	A	39	n.d.
	B	17	n.d.
	C	46	n.d.
EE2	A	25	2.3
	B	13	n.d.
	C	27	1.1
E3	A	16	5.8
	B	20	7.9
	C	14	4.8

4. Conclusions

In summary, MGO was successfully synthesized using the in situ chemical coprecipitation method. MGO can efficiently remove five typical estrogens from both ultrapure water and WWTP effluent at ppb levels. Sorption was decreased with increasing pH values, and E3 is the most sensitive to pH conditions compared to the other estrogens. Acidic and neutral conditions are favorable for estrogen adsorption onto MGO. Experimental data fit better with the pseudo-second-order kinetic model. Except for E3, sorption efficiency was increased with increasing temperatures. E3 expresses the opposite trend. MGO can be easily separated by using a powerful magnet and regenerated using 10 mL of ethanol. MGO exhibits an effective ability to adsorb the removal of estrogens from WWTP effluent. More than 90% of E2, E1, 17α-E2, and EE2 can be removed.

Author Contributions: X.W. and W.Y. conceived and designed the experiments; X.W., Z.L., and Z.Y. performed the experiments; X.W. and M.H. analyzed the data; Z.L. and L.Z. contributed analysis tools; W.Y. wrote the paper.

Funding: This work was funded by the National Natural Science Foundation of China (51508077, 51508079 and 51708094) and The Science and Technology Development Project of Jilin Province (20160520082JH and 20180520167JH).

Acknowledgments: We thank Min Ma, Haiyang Wang, Fengmin Yang for supporting us in the experiments.

Conflicts of Interest: The authors declare no conflict of interest.

References

1. China Urban Construction Statistics Yearbook. 2016. Available online: http://www.mohurd.gov.cn/xytj/tjzljsxytjgb/jstjnj/index.html (accessed on 5 January 2018).

2. Chen, W.; Lu, S.; Jiao, W.; Wang, M.; Chang, A.C. Reclaimed water: A safe irrigation water source? *Environ. Dev.* **2013**, *8*, 74–83. [CrossRef]

3. Lyu, S.; Chen, W.; Zhang, W.; Fan, Y.; Jiao, W. Wastewater reclamation and reuse in China: Opportunities and challenges. *J. Environ. Sci.* **2016**, *39*, 86–96. [CrossRef] [PubMed]

4. Wang, Y.; Hu, W.; Cao, Z.; Fu, X.; Zhu, T. Occurrence of endocrine-disrupting compounds in reclaimed water from Tianjin, China. *Anal. Bioanal. Chem.* **2005**, *383*, 857–863. [CrossRef] [PubMed]

5. Ma, X.Y.; Li, Q.; Wang, X.C.; Wang, Y.; Wang, D.; Ngo, H.H. Micropollutants removal and health risk reduction in a water reclamation and ecological reuse system. *Water Res.* **2018**, *138*, 272–281. [CrossRef] [PubMed]

6. Ma, W.; Sun, J.; Li, Y.; Lun, X.; Shan, D.; Nie, C.; Liu, M. 17α-Ethynylestradiol biodegradation in different river-based groundwater recharge modes with reclaimed water and degradation-associated community structure of bacteria and archaea. *J. Environ. Sci.* **2018**, *64*, 51–61. [CrossRef] [PubMed]

7. Estévez, E.; del Carmen Cabrera, M.; Molina-Díaz, A.; Robles-Molina, J.; del Pino Palacios-Díaz, M. Screening of emerging contaminants and priority substances (2008/105/EC) in reclaimed water for irrigation and groundwater in a volcanic aquifer (Gran Canaria, Canary Islands, Spain). *Sci. Total Environ.* **2012**, *433*, 538–546. [CrossRef] [PubMed]

8. Gavrilescu, M.; Demnerová, K.; Aamand, J.; Agathos, S.; Fava, F. Emerging pollutants in the environment: Present and future challenges in biomonitoring, ecological risks and bioremediation. *New Biotechnol.* **2015**, *32*, 147–156. [CrossRef] [PubMed]

9. Li, Z.; Xiang, X.; Li, M.; Ma, Y.; Wang, J.; Liu, X. Occurrence and risk assessment of pharmaceuticals and personal care products and endocrine disrupting chemicals in reclaimed water and receiving groundwater in China. *Ecotoxicol. Environ. Saf.* **2015**, *119*, 74–80. [CrossRef] [PubMed]

10. Plahuta, M.; Tišler, T.; Toman, M.J.; Pintar, A. Toxic and endocrine disrupting effects of wastewater treatment plant influents and effluents on a freshwater isopod *Asellus aquaticus* (Isopoda, Crustacea). *Chemosphere* **2017**, *174*, 342–353. [CrossRef] [PubMed]

11. Ting, Y.F.; Praveena, S.M. Sources, mechanisms, and fate of steroid estrogens in wastewater treatment plants: A mini review. *Environ. Monit. Assess.* **2017**, *189*, 178. [CrossRef] [PubMed]

12. Servos, M.R.; Bennie, D.T.; Burnison, B.K.; Jurkovic, A.; McInnis, R.; Neheli, T.; Schnell, A.; Seto, P.; Smyth, S.A.; Ternes, T.A. Distribution of estrogens, 17β-estradiol and estrone, in Canadian municipal wastewater treatment plants. *Sci. Total Environ.* **2005**, *336*, 155–170. [CrossRef] [PubMed]

13. Zhao, G.; Li, J.; Ren, X.; Chen, C.; Wang, X. Few-layered graphene oxide nanosheets as superior sorbents for heavy metal ion pollution management. *Environ. Sci. Technol.* **2011**, *45*, 10454–10462. [CrossRef] [PubMed]

14. Sitko, R.; Turek, E.; Zawisza, B.; Malicka, E.; Talik, E.; Heimann, J.; Gagor, A.; Feist, B.; Wrzalik, R. Adsorption of divalent metal ions from aqueous solutions using graphene oxide. *Dalton Trans.* **2013**, *42*, 5682–5689. [CrossRef] [PubMed]

15. Gao, Y.; Li, Y.; Zhang, L.; Huang, H.; Hu, J.; Shah, S.M.; Su, X. Adsorption and removal of tetracycline antibiotics from aqueous solution by graphene oxide. *J. Colloid Interface Sci.* **2012**, *368*, 540–546. [CrossRef] [PubMed]

16. Pavagadhi, S.; Tang, A.L.; Sathishkumar, M.; Loh, K.P.; Balasubramanian, R. Removal of microcystin-LR and microcystin-RR by graphene oxide: Adsorption and kinetic experiments. *Water Res.* **2013**, *47*, 4621–4629. [CrossRef] [PubMed]

17. Wang, J.; Chen, Z.; Chen, B. Adsorption of polycyclic aromatic hydrocarbons by graphene and graphene oxide nanosheets. *Environ. Sci. Technol.* **2014**, *48*, 4817–4825. [CrossRef] [PubMed]

18. Liu, S.; Zeng, T.H.; Hofmann, M.; Burcombe, E.; Wei, J.; Jiang, R.; Kong, J.; Chen, Y. Antibacterial Activity of Graphite, Graphite Oxide, Graphene Oxide, and Reduced Graphene Oxide: Membrane and Oxidative Stress. *ACS Nano* **2011**, *5*, 6971–6980. [CrossRef] [PubMed]

19. Ahmed, F.; Rodrigues, D.F. Investigation of acute effects of graphene oxide on wastewater microbial community: A case study. *J. Hazard. Mater.* **2013**, *256–257*, 33–39. [CrossRef] [PubMed]

20. Mehta, D.; Mazumdar, S.; Singh, S.K. Magnetic adsorbents for the treatment of water/wastewater—A review. *J. Water Process Eng.* **2015**, *7*, 244–265. [CrossRef]

21. Altıntıg, E.; Altundag, H.; Tuzen, M.; Sarı, A. Effective removal of methylene blue from aqueous solutions using magnetic loaded activated carbon as novel adsorbent. *Chem. Eng. Res. Des.* **2017**, *122*, 151–163. [CrossRef]

22. Sen, T.; Nomura, S.; Nishioka, H.; Sen, T. Synthesis and arsenic adsorption characteristics of a novel magnetic adsorbent. *J. Environ. Conserv. Eng.* **2017**, *46*, 156–162.

23. Zhu, S.; Dong, G.; Yu, Y.; Yang, J.; Yang, W.; Fan, W.; Zhou, D.; Liu, J.; Zhang, L.; Huo, M. Hydrothermal synthesis of a magnetic adsorbent from wasted iron mud for effective removal of heavy metals from smelting wastewater. *Environ. Sci. Pollut. Res.* **2018**, 1–15. [CrossRef] [PubMed]

24. Dhoble, R.M.; Maddigapu, P.R.; Rayalu, S.S.; Bhole, A.; Dhoble, A.S.; Dhoble, S.R. Removal of arsenic(III) from water by magnetic binary oxide particles (MBOP): Experimental studies on fixed bed column. *J. Hazard. Mater.* **2017**, *322*, 469–478. [CrossRef] [PubMed]

25. Drenkova-Tuhtan, A.; Schneider, M.; Franzreb, M.; Meyer, C.; Gellermann, C.; Sextl, G.; Mandel, K.; Steinmetz, H. Pilot-scale removal and recovery of dissolved phosphate from secondary wastewater effluents with reusable ZnFeZr adsorbent@ Fe_3O_4/SiO_2 particles with magnetic harvesting. *Water Res.* **2017**, *109*, 77–87. [CrossRef] [PubMed]

26. Lin, Y.; Xu, S.; Li, J. Fast and highly efficient tetracyclines removal from environmental waters by graphene oxide functionalized magnetic particles. *Chem. Eng. J.* **2013**, *225*, 679–685. [CrossRef]

27. Deng, J.-H.; Zhang, X.-R.; Zeng, G.-M.; Gong, J.-L.; Niu, Q.-Y.; Liang, J. Simultaneous removal of Cd(II) and ionic dyes from aqueous solution using magnetic graphene oxide nanocomposite as an adsorbent. *Chem. Eng. J.* **2013**, *226*, 189–200. [CrossRef]

28. Hu, X.-J.; Liu, Y.-G.; Wang, H.; Chen, A.-W.; Zeng, G.-M.; Liu, S.-M.; Guo, Y.-M.; Hu, X.; Li, T.-T.; Wang, Y.-Q.; et al. Removal of Cu(II) ions from aqueous solution using sulfonated magnetic graphene oxide composite. *Sep. Purif. Technol.* **2013**, *108*, 189–195. [CrossRef]

29. Liu, M.; Chen, C.; Hu, J.; Wu, X.; Wang, X. Synthesis of Magnetite/Graphene Oxide Composite and Application for Cobalt(II) Removal. *J. Phys. Chem. C* **2011**, *115*, 25234–25240. [CrossRef]

30. Bao, C.; Song, L.; Xing, W.; Yuan, B.; Wilkie, C.A.; Huang, J.; Guo, Y.; Hu, Y. Preparation of graphene by pressurized oxidation and multiplex reduction and its polymer nanocomposites by masterbatch-based melt blending. *J. Mater. Chem.* **2012**, *22*, 6088–6096. [CrossRef]

31. Yang, X.; Chen, C.; Li, J.; Zhao, G.; Ren, X.; Wang, X. Graphene oxide-iron oxide and reduced graphene oxide-iron oxide hybrid materials for the removal of organic and inorganic pollutants. *RSC Adv.* **2012**, *2*, 8821. [CrossRef]

32. Kumar, A.K.; Mohan, S.V.; Sarma, P. Sorptive removal of endocrine-disruptive compound (estriol, E3) from aqueous phase by batch and column studies: Kinetic and mechanistic evaluation. *J. Hazard. Mater.* **2009**, *164*, 820–828. [CrossRef] [PubMed]

33. Li, S.-H.; Liu, Q.-H.; Qi, L.; Liu, H.-H.; Wang, H.-Y. Progress in research on manganese dioxide electrode materials for electrochemical capacitors. *Chin. J. Anal. Chem.* **2012**, *40*, 339–346. [CrossRef]

34. Gabet-Giraud, V.; Miège, C.; Choubert, J.; Ruel, S.M.; Coquery, M. Occurrence and removal of estrogens and beta blockers by various processes in wastewater treatment plants. *Sci. Total Environ.* **2010**, *408*, 4257–4269. [CrossRef] [PubMed]

35. Nakada, N.; Yasojima, M.; Okayasu, Y.; Komori, K.; Tanaka, H.; Suzuki, Y. Fate of oestrogenic compounds and identification of oestrogenicity in a wastewater treatment process. *Water Sci. Technol.* **2006**, *53*, 51–63. [CrossRef] [PubMed]

36. Nie, Y.; Qiang, Z.; Zhang, H.; Ben, W. Fate and seasonal variation of endocrine-disrupting chemicals in a sewage treatment plant with A/A/O process. *Sep. Purif. Technol.* **2012**, *84*, 9–15. [CrossRef]

International Journal of
Environmental Research and Public Health

Article

Pollutant Removal from Synthetic Aqueous Solutions with a Combined Electrochemical Oxidation and Adsorption Method

Amin Mojiri, Akiyoshi Ohashi, Noriatsu Ozaki, Ahmad Shoiful and Tomonori Kindaichi *

Department of Civil and Environmental Engineering, Graduate School of Engineering, Hiroshima University, 1-4-1 Kagamiyama, Higashihiroshima 739-8527, Japan; amin.mojiri@gmail.com (A.M.); ecoakiyo@hiroshima-u.ac.jp (A.O.); ojaki@hiroshima-u.ac.jp (N.O.); d165199@hiroshima-u.ac.jp (A.S.)
* Correspondence: tomokin@hiroshima-u.ac.jp; Tel./Fax: +81-82-424-5718

Abstract: Eliminating organic and inorganic pollutants from water is a worldwide concern. In this study, we applied electrochemical oxidation (EO) and adsorption techniques to eliminate ammonia, phenols, and Mo(VI) from aqueous solutions. We analyzed the first stage (EO) with response surface methodology, where the reaction time (1–3 h), initial contaminant concentration (10–50 mg/L), and pH (3–6) were the three independent factors. Sodium sulfate (as an electrolyte) and Ti/RuO_2–IrO_2 (as an electrode) were used in the EO system. Based on preliminary experiments, the current and voltage were set to 50 mA and 7 V, respectively. The optimum EO conditions included a reaction time, initial contaminant concentration, and pH of 2.4 h, 27.4 mg/L, and 4.9, respectively. The ammonia, phenols, and Mo elimination efficiencies were 79.4%, 48.0%, and 55.9%, respectively. After treating water under the optimum EO conditions, the solution was transferred to a granular composite adsorbent column containing bentonite, limestone, zeolite, cockleshell, activated carbon, and Portland cement (i.e., BAZLSC), which improved the elimination efficiencies of ammonia, phenols, and molybdenum(VI) to 99.9%. The energy consumption value (8.0 kWh kg^{-1} N) was detected at the optimum operating conditions.

Keywords: adsorption; ammonia; electrochemical oxidation; molybdenum; phenols

---

1. Introduction

Disposal of industrial, agricultural, and municipal waste into lakes and rivers can result in environmental contamination [1], where various pollutants can have detrimental effects on human health. Major aquatic pollutants include ammonia, phenol, and heavy metals. Of these, ammonia nitrogen is one of the most common aquatic pollutants and it contributes to the enhanced eutrophication of rivers and lakes, depletion of dissolved oxygen, and fish toxicity in gaining water [2]. Proposed techniques for eliminating ammonia include air stripping, biological reactors, electrochemical oxidation (EO), ozonation, and adsorption [3,4]. The elimination of ammonia has attracted considerable attention due to the need to control nitrogen pollution and to prevent the eutrophication of water sources.

Meanwhile, phenols are among the most toxic contaminants in wastewater, and phenols and related compounds are prevalent organic contaminants in the wastewater of various chemical plants. Given the widespread prevalence of phenols in wastewater and their toxicity to human and animal life, even at low concentrations, their elimination from wastewater is essential. The efficient elimination of phenols from waste streams has gradually become a major environmental concern [5]. Numerous methods have been applied to phenol elimination in wastewater treatment, including biological treatment, reverse osmosis, adsorption, ion exchange, catalytic oxidation, electrochemical oxidation

(EO), and solvent extraction as common conventional methods for eliminating phenols and the related organic substances [6].

Finally, among inorganic pollutants, heavy metals have attracted substantial academic attention. The contamination of water by heavy metals during industrial wastewater disposal is an international environmental issue, as rapid industrialization worldwide has significantly contributed to the release of theoretically toxic heavy metals into aquatic systems [7]. Among heavy metals, Mo is highly toxic. Vital techniques for eliminating metals from water include physical/chemical methods, such as adsorption and EO.

EO is a physical/chemical method for treating water and wastewater, and its application to various types of wastewater has been investigated extensively in recent years. The EO process is a promising wastewater treatment method chiefly due to its effectiveness and the ease of operation [3]. Electrochemical technology, which represents an advanced oxidation process, is the most promising method for treating both organic contaminants and heavy metals. Such technology includes electrodialysis, electrocoagulation, electroflotation, anodic oxidation, and EO [8]. Several researchers have investigated the treatment of different types of wastewater with EO [9]. The process of electrochemical degradation can be divided into direct and indirect oxidation procedures. In direct oxidation, the contaminants are first adsorbed onto the surface of the anode, and then undergo the electron transfer reaction of hydroxyl radical ($\bullet$OH) formation, which is followed by strong oxidative free radical damage of the contaminant molecular structure, and finally, decomposition into CO_2. In indirect electrochemical oxidation, to produce strong oxidizing agents (e.g., hypochlorous acid/chlorine, ozone, H_2O_2, etc.), oxidation of contaminants and electrochemical oxidation occurs in the bulk solution [10]. For example, Chen et al. [8] investigated EO in the treatment of heavy metal wastewater. They expressed that EO could be effective in removing heavy metals. Meanwhile, Cossu et al. [11] used Ti/PbO_2 and Ti/SnO_2 anodes to eliminate ammonia nitrogen and chemical oxygen demand from landfill leachate by EO, and found that EO could remove a large proportion of chemical oxygen demand and ammonia.

Researchers often recommend combined techniques to treat the high pollutant concentrations in industrial wastewater. For example, adsorption is a common wastewater treatment method and is typically combined with other techniques [12]. Several researchers have investigated the use of adsorption to treat various types of wastewater [13,14], and several studies [15] have verified that adsorbents can eliminate considerable amounts of contaminants, especially heavy metals. Numerous adsorbents, such as activated carbon, limestone, shell, cement, and zeolite, have been studied in the literature [3,4]. Here, we used a new composite adsorbent, BAZLSC (i.e., bentonite, zeolite, cockleshell, limestone, activated carbon, and Portland cement) to simultaneously adsorb contaminants and to perform ion exchange.

In this study, we assessed a novel combination of EO and adsorption techniques to achieve 100% elimination efficacy. The aims of this research were to (1) evaluate the performance of a combined system incorporating EO and adsorption to remove ammonia, phenols, and Mo from aqueous solutions, (2) introduce a novel process, granular BAZLSC adsorption combined with EO, and (3) monitor the adsorption isotherms during adsorption treatment. There are no reports in the literature of reactors with the same design as our reactor with such a high performance. In addition, we introduced a novel composite adsorbent.

2. Materials and Methods

The treatment process in this study was divided into two stages. In the first stage, a synthetic aqueous solution was treated with EO. Statistical analysis and optimization were performed using response surface methodology (RSM). In the second stage, water was transferred to a fixed-bed adsorption column for further treatment. During this process, desorption isotherms were monitored. Figure 1 presents a schematic diagram of the reactor that was employed in this study.

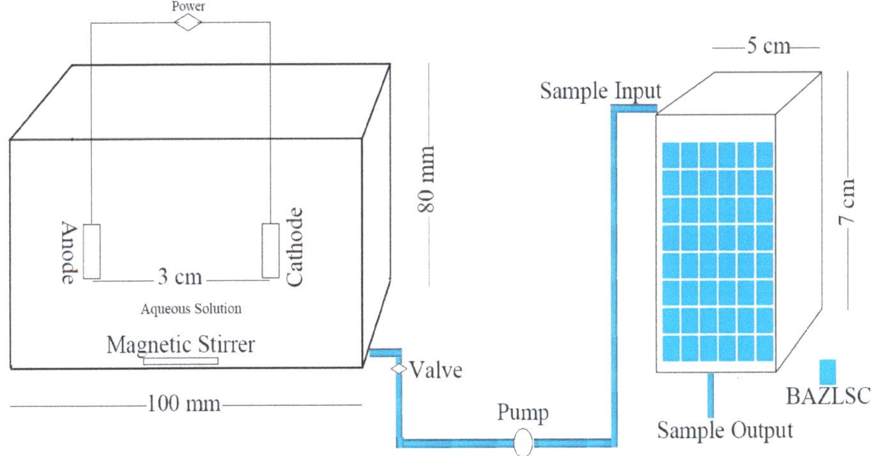

Figure 1. Schematic diagram of the EO reactor (**left**) and adsorption column (**right**) used in this study.

2.1. Synthetic Aqueous Solution Production

To created synthetic polluted water, tap water was spiked with three compounds, ammonia (NH_3-N), phenols, and Mo(IV). To create standard solutions of each contaminant, chemical-grade compounds were dissolved in water to the prescribed concentration. Aqueous ammonia was obtained by dissolving ammonium hydroxide in water [16]. The phenol had a purity of 99.9% and molecular mass of 94.11 g/mol based on laboratory analysis [17]. Finally, the standard solution of Mo(IV) was obtained by dissolving $Na_2MoO_4 \cdot 2H_2O$ in water [18].

The pH of the influent solution was measured with a pH meter and was adjusted with 0.1 M NaOH or 0.1 M HCl.

2.2. EO Reactor Characteristics

A reactor with a working capacity of 500 mL, width of 100 mm, length of 100 mm, and height of 80 mm was used for EO. An electrical current was employed via a constant-voltage/current-controlled DC power source. Ti/RuO_2–IrO_2 electrodes were employed as the anode and cathode. A plate anode and plate cathode of the same dimensions (3.5 cm × 3 cm × 1 cm, L × W × T) were arranged parallel to each other. Based on preliminary experiments and Koppad et al. [19], who used a distance of 30 mm between the anode and cathode during wastewater treatment with EO, we set the distance between the anode and cathode as 30 mm. A magnetic stirrer was placed at the bottom of the reactor for mixing. The experiments were completed at room temperature. As an electrolyte, 1 g/L of $Na_2S_2O_8$ added to the samples before each experiment [3]. The current and voltage were fixed at 50 mA and 7 V, and were set according to the ranges that were reported by Bashir et al. [3] and Mojiri et al. [12], respectively.

2.3. Statistical Analysis

We computed the elimination effectiveness using Equation (1).

$$Removal\ (\%) = \frac{(C_i - C_f)\,100}{C_i} \tag{1}$$

where C_i and C_f denote the preliminary and final concentrations of the contaminants, respectively.

A three-level factorial design was created with Design Expert ver. 10.0.7 software for the experimental design and data analysis. The three independent factors in this research were reaction time (1, 2, and 3 h), initial contaminant concentration (10, 30, and 50 mg/L), and pH (3, 4.5, and 6). Ammonia, phenols, and Mo(VI) removal were selected as the response parameters. We performed preliminary experiments to narrow the ranges of the variables before carrying out the full factorial experiments, which were based on previous studies. For example, Kearney et al. [20] set the pH to 4 for ammonia elimination by EO. Meanwhile, Asghar et al. [21] reported that a 1-h contact time was optimal to treat an aqueous solution by EO. Finally, Xu et al. [22] reported that a 4-h contact time was optimal for the removal of cyanide from water by EO. In addition, we considered increased and decreased ranges to improve the accuracy of the results. The three independent factors and their corresponding levels are presented in Table 1. Equation (2), which is an empirical second-order polynomial model, considers the performance of the scheme.

$$Y = \beta_0 + \sum_{j=1}^{k} \beta_j X_j + \sum_{j=1}^{k} \beta_{jj} X_j^2 + \sum_{i}\sum_{<j=2}^{k} \beta_{ij} X_i X_j + e_i, \tag{2}$$

where Y signifies the response; X_i and X_j denote the variables; β_0 is a fixed coefficient; β_j, β_{jj}, and β_{ij} denote the interface coefficients of the linear, quadratic, and second-order terms, respectively; k denotes the quantity of considered factors; and, e denotes the error. We used analysis of variance (ANOVA) to analyze the results using Design Expert.

Table 1. Independent variables of the three-level factorial design.

Level	Reaction Time (h)	Initial Pollutant Concentration (mg/L)	pH
−1	1	10	3
0	2	30	4.5
+1	3	50	6

We selected the initial concentrations of pollutants based on preliminary experiments and a literature review. For example, Li et al. [23] used 25 mg/L as the initial ammonia concentration during the application of an electrochemical ion-exchange reactor for ammonia removal, while Peings et al. [24] used 30 mg/L as the initial phenol concentration for the removal from an aqueous solution by advance oxidation.

2.4. Fixed-Bed Adsorption Column

After treating the synthetic aqueous solution with an EO reactor, the samples were transferred to an adsorption column with a water pump for further treatment.

2.4.1. Fixed-Bed Adsorption Column Preparation

Dynamic adsorption was performed in a 5–7-cm glass column filled with a 1-mm composite adsorbent, BAZLSC. Based on preliminary experiments, the contact time of water with the adsorption column was set to 10~15 min [25].

2.4.2. Composite Adsorbent (BAZLSC) Preparation

Bentonite, limestone, zeolite, cockleshell, activated carbon, and Portland cement were crushed, passed through a 300-μm mesh sieve, and then blended to obtain BAZLSC. The mixture was then carefully poured into a mold after adding water. The materials were removed from the mold after 24 h and immersed in water for approximately two days for the curing procedure. After letting the materials dry for three days, they were ground and passed through a sieve. Before using BAZLSC in the experiments, it was dried at 105 °C for 24 h. Table 2 and Figure 2 show the structures of the BAZLSC and x-ray diffraction analysis results, respectively. BAZLSC supported simultaneous adsorption and ion exchange [11].

Table 2. Powdered BAZLSC characteristics.

Characteristic	Value
Surface area (m^2/g)	288.6
External surface area (m^2/g)	246.7
Micropore area (m^2/g)	61.9
Micropore volume (cc/g)	0.08

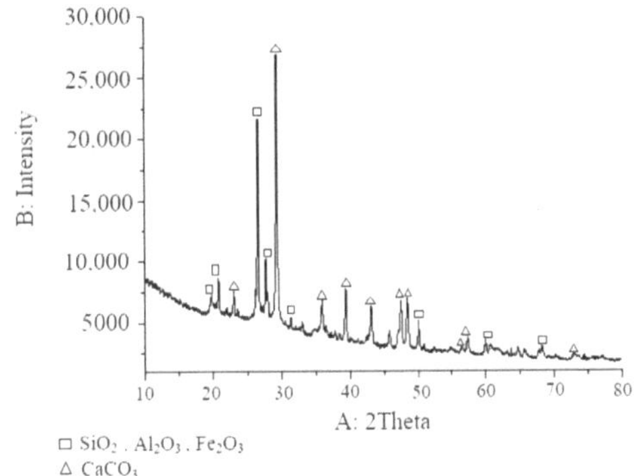

Figure 2. X-ray diffraction results of BAZLSC.

2.4.3. Adsorption Isotherm

Adsorption is the adhesion of atoms, ions, biomolecules, or molecules of gas, liquid, or solids onto a surface. The adsorption isotherm equation is (Equation (3)):

$$q_e = \frac{(C_0 - C_e)V}{M},\qquad(3)$$

where q_e represents the sum of the solute adsorbed per unit weight of the adsorbent (mg/g), C_0 denotes the preliminary adsorbate concentration, C_e denotes the equilibrium adsorbate concentration (mg/L), V denotes the volume of the solution (L), and M represents the mass of the adsorbent (g).

The Langmuir isotherm represents the foundation of a monolayer adsorbate on the outward surface of an adsorbent. Hence, this isotherm represents the equilibrium spreading of ions between the solid and liquid phases [26]. Meanwhile, the Langmuir isotherm is suitable for monolayer adsorption onto a surface comprising a confined quantity of identical positions. The Langmuir equation can be expressed as Equation (4) [27]:

$$\frac{x}{m} = \frac{abC_e}{(1 + bC_e)},\qquad(4)$$

where x/m denotes the mass of the adsorbate adsorbed per unit mass of adsorbent (mg/g), a and b denote the empirical fixed, and C_e is the equilibrium concentration of the adsorbate in the solution after adsorption (mg/L).

The Freundlich isotherm is regularly applied in order to justify the adsorption characteristics of heterogeneous surfaces [26]. The Freundlich equation can be expressed as Equation (5):

$$q_m = K_f C_e^{1/n},\qquad(5)$$

where K_f is a constant representative of the relative adsorption capability of the adsorbent (mg$^{1-(1/n)}$ L$^{1/n}$ g^{-1}) and n denotes a constant that is related to the adsorption intensity [28].

Freundlich and Langmuir isotherms were used to simplify the characteristics of BAZLSC adsorption. The adsorption isotherms were monitored in batch experiments. First, 20 mg/L of each contaminant were added to 200-mL beakers containing various concentrations (0–2.5 g/L) of adsorbent. Then, the beakers were shaken at 200 rpm for 30 min [12].

2.5. Analytical Methods

We followed the Standard Methods for the Investigation of Water and Wastewater [29] to analyze wastewater. A YSI 556 MPS (YSI Inc., Yellow Springs, OH, USA) was employed to record the temperature (°C), pH, electrical conductivity (mS/cm), oxidation–reduction potential (mV), and salinity (g/L). Inductively coupled plasma optical emission spectrometry (Varian 715; Varian Inc., Palo Alto, CA, USA,) and a spectrophotometer (HACH/2500; HACH, Loveland, CO, USA) were used to measure the components of the water. Phenol was tested using a HACH DR/2500 based on method 8047 and method 2540B, the 4-aminontipyrine method. Ammonia was tested by using a HACH DR/2500 that is based on method 8190, the Nessler method.

3. Results and Discussion

We investigated the elimination of ammonia, phenols, and Mo from contaminated wastewater via a combined EO and adsorption method. Tables 1 and 3 show the independent variables of the three-level factorial design and the response values for the experimental conditions, respectively. Table 4 presents the statistical results of the response parameters. Figure 3 displays three-dimensional surface plots of contaminant elimination.

Table 3. Response values under different experimental conditions.

Run	Contact Time (h)	Initial Concentration (mg/L)	pH	Ammonia Rem. * (%)	Phenols Rem. (%)	Mo Rem. (%)
13	0.8	30.0	4.5	74.17	45.17	49.64
6	1.0	10.0	3.0	72.11	51.95	49.87
8	1.0	10.0	6.0	82.62	35.12	40.13
4	1.0	50.0	3.0	69.53	35.11	44.00
11	1.0	50.0	6.0	81.11	28.71	34.18
17	2.0	6.0	4.5	74.31	44.11	46.11
2	2.0	30.0	4.5	77.00	50.86	58.95
3	2.0	30.0	6.3	90.11	33.00	38.00
5	2.0	30.0	2.7	71.17	43.84	47.11
9	2.0	54.0	4.5	72.00	42.92	47.11
10	2.0	30.0	4.5	76.92	49.97	58.74
12	2.0	30.0	4.5	77.11	50.81	59.49
15	2.0	30.0	4.5	77.12	50.57	58.00
16	2.0	30.0	4.5	77.00	50.89	58.76
18	2.0	30.0	4.5	76.93	51.18	59.00
20	2.0	30.0	4.5	76.93	51.11	59.40
22	2.0	30.0	4.5	77.71	50.35	59.10
7	3.0	10.0	6.0	84.13	37.19	43.45
21	3.0	10.0	3.0	73.46	52.17	48.69
14	3.0	50.0	3.0	71.64	41.50	50.13
1	3.0	50.0	6.0	90.95	29.97	37.64
19	3.2	30.0	4.5	73.18	46.18	53.90

* Abbreviation: Rem. means removal.

Table 4. Results of the analysis of variance of the response parameters.

Response	Final Equation in Terms of Actual Factor [a]	R^2	Adj. R^2	Adec. P.	SD	CV	PRESS
Ammonia	$94.27 + 1.032A - 12.849C + 1.655C^2$	0.9422	0.8988	18.19	1.80	2.40	334.32
Phenols	$3.532 - 0.039B + 21.99C - 0.008B^2 - 3.003C^2$	0.9297	0.8769	11.68	2.72	6.93	668.22
Mo(VI)	$-22.835 + 30.369C - 0.013B^2 - 3.634C^2$	0.9202	0.8604	11.98	3.06	6.65	712.52

Abbreviations: R^2: Coefficient of determination; Adj. R^2: Adjusted R^2; Adec. P.: Adequate precision; SD: Standard deviation; CV: Coefficient of variation; PRESS: Predicted residual error sum of squares; [a] In the final equations, A is the electrochemical oxidation reaction time (h), B is the initial concentration of pollutants (mg/L), and C is pH; Significant at 0.05.

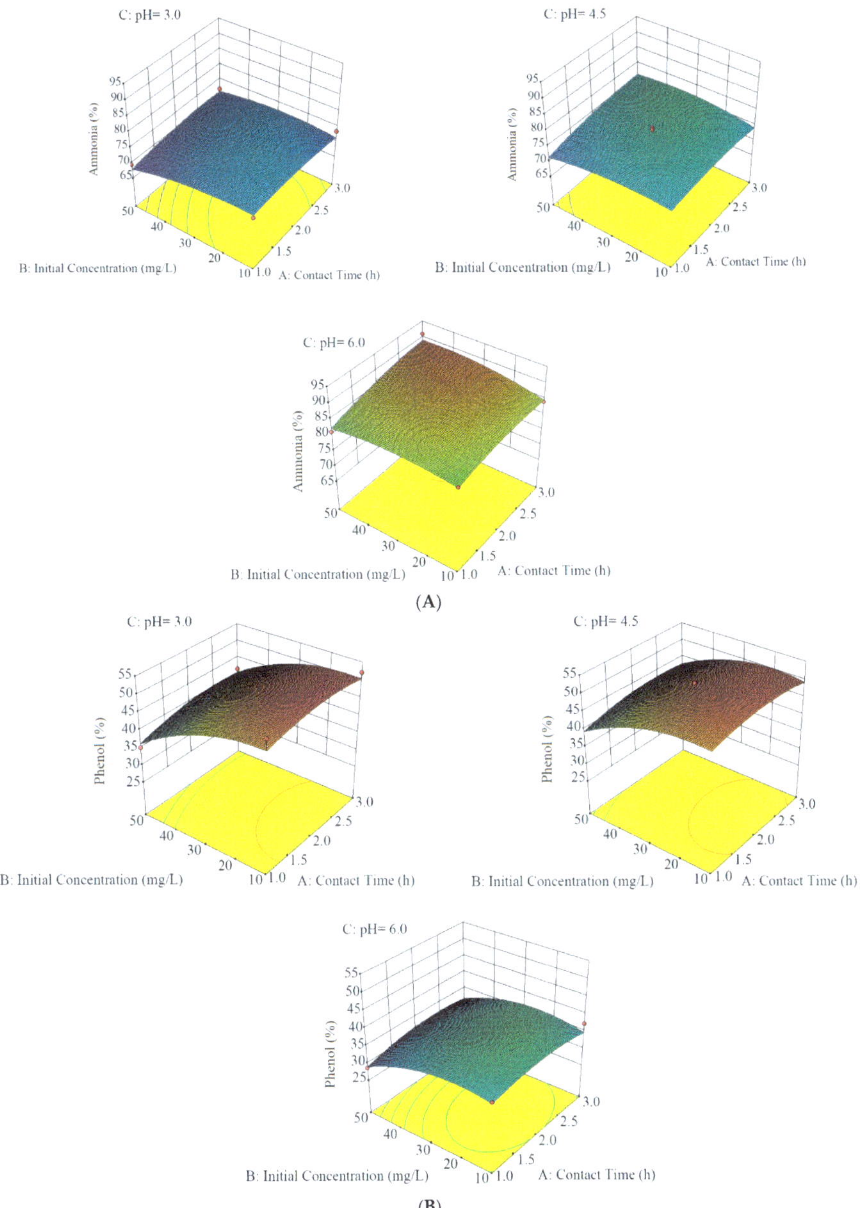

Figure 3. *Cont.*

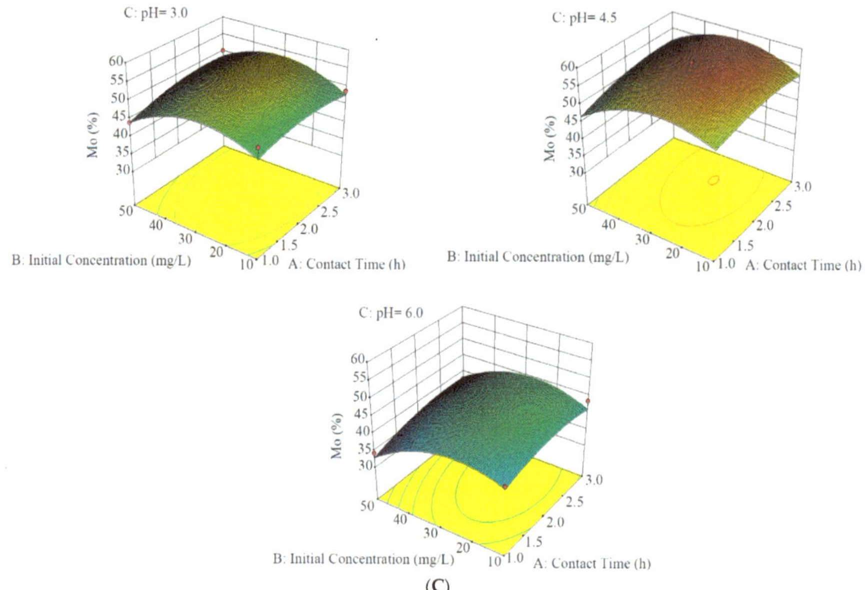

Figure 3. Three-dimensional surface plots of (**A**) ammonia, (**B**) phenol, and (**C**) Mo removal.

3.1. Ammonia, Phenol, and Mo Removal Using EO

The lowest elimination efficacy for ammonia was 69.5% (reaction time = 1 h, pH = 3, preliminary ammonia concentration = 50 mg/L) and the highest was 90.9% (contact time = 3 h, pH = 6, preliminary ammonia concentration = 50 mg/L) (Table 3 and Figure 3). Under the optimum conditions (contact time = 3 h, pH = 6, preliminary ammonia concentration = 44.8 mg/L), ammonia removal could reach approximately 91.0%. He et al. [30] investigated ammonia removal by EO at pH = 6.5 in the presence of Ru–Ir/TiO$_2$ and Na$_2$SO$_4$. They expressed that up to 80% of ammonia could be eliminated and be mostly transferred to N$_2$ in a powdered activated carbon (PAC) packed bed reactor under optimum conditions (pH = 6.5, I = 0.9 A, 2% Na$_2$SO$_4$, Cl$^-$ = 1500 mg/L, and inlet velocity = 0.8 L/h). In addition, Ding et al. [31] removed 90% of ammonia using Ti/RuO$_2$-Pt electrodes by the EO method. Similarly, Li et al. [23] reported an 89% ammonia removal at an initial concentration of 30 mg/L using a vermiculite-packed electrochemical reactor. Overall, the ammonia elimination efficiency in the current study was similar to those in previous studies.

The lowest phenol elimination efficacy was 28.7% (reaction time = 1 h, pH = 6, preliminary phenol concentration = 50 mg/L) and the highest was 52.1% (contact time = 3 h, pH = 3.5, initial phenol concentration = 10 mg/L) (Table 3 and Figure 3). Under the optimum conditions (contact time = 1.8 h, pH = 3, initial phenol concentration = 17.7 mg/L), the phenol elimination rate was approximately 52.6%. Saratale et al. [32] explored phenol elimination from wastewater by EO and reported an acidic pH and a voltage of 5 V for optimum phenol elimination. In addition, Wu et al. [33] investigated phenol removal by EO using N$_2$SO$_4$ as an electrolyte. Meanwhile, Tasic et al. [34] used two EO methods (direct and indirect oxidation) for the removal of organic contaminants. Direct oxidation of contaminants first leads to their adsorption onto the anode surface without the contribution of other substances in the solution, except for electrons, which are considered to be pure reagents. Direct electro-oxidation is ideally possible at low potential values before oxygen evolution, but the reactions are often slow and dependent on the electrocatalytic activity of the anode. Rapid electrochemical reaction is achieved while using noble metals and anodes that are based on metal oxides (e.g., IrO$_2$, TiO$_2$-Ru, and Ir-TiO$_2$)

for phenol elimination using the electrochemical technique. Two responses are probable for the direct anodic oxidation of organic pollutants [35].

(a) Electrochemical conversion, where the organic complexes are partly oxidized, according to the reaction (Equation (6)):

$$R \rightarrow RO + e^- \tag{6}$$

(b) Electrochemical combustion, where the organic complexes break down into CO_2, water, and other inorganic complexes (Equation (7)):

$$R \rightarrow CO_2 + H_2O + Salt + e^- \tag{7}$$

The lowest elimination effectiveness for Mo was 34.1% (reaction time = 1 h, pH = 6, initial Mo concentration = 50 mg/L) and the highest was 59.4% (contact time = 2 h, pH = 4.5, initial Mo concentration = 30 mg/L) (Table 3 and Figure 3). Under the optimum conditions (contact time = 2.4 h, pH = 4.1, initial Mo concentration = 28.5 mg/L), the Mo removal rate was approximately 59.5%. Tran et al. [36] investigated metal elimination from an aqueous solution by the EO method at a voltage of 10 V. Their investigation showed a high performance in removing metals, but the selected voltage (10 V) and reaction time (20 h) were higher than the our findings.

In electrochemical treatments, the two methods for the elimination of contaminants in the presence of $Na_2S_2O_8$ (as an electrolyte) are as follows:

(i) Direct oxidation, where metal cations (commonly heavy metals) are reduced at the cathode and organic contaminants are oxidized at the anode even without the connection of other chemical reagents [37].

(ii) Indirect electrolysis, where the concentration of $Na_2S_2O_8$ hastens the mineralization of organic compounds. In general, reasonable concentrations of Na_2SO_4 accelerate the mineralization of organic matter via indirect oxidation, as shown in the following reactions [38]; however, it should be mentioned that some researchers have applied heat or ultraviolet light + heat in order to improve the persulfate oxidation ability of phenols [39].

The probable reactions occurring at the anode, cathode, and in the bulk material are shown below (Equations (8)–(15)) [40].

At the anode (oxidation):

$$SO_4{}^{2-} + 2OH^- \leftrightarrow SO_4{}^{2-} + 2e^- + H_2O \tag{8}$$

$$SO_4{}^{2-} \leftrightarrow S_2O_8 + 2e^- \tag{9}$$

$$4OH^- \leftrightarrow O_2 + 2H_2O + 4e^- \tag{10}$$

At the cathode (reduction):

$$H_2O \leftrightarrow H^+ + OH^- \tag{11}$$

$$2H_2O + 2e^- \leftrightarrow H_2 + 2OH^- \tag{12}$$

$$HSO_4{}^- + OH^\circ \leftrightarrow SO_4{}^{-\circ} + H_2O \tag{13}$$

$$SO_4{}^{-\circ} + SO_4{}^{-\circ} \leftrightarrow S_2O_8{}^{2-} \tag{14}$$

$$Organics + S_2O_8{}^{2-} \rightarrow intermediates \rightarrow H_2O + CO_2 + \uparrow A^\circ \tag{15}$$

Figure 3 shows the pollutant removal under different pH conditions, initial pollutant concentrations, and contact times, and has been extracted based on the information in Tables 3 and 4. Figure 3A shows the effect of the operational parameters on ammonia elimination. Ammonia elimination increased with increasing initial concentration, pH, and contact time. With an influent pH of 6, initial concentration of 50 mg/L, and contact time of 90 min, the maximum ammonia removal was 90.95%. Figure 3B shows the effects of the variables on phenol elimination. Phenol elimination increased with an increasing initial concentration until 30 mg/L, pH until 3, and contact time until 2–3 h. Finally, Figure 3C displays the effects of the independent factors on Mo elimination. Mo elimination increased with increasing initial concentration until 30 mg/L, pH until 4.5, and contact time until 2 h.

3.2. Energy Consumption (EC; kWh/kg N)

EC (kWh kg^{-1} N) at optimum condition was calculated by Equation (16) [41]. EC states to the electrochemical treatment cost. The EC value was 8.0 (kWh kg^{-1}) at the optimum operating conditions. Christiaens et al. [41] reported EC = 13.9 (kWh kg^{-1}) during electrochemical ammonia recovery. EC in this contemporary research is lower than them, it displays the combined system could diminish energy consumption.

$$EC = \frac{UIt}{(N_0 - N_t)V},$$
(16)

where, N_0 and N_t present N at initial time and set time; U, I and t are voltage, current (A), and time (h), respectively; and, V is volume (L).

3.3. Ammonia, Phenol, and Mo Removal Using an Adsorption Column

The synthetic aqueous solution was initially treated in the EO reactor under the optimal conditions before being transferred to an adsorption column for the second stage of treatment. The performance of EO combined with adsorption enhanced the elimination efficiencies of ammonia, phenols, and Mo from 79.4% to 99.9%, 48.0% to 99.9%, and 55.9% to 99.9%, respectively. The composite adsorbent used in this study simultaneously performed adsorption and ion exchange, and it was produced from effective, low-cost materials, such as bentonite, zeolite, activated carbon, cockleshell, cement, and limestone. For example, Mazloomi and Jalali [42] reported that zeolite can be used to eliminate NH_4^+ from domestic and industrial wastewater. Meanwhile, Halim et al. [14] reported that activated carbon is effective in removing ammonia. Finally, Haseena et al. [43] investigated the potential use of bentonite in eliminating ammonia from aqueous solutions. Based on the literature [44], we speculated that bentonite and zeolite could be used to remove phenols and metals from water, which was verified by the results. As noted above, the composite adsorbent BAZLSC can perform ion exchange and adsorption, since it contains bentonite, zeolite, limestone, shell, cement, and activated carbon [45]. This has been confirmed in equilibrium and adsorption studies of composite adsorbents [46].

3.4. Adsorption Isotherms of Pollutant Removal by the Composite Adsorbent

Table 5 and Figure 4 show the Langmuir equation and isotherm regression for ammonia, phenols, and Mo, respectively.

Table 5. Langmuir equation for ammonia, phenols, and Mo.

Parameter	Q (mg/g)	b (L/mg)	R^2
Ammonia	1.027	0.240	0.9333
Phenols	0.554	0.087	0.8696
Mo	0.874	0.45	0.8051

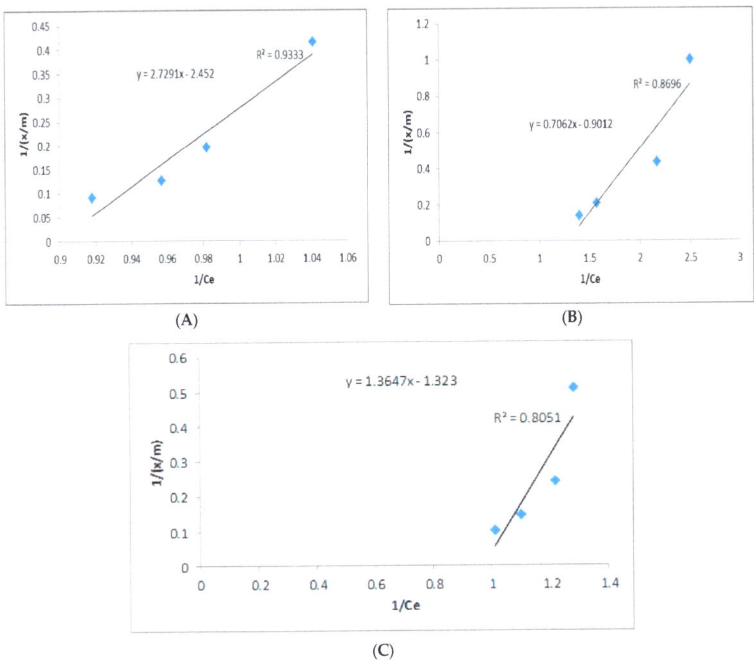

Figure 4. Langmuir isotherm regressions for (**A**) ammonia, (**B**) phenols, and (**C**) Mo.

The adsorption capacities (Q, Table 5) of ammonia nitrogen, phenols, and Mo were 6.198, 3.86, and 5.44 mg/g, respectively. Based on a study by Mojiri et al. [7], we determined that the composite adsorbent could reach Q = 0.70 mg/g for Fe adsorption. The energy of adsorption (b, Table 5) for ammonia nitrogen, phenol, and Mo removal were 0.240, 0.087, and 0.045 L/mg, respectively. By increasing the efficacy of elimination (i.e., increased C_e values), the value of x/m decreased [4]. Finally, the regression coefficient (R^2, Table 5 and Figure 4) was 0.9333, 0.8696, and 0.8051 for ammonia, phenols, and Mo, respectively. Similar values for R^2 have been reported for ammonia by zeolite, activated carbon, and composite adsorbent [14]. Moreover, similar R^2 values have been reported for phenol adsorption using granular activated carbon and Mo adsorption while using Pb–Fe-based adsorbent [47].

The Freundlich capability factors (K_f, Table 6) for ammonia, phenol, and Mo elimination were 0.014, 0.063, and 0.028 (mg/g (L/mg)$^{1/n}$), respectively. The R^2 values were 0.9795, 0.9641, and 0.9266 for ammonia, phenols, and Mo, respectively (Table 6 and Figure 5). These collective results indicated that the Freundlich isotherm is more suitable than the Langmuir isotherm for describing ammonia, phenol, and Mo removal by the adsorption column. Finally, greater capacities for adsorption are indicated by higher K values [7]. The collective 1/n values for ammonia, phenol, and Mo were 12.458, 3.1217, and 6.3689, respectively.

Table 6. Freundlich equation for ammonia, phenols, and Mo.

Parameter	K_f (mg/g (L/mg)$^{1/n}$)	1/n	n	R^2
Ammonia	0.014	12.458	0.080	0.9795
Phenols	0.063	3.121	0.320	0.9641
Mo	0.028	6.368	0.157	0.9266

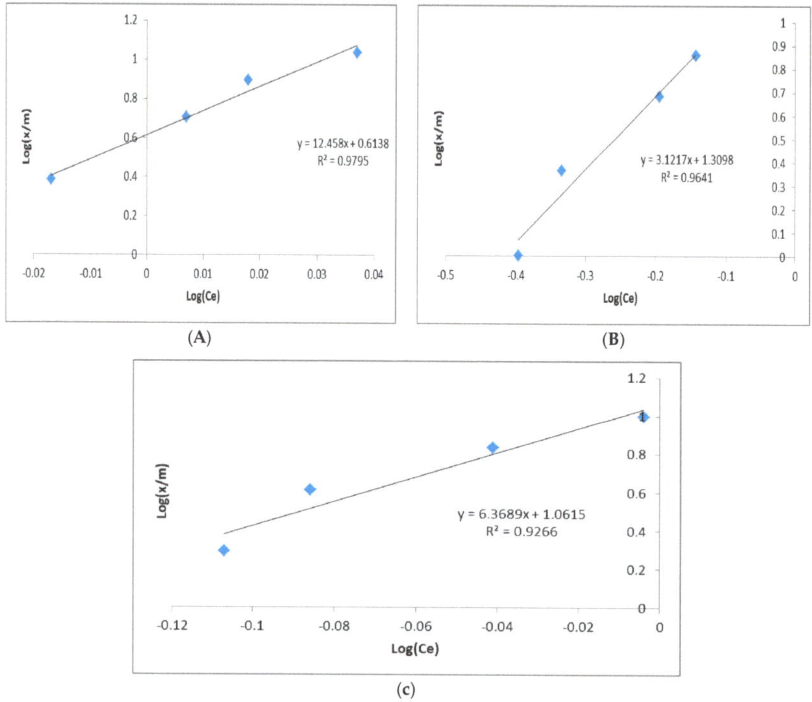

Figure 5. Freundlich isotherm regressions for (**A**) ammonia, (**B**) phenols, and (**C**) Mo.

4. Conclusions

In this study, we successfully removed ammonia, phenols, and Mo from aqueous solution via combined EO and adsorption column. RSM was used to design the experiments. The key outcomes of this research indicate that the EO reactor could remove 79.4%, 48%, and 55.9% of ammonia, phenols, and Mo, respectively, under optimum conditions (contact time = 2.4 h, pH = 4.9, initial pollutant concentration = 27.4 mg/L), with a fixed current and voltage of 50 mA and 7 V, respectively. Next, the water was moved to an adsorption column filled with a composite adsorbent, BAZLSC. Adsorption improved the removal efficiency to 99.99% for ammonia, phenols, and Mo(VI). Finally, adsorption isotherm analysis revealed that the adsorption of ammonia, phenols, and Mo by BAZLSC followed the Freundlich isotherm equation better than the Langmuir isotherm equation. More investigation about using different adsorbents, different electrolytes, and electrolytes function would be considered in future studies.

Author Contributions: A.M. was responsible for setting up the experiments, completing most of the experiments, and writing the initial draft of the manuscript. A.O., N.O., and T.K. modified the manuscript and contributed to the literature search. A.S. helped process the data.

Funding: This work was supported by a Japan Society for the Promotion of Science KAKENHI grant [Grant No. JP17F17375].

Acknowledgments: The authors would like to express their gratitude to the Japan Society for the Promotion of Science for providing a fellowship (reference No. P17375).

Conflicts of Interest: The authors declare no conflict of interest.

References

1. Barkhordar, B.; Ghiasseddin, M. Comparision of Langmuir and Freundlich Equilibriums in Cr, Cu and Ni Adsorption by Sargassum Iran. *J. Environ. Health Sci. Eng.* **2004**, *1*, 58–64.
2. Du, X. A review on the electrochemical treatment of the salty organic wastewater. *IOP Conf. Ser.* **2015**, *87*, 012037. [CrossRef]
3. Bashir, M.J.K.; Isa, M.H.; Kutty, S.R.M.; Awang, Z.B.; Aziz, H.A.; Mohajeri, S.; Farooqi, I.H. Landfill leachate treatment by electrochemical oxidation. *Waste Manag.* **2009**, *29*, 2534–2541. [CrossRef] [PubMed]
4. Aziz, S.Q.; Aziz, H.A.; Yusoff, M.S. Powdered activated carbon augmented double react-settle sequencing batch reactor process for treatment of landfill leachate. *Desalination* **2011**, *277*, 313–320. [CrossRef]
5. Caetano, M.; Valderrama, C.; Farran, A.; Cortina, J.L. Phenol removal from aqueous solution by adsorption and ion exchange mechanisms onto polymeric resins. *J. Colloid Interface Sci.* **2009**, *338*, 402–409. [CrossRef] [PubMed]
6. Rodrigues, L.A.; Pintoda Silva, M.L.C.; Alvarez-Mendes, M.O.; dos Reis Coutinho, A.; Thim, G.P. Phenol removal from aqueous solution by activated carbon produced from avocado kernel seeds. *Chem. Eng. J.* **2009**, *174*, 49–57. [CrossRef]
7. Mojiri, A.; Aziz, H.A.; Zaman, N.Q.; Aziz, S.; Zahed, M.A. Metals removal from municipal landfill leachate and wastewater using adsorbents combined with biological method. *Desalination Water Treat.* **2016**, *57*, 2819–2833. [CrossRef]
8. Chen, X.; Huang, G.; Wang, J. Electrochemical Reduction/Oxidation in the Treatment of Heavy Metal Wastewater. *J. Metall. Eng.* **2013**, *2*, 161–164.
9. Duan, F.; Li, Y.; Cao, Y.; Wang, Y.; Crittenden, J.C.; Zhang, Y. Activated carbon electrodes: Electrochemical oxidation coupled with desalination for wastewater treatment. *Chemosphere* **2015**, *125*, 205–211. [CrossRef] [PubMed]
10. Du, Q.; Liu, S.; Cao, Z.; Wang, Y. Ammonia removal from aqueous solution using natural Chinese clinoptilolite. *Sep. Purif. Technol.* **2005**, *44*, 229–234. [CrossRef]
11. Cossu, R.; Polcaro, A.M.; Lavagnolo, M.C.; Mascia, M.; Palmas, S.; Renoldi, F. Electrochemical treatment of landfill leachate: Oxidation at Ti/PbO_2 and Ti/SnO_2 anodes. *Environ. Sci. Technol.* **1998**, *32*, 3570–3573. [CrossRef]
12. Mojiri, A.; Ziyang, L.; Hui, W.; Ahmad, Z.; Tajuddin, R.M.; Abu Amr, S.S.; Kindaichi, T.; Aziz, H.; Farraji, H. Concentrated landfill leachate treatment with a combined system including electro-ozonation and composite adsorbent augmented sequencing batch reactor process. *Process. Saf. Environ. Prot.* **2017**, *111*, 253–262. [CrossRef]
13. Foo, H.Y.; Hameed, B.H. An overview of landfill leachate treatment via activated carbon adsorption process. *J. Hazard. Mater.* **2009**, *171*, 54–60. [CrossRef] [PubMed]
14. Halim, A.A.; Aziz, H.A.; Johari, M.A.M.; Ariffin, K.S.; Bashir, M.J.K. Semi-Aerobic Landfill Leachate Treatment Using Carbon–Minerals Composite Adsorbent. *Environ. Eng. Sci.* **2012**, *29*, 306–312. [CrossRef]
15. Risnawati, I.; Damanhuri, T.P. *Faculty Civil and Environmental Engineering SW5-1 to SW5-11*; Institute Technology Bandung: Kota Bandung, Indonesia, 2010.
16. Ashrafizadeh, S.N.; Khorasani, Z. Ammonia removal from aqueous solutions using hollow-fiber membrane contactors. *Chem. Eng. J.* **2010**, *162*, 242–249. [CrossRef]
17. Najafpoor, A.A.; Dousti, S.; Jafari, A.J.; Hosseinzadeh, A. Efficiency in phenol removal from aqueous solutions of pomegranate peel ash as a natural adsorbent. *Environ. Health Eng. Manag.* **2016**, *3*, 41–46.
18. Kafshgari, F.; Keshtkar, F.; Mousavian, M.A. Study of Mo (VI) removal from aqueous solution: Application of different mathematical models to continuous biosorption data. *Iran. J. Environ. Health Sci. Eng.* **2013**, *10*, 14. [CrossRef] [PubMed]
19. Koppad, V.B. Performance Evaluation of Electrochemical Oxidation System to Treat Domestic Sewage. *Int. J. Eng. Res. Technol.* **2014**, *3*, 765–771.
20. Kearney, D.; Dugauguez, O.; Bejan, D.; Bunce, N.J. Electrochemical Oxidation for Denitrification of Ammonia: A Conceptual Approach for Remediation of Ammonia in Poultry Barns. *ACS Sustain. Chem. Eng.* **2013**, *1*, 190–197. [CrossRef]
21. Asghar, H.M.A.; Ahmad, T.; Hussain, S.N.; Sattar, H. Electrochemical Oxidation of Methylene Blue in Aqueous Solution. *Int. J. Chem. Eng. Appl.* **2015**, *6*, 352–355. [CrossRef]

22. Xu, H.; Li, A.; Feng, L.; Cheng, X.; Ding, S. Destruction of Cyanide in Aqueous Solution by Electrochemical Oxidation Method. *Int. J. Electrochem. Sci.* **2012**, *7*, 7516–7525.
23. Li, L.; Yao, J.; Fang, X.; Huang, Y.; Mu, Y. Electrolytic ammonia removal and current efficiency by a vermiculite-packed electrochemical reactor. *Sci. Rep.* **2017**, *7*, 41030. [CrossRef] [PubMed]
24. Peings, V.; Frayret, J.; Pigot, T. Mechanism for the oxidation of phenol by sulfatoferrate (VI): Comparison with various oxidants. *J. Environ. Manag.* **2015**, *157*, 287–296. [CrossRef] [PubMed]
25. Biswas, S.; Mishra, U. Continuous Fixed-Bed Column Study and Adsorption Modeling: Removal of Lead Ion from Aqueous Solution by Charcoal Originated from Chemical Carbonization of Rubber Wood Sawdust. *J. Chem.* **2015**, *2015*, 907379. [CrossRef]
26. Dada, A.O.; Olalekan, A.P.; Olatunya, A.M.; Dada, O. Langmuir, Freundlich, Temkin and Dubinin–Radushkevich Isotherms Studies of Equilibrium Sorption of Zn^{2+} Unto Phosphoric Acid Modified Rice Husk. *IOSR J. App. Chem.* **2012**, *3*, 38–45.
27. Altig, J. *The Langmuir Adsorption Isotherm*; Revision 2.0; CHEM 331L; Physical Chemistry Laboratory: New Mexico, NM, USA, 2013; pp. 1–7. Available online: http://infohost.nmt.edu/~jaltig/Langmuir.pdf (accessed on 1 May 2018).
28. Hamdaoui, O.; Naffrechoux, E. Modeling of adsorption isotherms of phenol and chlorophenols onto granular activated carbon Part, I. Two-parameter models and equations allowing determination of thermodynamic parameters. *J. Hazard. Mater.* **2007**, *147*, 381–394. [CrossRef]
29. American Public Health Association (APHA). *Standard Methods for the Examination of Water and Wastewater*, 21st ed.; APHA: Washington, DC, USA, 2005; p. 541.
30. He, S.; Huang, Q.; Zhang, Y.; Wang, L.; Nie, Y. Investigation on Direct and Indirect Electrochemical Oxidation of Ammonia over $Ru-Ir/TiO_2$ Anode. *Ind. Eng. Chem. Res.* **2015**, *54*, 1447–1451. [CrossRef]
31. Ding, J.; Zhao, Q.L.; Wei, L.L.; Chen, Y.; Shu, X. Ammonium nitrogen removal from wastewater with a three-dimensional electrochemical oxidation system. *Water Sci. Technol.* **2013**, *68*, 552–559. [CrossRef] [PubMed]
32. Saratale, R.S.; Hwang, K.J.; Song, J.Y. Electrochemical Oxidation of Phenol for Wastewater Treatment Using Ti/PbO_2 Electrode. *J. Environ. Eng.* **2016**, *142*, 04015064. [CrossRef]
33. Wu, W.; Huang, Z.H.; Lim, T.T. Enhanced electrochemical oxidation of phenol using a hydrophobic TiO_2-NTs/SnO_2-Sb-PTFE electrode prepared by pulse electrodeposition. *RSC Adv.* **2015**, *5*, 32245–32255. [CrossRef]
34. Tasic, Z.; Gupta, V.K.; Antonijevic, M.M. The Mechanism and Kinetics of Degradation of Phenolics in Wastewaters Using Electrochemical Oxidation. *Int. J. Electrochem. Sci.* **2014**, *9*, 3473–3490.
35. Anglada, A.; Urtaga, A.; Ortiz, I. Contributions of electrochemical oxidation to waste-water treatment: Fundamentals and review of applications. *J. Chem. Technol. Biotechnol.* **2009**, *84*, 1747–1755. [CrossRef]
36. Tran, T.K.; Chiu, K.F.; Lin, C.Y.; Leu, H.J. Electrochemical treatment of wastewater: Selectivity of the heavy metals removal process. *Int. J. Hydrogen Energy* **2017**, *42*, 27741–27748. [CrossRef]
37. Deng, Y.; Englehardt, J.D. Electrochemical oxidation for landfill leachate treatment. *Waste Manag.* **2007**, *27*, 380–388. [CrossRef] [PubMed]
38. Mao, X.H.; Tian, F.; Gan, F.X.; Zhang, X.J.; Peng, T.W. A study of electrochemical degradation of azo-dye AO7 on Na_2SO_4 medium. *Fresenius Environ. Bull.* **2006**, *15*, 1307–1315.
39. Pan, X.; Yan, L.; Qu, R.; Wang, Z. Degradation of the UV-filter benzophenone-3 in aqueous solution using persulfate activated by heat, metal ions and light. *Chemosphere* **2018**, *196*, 95–104. [CrossRef] [PubMed]
40. Mao, X.H.; Wei, L.; Hong, S.; Zhu, H.; Lin, A.; Gan, F.X. Enhanced electrochemical oxidation of phenol by introducing ferric ions and UV radiation. *J. Environ. Sci.* **2008**, *20*, 1386–1391. [CrossRef]
41. Christiaens, M.E.R.; Gildemyn, S.; Matassa, S.; Ysebaert, T.; de Varieze, J.; Rabaey, K. Electrochemical Ammonia Recovery from Source-Separated Urine for Microbial Protein Production. *Environ. Sci. Technol. Lett.* **2017**, *51*, 13143–13150. [CrossRef] [PubMed]
42. Mazloomi, F.; Jalali, M. Ammonium removal from aqueous solutions by natural Iranian zeolite in the presence of organic acids, cations and anions. *J. Environ. Chem. Eng.* **2016**, *4*, 1664–1673. [CrossRef]
43. Haseena, P.V.; Padmavathy, K.S.; Rohit Krishnan, P.; Madhu, G. Adsorption of Ammonium Nitrogen from Aqueous Systems Using Chitosan-Bentonite Film Composite. *Procedia Technol.* **2016**, *24*, 733–740. [CrossRef]

44. Leili, M.; Faradmal, J.; Kosravian, F.; Heydari, M. A Comparison Study on the Removal of Phenol from Aqueous Solution Using Organo-modified Bentonite and Commercial Activated Carbon. *Avicenna J. Environ. Health Eng.* **2015**, *2*, e2698. [CrossRef]
45. Inglezakis, V.J.; Stylianou, M.; Loizidou, M. Ion exchange and adsorption equilibrium studies on clinoptilolite, bentonite and vermiculite. *J. Phys. Chem. Solids* **2010**, *71*, 279–284. [CrossRef]
46. Mojiri, A.; Hui, W.; Arshad, A.K.; Ridzuan, A.R.M.; Hamid, N.H.M.; Farraji, H.; Gholami, A.; Vakili, A. Vanadium(V) removal from aqueous solutions using a new composite adsorbent (BAZLSC): Optimization by response surface methodology. *Adv. Environ. Res.* **2017**, *6*, 173–187. [CrossRef]
47. Dodbiba, G.; Wu, I.C.; Lee, Y.C.; Matsuo, S.; Fujita, T. Adsorption of Molybdenum Ion in Nitric Acid Solution by Using a Pb-Fe Based Adsorbent. *Int. J. Soc. Mater. Eng. Resour.* **2010**, *17*, 28–34. [CrossRef]

International Journal of
*Environmental Research
and Public Health*

Article

Use of Sorption of Copper Cations by Clinoptilolite for Wastewater Treatment

Iveta Pandová [1,*], Anton Panda [1], Jan Valíček [2,3], Marta Harničárová [2,3], Milena Kušnerová [3] and Zuzana Palková [2]

[1]   Faculty of Manufacturing Technologies with a Seat in Prešov, Technical University of Košice, Bayerova 1, 080 01 Prešov, Slovak Republic; anton.panda@tuke.sk

[2]   Slovak University of Agriculture in Nitra, Technical Faculty, Tr. A. Hlinku 2, 949 76 Nitra, Slovakia; jan.valicek@vsb.cz (J.V.); marta.harnicarova@vsb.cz (M.H.); zuzana.palkova@uniag.sk (Z.P.)

[3]   Department of Mechanical Engineering, Faculty of Technology, Institute of Technology and Business in České Budějovice, Okružní 10, 370 01 České Budějovice, Czech Republic; milena.kusnerova@vsb.cz

*   Correspondence: iveta.pandova@tuke.sk; Tel.: +421-55-602-6316

Abstract: This paper from the field of environmental chemistry offers an innovative use of sorbents in the treatment of waste industrial water. Various industrial activities, especially the use of technological fluids in machining, surface treatment of materials, ore extraction, pesticide use in agriculture, etc., create wastewater containing dangerous metals that cause serious health problems. This paper presents the results of studies of the natural zeolite clinoptilolite as a sorbent of copper cations. These results provide the measurement of the sorption kinetics as well as the observed parameters of sorption of copper cations from the aquatic environment to the clinoptilolite from a promising Slovak site. The effectiveness of the natural sorbent is also compared with that of certain known synthetic sorbents.

Keywords: wastewater; environment; copper cations; sorption; zeolites

1. Introduction

Heavy metals and their compounds endanger the environment as no method yet exists which provides a natural method for their decomposition. The predominant instruments for EU legislation which regulates pollution reduction of the aquatic environment is Council Directive 76/464/EEC on pollution, which was precipitated by certain dangerous substances discharged into the aquatic environment, as well as Council Directive 98/83/EC, which refers to the quality of drinking water. In wastewater treatment plants, heavy metal ions are removed through a number of methods, including oxidation-reduction reactions, by coagulation and sedimentation, magnetic separation, activated carbon sorption, ion exchange within ion exchangers, and biochemical methods which use algae [1–3].

Currently, research is focused on three areas: (1) the application of natural and modified zeolite composites, (2) the use of natural materials that are more affordable than synthetic ones, and (3) the reduction of metal in water [3–5]. Copper is an essential element found in metalloenzymes, but it affects living organisms toxically at higher concentrations. Therefore, it is important to monitor the concentration of this element in water and, in cases of increased concentrations, its necessary adjustment. Conventional chemical methods can be costly, as they require the use of several chemicals; a further disadvantage is the production of waste by-products. Therefore, it is preferable to attach copper in the form of its cations to sorbents. Several types of synthetic zeolites are suitable for this purpose; however, their cost is disadvantageous in comparison with natural zeolite. Their sorption and ion exchange capabilities, which derive from their inherent structure, are interesting for the purposes of cleaning water. An uptake of zinc (Zn), copper (Cu), and lead (Pb) from aqueous solutions by ion

Int. J. Environ. Res. Public Health **2018**, *15*, 1364; doi:10.3390/ijerph15071364   www.mdpi.com/journal/ijerph

exchange on natural zeolitic tuff has been studied. The Croatian zeolite clinoptilolite from the Donje Jesenje deposit has been used as a natural ion exchanger. The efficiency of removal is higher for Pb and Cu than for Zn ions [5].

According to the literature, the sorption method used in water treatment processes has recently become a dominant method, especially in the use of locally available, economically undemanding, and natural materials. Such materials include natural zeolites, which, because of their chemical composition, represent inorganic alumino silicate cation exchangers. As a carrier matrix, zeolite has the necessary component to meet the demanding criteria to produce new composite materials. In contrast to amorphous organic ion exchangers, natural zeolites have a solid skeleton comprising silicon and aluminium polyoxides. Additionally, natural zeolites have a sufficiently large adsorption surface and are hydrophilic, polar, micro-porous, semi-resistant to thermal and radiation effects, affordable, and exhibit lower abrasive properties than activated carbon, which predisposes them to an appropriate hydrodynamic use in practice [6,7].

The three-dimensional zeolite structure consists of a regular arrangement of tetrahedral $[SiO_4]^{-4}$ and $[AlO_4]^{-5}$. In the $[SiO_4]^{-4}$ tetrahedra network, part of the silicon atoms is replaced isomorphically by aluminum atoms. The negative charge of the formed zeolitic grid is compensated by cations that are located in extra-grid positions. These are most often cations of the alkali metals and alkaline earth metals [6,8–21]. These cations can be replaced by other cations with the use of sorption and ion exchange. Natural zeolites are hydrated alumino-silicates, which are characterized by their ability to sorb heavy metal cations from aqueous solutions. Cations are immobilized on the zeolite by two mechanisms: ion exchange and chemisorption [22]. In this way, as in other micro-porous materials, the ion exchange sorption between the components of the liquid and solid phases is carried out according to the surface diffusion and the internal diffusion mechanism.

The application possibilities of natural zeolites ensue from their specific physicochemical properties, such as ion exchange, sorption, and the molecular sieve properties derived therefrom. Concurrently, the properties of natural zeolites offer possibilities of dehydration and hydration as well as the silicate structure itself, not to mention the micron dimensions of crystals with highly active specific surface areas. In terms of practical use, clinoptilolite deposits offer numerous interesting potential uses, for example, in the purification of gases, in the treatment of contaminated waters, and in agriculture regarding the neutralization of acidic soils. The ability of ion exchange and adsorption properties implicate a use of natural zeolites as effective carriers of herbicides, pesticides, as well as mineral fertilizers. After thermal activation, clinoptilolite has a relatively high sorption capacity for several gases, such as NH_3, CO_2, H_2S, SO_2, and NO_x [23]. In the United States and Japan, several water treatment stations use clinoptilolite for the protection of drinking water sources. The use of natural clinoptilolite for the removal of ammonium ions from water is also well known. After its regeneration by diluted sulphuric acid, an ammonium sulphate is obtained, which can be used as a fertilizer. In several locations around the world, clinoptilolite has demonstrated its ability to capture radioactive elements. In this respect, natural zeolites are also used for the purification of low-activity liquid wastes from nuclear power plants, predominantly for the capture of radioactive strontium and radioactive cesium. Zeolites have a high affinity for cesium. After the modification of natural clinoptilolite by the solution of cations of metals, clinoptilolite acquires catalytic properties and, accordingly, is used for the purification of gases [24].

The latest water treatment technologies comprise sorption on a solid layer. The sorption isotherm and the kinetic evolution of sorption must be known for the design of the cleaning process arrangement. This article represents a laboratory experiment for the determination of sorption parameters of sorption of cupric cations on natural zeolite-clinoptilolite, which is a low-cost and environmentally friendly sorption material. Experimental data obtained from equilibrium tests were analyzed using the Freundlich model.

2. Materials and Methods

2.1. Properties of Used Clinoptilolite

We tested the natural zeolite clinoptilolite, which was acquired at the Nižný Hrabovec site, to investigate the possibility of replaceing synthetic zeolites with natural ones. This Slovak deposit is economically significant, with annual mining output ranging from 40 to 50 thousand tons. Reserves of approximately 9,500,000 tons ensure the long-term availability of this natural material.

Clinoptilolite is a potassium—calcium type mineral with a pore size of 0.3–0.4 nm. Clinoptilolite composition is expressed by the formula [8]: $(Na, K)_4 Ca (Al_6 Si_{30} O_{72}) \times 24 H_2O$.

Chemical and physical properties of the clinoptilolite are presented in Tables 1–3.

Table 1. Clinoptilolite chemical composition [25].

Compound	Content	Compound	Content
SiO_2	65–71.3%	Fe_2O_3	0.7–1.9%
Al_2O_3	11.5–13.1%	MgO	0.6–1.2%
CaO	2.7–5.2%	Na_2O	0.2–1.3%
K_2O	2.2–3.4%	TiO_2	0.1–1.3%
P_2O_5	0.02%	Si/Al	4.5–5.4%

Table 2. Ion exchange properties [25].

Cation	Overall Interchange [$mol \cdot kg^{-1}$]
Ca^{+2}	0.64–0.98
Mg^{+2}	0.06–0.19
K^+	0.22–0.45
Na^+	0.01–0.19

Table 3. Physical properties [25].

Physical Property	Value
Temperature of softening	1 260 °C
Temperature of fusion	1 340 °C
Stability in acids	79.50 °C
Density	2200–2440 $kg \cdot m^{-3}$

We compared this natural material with several types of synthetic zeolites with respect to the cupric cation sorption, specifically, with nalsite, calcite, and y-site. Nalsite is a synthetic zeolite of the type 4 A with a pore size of 0.4 nm. Its chemical composition in a dehydrated state is expressed by the formula: $Na_2O \cdot Al_2O_3 \cdot 2 SiO_2$. Calsite is a synthetic zeolite of the type 5 A with a pore size of 0.5 nm. Its chemical composition in the dehydrated state is expressed by the formula: $CaO \cdot Na_2O \cdot Al_2O_3 \cdot 2 SiO_2$. Y-site is a synthetic zeolite of the Y type with a pore size of 0.9 nm. Its chemical composition in the dehydrated state is expressed by the formula: $Na_2O \cdot Al_2O_3 \cdot 4.5SiO_2$ [12,14]. Synthetic zeolites are at low pH (pH $\leq$ 3) subjected to hydrolysis and elution of alumina from the skeleton, offering a use case for natural clinoptilolite in industrial practice.

For a quantitative determination of the concentration of copper cations in aqueous solution at precise time intervals, absorption photometry was applied with the use of the equipment Optima DIGITAL COLORIMETER Model AC 114 (Optima, Tokyo, Japan). Measurements were realized at a wavelength of 620 nm using the calibration curve method. For plotting the calibration curve, we used a set of standard solutions with the concentration of copper cations ranging from 2.50 to 9.05 $g \cdot dm^{-3}$. After addition of ammonia with ammonium ion, the copper produced a blue-violet complex composition $[Cu(NH_3)_4]^{2+}$. For determination of the absorption maximum of this complex, we plotted the graphical dependence A = f (*L*) using a standard solution. Accordingly, we determined the adsorption maximum for the wavelength *L* of 620 nm. The analytical measurement of concentration of copper cations was based on the construction of the calibration curve in dependence on the absorbance from the concentration of the standard A = f (*c*).

Determination of dependence of the sorbed quantity on the sample concentration in the solution was performed by the method of the container experiments. To the weighed samples of sorbent (natural clinoptilolite) of 50 g, the same volume (0.1 dm^3) of an aqueous solution of copper cations was added. For this model, we used samples with an initial mass concentration of 2.56 g·dm^{-3}, 4.95 g·dm^{-3}, 6.72 g·dm^{-3}, 7.05 g·dm^{-3}, and 8.55 g·dm^{-3}. The samples were mixed and, at regular one-hour intervals, the concentration of copper cations was measured photometrically in the aken liquid phase. With the liquid phase, we also took the equivalent amount of the sorbent. On the basis of the measured values, the sorbed quantities were calculated for the individual time intervals. The measurements were performed until achievement of the equilibrium state, i.e., the state when the concentration of copper cations in the solutions ceased to change. The experiment was performed three times, with a standard deviation of 1.0049 calculated for the measured equilibrium concentrations.

2.2. Sorption of Cupric Cations from Aquatic Solutions

The process of sorption of a chemical substance from solution to solid matter can be expressed as a result of the reversible reaction, sorption and desorption, which achieves the resulting equilibrium between the concentrations of the chemical substance in both phases. This process is studied by evaluating the equilibrium concentration of a chemical substance in the sorbent as a function of the total equilibrium concentration in solution at a given temperature. This dependence is expressed by isotherms. The efficiency of the sorption of soluble matters on the solid matrix to the aqueous solution is most often expressed by the effective distribution coefficient K_R, which is the slope of a straight line of the linear sorption isotherm and which gives the share of the sorbed amount of the substance in the solid phase (c_s) to its equilibrium concentration in the solution (c_r) during the equilibrium state [26,27]. This parameter is a quantitative indicator of substance distribution between the solid and liquid phases $K_R = c_s/c_r$ [28,29]. The amount of the sorbed substance per sorbent unit increases linearly with the increasing concentration at low surface coverage, under three assumptions that must be met. The sorption energy must be the same for all sorption sites and it must be independent of the degree of coverage; sorption should take place only at localized sorption sites and without interaction between the sorbed molecules, the sorption capacity being a one-layer coating [30].

Assuming that the sorbed substance reaches the sorbent surface by molecular diffusion through a boundary diffusion layer, it is possible to generally express the concentration of the sorbed substance *c* at a time *t* by the Equation (1) [30].

$$c = \frac{\lambda \cdot c_r - \beta \cdot e^{\bar{\rho} \cdot t}}{\lambda - e^{-\bar{\rho} \cdot t}} \tag{1}$$

where λ, β, ρ are constants that are obtained from the measured values of concentration for individual time intervals ("β" and "λ" have a concentration dimension, "ρ" has a dimension of reciprocal value of time, c_0 is the initial concentration of copper cations in solution, c_r is the equilibrium concentration of copper cations in the solution).

$$\lambda = \frac{c_0 - \beta}{c_0 - c_r} \tag{2}$$

where:

$$\beta = \frac{2 \cdot \gamma \cdot c_1 - c_0 - c_2}{\gamma - 1} - c_r \tag{3}$$

$$\gamma = \frac{(c_0 - c_r) \cdot (c_2 - c_r)}{(c_1 - c_r)^2} \tag{4}$$

$$\bar{\rho} = \frac{1}{t} \cdot \ln \frac{(c_0 - c_r) \cdot (c - \beta)}{(c_0 - \beta) \cdot (c - c_r)} \tag{5}$$

3. Results and Discussion

Measurement Results and Their Evaluation

Laboratory measurements were focused on the sorption of cupric cations from the aquatic environment. The kinetic course of sorption of cupric cations on natural sorbent-clinoptilolite with a grain size of 2.5–5 mm (Figure 1) and on synthetic zeolites was investigated. From the measured concentration values, the efficiency of the individual sorbents was calculated. On the basis of the measured concentrations of cupric cations in solution, the effective distribution coefficients were calculated for individual sorbents. The degree of cleaning of the contaminated water was evaluated using the sorption efficiency parameter expressed as a percentage.

Figure 1. Zeolite clinoptilolite with a grain size from 2.5 to 5 mm.

In individual types of sorbents, the influence of duration of contact on sorption of cupric cations was recorded within 48 h. To define the time required to achieve chemical equilibrium, a dependence of the sorbent quantity on the duration of contact of the sorbent with sorbate was investigated. Individual types of sorbents weighing 50 g were used for the experiments. The sorbents were exposed to an aqueous solution of cupric cations with a volume of 0.25 dm^3 with an initial concentration of 2.54 g·dm^{-3}. In individual samples of sorbents, samples of solutions were taken at the exact time intervals until equilibrium was reached for analytical determination of the content of cupric cations. On the basis of the analysis, the sorption evolution with the use of individual sorbents was established (Table 4).

Table 4. The process of sorption in sorbents.

Time (h)	Sorbed Amount (g·dm^{-3})			
Sorben	Clinoptilolite	Calsite	Y-Site	Nalsite
1	1.03	2.42	2.22	2.54
2	1.42	2.53	2.29	0
3	1.73	0	2.38	
24	1.96		0	
48	2.2			
72	2.2			

The measured results show that, as a result of the sorption of cupric cation, the most rapid reduction of cations was on the nalsite, where the sorption capacity had been exhausted during the first hour. In terms of sorption rate, the second most rapid reduction was calsite, in which the reduction in concentration to almost zero was recorded after 120 min. Y-site was the third most rapid according to the sorption rate. Clinoptilolite, with sorption equilibrium for this sorbent, was reached after 48 h and showed the slowest sorption rate. The highest efficacy was recorded both in nalsite, which, after 60 min, showed 100% efficiency, and in calsite, with a reduction of cupric cations to 5% of the original concentration after one hour. Total efficacy at steady state was 94%. The efficiency of the y-site at steady state was 92%, and, after the first hour, the concentration of cupric cations was reduced to 13% of the original concentration. On the natural zeolite clinoptilolite, after the first hour, the cupric cations fell to 60% of the original concentration; the efficiency after 48 h was 81%. The efficacy was calculated by the formula $\eta = (c_1 - c_2/c_1) \cdot 100$ where c_1 is the initial concentration of cupric cations in the solution and c_2 is the concentration of cupric cations in steady-state solution.

The effective distribution coefficients recorded at steady state for the compared sorbents are presented in Table 5.

Table 5. Effective distribution coefficients calculated for various types of sorbents in the concentration of cupric cations of 2.54 g·dm^{-3}.

Type of Zeolite	c_s	c_r	K_R
Calsite	2.40	0.137	17.5
Y-site	2.35	0.182	5.5
Nalsite	2.54	0	2.54
Clinoptilolite	2.03	0.508	4.0

The results indicate the advantage of synthetic zeolites in comparison with natural clinoptilolite according to their faster evolution of sorption and higher efficiency. In contrast, the natural zeolite, due to its rich deposits, is more affordable. As a comparison, we provide the relative prices of individual sorbents: clinoptilolite-0.078 euro/kg, nalsite-4.813 euro/kg, calsite-5.411 euro/kg, and y-site-9.892 euro/kg. On the basis of these facts, we focused our subsequent experiments on clinoptilolite.

In the case of clinoptilolite, the constants were calculated from the measured concentration values for individual time intervals and were then used for the searched relationship $c = f(t)$. The values required for the calculation of the constants for the Equation (1) as well as the calculated values of the concentration of cupric cations in the solution are provided in Table 6.

Table 6. Temporal change of cupric cations and calculated parameters.

t (min)	c_m (g·dm^{-3})	σ (min^{-1})	c_v (g·dm^{-3})
60	1.524	0.00301	1.21
120	1.143	0.0029	0.874
180	0.82	0.0039	0.729
1440	0.601	0.0112	0.508
2880	0.508		0.508

For the calculation of the constants, an average value $\rho_p = 0.0053$ was used. The searched relation $c = f(t)$ was according to the Equation (1) with the use of the calculated constants according to the following

$$c_v = \frac{0.508 \cdot 1.234 - 0.02 \cdot e^{-0.0053 \cdot t}}{1.234 - e^{-0.0053 \cdot t}} \tag{6}$$

Standard deviation calculated according to the relation [27,30–34] had the value of 0.6.

$$S = \frac{\sqrt{\Sigma(c_m - c_v)^2}}{n - 2} \tag{7}$$

Figure 2 shows the experimentally determined values and the curve fitted to them, for which the parameters were calculated according to the Equation (6).

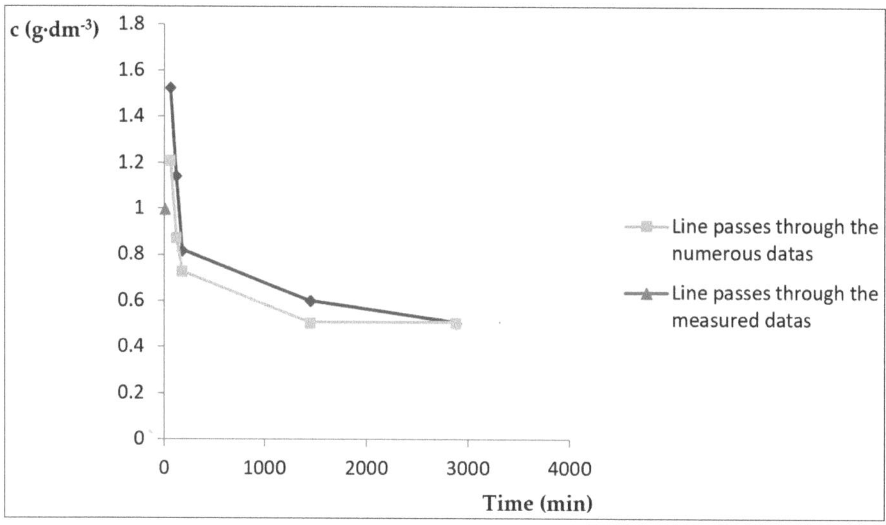

Figure 2. Dependence of the weight concentration c of cupric cations in solution according to time.

Further laboratory measurements were aimed at the evolution of the adsorption isotherm of sorption of cupric cations on the natural zeolite clinoptilolite with a grain size of 2.5–5 mm from Nižný Hrabovec. For the sorption of cuprous cations, we used model solutions with an initial weight concentration of c_0 2.56 g·dm^{-3}, 4.95 g·dm^{-3}, 6.72 g·dm^{-3}, 7.05 g·dm^{-3}, and 8.55 g·dm^{-3}. For determination of the time required to achieve balance in the system, we monitored the dependence of the sorbed quantity from the moment of contact of the sorbent with the adsorbate at a temperature of 25 °C. The results were processed graphically and mathematically with the use of the Freundlich adsorption isotherm. The evolution of sorption for all model samples was monitored at precise one-hour time intervals.

With prolonged contact time of the sorbent with the solution, the concentration of the cupric cations in solution asymptotically approached the equilibrium concentration c_r. The quantity of absorbed cupric cations was calculated according to the Equation (8) [27,31–33] as the difference between the initial concentration of c_0 and the concentration in solution in the equilibrium state c_r, where a is the sorption capacity [mg·g^{-1}], V is the volume of the solution, and m is the sorbent mass.

$$a = \frac{c_0 - c_r}{m} \cdot V \tag{8}$$

To analyze the equilibrium experimental data for adsorption, the Langmuir or Freundlich isothermal models were used. The Langmuir isothermal model is based on the assumption that the surface areas of the adsorbent are homogeneous, and that the maximum adsorption is limited to covering the surface of the monolayer; in contrast, the Freundlich isothermal model is based on the assumption of heterogeneous surface areas and multilayer surface coverage. If the dependence $a = f(c_r)$ can be expressed by the equation

$$a = \frac{a_m \cdot b \cdot c_r}{1 + b \cdot c_r} \tag{9}$$

The experimentally obtained values were plotted with the coordinates $X_i = c_{r,i}$, $Y_i = c_{r,i}/a_i$ to create a line. The Freundlich isotherm assumes that the adsorbate concentration on the surface of the adsorbent increases with the increase in adsorbate concentration. This isotherm is based on sorption on a heterogeneous surface, which is expressed by an exponential equation [9,30].

If the Freundlich isotherm satisfies the expression of the dependence $a = f(c_r)$, it is possible to fit the line with the experimental values plotted in the coordinates $X = \log c_r$, $Y = \log a$. From the obtained values of the angular coefficient of this line and from the section of the line on the Y-axis, it is possible to calculate the sought constants of the Freundlich isotherm. The experimentally obtained and calculated parameters are presented in Table 7.

Table 7. Obtained experimental parameters.

c_r (mg·dm^{-3})	a (mg·g^{-1})	$\log c_r$	$\log a$	c_0 (g·dm^{-3})
200	4.9	2.3	0.6148	2.56
300	6.18	2.47	0.79	4.95
700	11	2.8	1.04	6.72
800	12	2.9	1.079	7.05
820	15.7	3.07	1.2	8.55

The experimentally measured and computed values plotted at coordinates $X = c_{r,i}$, $Y = c_{r,i}/a_i$ did not conform to a straight line. For this reason, it was not possible to describe the dependence $a = f(c_r)$ by the Langmuir equation. We constructed the straight line after having calculated the logarithm of the quantities a and c_r and their plotting in coordinates $X = \log c_r$, $Y = \log a$. For the calculation of the searched quantities, we adopted the Freundlich model.

Assuming an adsorption on a non-homogeneous adsorption surface, a Freundlich adsorption isotherm was used with the following form [9,27,32–38]:

$$a = K \cdot c_r^{1/n} \tag{10}$$

where K and n are Freundlich constants, which indicate the adsorption capacity of the adsorbent and the adsorbent adsorbate affinity. Graphical representation of the Freundlich adsorption isotherm is shown in Figure 3.

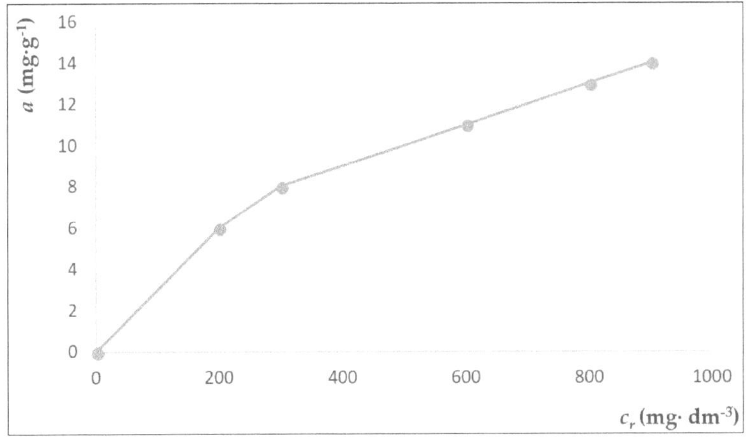

Figure 3. Sorption isotherm for sorption of cupric cations from solutions to clinoptilolite.

The constant values of the isotherm are determined by the least squares method and from the linearised Freundlich equation:

$$\log a = \log K + \frac{1}{n} \cdot \log c_r \tag{11}$$

The logarithmic shape of the isotherm is shown in Figure 4.

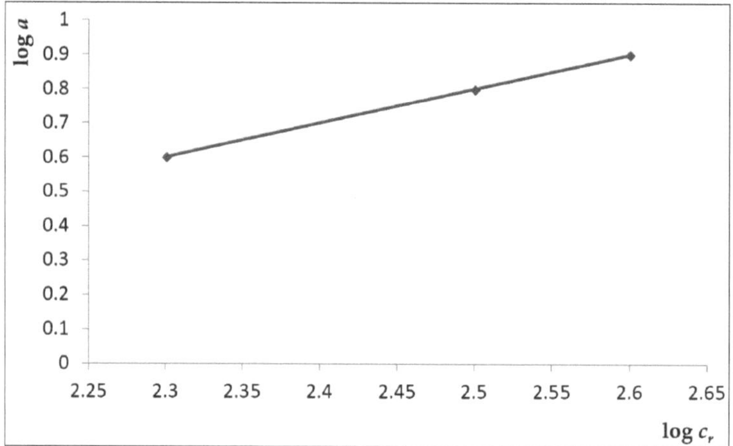

Figure 4. Logarithmic shape of the Freundlich isotherm.

The constant values of the isotherm are determined by the least squares method. By this method, we calculated for the constant, K, a value of 5.012 and for q a value of 0.67.

$$n = 1/k = 0.23$$

$$K = 10^q = 10^{0.67} = 5.012$$

Accordingly, the isotherm shown in Figure 2 can be expressed by the equation

$$a = 5.012 \cdot C_r^{1/0.23}.$$

The Freundlich isotherm [3] assumes that the adsorbate concentration on the surface of the adsorbent increases with an increase in the concentration of solution. Our experiment confirms this. This finding is consistent with the results of the experiments performed by the cited authors. An increase in a concentration generally results in an increase in the amount of copper adsorbed and the rate of adsorption. According to the results obtained by Zendelska et al., the adsorption capacity will increase with an increase in initial concentration until the system reaches a saturation point [38].

Because dependence of the sorbed quantity on the equilibrium concentrations in the logarithmic form was linear, the measured sorption isotherm conformed to the Freundlich sorption isotherm.

Adsorption isotherms tested under laboratory conditions can be used for preliminary investigation as to the potential technological use of natural zeolite in the sewage water treatment process. The isotherm parameters were calculated with the use of experimental results of the sorbed copper cations per gram of sorbent versus their equilibrium concentrations in solution. We compared the sorption capacity calculated for the sample of the natural zeolite used in our experiment with the sorption capacity determined in previous experiments [39]. In the case of monoionic form of sodium, a sample of approximately the same initial concentration, the sorption capacity of the unmodified sample reached half the capacity of the modified sample. When a natural sample was used, we reached equilibrium after 24 h; in contrast, with the modified sample, we reached equilibrium in three hours.

Peric et al. [5] observed the sorption behaviour of Zn, Cu, and Pb on the natural zeolite clinoptilolite from Croatia. Their results show that ion exchange capacity for Cu and Pb is twice the size of Zn under the same experimental conditions, when equilibrium for Cu was achieved after 72 h. In light of these facts and the results obtained, the sorption-ion-exchange method using the natural clinoptilolite is an efficient process for removing heavy metal ions from wastewater containing lower concentrations of contamination. To accelerate this process, it is appropriate to modify the natural clinoptilolite, for example, into the Na-form. According to the results obtained by Holub et al. [40], the sorption of copper cations can also be influenced by the sorbent grain size. More favourable results were obtained with the use of natural clinoptilolite of smaller granularity. Butnariu et al. [41] observed the effects of the sorption of environmental applications by various source materials of natural organic matter. The results suggest a potential for obtaining efficient and cost-effective engineered natural organic sorbents for environmental applications.

4. Conclusions

In emergency situations, the principle of chemical reactions, such as precipitation, can be used for immediate reduction of concentrations of cations of heavy and toxic metallic elements in water. However, it is required that the elimination of ongoing and prolonged contamination is performed by inexpensive concentration-reducing methods. This creates an opportunity for the use of natural zeolites. Although these have a lower sorption rate in comparison with synthetical zeolites, they are much cheaper. The Zeolite-based sorption technology does not require significant space or the use of expensive chemicals. Because this is a natural and easily accessible material, it is assumed that this method could be used in the future in greater extent for cleaning water from cupric cations and from other heavy metals in process plants where high rates of cleaning are not required.

On the basis of the above results, the ability of adsorbent based on the natural zeolite to remove cupric cations from the aqueous environment was confirmed, while sorption capacity of the sorbent increased with the initial concentration of cupric cations in the aqueous solution. We described the evolution of the sorption process by the Freundlich isotherm.

The contact filtration through a suitable material represents an economically acceptable and undemanding technology for removing cupric cations from water. The acquired findings of basic research on specific natural sorbents can, in perspective, provide an important information for technological processes of water purification. The use of sorbents in water purification processes can help the efforts to increase water reserves through a safe re-use of wastewater.

Author Contributions: I.P. and A.P. conceived, designed the experiments, and wrote the paper; J.V., M.H., M.K. performed the experiments and analyzed the data. Z.P. performed review research and made final corrections of the paper.

Funding: This research was funded by the KEGA grant agency 004TUKE-4/2017 and the programme Inter-Excelence LTC17051 European anthroposphere as a source of mineral raw materials.

Acknowledgments: We thank professor Helena Raclavska for supporting our research performed at the ENET centre.

References

1. Pagnanelli, F.; Mainell, S.; Veglio, F.; Toro, L. Heavy metal removal by olive pomace: Biosorbent and equilibrium modelling. *Chem. Eng. Sci.* **2003**, *58*, 4709–4717. [CrossRef]
2. Abu Al-Rub, F.A.; El-Naas, M.H.; Benyahia, F.; Ashour, I. Biosorption of Nickel on Blank Alginate Beads, free and Immobilized Algal cells. *Process Biochem.* **2004**, *39*, 1767–1773. [CrossRef]
3. Christian, P.; Von der Kammer, F.; Baalousha, M.; Hofmann, T. Nanoparticles: Structure, properties, preparation and behaviour in environmental media. *Ecotoxicology* **2008**, *17*, 326–343. [CrossRef] [PubMed]

4. Abdulkareem, S.A.; Muzenda, E.; Afolabi, A.S.; Kabuba, J. Treatment of Clinoptilolite as an Adsorbent for the Removal of Copper Ion from Synthetic Wastewater Solution. *Arab. J. Sci. Eng.* **2013**, *38*, 2263–2272. [CrossRef]

5. Peric, J.; Trgo, M.; Vukojevic Medvidovi, N. Removal of zinc, copper and lead by natural zeolite—A comparison of adsorption isotherms. *Water Res.* **2004**, *38*, 1893–1899. [CrossRef] [PubMed]

6. Sabová, L.; Chmielewská, E.; Gáplovská, K. Preparation and utilization of combined adsorbents on zeolite base for removal of the oxyanion pollutants from water. *Chem. Listy* **2010**, *104*, 243–250.

7. Genç, N.; Dogan, E.C.; Yurtsever, M. Bentonite for ciprofloxacin removal from aqueous solution. *Water Sci. Technol.* **2013**, *68*, 848–855. [CrossRef] [PubMed]

8. Obalová, L.; Bernauer, B. Catalytic decomposition of nitrogen monoxide. *Chem. Listy* **2003**, *97*, 255–259.

9. Chmielewska, E. Vývoj novej generácie environmentálnych adsorbentov a biokompozitov na báze prírodných nanomateriálov. *Chem. Listy* **2008**, *102*, 124–130.

10. Breck, D.W. *Zeolite Molecular Sieves: Structure, Chemistry and Use*; Wiley: New York, NY, USA, 1974.

11. Sovová, T.; Kočí, V. Ekotoxikologie nanomateriálů. *Chem. Listy* **2012**, *106*, 82–87.

12. Van Bekkum, H.; Flanigen, E.M.; Jansen, J.C. *Introduction to Zeolite Science and Practice*; Elsevier Science Publishers: Amsterdam, The Netherlands, 2001.

13. Jacobs, P.A. *Zeolite Chemistry and Catalysis*; Elsevier: Amsterdam, The Netherlands; Oxford, UK; New York, NY, USA, 1991.

14. Wan Yaacob, W.Z.; Samsudin, A.R.; Kong, T.B. The Sorption Distribution Coefficient of Lead and Cooper on the Selected Soil Samples from Selangor. *Bull. Geol. Soc. Malays.* **2008**, *54*, 21–25.

15. Beausse, J. Selected drugs in solid matrices: A review of environmental determination, occurrence and properties of principal substances. *Trends Anal. Chem.* **2004**, *23*, 753–761. [CrossRef]

16. Phan, N.T.S.; van Der Sluta, M.; Jones, C.W. A Palladium (II) Catalyst Supported on Polyacetylene Nanoparticles Combining the Advantages of Homogeneous and Heterogeneous Catalysts. *Adv. Synth. Catal.* **2006**, *348*, 609–617. [CrossRef]

17. Panda, A.; Jurko, J.; Pandová, I. *Monitoring and Evaluation of Production Processes. An Analysis of the Automotive Industry*; Monograph; Springer International Publishing: Cham, Switzerland, 2016.

18. Panda, A.; Jurko, J.; Džupon, M.; Pandová, I. Optimalization of heat treatment bearings rings with goal to eliminate deformation of material. *Chem. Listy* **2011**, *105*, 459–461.

19. Prislupčák, M.; Panda, A.; Jančík, M.; Pandová, I.; Orendáč, P.; Krenický, T. Diagnostic and Experimental Valuation on Progressive Machining unit. *Appl. Mech. Mater.* **2014**, *616*, 191–199. [CrossRef]

20. Jenne, E.A. *Adsorption of Metals by Geomedia: Variables, Mechanism and Model Applications*; Academic Press: San Diego, CA, USA, 1998.

21. Sieczka, A.; Koda, E. Kinetic and Equilibrium Studies of Sorption of Ammonium in the Soil-Water Environment in Agricultural Areas of Central Poland. *Appl. Sci.* **2016**, *6*, 269. [CrossRef]

22. Strnadová, N.; Matějková, D. Adsorption of Copper and Zinc from Aqueous Solution on $Mg(OH)_2$. *Chem. Listy* **2006**, *100*, 803–808.

23. Čejka, J.; Žilková, N. Synthesis and Structure of the Zeolites. *Chem. Listy* **2000**, *94*, 278–287.

24. Pandová, I.; Panda, A.; Jurko, J. Clinoptilolite Testing as Sorbent Nitrogen Oxides from Combustion Engines Gases. *Chem. Listy* **2011**, *105*, 609–611.

25. We Know How to Seduce Nature. Available online: http://www.ZEOCEM.sk (accessed on 16 February 2001).

26. Delle, S.A. Factors affecting sorption of organic compounds in natural sorbent/water systems and sorption coefficients for selected pollutants. A Review. *J. Phys. Chem. Ref. Data* **2001**, *30*, 187–439. [CrossRef]

27. Navrátilová, M.; Sporka, K. Modified Zeolitic Catalysts for the Adamantane Synthesis. *Chem. Listy* **2000**, *94*, 445–448.

28. Hiller, E.; Krascsenits, Z.; Bartal, M. Environmental Fate of Selected Pharmaceuticals: Sorption and Desorption of Sediment. *Acta Environ. Univerzitatis Comen.* **2007**, *15*, 16–25.

29. Boivin, A.; Cherrier, R.; Schiavon, M. Bentazone adsorption and desorption on agricultural soils. *Agron. Sustain. Dev.* **2005**, *25*, 309–315. [CrossRef]

30. Pitter, P.; Tuček, F.; Chudoba, J.; Žáček, I. *Laboratorní Metody v Technologii Vody [Laboratory Methods in Water Treatment Technology]*; SNTL Alfa: Praha, Czech, 1983.

31. Ghobadi, N.Z.; Borghei, S.M.; Yaghmaei, S. Kinetic studies of Bisphenol A in aqueous solutions by enzymatic treatment. *Int. J. Environ. Sci. Technol.* **2018**, 1–12. [CrossRef]

32. Thirugnanasambandham, K.; Sivakumar, V. Enzymatic catalysis treatment method of meat industry wastewater using lacasse. *J. Environ. Health Sci. Eng.* **2015**, *13*, 86. [CrossRef] [PubMed]

33. Roca Jalil, M.I.; Baschini, M.; Sapag, K. Removal of Ciprofloxacin from Aqueous Solutions Using Pillared Clays. *Materials* **2017**, *10*, 1345. [CrossRef] [PubMed]

34. Parker, J.K.; Lignou, S.; Shankland, K.; Kurwie, P.; Griffiths, H.D.; Baines, D.A. Development of a Zeolite Filter for Removing Polycyclic Aromatic Hydrocarbons (PAHs) from Smoke and Smoked Ingredients while Retaining the Smoky Flavor. *J. Agric. Food Chem. Am. Chem. Soc.* **2018**, *66*, 2449–2458. [CrossRef] [PubMed]

35. Wang, C.J.; Li, Z.; Jiang, W.T. Adsorption of ciprofloxacin on 2:1 dioctahedral clay minerals. *Appl. Clay Sci.* **2011**, *53*, 723–728. [CrossRef]

36. Putra, E.K.; Pranowo, R.; Sunarso, J.; Indraswati, N.; Ismadji, S. Performance of activated carbon and bentonite for adsorption of amoxicillin from wastewater: Mechanisms, isotherms and kinetics. *Water Res.* **2009**, *43*, 2419–2430. [CrossRef] [PubMed]

37. Azevedo, D.C.S.; Cardoso, D.; Fraga, M.A.; Pastore, H.O. Zeolites for a Sustainable World. *Microporous Mesoporous Mater.* **2017**, *254*, 1–2. [CrossRef]

38. Zendelska, A.; Golemeova, M.; Blazev, K.; Krstev, A. Adsorption of copper ions from aqueos solutions on natural zeolite. *Environ. Prot. Eng.* **2015**, *41*, 17–36.

39. Pandová, I.; Oravec, P.; Matisková, D. Sorption Characteristics of Cooper Sorption on the Clinoptilolite Measurement. *MM Sci. J.* **2017**, *2017*, 1977–1980. [CrossRef]

40. Holub, M.; Balintová, M.; Pavliková, P.; Palascaková, L. Study of Sorption Properties of Zeolite in Acidic Conditions in Dependence on Particle Size. *Chem. Eng. Trans.* **2013**, *32*, 559–564.

41. Butnariu, M.; Negrea, P.; Lupa, L.; Ciopec, M.; Negrea, A.; Pentea, M.; Sarac, I.; Samfira, I. Remediation of Rare Earth Element Pollutants by Sorption Process Using Organic Natural Sorbents. *Int. J. Environ. Res. Public Health* **2015**, *12*, 11278–11287. [CrossRef] [PubMed]

International Journal of
*Environmental Research
and Public Health*

Article

Influence of Ammonium Ions, Organic Load and Flow Rate on the UV/Chlorine AOP Applied to Effluent of a Wastewater Treatment Plant at Pilot Scale

Eduard Rott *, Bertram Kuch, Claudia Lange, Philipp Richter and Ralf Minke

Institute for Sanitary Engineering, Water Quality and Solid Waste Management, University of Stuttgart, Bandtäle 2, 70569 Stuttgart, Germany; bertram.kuch@iswa.uni-stuttgart.de (B.K.); claudia.lange@dekra.com (C.L.); philipp.richter@iswa.uni-stuttgart.de (P.R.); ralf.minke@iswa.uni-stuttgart.de (R.M.)
* Correspondence: eduard.rott@iswa.uni-stuttgart.de; Tel.: +49-711-685-60497

Abstract: This work investigates the influence of ammonium ions and the organic load (chemical oxygen demand (COD)) on the UV/chlorine AOP regarding the maintenance of free available chlorine (FAC) and elimination of 16 emerging contaminants (ECs) from wastewater treatment plant effluent (WWTE) at pilot scale (UV chamber at 0.4 kW). COD inhibited the FAC maintenance in the UV chamber influent at a ratio of 0.16 mg FAC per mg COD ($k_{\text{HOCl-COD}} = 182\ \text{M}^{-1}\text{s}^{-1}$). An increase in ammonium ion concentration led to a stoichiometric decrease of the FAC concentration in the UV chamber influent. Especially in cold seasons due to insufficient nitrification, the ammonium ion concentration in WWTE can become so high that it becomes impossible to achieve sufficiently high FAC concentrations in the UV chamber influent. For all ECs, the elimination effect by the UV/combined Cl_2 AOP (UV/CC) was not significantly higher than that by sole UV treatment. Accordingly, the UV/chlorine AOP is very sensitive and loses its effectiveness drastically as soon as there is no FAC but only CC in the UV chamber influent. Therefore, within the electrical energy consumption range tested (0.13–1 kWh/m^3), a stable EC elimination performance of the UV/chlorine AOP cannot be maintained throughout the year.

Keywords: ammonium; emerging contaminants; pilot plant; UV/chlorine AOP; UV/HOCl; wastewater treatment

---

1. Introduction

The prevention of the emission of anthropogenic emerging contaminants (ECs) in surface waters is becoming increasingly important, as such compounds can be endocrine disrupting [1] and carcinogenic [2]. Since these compounds are mainly introduced into the environment via wastewater treatment plant effluents (WWTE), additional treatment steps, such as the advanced oxidation process (AOP), are provided for in wastewater treatment plants (WWTPs). One such method is the UV/chlorine AOP. The principle of this process is the transformation of free available chlorine (FAC), e.g., in the form of hypochlorous acid (HOCl) or the hypochlorite anion (OCl$^-$) (pK$_a$ = 7.5), by UV radiation into highly reactive radicals (Equations (1)–(3)), with the aim of oxidizing the ECs to CO_2 and H_2O or at least rendering them biodegradable [3–6]:

$$\text{HOCl} + \text{UV photons} \rightarrow \bullet\text{OH} + \text{Cl}\bullet \tag{1}$$

$$\text{ClO}^- + \text{UV photons} \rightarrow \bullet\text{O}^- + \text{Cl}\bullet \tag{2}$$

$$\bullet\text{O}^- + \text{H}_2\text{O} \rightarrow \bullet\text{OH} + \text{OH}^- \tag{3}$$

In a study with real effluent of a WWTP (continuous operation with 1 m^3/h, medium pressure UV lamp operated at 0.4–1.0 kW) by Rott et al. [7] it was shown at pilot scale that the UV/chlorine AOP is superior to the UV/H$_2$O$_2$ AOP [8] in terms of the elimination of ECs, the bacterial count and the total estrogenic activity, as much lower mass concentrations of oxidant are required. All investigations in this study were carried out at NH$_4^+$-N concentrations <0.1 mg/L.

A major concern associated with the UV/chlorine AOP is the formation of potentially toxic and lipophilic halogenated degradation by-products such as adsorbable organohalogens (AOX) [7,9,10]. Side reactions contribute to the fact that the dosed chlorine immediately reacts to form combined chlorine (CC) or decomposes into chloride. Ammonium ions belong to the most important compounds in WWTE making it difficult to maintain free chlorine in such a form of wastewater. For example, chlorine reacts with ammonium ions preferably to form chloramines, as shown in Equation (4) in the form of monochloramine [11]:

$$HOCl + NH_4^+ \rightarrow NH_2Cl + H_2O + H^+ \tag{4}$$

An important task of WWTPs is the removal of nitrogen. Ammonium ions in the feed are oxidized to nitrate (NO$_3^-$) by aerobic, autotrophic, nitrifying microorganisms. This, in turn, is converted by predominantly heterotrophic, denitrifying bacteria under anaerobic conditions to gaseous, elementary nitrogen (N$_2$), which escapes into the atmosphere. The speed of both processes is severely impaired at low temperatures and comes to a standstill at 8 °C. In the temperate zone, where very cold and warm seasons alternate, the elimination of ammonium ions in WWTPs is thus not guaranteed throughout. Figure 1 shows the ammonium ion concentration in the effluent of the WWTP (Lehr- und Forschungsklärwerk, LFKW, Stuttgart, Germany), the effluent of which was used in the investigations by Rott et al. [7] and this work. It becomes clear that NH$_4^+$-N concentrations of up to 10 mg/L can occur in cold seasons. In this year, the wastewater temperature varied between 8 and 21 °C, with 125 days of it being <13 °C. It is therefore necessary to find out to what extent the UV/chlorine AOP is influenced by high NH$_4^+$ concentrations in the effluent of the WWTP when it is used for the elimination of ECs.

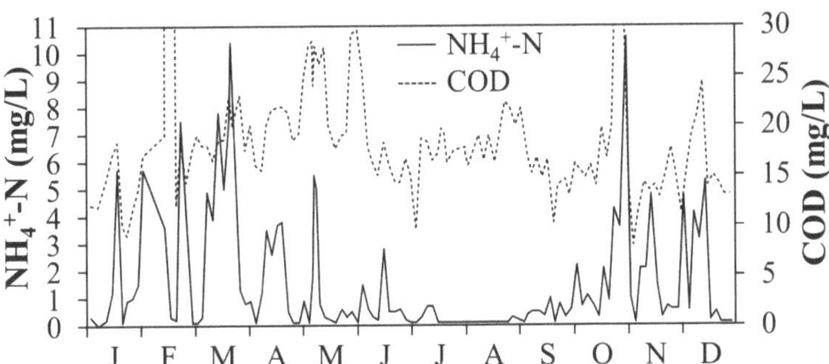

Figure 1. Ammonium ions and chemical oxygen demand (COD) concentrations determined in WWTE of the LFKW (x-axis: months).

The formation of chloramines in the UV/chlorine process does not necessarily lead to the absence of oxidation of ECs. It is known that in the presence of UV radiation chloramines are also converted to reactive radicals (e.g., aminyl and chlorine radicals) (Equation (5)) [6,9]:

$$NH_2Cl + UV \text{ photons} \rightarrow NH_2\bullet + Cl\bullet \tag{5}$$

Only a few studies investigating the UV/chlorine AOP involve high ammonium ion concentrations in their experiments [9,12,13]. Generally, these investigations are only based on synthetic wastewater on a laboratory scale often only simulating the organic carbon content by the dosage of a specific compound such as *tert*-butanol or citric acid and urea [14,15]. However, real conditions in WWTE can be completely different. The aim of this work is therefore to investigate the influence of ammonium ions and the organic load (COD, Figure 1) in wastewater under realistic conditions, i.e., at pilot scale in continuous operation with real WWTE.

The article discussed here is to be understood as a continuation of the article by Rott et al. [7]. In this article, experiments with the pilot plant used here were carried out with WWTE of negligible ammonium ion concentrations of the same WWTP.

2. Materials and Methods

2.1. Electrical Energy Consumption

The electrical energy consumption E (in kWh/m^3, Equation (6) with electrical power input P (in kW) and the flow rate F of the pilot plant (in m^3/h)) is correlated to the running costs of a flow-through plant and is therefore an important factor for the technical applicability of the process [14]. Moreover, assuming first-order kinetics, the electrical energy consumption per order of compound removal (E$_{EO}$) can be calculated using Equation (7), where c_0 is the initial concentration of the compound and c is the concentration of the compound after treatment [16]:

$$E = \frac{P}{F} \tag{6}$$

$$E_{EO} = \frac{P}{F \times \log\left(\frac{c_0}{c}\right)} \tag{7}$$

2.2. Overview of Experiments

Four experiments were carried out. In these experiments, either tap water spiked with diclofenac and carbamazepine (these ECs are among the most frequently found pharmaceuticals in water bodies and they are ineffectively removed in WWTPs [17]) or the effluent of a WWTP was treated with a pilot plant equipped with a medium pressure UV chamber. The WWTE was examined for 16 different ECs. The individual parameters were varied as follows:

- Exp. A: Variation of NH$_4^+$-N concentration (0.5, 1.0, 1.5 mg/L) in spiked tap water (6.9 mg/L dosed free Cl$_2$, 0.4 kW UV power, 1 m^3/h flow rate)
- Exp. B: Variation of WWTE dilutions with spiked tap water (6.9 mg/L dosed free Cl$_2$, 0.4 kW UV power, 1 m^3/h flow rate)
- Exp. C: Variation of CC concentration (1–5 mg/L CC in UV chamber influent) on WWTE (0.0 and 0.4 kW UV power, 1 m^3/h flow rate)
- Exp. D: Variation of flow rate (1, 2, 3 m^3/h WWTE) at 3 mg/L FAC dosage in UV chamber influent and 0.4 kW UV power (0.13, 0.20, 0.40 kWh/m^3 electrical energy consumption)

2.3. Chemicals and Reagents

NaOCl solution (14% active chlorine) and hydrochloric acid (HCl, 32%, AnalaR Normapur) were purchased from VWR International (Radnor, PA, USA). The used H$_2$O$_2$ solution (35%) provided by Siemens Water Technologies (Günzburg, Germany) was of technical grade. Carbamazepine (99%) and diclofenac sodium salt (>98.5%) were purchased from Sigma-Aldrich (St. Louis, MO, USA). NH$_4$Cl (p.a.) was purchased from Merck KGaA (Darmstadt, Germany). Crystalline sodium thiosulfate pentahydrate (Na$_2$S$_2$O$_3$·5H$_2$O, ≥99%) was purchased from Carl Roth (Karlsruhe, Germany). Dichloromethane (CH$_2$Cl$_2$, >99.8%) was purchased from Bernd Kraft GmbH (Duisburg, Germany). N,N-diethyl-p-phenylenediamine (DPD) was contained in powder pillows obtained from Hach (Berlin, Germany).

2.4. Tap Water (TW) and Wastewater Treatment Plant Effluent (WWTE)

The tap water used was analyzed for the following parameters the concentrations of which were mainly below the limit of detection: <0.1 mg/L NH_4^+-N, <5.0 mg/L COD, <1.5 mg/L DOC (dissolved organic carbon), 300–350 µS/cm electrical conductivity, <0.02 mg/L free Cl_2, <0.02 mg/L total Cl_2. Thus, the tap water had not been chlorinated in the waterworks when the experiments were conducted.

The municipal Treatment Plant for Education and Research (LFKW, Lehr- und Forschungsklärwerk) with a capacity of 30 L/s treats an average amount of 900,000 m^3 per annum (9000 population equivalents). Its raw wastewater is composed of domestic wastewater and wastewater from the university grounds mainly of industrial effluents. After the primary clarifier, the wastewater is treated in separated denitrification and nitrification tanks (simultaneous P precipitation). The aerated sludge is separated in a secondary clarifier the effluent of which is additionally separated from particles by micro sieves (15–20 µm pore size).

In experiments with tap water and dilutions of WWTE, carbamazepine and diclofenac were spiked and analyzed. The initial concentrations of these compounds in raw samples of all experiments and important parameters characterizing the wastewater composition are given in Table 1. The temperature of the wastewater was between 14 and 19 °C. The pH varied slightly between 6.9 and 8.2. The WWTE was mainly composed of 5–8 mg/L DOC, approx. 20–30 mg/L COD and approx. 1000 µS/cm electrical conductivity. The NH_4^+-N concentration in WWTE could vary strongly between <0.1 and 6.7 mg/L (Table 1). In experiments with pure WWTE directly drained from the micro sieves effluent, fourteen other emerging contaminants presented in Figure 2 were analyzed. Their initial concentrations are given in the Supplementary Material (Table S1) and varied between 0.02 and 2.19 µg/L.

Table 1. Initial parameter values c_0 measured in the reference samples.

	Experiment	T	pH	COD	DOC	NH_4^+-N	Cond.	CBZ	DCF
		°C	-	mg/L	mg/L	mg/L	µS/cm	µg/L	µg/L
A	<0.1 mg/L NH_4^+-N	16.3	7.4	<5.0	<1.5	<0.1	334 ± 0	0.25 ± 0.00	1.70 ± 0.01
	0.5 mg/L NH_4^+-N	16.0	8.0	<5.0	<1.5	0.5	341 ± 0	0.94 ± 0.02	1.66 ± 0.18
	1.0 mg/L NH_4^+-N	16.7	8.0	<5.0	<1.5	1.0	348 ± 3	0.59 ± 0.03	1.90 ± 0.03
	1.5 mg/L NH_4^+-N	15.5	8.0	<5.0	<1.5	1.5	350 ± 0	1.07 ± 0.04	1.40 ± 0.25
B	<5 mg/L COD	16.3	7.4	<5.0	<1.5	<0.1	334 ± 0	0.25 ± 0.00	1.70 ± 0.01
	10 mg/L COD	17.7	8.2	12.1 ± 1.6	3.1 ± 0.2	<0.1	520 ± 1	0.40 ± 0.00	3.08 ± 0.07
	16 mg/L COD	17.1	8.2	16.2 ± 0.1	4.2 ± 0.4	<0.1	717 ± 4	0.48 ± 0.07	3.01 ± 0.37
	22 mg/L COD	18.2	8.2	21.0 ± 1.5	5.1 ± 0.1	<0.1	909 ± 1	0.48 ± 0.01	2.08 ± 0.51
C	1 mg/L CC	13.9	6.9	31.2 ± 1.0	8.3 ± 0.9	1.58 ± 0.02	920 ± 0	0.68 ± 0.05	2.67 ± 0.18
	3 mg/L CC	14.8	7.0	23.0 ± 0.8	5.9 ± 0.3	6.26 ± 0.05	973 ± 0	0.57 ± 0.02	1.98 ± 0.25
	5 mg/L CC	14.2	7.0	24.8 ± 1.5	6.5 ± 0.9	6.73 ± 0.85	1048 ± 9	0.57 ± 0.01	3.87 ± 0.45
D	1 m^3/h flow rate	18.3	6.9	21.9 ± 0.1	4.5 ± 0.0	0.13 ± 0.00	823 ± 0	0.54 ± 0.02	1.31 ± 0.02
	1 m^3/h flow rate	18.3	6.9	20.6 ± 2.3	n.m.	0.57 ± 0.00	822 ± 6	0.57 ± 0.02	1.42 ± 0.14

n.m.: not measured, T: temperature, COD: chemical oxygen demand, DOC: dissolved organic carbon, Cond.: electrical conductivity, CBZ: carbamazepine, DCF: diclofenac, FAC: free available chlorine, CC: combined Cl_2.

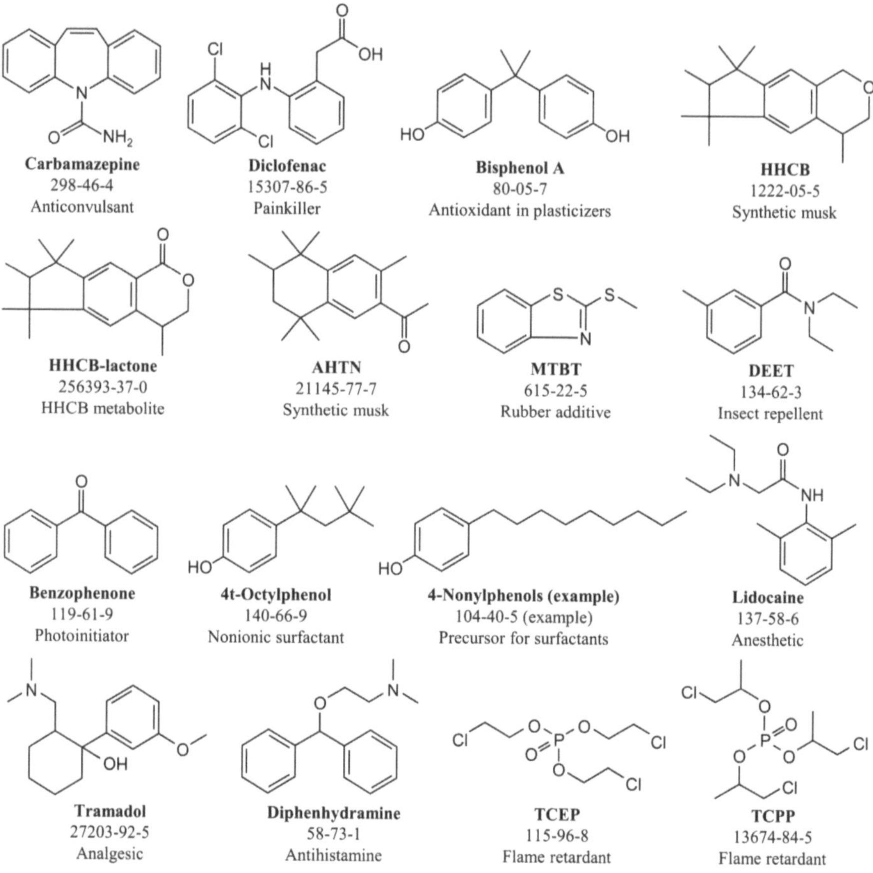

Figure 2. Emerging contaminants analyzed in WWTE samples with CAS numbers (Rott et al. [7] based on [18]).

2.5. Pilot Plant

The UV/AOP pilot plant (Figure 3, see Rott et al. [7] for more details) was placed next to the micro sieves of the WWTP. The plant was fed by means of an eccentric screw pump. The flow rate (1–3 m^3/h) was controlled using a variable area flowmeter. A tap at the inlet of the plant allowed for the sampling of untreated sample (c_0). NaOCl stock solution was dosed into the feed stream using a peristaltic pump (0.08–4 L/h). Subsequent to a static mixer, a portion of the feed stream was directed to a measuring cell where the temperature, pH (single junction, combination electrode sensor, Wallace & Tiernan, Günzburg, Germany) and free available Cl$_2$ (FAC) (potentiostatic electrode amperometry sensor, Wallace & Tiernan, Günzburg, Germany) were analyzed. The medium pressure UV lamp (type: WTL 1000, 1 kW maximum power, 230 mm length × 22 mm diameter, Wallace & Tiernan, Günzburg, Germany), protected by a quartz sleeve with a thickness of 1 mm and cut-off at 200 nm wavelength, was installed in a stainless steel chamber (Wallace & Tiernan Barrier M35, 300 mm assembly dimension × 214 mm height × 600 mm length) (approx. contact time in the UV chamber: 6–10 s). The irradiance was visualized by a 4–20 mA UV sensor (signal in W/m^2) on the cabinet. The UV chamber effluent could be mixed with H$_2$O$_2$ via a second peristaltic pump in order to quench residual free Cl$_2$ (RFC). This study focused on the technical feasibility of the UV/chlorine process. Therefore, this peristaltic pump was mainly operated in automatic mode, i.e., the H$_2$O$_2$ dosage was automatically controlled by means of a chemical feed analyzer (via a further static mixer, a partial stream was passed into a further measuring cell where the RFC concentration could be measured) and process controller (MFC Analyzer/Controller) from Wallace & Tiernan. Since the FAC concentration could vary during an experiment while the H_2O_2 dosage was running and the experiments were limited in time, it was therefore not possible to determine the RFC concentration on a regular basis in case of missing H_2O_2 dosage. This aspect is therefore not addressed in this article. The contact time of the quenching agent from its dosage point to the effluent of the pilot plant was 4–6 s. The pilot plant effluent (treated sample (c)) could be sampled via a sampling tap at the outlet of the pilot plant (upper sampling tap). At a flow rate of 1 m^3/h, the flow time in the pilot plant was 25–28 s.

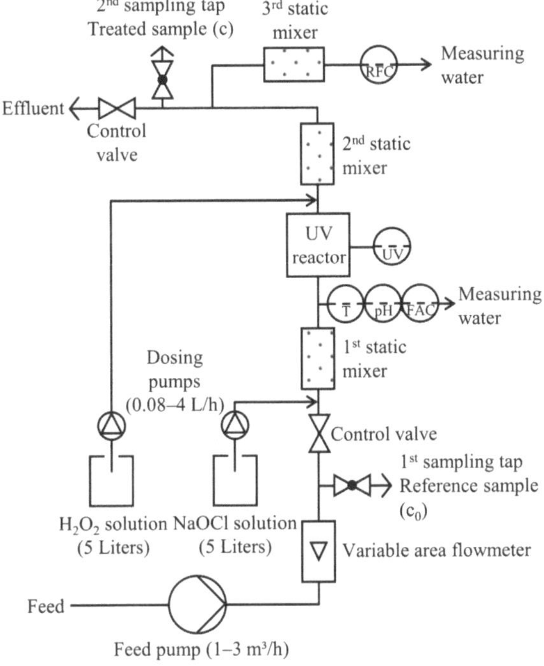

Figure 3. Technical scheme of the UV/chlorine AOP pilot plant (with changes from Rott et al. [7]).

2.6. Experimental Procedure

2.6.1. Preparations

Prior to some experiments, carbamazepine, diclofenac and ammonium chloride (NH_4Cl) were dissolved in pure water in separate 1 L flasks (stock solutions). In order to obtain similar carbamazepine and diclofenac concentrations in all spiked tap water experiments (Exp. A, B and C), at first two 1 L samples were collected from continuously circulated WWTE in a big tank (40 m^3). In these samples, the concentrations of both compounds were analyzed. By adding a quantity of the abovementioned stock solutions matched to this concentration to 800 L of sample, an attempt was made to obtain similar initial EC concentrations in these experiments. As it can be seen in Table 1, only in some cases similar concentrations could not be achieved in all batches showing that an exact adjustment of the EC concentration in the µg/L range on this scale was challenging. This can be attributed to very fine residual pollution contaminated with the ECs in the 800 L sample tank despite meticulous cleaning of the tank prior to the experiment. Furthermore, the spiked compounds could adsorb on such deposits and the analyzed dissolved concentration of the ECs could therefore be lower than expected. However, previous experiments [7] had shown that as long as the initial concentrations of the ECs are in a similar range, comparable results of the c/c_0-ratio can be determined.

2.6.2. General Procedure

The feed pump (1 m^3/h flow rate), UV lamp (operated at 0.4 kW), the NaOCl dosing pump and quenching agent dosing pump were switched on consecutively. After 10 min, the samples were collected. First two or three sample bottles (i.e., duplicate or triplicate samples) were filled with reference sample (c_0). Subsequently, two or three sample bottles were filled with treated sample (c).

2.6.3. Exp. A: Variation of NH_4^+-N Concentration in Spiked Tap Water

Four 800 L tap water batches with different concentrations of NH_4^+ were prepared (0.0, 0.5, 1.0 and 1.5 mg/L NH_4^+-N) and treated separately as described as follows. When the feed tank was filled with 800 L of tap water, the carbamazepine, diclofenac and NH_4^+ stock solutions were added to the tank. In order to achieve a good homogenization, the tank was stirred for 1 h. For each of the four batches, separately the general procedure described in Section 2.6.2 was performed (duplicate samples taken for the analysis). For all batches, the flow rate of the NaOCl dosing pump was set adjusting a concentration of 6.9 mg/L dosed free Cl_2. The H_2O_2 concentration (quenching agent) was around 3.2 mg/L. The UV sensor signal was 224 ± 21 W/m^2.

2.6.4. Exp. B: Variation of WWTE Dilutions with Spiked Tap Water

40 m^3 WWTE were collected in a tank in the micro sieves hall (Figure 4). With this WWTE, three 800 L batches with different tap water (TW) to WWTE ratios were prepared (530 L TW and 270 L WWTE, 270 L TW and 530 L WWTE, 800 L WWTE) and treated separately as described as follows (the batch regarding sole TW was already investigated in Exp. A (0 mg/L NH_4^+-N)). At first, the stirred tank was filled with WWTE and then with tap water. At the same time, the carbamazepine and diclofenac stock solutions were added to the tank. In order to achieve a good homogenization, the tank was stirred for 1 h. For each of the three batches, the general procedure described in Section 2.6.2 was performed separately (duplicate samples taken for the analysis). For all batches, the flow rate of the NaOCl dosing pump was set adjusting a concentration of 6.9 mg/L dosed free Cl_2. The H_2O_2 concentration (quenching agent) was around 3.2 mg/L. The UV sensor signals were 243 W/m^2 (800 L TW), 175 ± 11 W/m^2 (530 L TW and 270 L WWTE), 128 ± 8 W/m^2 (270 L TW and 530 L WWTE), 101 ± 8 W/m^2 (800 L WWTE).

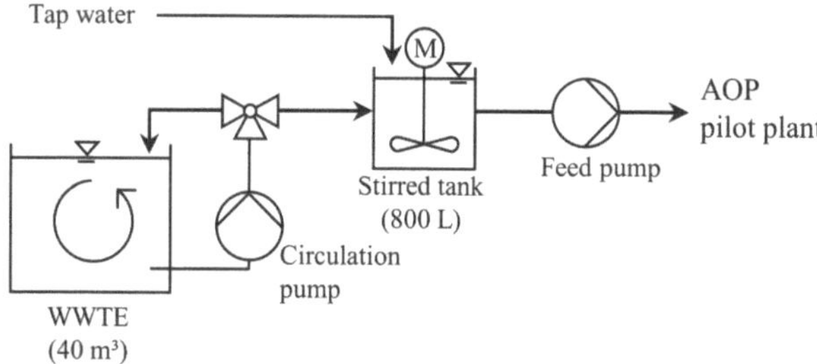

Figure 4. Scheme of the process setup in UV/chlorine AOP Exp. B with different WWTE dilutions.

2.6.5. Exp. C: Variation of CC Concentration on WWTE

In this experiment, NH_4^+-loaded WWTE was withdrawn from the micro sieves effluent directly. At first, the general procedure was performed as described in Section 2.6.2, but with the UV lamp switched off. Due to very high NH_4^+ concentrations in the WWTE (Table 1), free Cl_2 from the dosed NaOCl solution reacted immediately to form CC, so no FAC could be detected in the UV chamber influent. Thus, a 1 mg/L total Cl_2 concentration (1 mg/L CC, 0.0 kWh/m^3) in the UV chamber influent was adjusted. Now, the UV lamp was switched on and set to 0.4 kW (75 $\pm$ 5 W/m^2). After 10 min, further treated samples were collected. Now, the quenching agent H_2O_2 dosage via the second dosing pump was switched on (3.0–4.5 mg/L). After 10 min, the next treated samples were taken from the upper sampling tap. This procedure was repeated for 3 mg/L CC and 5 mg/L CC (both with and without 0.4 kW UV power, both with and without quenching agent dosage) on two different days. Each time triplicate samples were taken for the analysis.

2.6.6. Exp. D: Variation of Flow Rate

In this experiment, WWTE was withdrawn directly from the micro sieves effluent. Through the whole experiment, a FAC concentration of 3 mg/L in the UV chamber influent was set. At first, the general procedure was performed as described in Section 2.6.2 (0.4 kW, 1 m^3/h, 0.40 kWh/m^3, 106 W/m^2). Next, the flow rate was increased to 2 m^3/h, subsequently repeating the general procedure (0.4 kW, 0.20 kWh/m^3, 98 W/m^2, no reference sample taken). The same procedure was repeated with a flow rate of 3 m^3/h (0.4 kW, 0.13 kWh/m^3, 92 W/m^2). Each time triplicate samples were taken for the analysis. The concentration of quenching agent was 3.8–5.8 mg/L H_2O_2.

2.7. Analytical Methods

2.7.1. Free Cl_2 (FAC, RFC), Combined Cl_2 (CC), Total Cl_2

For the on-site determination of free Cl_2 and total Cl_2 equivalent concentrations, a DPD powder pillow method was used (Hach, photometer SQ 118, Merck) (DPD: *N,N*-diethyl-*p*-phenylenediamine). The concentration of dosed Cl_2 was calculated from the flow rates of the feed pump, the dosing pump and the NaOCl stock solution concentration [7]. With 'other Cl-containing reaction products' (OCRP) the difference between dosed Cl_2 and measured total Cl_2 is described (e.g., OCRP can be chloride). During all experiments, Cl_2 measurements were carried out as soon as a certain state of equilibrium was achieved.

2.7.2. Emerging Contaminants

Each 1 liter sample was pretreated with 15 mg of the reducing agent sodium thiosulfate ($Na_2S_2O_3$). The determination of ECs was performed via gas chromatography directly coupled with a mass selective spectrometer (5890N Series II GC, Hewlett Packard, Palo Alto, CA, United States, Hewlett Packard 5972 Series detector, column: VF-Xms, length: 30 m, diameter: 0.25 mm, film thickness: 0.25 µm, Varian, Palo Alto, CA, United States). After the addition of internal standards, the samples were liquid-liquid extracted (dichloromethane, 2 $\times$ 40 mL) and evaporated to 100 µL. Quantification was done using the isotope dilution method and external calibration. The limit of quantification (LOQ) was 1 ng/L.

2.7.3. Other Parameters

The temperature and pH value were measured on-site in measuring cells of the pilot plant using a single junction combination electrode sensor by Wallace & Tiernan. The electrical conductivity was measured by means of a WTW TetraCon 325 conductivity detector and a WTW Multi 350i device. The $NH_4{}^+$-N concentration (Hach LCK 304) and COD (Hach LCK 414, 5 mg/L LOQ) were determined with cuvette rapid tests without prior treatment. The COD cuvettes were heated in a thermostat (Hach HT 200S) for 2 h at 148 °C. The DOC concentrations (1.5 mg/L LOQ) were measured by means of the thermo-catalytic UV oxidation method implemented in the multiN/C 3000 device (Analytik Jena, Jena, Germany). Prior to this analysis, each sample was acidified by hydrochloric acid (pH 2) and filtered (cellulose nitrate, 0.45 µm pore size).

2.7.4. Number of Measurements

The given values in diagrams or tables are mean values calculated from two or three equivalent samples taken consecutively (see experiment descriptions). Error bars in diagrams as well as numbers after the $\pm$ symbol in tables correspond to the calculated standard deviation.

3. Results and Discussion

3.1. Chlorine Species

3.1.1. Exp. A: Variation of NH_4^+-N Concentration in Spiked Tap Water

In Figure 5, the left columns depict the measured concentrations of FAC, CC and OCRP in the UV chamber influent of Exp. A and B. The right columns show the concentrations of these chlorine species in the pilot plant effluent after quenching with H_2O_2.

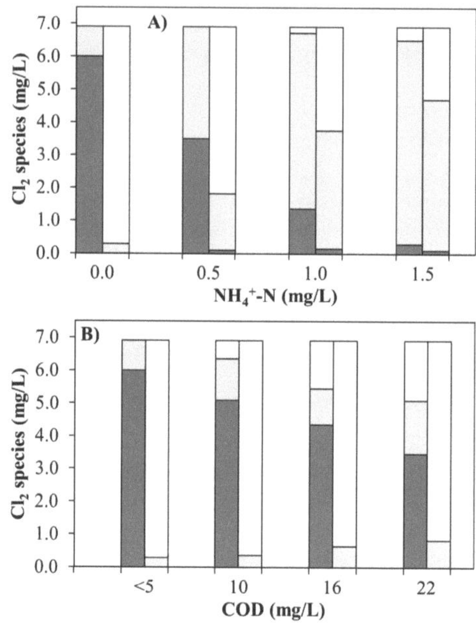

Figure 5. FAC/RFC (■), CC (□) and OCRP (□) in the UV chamber influent (left columns) and after UV treatment at 0.4 kW and subsequent quenching with H_2O_2 (right columns) by means of an AOP pilot plant. (**A**) Exp. A (1 m³/h TW, mainly <5 mg/L COD) and (**B**) Exp. B (1 m³/h, dilutions of WWTE, <0.1 mg/L NH_4^+-N).

With RFC concentrations <0.1 mg/L in the pilot plant effluent, it is evident from all experiments that by quenching with H_2O_2 a good removal of that proportion of chlorine that had not reacted in the UV reactor (not determined) was achieved. As a rule, this required a concentration of around 3.2 mg/L H_2O_2, which corresponds to 95 µM. For the complete reaction of free Cl_2, the stoichiometric equivalent of H_2O_2 is theoretically sufficient [19], which corresponds to 6.7 mg/L Cl_2. Thus, since the maximum concentration in the pilot plant influent was only 6 mg/L Cl_2, the H_2O_2 dosage of 3.2 mg/L H_2O_2 was theoretically sufficient. The reaction rate for this case ($k_{HO2^- \text{-HOCl}} = 4.4 \times 10^7$ $M^{-1}s^{-1}$ [19]) is so high that the quenching can take place within a few milliseconds. A contact time of about 4–6 s between the H_2O_2 dosing point and the pilot plant effluent was therefore more than sufficient.

Although Exp. A was carried out with tap water, which in traces had only been spiked with carbamazepine and diclofenac, and the DOC of which was below the LOQ of 1.5 mg/L, in the case of no ammonium chloride spiking, a dosage of 6.9 mg/L Cl_2 was required to obtain a concentration of 6 mg/L FAC in the UV chamber influent (Figure 5A). The slightly higher dosage was due to hardly noticeable organic impurities that were present in the tap water or e.g., residual impurities in the stirring tank, the pump hoses or in the static mixers (the plant was thoroughly flushed before each experiment, however, a 100 percent cleaning was challenging). For all NH_4^+ concentrations examined, the same dosage concentration of 6.9 mg/L Cl_2 was used. As expected, the increase in ammonium ion concentration led to a decrease of the FAC concentration in the UV chamber influent. At all NH_4^+ concentrations studied, furthermore, lost free chlorine was found almost entirely in the form of combined chlorine, i.e., in the form of chloramines. Thus, a concentration of 1 mg/L NH_4^+-N (71 µM) already reduced the achievable FAC concentration by 75%. The associated loss of 4.7 mg/L free Cl_2 (66 µM) was quasi-equimolar with the NH_4^+-N concentration of 71 µM. Accordingly, a stoichiometric inhibition of the UV/chlorine AOP is to be expected in tap water by ammonium ions (inhibition ratio of 4.7 mg FAC per mg NH_4^+-N between 0 and 1 mg/L NH_4^+-N) (Figure S1).

Margerum et al. [20] as well as Qiang and Adams [21] found an apparent rate constant of 1.3×10^4 $M^{-1}s^{-1}$ for the reaction of HOCl with NH_4^+ at 25 °C and pH 7 [22]. Using this value, a contact time of 5.6 s between the dosage point of chlorine and the Cl_2 measuring cell could be determined with the least squares method (Table S2 and Figure S2). Accordingly, on the basis of the diameter of the pipes, a contact time of 6.1 s between the chlorine dosage point and the UV chamber influent was calculated (Table S3). The difference in contact time was therefore only slight, so that the FAC concentration between the UV chamber influent and the measuring cell differed only by a maximum of 0.05 mg/L (Table S2).

Furthermore, it can be seen from Figure 5A that the CC concentration in the pilot plant effluent was always lower than in the UV chamber influent for all tested batches (with an increase in NH_4^+ concentration from left to right, the degree of CC elimination changed as follows: 68, 50, 33, 26%). A specific proportion of the chloramines present in the UV chamber influent was therefore degraded in the UV chamber. The elimination of CC by UV light is a familiar phenomenon. Yang et al. [9] found a similar decrease from 2.1 to 1.6 mg/L monochloramine in ammonium-rich wastewater (pH 7) by UV light (10 W). Chuang et al. [23] also found that at pH 7 NH_2Cl is reduced up to 50% at fluences of up to 3000 mJ cm^{-2}. The weakly pronounced falsification of the result of the CC concentration in the pilot plant effluent by the quenching agent H_2O_2 as quantified as 0.0388 mg total Cl_2/mg H_2O_2 [7] is estimated to be very low.

3.1.2. Exp. B: Variation of WWTE Dilutions with Spiked Tap Water

Since the COD is an adequate parameter to describe the cumulative organic load of WWTE, all dilutions in Exp. B are classified by their initial COD (Figure 5B). In Table 1, the exact measured COD values of the dilutions can be seen. For simplification reasons, these COD values were simplified to 10, 16 and 22 mg/L COD in Figure 5B. The COD of pure tap water could therefore be calculated to approx. 4 mg/L (<5 mg/L). In all batches, NH_4^+-N was always <0.1 mg/L, which allowed the investigation of the sole influence of COD, i.e., organic and some inorganic compounds in WWTE, on the UV/chlorine process. As in Exp. A, for all of the four batches in Exp. B, always the same NaOCl stock solution dosage of 6.9 mg/L Cl_2 was applied. The FAC concentration obtained in the UV chamber influent decreased linearly proportional to the COD up to 22 mg/L COD at a ratio of 0.16 mg FAC per mg COD (Figure S3). As already reported by Rott et al. [7] in the case of undiluted effluent from a WWTP, therefore, to obtain the desired FAC concentration in the UV chamber influent approximately the double dosage was necessary.

CC increased slightly with an increase in COD (approx. 0.03 mg CC/mg COD). In the UV chamber, however, CC was eliminated between 40 and 70%. Compared to Exp. A, a far greater proportion of OCRP was found in the UV chamber influent, which also increased at a significantly higher ratio of

0.1 mg OCRP/mg COD. This is obvious, as in Exp. B chlorine reacted predominantly with organic compounds, not all chlorinated products of which can necessarily be detected as CC using the DPD method [7].

Knowing the exact contact time of 5.6 s between the dosing point of chlorine and the Cl_2 measuring cell from Exp. A and the recorded FAC concentrations in this measuring cell at known COD concentrations, the least squares method could be used to determine the rate constant between HOCl and COD to be 182 $M^{-1}s^{-1}$ (the COD is not the actual reaction partner of HOCl, but represents the sum of all organic compounds in the sample, Table S4 and Figure S4). The COD of the investigated WWTE of 22 mg/L was typical for the investigated WWTP and thus representative, albeit slightly above the annual mean value of 19.7 mg/L (Figure 1). Since COD limit values usually depend on the size class of WWTPs and can even be in the three-digit mg/L range, it should be considered that such WWTPs would require relatively high Cl_2 doses. Assuming the abovementioned rate constant of HOCl with COD, for instance, at a COD of 80 mg/L in WWTE (neglecting NH_4^+) about 34 mg/L of dosed Cl_2 would be required to obtain a desired FAC concentration of 3 mg/L in the UV chamber influent (Table S5).

Whether ammonium ions or organic pollution play a major role in FAC inhibition during the entire operating year of a WWTP, is very case-specific. The following calculation intends to solve this question for the year of operation of the LFKW shown in Figure 1. In this year, the annual mean value of the NH_4^+-N concentration in WWTE was 1.56 mg/L and the COD average was 19.7 mg/L. Based on the abovementioned inhibition ratios, because of ammonium ions the required Cl_2 dosage to obtain 3 mg/L FAC on average in the UV chamber influent would have been 10.3 mg/L, with 7.3 mg/L of it being inhibited by ammonium ions on average. Due to the organic constituents (COD), the required average Cl_2 dosage to obtain 3 mg/L FAC in the UV chamber influent would have been 6.2 mg/L, with 3.2 mg/L FAC of it being inhibited on average (under the simplified assumption that the required Cl_2 dosage to obtain a specific FAC concentration is linearly proportional to the COD at $\leq$30 mg/L COD (Figure S5)—over 95% of the year, this concentration range prevailed in the WWTE). Thus, under very simplified assumptions, this would have resulted in a required annual average Cl_2 dosage concentration of at least 13.5 mg/L. This shows that on average ammonium ions in the WWTE would have inhibited dosed chlorine more strongly than organic components.

NH_4^+-N concentrations of 5–10 mg/L, for example, would result in minimum dosages of 27–54 mg/L Cl_2 to obtain a FAC concentration of 3 mg/L in the UV chamber influent (Table S5). However, such high dosing quantities are highly questionable with regard to the formation of critical by-products. For periods in which such high ammonium ion concentrations prevail, it would therefore be decisive for the applicability of the UV/chlorine AOP whether CC also causes a sufficiently efficient EC elimination due to its activation with UV light.

3.1.3. Exp. C: Variation of CC Concentration on WWTE

In the experiment investigating the efficiency of the UV/CC AOP, a sufficiently high NH_4^+ concentration was present in the WWTE at all three CC concentrations tested (1, 3, 5 mg/L). Consequently, the dosed free Cl_2 reacted quickly to form CC and was thus almost completely detected in the UV chamber influent as CC (OCRP concentrations in the UV chamber influent at 1, 3, 5 mg/L CC were: 0.6 mg/L, 0.2 mg/L, 0.1 mg/L (not shown in Figure 6)). The COD of the raw samples varied only slightly between 24 and 31 mg/L. When the UV lamp was off, the CC concentration between the UV chamber influent and the pilot plant effluent did not change significantly (Figure 6). Only when the UV lamp was on (operated at 0.4 kW), the CC concentration dropped between 5 and 20%, indicating activation/decay of chloramines possibly according to Equation (5).

Within the scope of this work, it was not investigated which compounds exactly made up the CC. However, since the rate constant of chlorine with ammonium ions is almost one hundred times greater than the rate constant of chlorine with COD (see Sections 3.1.1 and 3.1.2), it can be assumed that mainly inorganic chloramines were formed in the presence of ammonium ions. When H_2O_2

quenching was carried out additionally, the CC concentrations found in the pilot plant effluent were slightly higher than those without quenching. This is because H_2O_2 leads to a slight falsification of the total Cl_2 determination method with DPD [7,24]. However, this falsification is not significant enough to interfere with the conclusion that CC cannot be quenched with H_2O_2. Assuming that most of the CC was composed of inorganic chloramines, this is obvious since the rate constants of monochloramine ($k_{NH2Cl-H2O2} = 2.76 \times 10^{-2}$ $M^{-1}s^{-1}$ [25]) and dichloramine ($k_{NHCl2-H2O2} = 3.60 \times 10^{-6}$ $M^{-1}s^{-1}$ [25]) with H_2O_2 are very low.

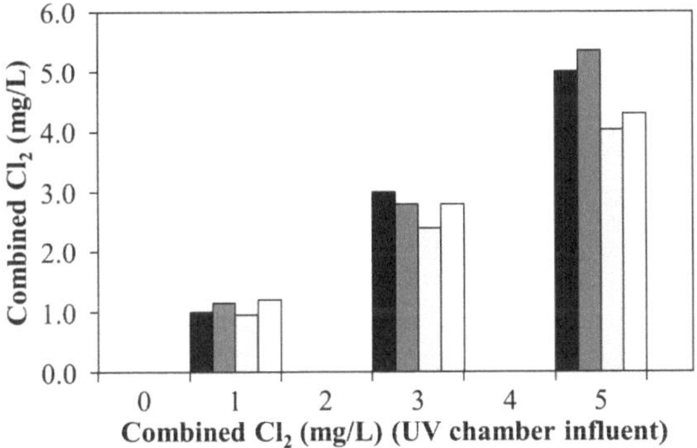

Figure 6. Combined Cl_2 in the UV chamber influent (■) and pilot plant effluent after treatment of 1 m^3/h NH_4^+-loaded WWTE with CC (no UV) (■), UV/CC (0.4 kW, no quenching) (□) and UV/CC (0.4 kW, 3.0–4.5 mg/L H_2O_2 quenching) (□) by means of an AOP pilot plant (Exp. C).

Chlorinated compounds are considered environmentally critical and should not be simply discharged with the WWTE into the receiving water. The fact that CC was only slightly eliminated when the UV lamp was on and CC could not be sufficiently removed by quenching with H_2O_2 shows that high NH_4^+ concentrations in the WWTE make the UV/CC AOP seem impractical at the investigated UV power range.

3.1.4. Exp. D: Variation of Flow Rate

From Figure 7 it becomes apparent that Exp. D was carried out at a time when the NH_4^+-N concentration slowly increased from 0.13 to 0.57 mg/L during the experiment. Thus, at a flow rate of 1 m^3/h, approximately twice the Cl_2 dosage amount was required to obtain 3 mg/L FAC in the UV chamber influent, whereas at 3 m^3/h this was only the case at four times the amount. This was accompanied by an increasing CC and OCRP concentration with increasing flow rate. The CC concentration was hardly reduced by the UV irradiation, at 3 m^3/h it even increased slightly. The lack of CC elimination at higher flow rates indicates that at 3 m^3/h the wastewater passed the UV chamber too quickly (2–3 s) resulting in no sufficient time for CC photolysis. The slight increase can be attributed to a slight falsification of the DPD method by H_2O_2.

3.2. Emerging Contaminants

3.2.1. Exp. A: Variation of NH_4^+-N Concentration and Exp. B: Variation of WWTE Dilutions

Figure 8A shows the residual concentrations of carbamazepine (CBZ) and diclofenac (DCF) as a function of the ammonium ion concentration in tap water matrix treated with 6.9 mg/L dosed Cl_2 at 0.4 kW UV power. Figure 8B shows the dependence of the residual concentrations on different dilutions of WWTE (22 mg/L COD) with tap water (approx. 4 mg/L COD). The elimination of CBZ from tap water without ammonium ions was approx. 84%, whereas DCF was eliminated in this matrix at approx. 99.4%. An increase in NH_4^+-N to 1.5 mg/L resulted in a similar deterioration of the degree of elimination for both ECs as an increase in the COD concentration to around 22 mg/L. Accordingly, with the highest NH_4^+-N concentration and COD tested, the CBZ elimination was only 33–34% and the DCF elimination was 82–86%. The trend lines could be represented well predominantly by means of square equations.

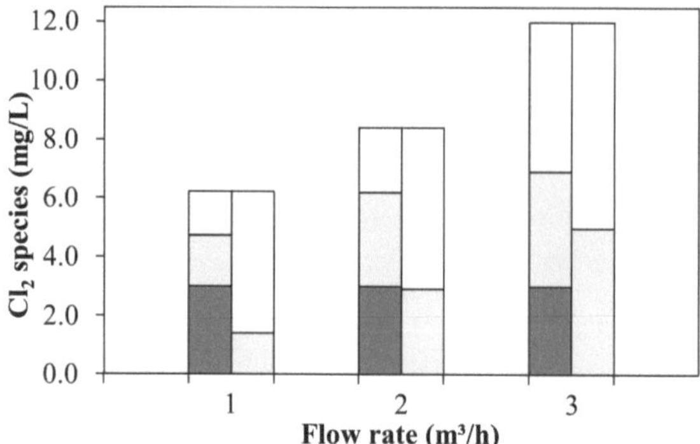

Figure 7. FAC/RFC (■), CC (□) and OCRP (□) in the UV chamber influent (left columns) and after UV treatment and subsequent quenching with H_2O_2 (right columns) by means of an AOP pilot plant at 0.4 kW (Exp. D: WWTE, 0.13–0.57 mg/L NH_4^+-N).

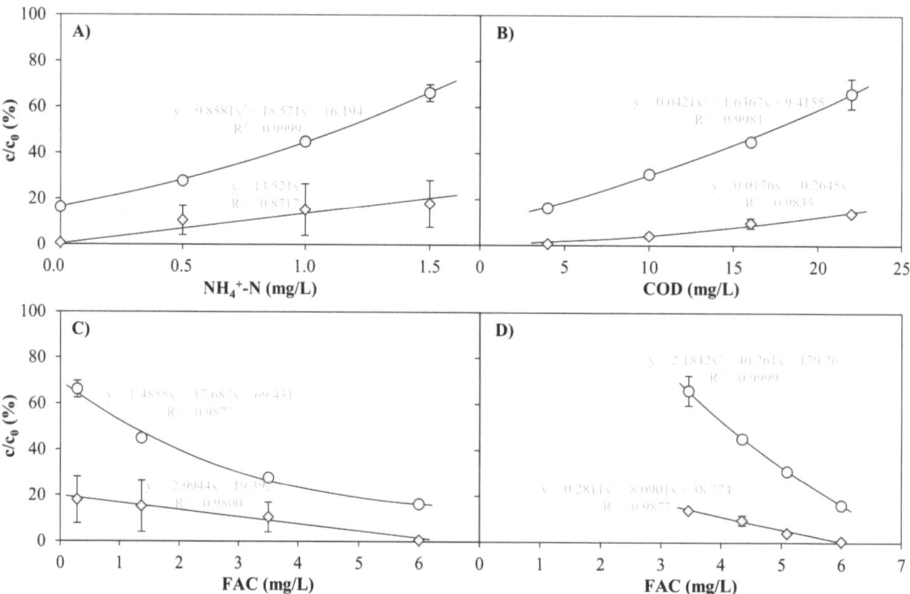

Figure 8. Carbamazepine (○) and diclofenac (◇) after treatment of tap water spiked with NH_4Cl (Exp. A, <5 mg/L COD) and dilutions of WWTE with tap water (Exp. B, <0.1 mg/L NH_4^+-N) by means of an AOP pilot plant at 1 m³/h, 6.9 mg/L dosed Cl_2, 0.4 kW UV lamp power (0.4 kWh/m³ electrical energy consumption) and subsequent quenching with around 3.2 mg/L H_2O_2 as functions of NH_4^+-N (**A**); COD (**B**) and FAC concentration as detected in the UV chamber influent (**C**: Exp. A; **D**: Exp. B). For E_{EO} values, see Table S6.

DCF is very susceptible to photolysis [26]. In experiments with the same UV pilot plant as in this study [7], the UV photolysis at 0.4 kWh/m^3 in WWTE matrix resulted in 81–90% elimination of this compound, whereas with 3 mg/L FAC in the UV chamber influent (7.3 mg/L dosed Cl$_2$) the elimination was only 52–53% (without UV). CBZ, on the other hand, was not eliminated at all with sole FAC treatment, but was removed by 18–22% with UV light only at 0.4 kWh/m^3 electrical energy consumption (no chlorine dosage) in WWTE matrix. It can therefore be assumed that the degradation of the latter compound as found in Figure 8 was mainly caused by radicals, whereas for DCF UV light was sufficient to degrade the molecule, and chlorine radicals but also free chlorine only slightly contributed to an improved elimination.

Soufan et al. [27] observed a third-order reaction between CBZ and HOCl at pH 7 (145 M^{-2}s^{-1}), whereas the reaction between DCF and HOCl was found to be of second-order (3.5 M^{-1}s^{-1}) [28]. These kinetic rate constants were determined at initial EC concentrations of 10 μM (3 mg/L DCF, 2.4 mg/L CBZ). The investigated [HOCl]/[CBZ] ratio was between 37 and 550, whereas the one of [HOCl]/[DCF] was 17–33. In this work, however, the EC concentrations were so low that the [HOCl]/[EC] ratio was between 10,000 and 25,000. Accordingly, it is not surprising that DCF half-lives of more than 30 min (and DCF is the more reactive EC), as calculated by using the rate constant 3.5 M^{-1}s^{-1}, were not applicable to the study presented here. For the same reason, the comparatively low rate constant of k_{obs} = 0.78 min^{-1} for CBZ degradation by UV/chlorine (2 mg/L CBZ (8.5 μM), 280 μM Cl$_2$, 1.48 mW/cm^2 (41 W), pH 7, in pure water) as found by Wang et al. [15] was not transferable to the results of this study as well.

Figure 8C,D clearly show that the similar degrees of elimination of the two ECs between Exp. A and Exp. B did not correlate with the FAC concentration in the UV chamber influent. This indicates that the elimination of ECs cannot be traced back to FAC alone. The CC concentrations in the UV chamber influent resulting from the different NH$_4$$^+$-N concentrations were considerably higher (0.9–6.2 mg/L CC) than those resulting from the different COD concentrations (0.9–1.6 mg/L CC). On the other hand, the FAC concentrations were considerably lower. Despite these lower FAC concentrations in the presence of NH$_4$$^+$, the elimination of ECs was similar in the investigated measuring range with both ammonium ions and COD. It is obvious that in Exp. A CC was composed of inorganic chloramines, which decompose to radicals by UV light (Equation (5) [6,9]). Figure 5A also shows that CC was degraded partially in the presence of UV irradiation. It can therefore be assumed that this conversion of chloramines into radicals by UV light also contributed to the degradation of the ECs.

3.2.2. Exp. C: Variation of CC Concentration on WWTE

In Table 1 and Table S1, the initial concentrations of the ECs in each reference sample of Exp. C can be seen. Thus, for the vast majority of ECs analyzed, the initial concentrations did not differ significantly. In Figure 9, the results of Exp. C (sole CC and UV/CC treatment) are compared to those obtained in an experiment with effluent of the same WWTP with varied FAC concentrations in the UV chamber influent of the same UV pilot plant at 0.4 kWh/m^3 (UV/FAC) [7]. Furthermore, a solid gray line demonstrates the residual EC concentrations after sole UV treatment (0.4 kWh/m^3) (range of standard deviation taken from Rott et al. [7]).

When no UV radiation was applied and only CC was present in the UV chamber influent, there was no elimination of the ECs in the WWTE. It is evident that the dosed Cl$_2$ reacted quickly with the ammonium ions in the wastewater (k = 1.3 × 10^4 M^{-1}s^{-1} [20,21]). These ammonium ions thus competed with the ECs for free Cl$_2$ [29]. Chloramines (it can be assumed that the majority of CC consisted of inorganic chloramines, see Section 3.1.3) can also react with ECs, but this reaction is significantly slower than with free Cl$_2$ [29]. This clearly shows that the oxidizing ability of inorganic chloramines is not sufficient for compounds that are present in traces to be significantly degraded within the very short contact time of less than 30 s prevailing in the pilot plant.

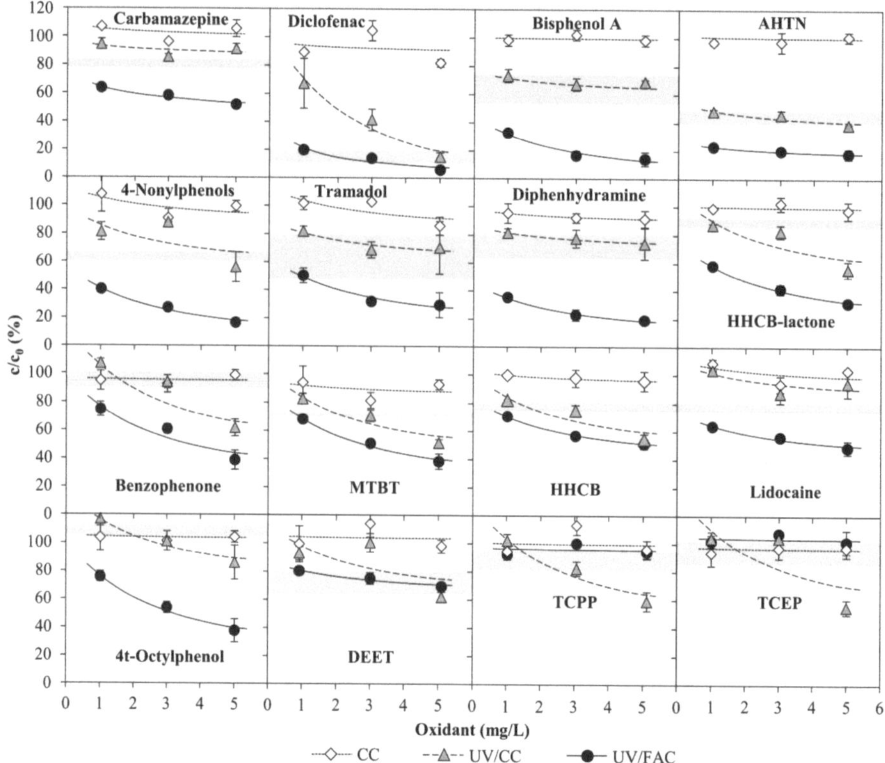

Figure 9. Emerging contaminants found in WWTE after treatment with the UV/FAC AOP [7], the UV/CC AOP (Exp. C) at 0.4 kWh/m^3 electrical energy consumption (0.4 kW) and with CC alone (no UV, Exp. C) as a function of the oxidant concentration (1 m^3/h flow rate). Gray line: sole UV treatment at 0.4 kWh/m^3 (1 m^3/h flow rate) [7]. For E_{EO} values, see Table S7.

In order to be able to assess the actual elimination effect of the AOPs, knowledge of the elimination performance of the 16 ECs by treatment with Cl$_2$ alone is required. In an experiment with 3 mg/L FAC at 1 m^3/h [7], only the following compounds were eliminated with sole FAC treatment: 4t-octylphenol (44% residual concentration), MTBT (64%), tramadol (64%), DCF (47%), diphenhydramine (35%), bisphenol A (28%) and 4-nonylphenols (20%). All other compounds (including CBZ) were not significantly degraded by 3 mg/L FAC. Thus, particularly the elimination of ECs such as CBZ, AHTN, HHCB, HHCB-lactone, benzophenone and lidocaine, which were not eliminated significantly by Cl$_2$ alone and the degree of elimination of which differed markedly between sole UV and UV/chlorine treatment, can be traced back to reaction with radicals.

For all compounds, the elimination effect of 1–3 mg/L CC with simultaneous UV treatment (UV/CC) was not significantly higher than the elimination effect by sole UV treatment. The fact that ammonium ions were present in the pilot plant influent thus had a decreasing effect on the EC elimination performance. Accordingly, due to the rapid reaction of free Cl$_2$ with ammonium ions, the •OH radical yield was considerably reduced [30]. Furthermore, the oxidizing ability of chloramines is significantly lower than that of HOCl [30]. In addition, radicals can also be consumed for the oxidation of ammonium ions to nitrite and nitrate ions [6,17]. On the other hand, HHCB, HHCB-lactone, benzophenone, MTBT, TCEP and TCPP seemed to be affected by a CC concentration of 5 mg/L at UV/CC (still, UV/CC was less effective than UV/FAC except for TCEP, TCPP). Especially for the latter ECs, however, it is questionable why UV/CC worked better than UV/FAC, although HOCl ($\Phi_{254\,nm}$ = 1.5, $\Phi_{200-350\,nm}$ = 3.3–4.0) has a significantly higher quantum yield than monochloramine ($\Phi_{254\,nm}$ = 0.3, $\Phi_{200-350\,nm}$ = 0.7) [6]. During the experiment, TCEP and TCPP were the ECs with the strongest variation in initial concentration (e.g., 0.37 µg/L TCEP at 3 and 5 mg/L CC and 2.19 µg/L TCEP at 1 mg/L CC). In contrast, the initial concentrations of the other ECs were usually on a similar scale. Thus, the supposedly better elimination can possibly be traced back to the changing wastewater composition and considerably different initial concentration during the course of the experiment. The very fact that the ammonium ion concentration in the WWTE varied considerably within a few minutes indicates that the nitrification of the WWTP did not function optimally at this moment. A non-functioning nitrification can also indicate a non-functioning elimination of other organic or inorganic compounds and thus a very different wastewater matrix as compared to the regular operation. Based on the few examples, a better efficiency of UV/CC compared to UV/FAC should therefore not be concluded for the organophosphoric acid esters without further research.

The effect of the changing wastewater composition could also be observed very well with DCF, 4-nonylphenols, lidocaine and DEET. Here, in parts the UV/CC combination was even less effective than sole UV treatment. The question now arises as to whether the changing wastewater composition during the experiment had a negative effect on the reliability of the results. It must be noted that high ammonium ion concentrations in WWTE are the exception in WWTP operation. Since nitrification obviously does not function reliably when the ammonium ion concentration is elevated, the ammonium ion concentration changes steadily, i.e., an equilibrium state of WWTE cannot be established for such experiments. Furthermore, a changing NH_4^+ concentration also indicates a poorer elimination of other compounds in the wastewater (e.g., occurring solids or other nitrogenous compounds may react with chlorine; color change of the wastewater may lead to a stronger absorption of UV light), so that other conditions may prevail for the pilot plant. This experiment should cover this exceptional case and is therefore representative.

In the experiments by Yang et al. [9], the elimination of some pharmaceuticals spiked in ammonium-rich wastewater (3.14 mg/L NH_4^+-N) was investigated. For example, at a dosage of 5 mg/L Cl_2, 10 W UV power, pH 7 and a contact time of 1.5 min, with 30% the elimination of CBZ was significantly lower than in wastewater with less than 0.03 mg/L NH_4^+-N. However, the degradation could be mainly attributed to chlorine radicals, which disagreed with the findings of this work for CBZ (no significant difference in elimination between UV and UV/CC). The comparison of both results demonstrates that the contact time in the UV chamber of 6–10 s is not sufficient for EC elimination when instead of FAC only CC is present in the UV chamber influent.

Compared to UV, CC, and UV/CC treatment, the UV/FAC process was the most effective method. Here, for many compounds (e.g., CBZ, HHCB-lactone, HHCB, benzophenone, MTBT) the degree of elimination differed significantly from the degree of elimination by sole UV and sole FAC treatment. It was already worked out by Rott et al. [7] that, in the dosing range of 1–6 mg/L oxidant, the UV/H_2O_2 process is significantly less effective than the UV/FAC process in terms of EC elimination. However, it is important to point out that especially in cold seasons due to insufficient nitrification, the ammonium ion concentration in the WWTE can become so high that it becomes impossible to achieve sufficiently high FAC concentrations in the UV chamber influent. In such cases, the UV/chlorine AOP becomes the UV/CC AOP. The fact that the degrees of elimination during UV/CC treatment differed insignificantly from sole UV treatment clearly indicates that the UV/chlorine AOP is very sensitive and loses its effectiveness drastically as soon as there is no FAC but only CC in the UV chamber influent. At lower ammonium ion concentrations, this may be compensated by an increased dosage of Cl_2. Considering by-product formation, however, at very high ammonium ion concentrations it is questionable whether the required Cl_2 dosage goes hand in hand with the concept of environmentally friendly wastewater treatment. Furthermore, it was shown that CC cannot be quenched with H_2O_2 (Section 3.1.3). The increased by-product emission can therefore not be met with H_2O_2 quenching.

3.2.3. Exp. D: Variation of Flow Rate

Figure 10 summarizes the residual concentrations of ECs in WWTE from two different experiments.

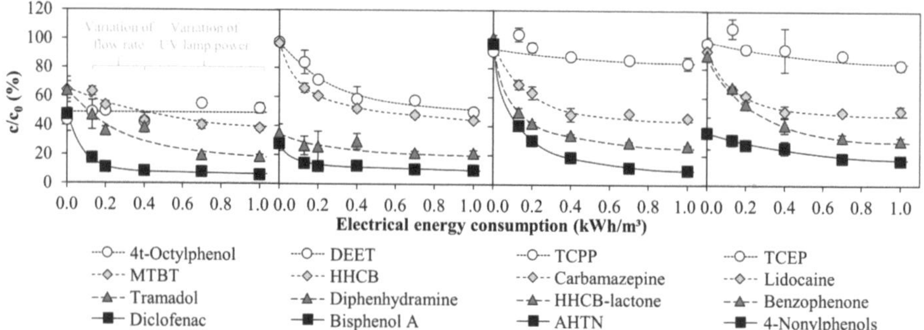

Figure 10. Emerging contaminants after treatment of WWTE by means of an AOP pilot plant as a function of the electrical energy consumption at 3 mg/L FAC in the UV chamber influent and subsequent quenching with H_2O_2. The results of two experiments are shown: variation of flow rate between 1 and 3 m³/h at 0.4 kW UV lamp power (Exp. D) and variation of UV lamp power between 0 and 1 kW at a flow rate of 1 m³/h (taken from Rott et al. [7]). For E_{EO} values, see Table S8.

In an experiment by Rott et al. [7] with the same pilot plant applying a flow rate of 1 m³/h and 3 mg/L FAC in the UV chamber influent, the UV power was varied between 0 and 1 kW. In Exp. D of this study, both the UV power and the FAC concentration in the UV chamber influent were kept constant at 0.4 kW and 3 mg/L, whereas the flow rate of the pilot plant was varied between 1 and 3 m³/h. The results of both experiments are shown as a function of the electrical energy consumption.

For many of the ECs investigated, it can be seen very well that the degree of elimination was negatively influenced by higher flow rates, i.e., lower electrical energy consumption. This is obvious, since doubling or tripling the flow rate is accompanied by shortening the contact time in the UV reactor by half and to one third, respectively. Increased UV power at a constant flow rate, i.e., higher electrical energy consumption, had a positive effect on the degree of EC elimination. Thus, applied over the electrical energy consumption, the results of two relatively different experiments proved to fit well together.

Compared to sole FAC dosage of 3 mg/L (0 kWh/m³), a significant improvement in the elimination of several ECs was achieved by addition of UV light. The ECs AHTN, HHCB-lactone, benzophenone, HHCB and CBZ were best eliminated with >50 percentage points (p.p.) difference between sole Cl_2 and UV/Cl_2 treatment. 40–50 p.p. differences were achieved for DEET, tramadol, lidocaine and DCF. Between 0.13 and 1.00 kWh/m³, the differences in the elimination of most ECs were small. Within this range, the variation of electrical energy consumption had the greatest effect on the ECs benzophenone, DEET and AHTN with more than 30 p.p. difference between the minimum and maximum degree of elimination. The smallest influence with less than 12 p.p. difference was found for diphenhydramine, DCF, and bisphenol A. However, this number is so low because even at very low electrical energy consumption high degrees of elimination already prevailed and a large leap to 100% elimination was therefore not possible.

The question remains whether the slightly increased CC concentration of up to 3.9 mg/L due to the increase of NH_4^+-N concentration in WWTE up to 0.57 mg/L (Table 1) towards the end of Exp. D influenced the EC elimination yields. Although in Exp. A for tap water matrix it was shown that chloramines may contribute to a greater EC elimination, in Exp. C it was observed for WWTE matrix with almost all ECs that CC in combination with UV light did not contribute significantly to a greater EC elimination.

4. Conclusions

Many influencing factors such as increased ammonium ion concentrations or increased COD values in WWTE have a negative effect on the maintenance of FAC in the UV chamber influent, which, as experiments of this work showed, is essential for the UV/chlorine AOP to effectively eliminate ECs. Within the electrical energy consumption range tested (0.13–1 kWh/m^3), a stable EC elimination performance of the UV/chlorine AOP can therefore not be maintained regularly throughout the year. To meet this problem, additional treatment steps would be a way of maintaining a good elimination performance. One possibility is to operate the UV AOP system with NaOCl only in the case of low ammonium ion concentrations and to dose H$_2$O$_2$ (UV/H$_2$O$_2$ AOP) instead of NaOCl at elevated ammonium ion concentrations, although in this case the EC elimination yields may decrease significantly [7]. In any case, the installation of an activated carbon stage downstream of the UV/chlorine AOP is recommended due to significant formation of adsorbable organohalogens (AOX) in the UV/chlorine AOP [7]. Since the entire process technology strongly depends on the fluctuating wastewater composition, for an immediate reaction to these changes, a very advanced control technology and process engineering would be required. Despite these issues, the UV/chlorine AOP has advantages over the UV/H$_2$O$_2$ AOP in terms of EC removal, hygienization and total estrogenic activity elimination [7]. Hence, further research at pilot scale would have to investigate whether shorter contact times between the chlorine dosing point and the UV chamber or higher UV lamp powers might contribute to weaker by-product formation and a more stable EC elimination performance.

Supplementary Materials: Supplementary calculations, figures and tables are available online at http://www.mdpi.com/1660-4601/15/6/1276/s1. Table S1: Initial concentrations of emerging contaminants in all experiments with WWTE. Table S2: Results of least squares method to determine the contact time between Cl$_2$ dosage point and Cl$_2$ measuring cell. Table S3: Calculation of contact time between Cl$_2$ dosage point and UV chamber influent. Table S4: Results of least squares method to determine the kinetic rate constant $k_{HOCl-COD}$. Table S5: Determination of regression curves to determine the required Cl$_2$ dosage for 3 mg/L FAC in the UV chamber influent as a function of NH$_4^+$-N or COD. Table S6: Electrical energy consumption per order of compound removal E$_{EO}$ as calculated from the data given in Figure 8. Table S7: Electrical energy consumption per order of compound removal E$_{EO}$ (kWh/m^3/order) as calculated from the data given in Figure 9. Table S8: Electrical energy consumption per order of compound removal E$_{EO}$ (kWh/m^3/order) as calculated from the data given in Figure 10. Figure S1: Measured FAC concentrations in the Cl$_2$ measuring cell as a function of the NH$_4^+$ concentration in the pilot plant influent in tap water matrix. Figure S2: Results of least squares method to determine the contact time between Cl$_2$ dosage point and Cl$_2$ measuring cell. Approximation to measured data by changing the contact time at a fix second order rate constant: $k_{HOCl,NH3} = 1.3 \times 10^4$ M^{-1}s^{-1}. Figure S3: Measured FAC concentrations and calculated FAC concentrations ($k = 182$ M^{-1}s^{-1}) in the Cl$_2$ measuring cell as a function of the COD concentration in the pilot plant influent (dilutions of wastewater treatment plant effluent). Figure S4: Results of least squares method to determine the kinetic rate constant $k_{HOCl-COD}$. Approximation to measured data by changing the kinetic rate constant at a fix contact time of 5.605 s. Figure S5: Required Cl$_2$ dosage to obtain 3 mg/L FAC in the UV chamber influent (6.112 s contact time) in the presence either of ammonium ions or COD (the results may differ when both are present at the same time) calculated either using the kinetic rate constants ($k_{HOCl,NH3} = 1.3 \times 10^4$ M^{-1}s^{-1} and $k_{HOCl-COD} = 182.1$ M^{-1}s^{-1}) or the inhibition ratios (4.7 mg FAC per mg NH$_4^+$-N and 0.16 mg FAC per mg COD).

Author Contributions: E.R. and B.K. conceived and designed the experiments; E.R. performed the experiments; E.R., B.K. and C.L. conducted the analyses; E.R., B.K., P.R. and R.M. analyzed the data; E.R. and P.R. wrote the paper.

Conflicts of Interest: The authors declare no conflict of interest.

References

1. Auriol, M.; Filali-Meknassi, Y.; Tyagi, R.D.; Adams, C.D.; Surampalli, R.Y. Endocrine disrupting compounds removal from wastewater, a new challenge. *Process Biochem.* **2006**, *41*, 525–539. [CrossRef]
2. Bolong, N.; Ismail, A.F.; Salim, M.R.; Matsuura, T. A review of the effects of emerging contaminants in wastewater and options for their removal. *Desalination* **2009**, *239*, 229–246. [CrossRef]
3. Buxton, G.V.; Subhani, M.S. Radiation chemistry and photochemistry of oxychlorine ions. Part 2—Photodecomposition of aqueous solutions of hypochlorite ions. *J. Chem. Soc. Faraday Trans. 1* **1972**, *68*, 958–969. [CrossRef]

4. Feng, Y.; Smith, D.W.; Bolton, J.R. Photolysis of aqueous free chlorine species (HOCl and OCl$^-$) with 254 nm ultraviolet light. *J. Environ. Eng. Sci.* **2007**, *6*, 277–284. [CrossRef]

5. Jin, J.; El-Din, M.G.; Bolton, J.R. Assessment of the UV/chlorine process as an advanced oxidation process. *Water Res.* **2011**, *45*, 1890–1896. [CrossRef] [PubMed]

6. Watts, M.J.; Linden, K.G. Chlorine photolysis and subsequent OH radical production during UV treatment of chlorinated water. *Water Res.* **2007**, *41*, 2871–2878. [CrossRef] [PubMed]

7. Rott, E.; Kuch, B.; Lange, C.; Richter, P.; Kugele, A.; Minke, R. Removal of Emerging Contaminants and Estrogenic Activity from Wastewater Treatment Plant Effluent with UV/Chlorine and UV/H$_2$O$_2$ Advanced Oxidation Treatment at Pilot Scale. *Int. J. Environ. Res. Public Health* **2018**, *15*, 935. [CrossRef] [PubMed]

8. Shu, Z.; Bolton, J.R.; Belosevic, M.; El Din, M.G. Photodegradation of emerging micropollutants using the medium-pressure UV/H$_2$O$_2$ Advanced Oxidation Process. *Water Res.* **2013**, *47*, 2881–2889. [CrossRef] [PubMed]

9. Yang, X.; Sun, J.; Fu, W.; Shang, C.; Li, Y.; Chen, Y.; Gan, W.; Fang, J. PPCP degradation by UV/chlorine treatment and its impact on DBP formation potential in real waters. *Water Res.* **2016**, *98*, 309–318. [CrossRef] [PubMed]

10. Zhou, S.; Xia, Y.; Li, T.; Yao, T.; Shi, Z.; Zhu, S.; Gao, N. Degradation of carbamazepine by UV/chlorine advanced oxidation process and formation of disinfection by-products. *Environ. Sci. Pollut. Res. Int.* **2016**, *23*, 16448–16455. [CrossRef] [PubMed]

11. Pressley, T.A.; Bishop, D.F.; Roan, S.G. Ammonia-nitrogen removal by breakpoint chlorination. *Environ. Sci. Technol.* **1972**, *6*, 622–628. [CrossRef]

12. Wu, Q.-Y.; Hu, H.-Y.; Zhao, X.; Sun, Y.-X. Effect of Chlorination on the Estrogenic/Antiestrogenic Activities of Biologically Treated Wastewater. *Environ. Sci. Technol.* **2009**, *43*, 4940–4945. [CrossRef] [PubMed]

13. Wu, Z.; Guo, K.; Fang, J.; Yang, X.; Xiao, H.; Hou, S.; Kong, X.; Shang, C.; Yang, X.; Meng, F.; et al. Factors affecting the roles of reactive species in the degradation of micropollutants by the UV/chlorine process. *Water Res.* **2017**, *126*, 351–360. [CrossRef] [PubMed]

14. Sichel, C.; Garcia, C.; Andre, K. Feasibility studies: UV/chlorine advanced oxidation treatment for the removal of emerging contaminants. *Water Res.* **2011**, *45*, 6371–6380. [CrossRef] [PubMed]

15. Wang, W.-L.; Wu, Q.-Y.; Huang, N.; Wang, T.; Hu, H.-Y. Synergistic effect between UV and chlorine (UV/chlorine) on the degradation of carbamazepine: Influence factors and radical species. *Water Res.* **2016**, *98*, 190–198. [CrossRef] [PubMed]

16. Bolton, J.R.; Bircher, K.G.; Tumas, W.; Tolman, C.A. Figures-of-merit for the technical development and application of advanced oxidation technologies for both electric- and solar-driven systems. *Pure Appl. Chem.* **2001**, *73*, 627–637. [CrossRef]

17. Zhang, Y.; Geissen, S.-U.; Gal, C. Carbamazepine and diclofenac: Removal in wastewater treatment plants and occurrence in water bodies. *Chemosphere* **2008**, *73*, 1151–1161. [CrossRef] [PubMed]

18. American Chemical Society Database. Available online: https://scifinder.cas.org (accessed on 21 March 2018).

19. Held, A.M.; Halko, D.J.; Hurst, J.K. Mechanisms of chlorine oxidation of hydrogen peroxide. *J. Am. Chem. Soc.* **1978**, *100*, 5732–5740. [CrossRef]

20. Margerum, D.W.; Gray, E.T.; Huffman, R.P. Chlorination and the Formation of N-Chloro Compounds in Water Treatment. In *Organometals and Organometalloids*; Brinckman, F.E., Bellama, J.M., Eds.; American Chemical Society: Washington, DC, USA, 1979; pp. 278–291.

21. Qiang, Z.; Adams, C.D. Determination of Monochloramine Formation Rate Constants with Stopped-Flow Spectrophotometry. *Environ. Sci. Technol.* **2004**, *38*, 1435–1444. [CrossRef] [PubMed]

22. Deborde, M.; von Gunten, U. Reactions of chlorine with inorganic and organic compounds during water treatment—Kinetics and mechanisms: A critical review. *Water Res.* **2008**, *42*, 13–51. [CrossRef] [PubMed]

23. Chuang, Y.-H.; Chen, S.; Chinn, C.J.; Mitch, W.A. Comparing the UV/Monochloramine and UV/Free Chlorine Advanced Oxidation Processes (AOPs) to the UV/Hydrogen Peroxide AOP Under Scenarios Relevant to Potable Reuse. *Environ. Sci. Technol.* **2017**, *51*, 13859–13868. [CrossRef] [PubMed]

24. National Research Council. *Drinking Water and Health, Volume 7: Disinfectants and Disinfectant By-Products*; National Academies Press: Washington, DC, USA, 1987.

25. McKay, G.; Sjelin, B.; Chagnon, M.; Ishida, K.P.; Mezyk, S.P. Kinetic study of the reactions between chloramine disinfectants and hydrogen peroxide: Temperature dependence and reaction mechanism. *Chemosphere* **2013**, *92*, 1417–1422. [CrossRef] [PubMed]

26. Keen, O.S.; Thurman, E.M.; Ferrer, I.; Dotson, A.D.; Linden, K.G. Dimer formation during UV photolysis of diclofenac. *Chemosphere* **2013**, *93*, 1948–1956. [CrossRef] [PubMed]

27. Soufan, M.; Deborde, M.; Delmont, A.; Legube, B. Aqueous chlorination of carbamazepine: Kinetic study and transformation product identification. *Water Res.* **2013**, *47*, 5076–5087. [CrossRef] [PubMed]

28. Soufan, M.; Deborde, M.; Legube, B. Aqueous chlorination of diclofenac: Kinetic study and transformation products identification. *Water Res.* **2012**, *46*, 3377–3386. [CrossRef] [PubMed]

29. Li, B.; Zhang, T. Different removal behaviours of multiple trace antibiotics in municipal wastewater chlorination. *Water Res.* **2013**, *47*, 2970–2982. [CrossRef] [PubMed]

30. Li, M.; Xu, B.; Liungai, Z.; Hu, H.-Y.; Chen, C.; Qiao, J.; Lu, Y. The removal of estrogenic activity with UV/chlorine technology and identification of novel estrogenic disinfection by-products. *J. Hazard. Mater.* **2016**, *307*, 119–126. [CrossRef] [PubMed]

International Journal of
*Environmental Research
and Public Health*

Review

Vibrio Species in Wastewater Final Effluents and Receiving Watershed in South Africa: Implications for Public Health

Allisen N. Okeyo [1,2,3,*], Nolonwabo Nontongana [1,2,3], Taiwo O. Fadare [1,2,3] and Anthony I. Okoh [1,2,3]

[1]   SAMRC Microbial Water Quality Monitoring Centre, University of Fort Hare, Alice 5700, South Africa; n.nontongana@gmail.com (N.N.); tosinfadare@yahoo.com (T.O.F.); aokoh@ufh.ac.za (A.I.O.)
[2]   Applied and Environmental Microbiology Research Group (AEMREG),
     Department of Biochemistry and Microbiology, University of Fort Hare, Alice 5700, South Africa
[3]   Department of Biochemistry and Microbiology, University of Fort Hare, P/Bag X1314, Eastern Cape,
     Alice 5700, South Africa
*   Correspondence: aokeyo@gmail.com; Tel.: +27-(0)71-2055-888

Abstract: Wastewater treatment facilities in South Africa are obliged to make provision for wastewater effluent quality management, with the aim of securing the integrity of the surrounding watersheds and environments. The Department of Water Affairs has documented regulatory parameters that have, over the years, served as a guideline for quality monitoring/management purposes. However, these guidelines have not been regularly updated and this may have contributed to some of the water quality anomalies. Studies have shown that promoting the monitoring of the current routinely monitored parameters (both microbial and physicochemical) may not be sufficient. Organisms causing illnesses or even outbreaks, such as *Vibrio* pathogens with their characteristic environmental resilience, are not included in the guidelines. In South Africa, studies that have been conducted on the occurrence of *Vibrio* pathogens in domestic and wastewater effluent have made it apparent that these pathogens should also be monitored. The importance of effective wastewater management as one of the key aspects towards protecting surrounding environments and receiving watersheds, as well as protecting public health, is highlighted in this review. Emphasis on the significance of the *Vibrio* pathogen in wastewater is a particular focus.

Keywords: wastewater effluent; *Vibrio* pathogens; wastewater monitoring; public health

---

1. Introduction

Water is a critical element and an important core for sustainable socio-economic development, along with the eradication of poverty and health discrepancies. The need for safe drinking water and proper sanitation management is a perennial concern for sustainable life globally [1]. Despite significant progress that has been made in this regard, discrepancies still exist. Globally, there are approximately 2.1 billion people who lack access to quality drinking water sources, and 4.5 billion that lack proper sanitation [2]. There are significant differences between more affluent and poorer communities, particularly in the developing world. This crisis is further exacerbated by increasing poverty, rapid urbanization and population growth as well as climate change, which can further stress an often-deteriorating water and sanitation infrastructure, putting many at risk of water and sanitation related discrepancies [3,4].

Proper management of a water treatment facility is pivotal to bettering the quality of water consumed (drinking water) or released (wastewater) into the environment [5]. In conjunction with the many determinants used to indicate the holistic health or quality of an area, wastewater treatment systems may suffice as a positive indicator of development [6]. The release of poorly treated wastewater into nearby watersheds directly threatens the macro and micro flora and fauna present; retarding the provision of good quality water required for societal functions [3].

Many wastewater treatment plants in South Africa still release final effluent containing significant amounts of enteric pathogens such as the *Vibrio* genus, known for its environmental resilience and relation to disease outbreaks. This results in the impairment of the surrounding receiving water bodies [5,7,8]. Wastewater facilities are obliged to make provision for quality management of their wastewater effluents prior to release into surrounding water bodies. There are microbial and physicochemical parameters which are required to be routinely monitored, indicated in the water quality guidelines; these have been the basis of water quality management and research over the years. However, evidence suggests that the organisms causing illnesses and outbreaks are not necessarily the ones which are routinely monitored. Studies have already indicated the need to monitor not only the classical pollution indicators, i.e., culturable total or faecal coliforms, but also viral pathogens, toxigenic *E. coli*, and highly infectious bacterial pathogens such as *Vibrio*. [5,8–11].

This paper focuses on sustainable wastewater management as a key approach towards protecting receiving watersheds and ultimately public health. Moreover, it highlights the significance of the *Vibrio* pathogen in wastewater, emphasizing their impact on public health and justifying the need to monitor for this pathogen.

2. Wastewater Complexities

According to Corcoran [12], wastewater is defined as a combination of one or more of the following: (1) domestic effluent (blackwater and greywater comprising of excreta, urine, faecal sludge, bathing and kitchen wastewater); (2) water from institutions such as hospitals, industries and other commercial establishments; (3) storm water and other types of urban run-off; (4) dissolved or suspended agricultural, horticultural and aquaculture waste. The composition of wastewater in its entirety is dependent primarily on its source, characterized by both its physicochemical content (temperature, turbidity, pH, colour, odour, suspended solids, total dissolved solids, dissolved oxygen (DO), biological oxygen demand (BOD), nutrients and toxic substances, organics, alkalinity, metals and chlorides) and biological or microbial content (plants and animals, bacteria, protozoa, helminths, viruses) [3,13,14].

The degree of environmental trauma of wastewater is directly related to its composition. The diversity of wastewater is brought about by factors affecting it, e.g., the natural environment, population, and/or the surrounding recreational, domestic and industrial related activities [13]. Municipal wastewater, for instance, is a combination of different inflows which include human excrement (sewage), suspended solids, debris and an array of chemicals originating from residential channels, industrial and commercial activities [15]; Table 1 shows some wastewater related contaminants. These factors also subsequently affect the discharge patterns of the wastewater and could impact on the chemical and microbial status of treated wastewater final effluent, in turn impacting on the surrounding water bodies [3,16].

Table 1. Wastewater-related contaminants.

Contaminants	Impact on Aquatic Environments
Pathogenic Organism	In wastewater they are notably detrimental, responsible for health-related discrepancies.
Suspended Solids (SS)	Suspended solids in untreated wastewater, when accumulated, may lead to the development of sludge deposits. These deposits can notably increase anaerobic conditions in aquatic environments.
Biodegradable organics	Commonly measured as BOD and COD, biodegradable organics consist of proteins, carbohydrates and fats. The discharge of these organics into receiving water bodies (rivers, lakes etc.,) may interfere with biological stabilization, for example, depleting the natural oxygen resources causes septic conditions detrimental to aquatic species.
Priority pollutants	Some organic and inorganic compounds present in wastewater are highly toxic, carcinogenic, mutagenic or teratogenic.
Refractory organics (surfactants, phenols and agricultural pesticides)	These are organics that tend to resist conventional waste-water treatment. Their accumulation in the environment may cause severe problems for the environment for instance environmental poisoning.
Heavy metals (arsenic, lead, mercury, cadmium, chromium, copper, nickel, silver, and zinc)	Usually added by commercial and industrial activities, they are notably the most persistent pollutants in wastewater. The release of high levels of heavy metals into receiving water bodies may cause serious health and environmental complications.
Dissolved inorganics (calcium, sodium and sulphate)	These are often present in domestic waste and must be removed, especially if the wastewater is intended for reuse, e.g., for irrigation purposes. Dissolved inorganics may have a long-term impact on the environment that increases with the continued use of wastewater.

Source: [17].

2.1. Wastewater Treatment

The implementation of a proper treatment plant design or the selection of effective treatment technologies depends on the nature of the wastewater to be treated [18]. Wastewater treatment was conceptualised with the aim to enable the disposal of wastewater safely, without this water polluting the receiving water bodies, being a danger to public health, or causing other water related anomalies [19]. Attempted first in the 1900s, wastewater treatment is now implemented worldwide and is continually being improved on. Globally, the treatment of wastewater is considered paramount. It is one of the key aspects for water quality management and is deemed to provide a genuine solution to two major challenges: (1) protecting water resources, and (2) access to sustainable water and sanitation services [18,20]. Typically, wastewater treatment processes involve the use of physical, chemical and biological unit operations. A summary of the operations employed is illustrated below in Figure 1:

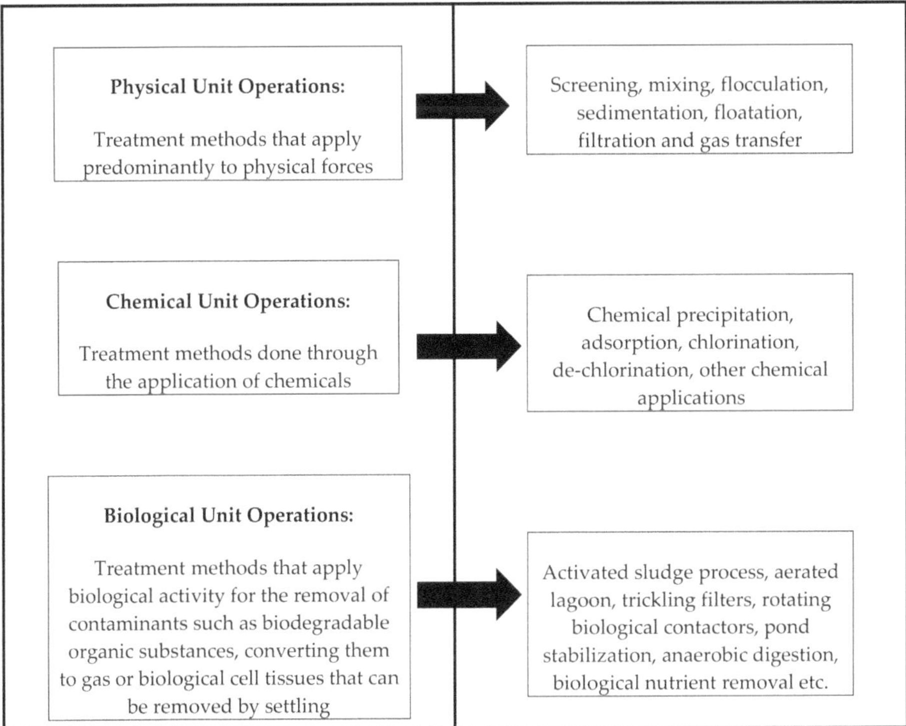

Figure 1. Summary of wastewater treatment unit operations. Source: [18,21].

Expanding on the above unit operations, wastewater treatment undergoes the following fundamental stages: preliminary, primary, secondary and tertiary stages [22].

2.1.1. Preliminary Treatment Process

Wastewater entering a treatment plant typically goes through preliminary treatment as the first process of treatment. The removal of larger debris such as paper, plastic and other large floating objects, for example, grit, and silt is performed at this stage. This stage prevents the accumulation of debris that may damage or clog the pumps, small pipes, other equipment and downstream processes in the treatment plant, is done at this stage. The screens are constructed from steel or iron bars and vary from coarse, with openings of about half an inch, to fine, generally mesh screens with much smaller openings. Materials removed by the screens are often dangerous and are safely disposed of using appropriate methods to a particular plant [6,23].

2.1.2. Primary Treatment Process

The primary treatment process is comprised predominantly of primary settling and sedimentation. At this stage, wastewater still contains dissolved organic and inorganic components, as well as some suspended solids (i.e. sand, grease, fats, oils and grit). The removal of these components at this stage is key to avoid excessive amounts of large solids from interfering with other processes [23]. Wastewater at this stage enters sedimentation tanks which reduce the flow rate. The flow of the wastewater becomes almost stagnant and is held in the tank for several hours. The formation of primary sludge is apparent at this stage, with most of the heavy solids falling to the bottom of the settling and sedimentation tank, separating suspended solid content from the liquid component of the wastewater. In addition, any surface floating materials can be drained off [24].

2.1.3. Secondary Treatment Process

During the secondary wastewater treatment process, the remaining suspended solids are decomposed and the microbial load is greatly reduced [25]. Typically, this is achieved by bringing the sewage, heterotrophic bacteria and oxygen together in trickling filters or within an activated sludge process, where mixture occurs by mechanical agitation or though mixing by air diffusers. This process utilizes oxygen and bacteria to stabilize the sewage, removing 85–90% of the Suspended Solids (SS) and Biological Oxygen Demand (BOD) [26–28]. Table 2 gives an overview of commonly used secondary wastewater treatment processes in South Africa.

2.1.4. Tertiary Treatment Process

Tertiary treatment further removes suspended solids, organic ions and nutrients such as phosphorus and nitrogen from the wastewater prior to release into a receiving watercourse [28]. This treatment process may incorporate procedures such as filtration, phosphorus removal, ammonia stripping and other special treatments which remove specific constituents from the wastewater. Other modes of treatment that may be used include sand filtration, wetlands or other advanced treatment processes. The process of disinfection improves the microbial quality of the wastewater (ideally bringing it to standard) before release into the environment. Some well-known methods of disinfection used in South Africa are: (1) Chemical, e.g., chlorination and ozonation; (2) Physical, e.g., ultraviolet radiation and microfiltration; and (3) Biological, e.g., detention ponds [29].

Table 2. Overview of commonly used secondary wastewater treatment processes in South Africa.

Treatment	Description
Trickling filters (bio-filters)	Organic matter is removed from wastewater using trickling or bio-filters. This is an aerobic treatment system that utilizes microorganism populations (bacteria, fungi, algae, and protozoa) attached to a medium forming a biological film approximately 0.1 to 0.2 mm thick to remove organic matter from wastewater. Wastewater passing through the medium with the microorganisms gradually attaches to the rock or plastic surface of the filters forming a film; organic material is degraded by the aerobic microorganisms in the outer part of the slime layer.
Rotating biological contractors	Rotating biological contractors are man-made aerobic attached-growth treatment disk systems, which are attached to shafts mounted over the wastewater to be treated. During treatment the shaft needs to rotate slowly so that the disks are immersed in the wastewater for a short period of time before returning to the air. This ensures the development of biological slime on the disks similar to that of the bio-filter. The developed slime falls back off into the wastewater where it will settle out and be removed or recycled.
Activated sludge processes	The activated sludge process removes organic matter from the wastewater by utilizing the high concentrations of microorganisms (mostly bacteria and some protozoa) present as floc. The floc is kept suspended in the wastewater through agitation. The main processes in the removal of the organic material are adsorption, carbonaceous oxidation, and nitrification. The key components for an activated sludge process are (1) Wastewater passes through a reactor (aeration tank), brought into contact with the present microorganisms; (2) The process transfers oxygen to the microorganisms; (3) The suspension is agitated; (4) The system separates the treated water from the microorganisms; (5) Live microorganisms are put back into the reactor and dead ones are removed.
Sedimentation tanks/Clarifier	The process of secondary sedimentation/clarification is necessary to remove high concentrations of sloughed biomass accumulated from the activated sludge process; separating it from the liquid.

Source: [19,25].

3. Overview of Some South African Wastewater Regulatory Legislation

"The Constitution of South Africa (Act 108, 1996) guarantees everyone the right to an environment that is not harmful to their health or wellbeing and guarantees the right to have the environment protected, for the benefit of present and future generations" [30]. This lays the foundation for a more equitable society through reasonable legislative frameworks. It is the basis of all regulatory procedures/policies that govern wastewater management, from the construction of a wastewater plant, wastewater treatment, to the release of the final effluent [6].

The improvement of supporting legislation or strategies at par with the provision of water services in South Africa is related directly to the primary impact of potable and wastewater on public health [6]. Some of the key legislation related to wastewater management addresses issues such as setting up a treatment plant and release of final effluent. The regulatory legislation has been compiled in Table 3.

Table 3. South African wastewater regulatory legislation.

Acts and Regulatory Legislations	Description
Environment Conservation Act (ECA) (No. 73 of 1989)	In January 1994, the Environmental Conservation Act was adopted to provide for economic growth and social welfare which is environmentally friendly (without negatively influencing, overstraining or irreversibly harming the natural environment and natural resources). A further consequential principle towards the polluter was incorporated in September 1994, stipulating the charges against the polluter for the negative environmental consequences of disposal or discharge actions.
Occupational Health and Safety Act (OHSA) (No. 85 of 1993)	A wastewater plant is required to comply with OHSA in its design and treatment requirements.
National Environmental Management (NEMA) Act (No. 107 of 1998)	NEMA contains internationally accepted principles of sustainability that concur with the South African Constitution (Section 24). Taking these principles into consideration is a legal requirement in all decisions that may affect the environment. Therefore, this is a prerequisite for intergovernmental co-ordination and harmonisation of policies relating to the environment. The Best Practical Environmental Option (BPEO) is defined in NEMA as "the option that provides the most benefit or causes the least damage to the environment as a whole, at a cost acceptable to society, in the long term as well as the short term".
The Environmental Management Plan (EMP)	This is recognised as the tool within NEMA that provides the assurance that any environmental related project makes suitable provisions for mitigation (Environmental Impact Assessment). An Environmental Impact Assessment (EIA) provides description of methods and procedures for mitigating and monitoring impacts through appropriate objectives. These methods take into consideration the various role players and responsibilities, timescales and cost.
The Water Services Act (No. 108 of 1997)	The Water Service Act designates the role of a Municipality as one of the major role players in Water and Sanitation Management providing an institutional framework for disseminating national norms and standards for providing water services. This simply means the authorisation of a service must be in harmony with funding mechanisms in place and EIA regulations.
The National Water Act (No. 36 of 1998)	This act conceptualises the management of water resources in South Africa based on constitutional rights. It positions the National Government as the custodian of water as a national resource. Provisions for the protection, use, development, conservation, management and control of water resources in the country are affirmed by the act. Setting up a new wastewater treatment plant requires a Water Act license from the Department of Water Affairs (DWA) and a Waste Act license from the Department of Environmental Affairs (DEA).

Source: [6,31].

4. Wastewater Management and Challenges

The paramount aim of wastewater treatment is to enable the disposal of wastewater that is not detrimental to the environment. The implementation of proper wastewater management strategies is not an option but an imperative in order to drive compliance with the set discharge standards. Wastewater management strategies safeguard the sustainable quality of water sources, reducing cost implications involved in drinking water treatment, and deterring or eradicating waterborne diseases [8].

The Department of Water Affairs and other related Water Quality Management units/stakeholders in South Africa, have the responsibility to ensure water quality management [32]. The Department is accountable for the initiation and implementation of water quality policies and documentation, such as the South African Water Quality Guidelines. The guidelines serve as the source for developing materials to advise water users about the physico-chemical, aesthetic and biological properties of water, and consist of the criteria for water quality, Target Water Quality Range (TWQR), and other relevant supporting data [33]. Table 4 shows the indicators for wastewater quality management, highlighting the limit values relevant to discharge of wastewater into a receiving water body [34].

A properly monitored and operational wastewater treatment plant is able to release effluent well within standards, eliminating up to 90% of bacterial and viral pathogens [35]. However, despite significant progress that has been made in relation to water quality management, anomalies in the area of wastewater treatment still exist, with there being significant differences between those in affluent and poorer communities [36]. Owing to factors such as the poorly functioning state and inadequate maintenance of some wastewater treatment plants, as well as a lack of facilities to monitor micro-pollutant content of effluent, particularly in poorer communities, the fundamental aim to produce standardized wastewater final effluent remains unrealized in this country. Studies have reported that several wastewater treatment plants still release effluent containing significant amounts of enteric pathogens, in turn resulting in the impairment of the surrounding receiving water bodies and thus posing a serious threat to public health [5,7,8,36]. Enteric pathogens are known to be resilient in wastewater, surviving treatment processes and also being able to develop resistance to chlorine [37].

Table 4. South African National Water Act waste discharge standard guidelines.

Variables and Substances	General Standards
Chemical oxygen demand	75 mg/L
Colour, odour or taste	No substance capable of producing the variables listed
Ionised and unionised ammonia (free and saline ammonia)	3 mg/L
Nitrate	15 mg/L
pH	5.5–9.5
Phenol index	0.1 mg/L
Residual chlorine (Cl)	0.25 mg/L
Suspended solids	25 mg/L
Total Aluminium (Al)	-
Total Cyanide (Cn)	0.02 mg/L
Total Arsenic (As)	0.02 mg/L
Total Boron (B)	1 mg/L
Total Cadmium (Cd)	0.005 mg/L
Total Chromium III (CrIII)	-
Total Chromium VI (CrVI)	0.05 mg/L
Total Copper (Cu)	0.01 mg/L
Total Iron (Fe)	0.3 mg/L
Total Lead (Pb)	0.01 mg/L
Total Mercury (Hg)	0.005 mg/L
Total Selenium (Se)	0.02 mg/L
Total Zinc (Zn)	0.1 mg/L
Faecal Coliform	1000 cfu/100 mL

Source: [38].

As mentioned in the introduction, routinely monitored microbial and physicochemical parameters promoted by current water quality guidelines the basis of water quality management and research may not include the organisms causing current illnesses or outbreaks. Studies have already indicated the need to monitor not only the classical pollution indicators but also the highly infectious ones [10]. In South Africa for instance, studies conducted on the occurrence of *Vibrio* pathogens in domestic water and wastewater effluent have made this apparent. A few recent examples are mentioned below.

In a study assessing the prevalence of disease-causing enteric pathogens in rural communities of Nkonkobe, South Africa, 25% of the bacterial isolates obtained from both ground water and surface water samples were confirmed to have toxigenic *Vibrio cholerae* [39]. A previous study on distribution of diarrhoea and microbial quality of domestic water in Khandanama River, Tshikuwi, South Africa, reported indicator microbial counts, which included heterotrophic bacteria, enterococci, total and faecal coliforms [9], exceeding the limits of no risk as indicated by the South African water quality guidelines for domestic use. Further analysis conducted also revealed a high presence of *Shigella*, *Vibrio* and *Salmonella* species in the Khandanama River.

Dungeni et al. [8] further reported on four wastewater treatment plants located in Gauteng Province which had recordings of *Vibrio cholerae* among other related pathogenic bacteria such as *Escherichia coli* and *Salmonella typhimuriam*. The presence of *Vibrio* pathogens in wastewater final effluents has also been found in studies conducted and published by members of the Applied and Environmental Microbiology Research Group (AEMREG) from the University of Fort Hare, South Africa. In a study published in 2010, the *Vibrio* strains *V. parahaemolyticus*, *V. metschnikovii*, *V. fluvialis* and *V. vulnificus* were obtained from wastewater final effluent from a rural community in the Eastern Cape of South Africa [40]. In a comprehensive study conducted in 2012 by Nongogo and Okoh [5], on the occurrence of *Vibrio* pathogens in final effluents from 5 wastewater treatment plants, in the Chris Hani District area in the Eastern Cape Province, South Africa, different species of *Vibrio* were confirmed from 310 isolates, including *V. parahaemoliticus*, *V. vulnificus* and *V. fluvialis*. A study on the presence of *Vibrio* pathogens in final effluents conducted on 14 wastewater treatment plants in the Chris Hani and Amathole district Municipalities, South Africa, reported the presence of *Vibrio* pathogens in most of the wastewater effluent samples in all seasons. Up to 66.8% of the isolates obtained were confirmed to belong to the *Vibrio* genus [11].

5. Some *Vibrio* Pathogens in Wastewater Final Effluents in South Africa

The *Vibrio* genus belongs to the family Vibrionaceae, which consists of opportunistic pathogens that affect humans and animals. Common inhabitants of marine coastal ecosystems, it is documented that their changing populations result from changes in seawater temperature in conjunction with warmer temperatures and algal blooms which decline with cooler temperatures. However, their adaptability to adverse conditions is also notably responsible for their presence in other environments such as sewage [8,41,42]. The wide distribution of the Vibrio genus in effluent environments, particularly those associated with domestic sewage has been strongly promoted by their ability to adapt to adverse conditions [43]. Studies have confirmed the association with and resilience of the *Vibrio* genus in wastewater, as they survive and thrive in the harsh conditions of the effluent (despite the various chemical, biological and physical characteristics); this in turn often causes a ripple-like effect of inconsistencies in the receiving water-bodies and the environment as a whole [11,44,45].

Vibrio cholerae is responsible for the most common form of *Vibrio* pathology, and results in the gastrointestinal disease cholera through the release of enterotoxins, which causes an efflux of important cell nutrient including sodium and water, leading to diarrhoea and dehydration. *V. cholera* is classified into two serotypes, O1 and nonO1 [46]. The O1 strain is further divided into two biotypes, Classical and ElTor. The biotype ElTor is known to be an emerging causative for *V. cholerae* in humans. Transmitted mostly via the faecal-oral route, cholera has two distinctive features: (1) its propensity to appear and cause sporadic outbreaks/epidemics, and (2) its ability to cause pandemics that spread over many countries over a period of many years [47]. Griffith et al. [48] report that the World Health Organisation receives hundreds of thousands of case reports every year, with these cholera-related epidemics reportedly mostly threatening the developing world. Contaminated wastewater has been associated with the spread of cholera. This is further validated by the high incidence of *V. cholerae* in water samples from various studies, indicating that they are natural inhabitants of aquatic environments [5,11,49].

Vibrio vulnificus is known to be extremely virulent and causes three types of infections: (1) severe gastroenteritis from consumption of raw or undercooked seafood; (2) necrotizing wound infections when injured skin is exposed to contaminated water for instance, marine water; and (3) intrusive septicaemia, which is caused when the bacterium invades the blood stream (this is 80 times more likely in immune-compromised individuals such as those with liver disease, such as haemochromatosis [50–53]. Igbinosa et al. [44] reports that *V. vulnificus* can survive in wastewater effluents even after chlorination, supporting the likelihood of finding it in the receiving waterbodies and surrounding environments after treatment.

Vibrio fluvialis causes cholera-like bloody diarrhoea, wound infection and primary septicaemia (prevalent in individuals with a weakened immune system) world-wide [45]. *V. fluvialis* infections are mostly found in areas exposed to increased levels of faecal contaminated food such as contaminated seafood products and water supplies [54]. It may also be transmitted by person-to-person contact [55]. Okoh et al. [40] reports that the association of *V. fluvialis* with low standards of living, inadequate water supply, and poor sanitary conditions have become significant over time.

Vibrio mimicus is a *Vibrio* species that mimics *V. cholerae*. It also is known to cause gastroenteritis transmitted from eating raw sea food. In very rare occurrences, in instances where *V. mimicus* is carrying the genes encoding for cholera toxin, it can cause severe watery diarrhoea [55]. *Vibrio alginolyticus* is known for its prevalence in sea water, causing otitis (ear infection), wound infection (skin ulcers), gastroenteritis, blood poisoning (septicaemia), food intoxication and haemorrhaging in humans [56]. *Vibrio parahaemolyticus* has been indicated in the literature to cause acute gastroenteritis, diarrhoea and abdominal pain in individuals who consume unclean seafood (usually raw or undercooked sea food) and, less commonly, wound infections through exposure to sea water.

As mentioned earlier, the presence of *Vibrio* pathogens in wastewater, even those cases documented in predominantly salt water or *Vibrio* pathogens in sea food-related cases, is notable due to their ability to acclimatise quickly to fluctuating environmental conditions. This positions *Vibrio* as an emerging pathogen with implications for public health [57].

6. Monitoring of Pathogenic *Vibrio* in Wastewater

The significance of monitoring wastewater effluents has over time become evident due to the presence of microbial pathogens that are considered environmental contaminants or precursors to health-related discrepancies. The process of wastewater treatment may remove most pathogens, but many still remain and are discharged into the effluent, which in turn affects the receiving water bodies and the environment as a whole [58]. In South Africa, the presence of environmental contaminants in wastewater effluents is not unusual, with notable significant differences between effluent in affluent and poorer communities [36].

In conjunction with classical microbiological indicators, the need to monitor outbreak-related pathogens, for example, viral pathogens, protozoa, toxic *E. coli*, highly infectious bacterial pathogens such as *Vibrio* spp., etc., as well as the need to perform direct analysis of specific pathogens of concern is paramount. These pathogens are notably more resilient in the environment with devastating effects. In wastewater for instance, these pathogens are able to withstand wastewater treatment processes, as well as lie dormant or survive for long periods of time. [36].

6.1. Conventional Methods

Phenotypic techniques are often used for identifying and enumerating *Vibrio* species. Culture-based techniques may include the pre-enrichment of samples in selective media, plating on to selective media followed by morphological and biochemical characterization [59,60]. For wastewater, for instance, the membrane filter technique is often used. In this technique, a water sample is filtered using a sterile filter with a 0.45 mm pore size, which will help retain bacteria, followed by placing of the filter on a selective medium such as thiosulphate citrate bile salts sucrose (TCBS) agar and appropriate incubation (at 37 °C for 24–48 h), after which typical colonies (green and yellow) will be enumerated on the filter [61]. Table 5 indicates the typical morphology of *Vibrio* pathogens when grown on TCBS agar. A predominant flaw in the membrane filter technique is that it is not able to recover injured bacteria, as a number of chemical and physical factors in the wastewater treatment processes, as well as disinfection, can cause sub-fatal injury to pathogens of interest, resulting in the lack of formation of clear and visible colonies on selective media. This shortcoming is often curbed by a pre-enrichment step using a broth, like alkaline peptone [61].

Table 5. Typical morphology of presumptive *Vibrio* pathogens grown on thiosulphate citrate bile salts sucrose (TCBS) agar.

Vibrio Pathogen	Colony Morphology (Colour)
Vibrio cholerae	Large yellow colonies
Vibrio vulnificus	Yellow-Greenish yellow colonies
Vibrio fluvialis	Yellow colonies
Vibrio mimicus	Green colonies
Vibrio parahaemolyticus	Blue colonies with green centres
Vibrio alginolyticus	Large yellow colonies

Source: [62].

6.2. Molecular Methods

The use of molecular methods for identification and quantification of pathogens, as well as rapid and sensitive characterization of bacteria with unique growth or biochemical requirements, may be seen as a valid alternative approach to the conventional or culture-based methods. Studies have shown that molecular methods are able to detect even the lowest mass of bacteria, and even those that are viable but non-culturable [63,64]. These techniques are also reportedly sensitive towards new emergent strains and indicators, such as *Vibrio* pathogens [58,65,66]. Considering the increasing importance of *Vibrio* spp. in conjunction to the environment and public health as a whole, the molecular method is an acceptable and reliable alternative technique for routine microbial screening and monitoring of environmental and food samples. These methods may also be used in the refinement and evaluation of disease and non-disease causing bacterial strains [58].

The Polymerase Chain Reaction (PCR)-based technique is one of the well-known molecular techniques used to amplify specific DNA sequences; however, PCR-based assays cannot differentiate between live and dead cells. Other regularly used molecular approaches for *Vibrio* species identification are Real-Time PCR, Restriction Fragment Length Polymorphism (RFLP), Microarrays, Multilocus Enzyme Electrophoresis (MLEE), Fluorescence in Situ Hybridization, Multilocus Sequence Typing (MLST), Ribotyping and Amplified Fragment Length Polymorphism (AFLP) [66].

7. Implications of Pathogenic Vibrio for Public Health

Enteric pathogens have been documented to cause many of the disease outbreaks that occur worldwide. Pathogenic *Vibrio*, for instance, is becoming a cause for great concern, especially considering the extent to which infections by these organisms are increasing globally. This is accentuated by the constant environmental changes like warmer waters due to climate change. *Vibrio*-related infections are responsible for some of the most deadly and costly food- and water-borne diseases [48].

Due to the severity of disease caused by *Vibrio cholerae*, it has been the focus of recent *Vibrio* species water related research [67]. However, over the last decade the relevance of the more minor *Vibrio* species has become significant. Some studies have reported a few *Vibrio* species of medical importance, which are being described as emerging pathogens that are able to cause mild to severe human diseases [60].

Brief Epidemiology of Diarrheal/Vibrio Pathogen-Related Health Cases in South Africa

Documented cholera outbreaks in South Africa date back as far as the 1980s, when sporadic outbreaks occurred. The beginning of the worst cholera epidemic, however, was experienced in August 2000, culminating to the deaths of 232, and the infection of about 106,389 people nationwide. The epidemic traversed Kwazulu-Natal, Gauteng, Mpumalanga and the Northern Province. The 2000/2001 epidemic subsided with 3901 reported cases and 45 deaths in Mpumalanga, Eastern Cape and Kwazulu-Natal in 2003. In 2004, 1773 cases of cholera infections and 29 deaths were reported in Mpumalanga's Nkomazi Region. In the same year, the Eastern Cape had reports of 738 people who were diagnosed with cholera, with four fatalities. Two deaths were also reported in the North West, with 260 cases of infection [68–70]. In 2007, 80 diarrhoea-related child deaths occurred in the Eastern Cape, in the Greater Barkly East Area, Ukhahlamba District Municipality [71]. Cholera was confirmed as being endemic in South African water resources by a spokesperson from the South African ministry of health in 2009 [72].

The Limpopo Province is prone to disease outbreaks, especially during rainy seasons, with reports of cholera and malaria being predominant. In 2014, it was reported that at least 45 people were admitted to Voortrekker Hospital after contracting diarrhoea. Nine of those admitted to hospital were in a critical condition. Contaminated water or food was suspected to be the source of the outbreak [73].

During the same year, an outbreak of diarrhoea was reported in Fort Beaufort and the surrounding areas, with some deaths and many hospitalizations; residents credited the incident to poor water quality supplied by the municipality [74]. Another outbreak causing infant mortalities was reported in Upington in the same year [75].

There was a contaminated water related scare in Chris Hani District Municipality in the Cradock area in 2015, with thirteen people, including two four-year-olds, treated at the Cradock hospital for gastroenteritis. Although cause of disease was not confirmed, the Health24 resident doctor Dr Heidi van Deventer was of the opinion that the disease could have been caused by either viral or bacterial sources [76].

8. Conclusions

Public health discrepancies related directly to water and sanitation in South African are taking centre stage. Emerging outbreak-related pathogens such as *Vibrio* are of importance due to their catastrophic health related implications. It is apparent that there is a need to monitor these pathogens. In the context of wastewater management, the need to perform direct analysis of specific pathogens of concern over and above the routinely monitored classical microbiological indicators is evident. Many of the outbreak-related pathogens present in wastewater do not form part of the routinely monitored indicators. Identifying key techniques/methods that can be used for the detection of *Vibrio* pathogens is needed for good public health. Recommendations have been made, following the detection of pathogenic vibrio species in wastewater treatment plants, to enforce environmental regulations and safeguard the impairment of receiving water bodies. However, proactive strategies should be employed to ensure effective monitoring of pathogenic vibrio species in the plants through efficient management that complies with set guidelines.

Author Contributions: All authors contributed to this review article, A.N.O. drafted the first manuscript. N.N. and T.O.F. and A.I.O. read and approved the review article.

Funding: This research was funded by South Africa Medical Research Council and the Water Research Commission of South Africa

Acknowledgments: We thank the South Africa Medical Research Council and the Water Research Commission of South Africa for financial support.

Conflicts of Interest: The authors declare no conflict of interest.

References

1. A Post-2015 Global Goal for Water: Synthesis of Key Findings and Recommendations from UN-Water. Available online: http://www.un.org/waterforlifedecade/pdf/27_01_2014_un-water_paper_on_a_post2015_global_goal_for_water.pdf (accessed on 24 October 2017).
2. The United Nations Children's Fund (UNICEF). 2.1 Billion People Lack Safe Drinking Water at Home, More than Twice as Many Lack Safe Sanitation. 2017. Available online: https://www.unicef.org/eca/press-releases/21-billion-lack-water-sanitation (accessed on 7 April 2018).
3. Naidoo, S.; Olaniran, A.O. Treated wastewater effluent as a source of microbial pollution of surface water resources. *Int. J. Environ. Res. Public Health* **2014**, *11*, 249–270. [CrossRef] [PubMed]
4. World Health Organisation (WHO). 2.1 Billion People Lack Safe Drinking Water at Home, More than Twice as Many Lack Safe Sanitation. 2017. Available online: http://www.who.int/mediacentre/news/releases/2017/water-sanitation-hygiene/en/ (accessed on 18 October 2017).
5. Nongogo, V.; Okoh, A.I. Occurrence of *Vibrio* Pathotypes in the Final Effluents of Five Wastewater Treatment Plants in Amathole and Chris Hani District Municipalities in South Africa. *Int. J. Environ. Res. Public Health* **2014**, *11*, 7755–7766. [CrossRef] [PubMed]
6. Department of Water Affairs, South Africa (DWA). Green Drop Handbook-Version I. 2011. Available online: http://www.dwaf.gov.za/dir_ws/GDS/Docs/DocsDefault.aspx (accessed on 21 September 2013).
7. Mara, D.; Horan, N.J. *Handbook of Water and Wastewater Microbiology*; Academic Press: Cambridge, MA, USA, 2003.
8. Dungeni, M.; van Der Merwe, R.R.; Momba, M.N.B. Abundance of pathogenic bacteria and viral indicators in chlorinated effluents produced by four wastewater treatment plants in the Gauteng Province, South Africa. *Water SA* **2010**, *36*, 607–614. [CrossRef]
9. Bessong, P.O.; Odiyo, J.O.; Musekene, J.N.; Tessema, A. Spatial distribution of diarrhoea and microbial quality of domestic water during an outbreak of diarrhoea in the Tshikuwi community in Venda, South Africa. *J. Health Popul. Nutr.* **2009**, *27*, 652–659. [CrossRef] [PubMed]
10. Figueras, M.; Borrego, J.J. New perspectives in monitoring drinking water microbial quality. *Int. J. Environ. Res. Public Health* **2010**, *7*, 4179–4202. [CrossRef] [PubMed]
11. Okoh, A.I.; Sibanda, T.; Nongogo, V.; Adefisoye, M.; Olayemi, O.O.; Nontongana, N. Prevalence and characterisation of non-cholerae *Vibrio* spp. in final effluents of wastewater treatment facilities in two districts of the Eastern Cape Province of South Africa: Implications for public health. *Environ. Sci. Pollut. Res.* **2015**, *22*, 2008–2017. [CrossRef] [PubMed]
12. Corcoran, E.; Nellemann, C.; Baker, E.; Bos, R.; Osborn, D.; Savelli, H. Sick Water? The Central Role of Wastewater Management in Sustainable Development. A Rapid Response Assessment. United Nations Environment Programme, UN-HABITAT, GRID-Arendal, 2010. Available online: https://www.grida.no/publications/218 (accessed on 18 October 2017).
13. Nagulapally, S.R. Antibiotic Resistance Patterns in Municipal Wastewater Bacteria. Ph.D. Thesis, Kansas State University, Manhattan, KS, USA, 2007. Available online: https://krex.k-state.edu/dspace/bitstream/handle/2097/331/SujathaNagulapally2007.pdf?sequence=1 (accessed on 24 October 2017).
14. Spellman, F.R.; Drinan, J. *Wastewater Treatment Plant Operations Made Easy: A Practical Guide for Licensure*; DEStech Publications, Inc.: Lancaster, PA, USA, 2003.
15. Argaw, N. Wastewater Sources and Treatment. In *Renewable Energy in Water and Wastewater Treatment Applications*; National Renewable Energy Laboratory, US Department of Energy Laboratory: Golden, CO, USA, 2004; Chapter 6, pp. 38–46.
16. The Consortium of Institutes for Decentralized Wastewater Treatment (CIDWT). Decentralized Wastewater Glossary. 2009. Available online: http://www.onsiteconsortium.org/Glossary2009.pdf (accessed on 10 October 2017).
17. Metcalf and Eddy, Inc. *Wastewater Engineering: Treatment Disposal and Reuse*, 3rd ed. McGraw-Hill: New York, NY, USA, 1991.

18. Topare, N.S.; Attar, S.J.; Manfe, M.M. Sewage/wastewater treatment technologies: A review. *Sci. Rev. Chem. Commun.* **2011**, *1*, 18–24.
19. Templeton, M.R.; Butler, D. Introduction to Wastewater Treatment. Bookboon, 2011. Available online: http://bookboon.com/en/introduction-to-wastewater-treatment-ebook (accessed on 18 October 2017).
20. Suez E-Mag. The Treatment of Wastewater: A Global Public Health and Environmental Protection Challenge. 2013. Available online: http://www.emag.suez-environnement.com/en/treatment-wastewater-global-public-health-environmental-protection-challenge-11126 (accessed on 22 November 2016).
21. ESCWA/UN. Waste-Water Treatment Technologies—A General Review. 2003. Available online: http://www.igemportal.org/Resim/Wastewater%20Treatment%20Technologies_%20A%20general%20rewiev.pdf (accessed on 10 October 2017).
22. Eslamian, S. *Urban Water Reuse Handbook*; CRC Press: Boca Raton, FL, USA, 2016; ISBN 978-1-4822-2914-1.
23. Sonune, A.; Ghate, R. Developments in wastewater treatment methods. *Desalination* **2004**, *167*, 55–63. [CrossRef]
24. United States Environmental Protection Agent (USEPA). Primer for Municipal Wastewater Treatment Systems. Office of Water and Wastewater Management, 2004. Available online: https://www3.epa.gov/npdes/pubs/primer.pdf (accessed on 10 October 2017).
25. Water Research Commission (WRC). Wastewater Treatment Technologies—A Basic Guide; WRC Project No. K8/1106; WRC Report No. TT 651/15. 2015. Available online: http://www.wrc.org.za/Lists/Knowledge%20Hub%20Items/Attachments/11606/WRC_Booklet_Interactive.pdf (accessed on 14 November 2017).
26. Horan, N.J. *Biological Wastewater Treatment Systems: Theory and Operation*; Baffins Lane, West Sussex PO 191 UD; John Wiley and Sons Ltd.: Chickester, UK, 1990.
27. Environmental Protection Agency (EPA), Ireland. Wastewater Treatment Manuals—Primary, Secondary and Tertiary Treatment. 1997. Available online: https://www.epa.ie/pubs/advice/water/wastewater/EPA_water_%20treatment_manual_primary_secondary_tertiary1.pdf (accessed on 21 September 2017).
28. Abdel-Raouf, N.; Al-Homaidan, A.A.; Ibraheem, I.B.M. Microalgae and wastewater treatment. *Saudi J. Biol. Sci.* **2012**, *19*, 257–275. [CrossRef] [PubMed]
29. WAMtechnology eBook. Water Treatment Processes. Stellenbosch, South Africa. Available online: http://anyflip.com/idpr/hhuv/basic (accessed on 24 November 2016).
30. Republic of South Africa. Constitution of the Republic of South Africa, Act 108 of 1996. *Gov. Gaz.* **1996**, *378*, 147.
31. Manxodidi, T.; Mackintosh, G.; Wensley, A.; Manus, L.; Gadzikwa, S. The Role of Regulatory and Co-Operative Governance in Ensuring the Quality of Water Services in South Africa. Available online: http://www.ewater.co.za/literature/files/107%20Manxodidi.pdf (accessed on 24 October 2017).
32. Department of Water Affairs (DWA). National Water Resource Strategy: Water for an Equitable and Sustainable Future. 2013. Available online: http://www.wrc.org.za/SiteCollectionDocuments/Acts%20for%20govenance%20page/DWS%20National%20Water%20Resources%20Strategy%202LinkClick.pdf (accessed on 10 October 2017).
33. Department of Water Affairs and Forestry (DWAF). *South African Water Quality Guidelines*, 2nd ed.; Volume 1: Domestic Water Use; The Department of Water Affairs and Forestry: Pretoria, South Africa, 1996. Available online: http://www.dwa.gov.za/iwqs/wq_guide/Pol_saWQguideFRESH_vol1_Domesticuse.PDF (accessed on 10 October 2017).
34. Jiménez, B.; Barrios, J.A.; Mendez, J.M.; Diaz, J. Sustainable sludge management in developing countries. *Water Sci. Technol.* **2004**, *49*, 251–258. [CrossRef] [PubMed]
35. Mema, V. Impact of Poorly Maintained Wastewater and Sewage Treatment Plants: Lessons from South Africa. *ReSource* **2010**, *12*, 60–65. Available online: https://pdfs.semanticscholar.org/afa6/d6ff3a25061a5c9b8f2194680f1b5c682fbc.pdf (accessed on 24 October 2017).
36. Bolong, N.; Ismail, A.F.; Salim, M.R.; Matsuura, T. A review of the effects of emerging contaminants in wastewater and options for their removal. *Desalination* **2009**, *239*, 229–246. [CrossRef]
37. Donlan, R.M. Biofilms: Microbial life on surfaces. *Emerg. Infect. Dis.* **2002**, *8*, 881–890. [CrossRef] [PubMed]
38. Department of Water Affairs and Forestry (DWAF). Government Gazette NO. 20526 8 October 1999. Available online: http://www.dwaf.gov.za/Documents/Notices/Gen_AuthPublished-eng.pdf (accessed on 12 June 2018).

39. Momba, M.N.; Malakate, V.K.; Theron, J. Abundance of pathogenic *Escherichia coli, Salmonella typhimurium* and *Vibrio cholerae* in Nkonkobe drinking water sources. *J. Water Health* **2006**, *4*, 289–296. [CrossRef] [PubMed]

40. Okoh, A.I.; Igbinosa, E.O. Antibiotic susceptibility profiles of some Vibrio strains isolated from wastewater final effluents in a rural community of the eastern cape province of South Africa. *BMC Microbiol.* **2010**, *10*, 143. [CrossRef] [PubMed]

41. Lobitz, B.; Beck, L.; Huq, A.; Wood, B.; Fuchs, G.; Faruque, A.S.G.; Colwell, R. Climate and infectious disease: Use of remote sensing for detection of *Vibrio cholerae* by indirect measurement. *Proc. Natl. Acad. Sci. USA* **2000**, *97*, 1438–1443. [CrossRef] [PubMed]

42. Worden, A.Z.; Seidel, M.; Smriga, S.; Wick, A.; Malfatti, F.; Bartlett, D.; Azam, F. Trophic regulation of *Vibrio cholerae* in coastal marine waters. *Environ. Microbiol.* **2006**, *8*, 21–29. [CrossRef] [PubMed]

43. Mezrioui, N.; Oufdou, K. Abundance and antibiotic resistance of non-O1 *Vibrio cholerae* strains in domestic wastewater before and after treatment in stabilization ponds in an arid region (Marrakesh, Morocco). *FEMS Microbiol. Ecol.* **1996**, *21*, 277–284. [CrossRef]

44. Igbinosa, E.O.; Obi, L.C.; Okoh, A.I. Occurrence of potentially pathogenic vibrios in final effluents of a wastewater treatment facility in a rural community of the eastern cape province of South Africa. *Res. Microbiol.* **2009**, *160*, 531–537. [CrossRef] [PubMed]

45. Igbinosa, E.O.; Obi, L.C.; Tom, M.; Okoh, A.I. Detection of potential risk of wastewater effluents for transmission of antibiotic resistance from *Vibrio* species as a reservoir in a peri-urban community in South Africa. *Int. J. Environ. Health Res.* **2011**, *21*, 402–414. [CrossRef] [PubMed]

46. Chatterjee, S.N.; Maiti, M. Vibriophages and vibriocins: Physical, chemical, and biological properties. *Adv. Virus Res.* **1984**, *29*, 263–312. [CrossRef] [PubMed]

47. Kaper, J.B.; Morris, J.G., Jr.; Levine, M.M. Cholera. *Clin. Microbiol. Rev.* **1995**, *8*, 48–86. [PubMed]

48. Griffith, D.C.; Kelly-Hope, L.A.; Miller, M.A. Review of reported cholera outbreaks worldwide, 1995–2005. *Am. J. Trop. Med. Hyg.* **2006**, *75*, 973–977. [CrossRef] [PubMed]

49. Maheshwari, M.; Krishnaiah, N.; Ramana, D. Evaluation of Polymerase Chain Reaction for the detection of *Vibrio cholerae* in Contaminants. *Ann. Biol. Res.* **2011**, *2*, 212–217.

50. Tacket, C.O.; Brenner, F.; Blake, P.A. Clinical features and an epidemiological study of *Vibrio vulnificus* infections. *J. Infect. Dis.* **1984**, *149*, 558–561. [CrossRef] [PubMed]

51. Levine, W.C.; Griffin, P.M.; Gulf Coast *Vibrio* Working Group. *Vibrio* infections on the Gulf Coast: Results of first year of regional surveillance. *J. Infect. Dis.* **1993**, *167*, 479–483. [PubMed]

52. Horseman, M.A.; Surani, S. A comprehensive review of *Vibrio vulnificus*: An important cause of severe sepsis and skin and soft-tissue infection. *Int. J. Infect. Dis.* **2011**, *15*, e157–e166. [CrossRef] [PubMed]

53. Cabral, J.P. Water microbiology. Bacterial pathogens and water. *Int. J. Environ. Res. Public Health* **2010**, *7*, 3657–3703. [CrossRef] [PubMed]

54. Igbinosa, E.O.; Okoh, A.I. Emerging *Vibrio* species: An unending threat to public health in developing countries. *Res. Microbiol.* **2008**, *159*, 495–506. [CrossRef] [PubMed]

55. Hasan, N.A.; Grim, C.J.; Haley, B.J.; Chun, J.; Alam, M.; Taviani, E.; Hoq, M.; Munk, A.C.; Saunders, E.; Brettin, T.S.; et al. Comparative genomics of clinical and environmental *Vibrio mimicus*. *Proc. Natl. Acad. Sci. USA* **2010**, *107*, 21134–21139. [CrossRef] [PubMed]

56. Qian, R.; Chu, W.; Mao, Z.; Zhang, C.; Wei, Y.; Yu, L. Expression, characterization and immunogenicity of a major outer membrane protein from *Vibrio* alginolyticus. *Acta Biochim. Biophys. Sin.* **2007**, *39*, 194–200. [CrossRef] [PubMed]

57. Denner, E.B.; Vybiral, D.; Fischer, U.R.; Velimirov, B.; Busse, H.J. *Vibrio calviensis* sp. nov.; a halophilic, facultatively oligotrophic 0.2 microm-filterable marine bacterium. *Int. J. Syst. Evol. Microbiol.* **2002**, *52*, 549–553. [CrossRef] [PubMed]

58. Girones, R.; Ferrus, M.A.; Alonso, J.L.; Rodriguez-Manzano, J.; Calgua, B.; de Abreu Corrêa, A.; Hundesa, A.; Carratala, A.; Bofill-Mas, S. Molecular detection of pathogens in water—The pros and cons of molecular techniques. *Water Res.* **2010**, *44*, 4325–4339. [CrossRef] [PubMed]

59. Colwell, R.R.; Huq, A. Vibrios in the environment: Viable but nonculturable *Vibrio cholerae*. *Am. Soc. Microbiol.* **1994**, 117–133. [CrossRef]

60. Tantillo, G.M.; Fontanarosa, M.; Di Pinto, A.; Musti, M. Updated perspectives on emerging *Vibrios* associated with human infections. *Lett. Appl. Microbiol.* **2004**, *39*, 117–126. [CrossRef] [PubMed]

61. American Public Health Association (APHA). *Standard Methods for the Examination of Water and Wastewater*; Water Environmental Federation and American Public Health Association: Washington, DC, USA, 2005.

62. Mustapha, S.; Mustapha, E.M.; Nozha, C. *Vibrio alginolyticus*: An emerging pathogen of foodborne diseases. *Int. J. Sci. Technol.* **2013**, *2*, 302–309.

63. Faruque, S.M.; Siddique, A.K.; Saha, M.N.; Rahman, M.M.; Zaman, K.; Albert, M.J.; Sack, D.A.; Sack, R.B. Molecular Characterization of a New Ribotype of *Vibrio cholerae* O139 Bengal Associated with an Outbreak of Cholera in Bangladesh. *J. Clin. Microbiol.* **1999**, *37*, 1313–1318. [PubMed]

64. Faruque, S.M.; Chowdhury, N.; Kamruzzaman, M.; Dziejman, M.; Rahman, M.H.; Sack, D.A.; Nair, G.B.; Mekalanos, J.J. Genetic diversity and virulence potential of environmental *Vibrio cholerae* population in a cholera-endemic area. *Proc. Natl. Acad. Sci. USA* **2004**, *101*, 2123–2128. [CrossRef] [PubMed]

65. Arias, C.R.; Garay, E.; Aznar, R. Nested PCR method for rapid and sensitive detection of *Vibrio vulnificus* in fish, sediments, and water. *Appl. Environ. Microbiol.* **1995**, *61*, 3476–3478. [PubMed]

66. Faruque, S.M.; Albert, M.J.; Mekalanos, J.J. Epidemiology, Genetics, and Ecology of Toxigenic *Vibrio cholerae*. *Microbiol. Mol. Biol. Rev.* **1998**, *62*, 1301–1314. [PubMed]

67. Mishra, M.; Farah, M.; Akulwar, S.L.; Katkar, V.J. Re-emergence of El Tor *Vibrio* in outbreak of cholera in & around Nagpur. *Indian J. Med. Res.* **2004**, *120*, 478–480. [PubMed]

68. Le Roux, C.S. Dealing with cholera: Exclusively the domain of environmental health practitioners? *Health SA Gesondheid* **2004**, *9*, 55–65. [CrossRef]

69. Health 24. Recent Cholera Outbreaks in SA. Available online: http://www.health24.com/Medical/Cholera/Recent-cholera-outbreaks/Recent-cholera-outbreaks-in-SA-20120721 (accessed on 24 October 2017).

70. Olaniran, A.O.; Naicker, K.; Pillay, B. Toxigenic *Escherichia coli* and *Vibrio cholerae*: Classification, pathogenesis and virulence determinants. *Biotechnol. Mol. Biol. Rev.* **2011**, *6*, 94–100.

71. Bateman, C. Up to its eyeballs in sewage: Government pleads for help. *S. Afr. Med. J.* **2009**, *99*, 556–560. [PubMed]

72. Mail and Gardian. Cholera Death Toll in South Africa Rises. 2009. Available online: https://mg.co.za/article/2009-01-14-cholera-death-toll-in-south-africa-rises (accessed on 10 October 2017).

73. Diarrhoea Outbreak in Limpopo. Available online: http://www.news24.com/SouthAfrica/News/Diarrhoea-outbreak-in-Limpopo-20140127 (accessed on 14 November 2017).

74. SABC Digital News. Available online: https://www.youtube.com/watch?v=TeSBwEJSmp0 (accessed on 17 November 2017).

75. ENCA News. Minister to Visit Bloemhof after Diarrhoea Deaths. 2014. Available online: https://www.enca.com/south-africa/minister-visit-bloemhof (accessed on 17 November 2017).

76. Toxic Water Scare Rocks Cradock as Several Fall Ill. Available online: http://www.health24.com/News/Public-Health/Toxic-water-scare-rocks-Cradock-as-several-falls-ill-20151104 (accessed on 14 November 2017).

International Journal of
*Environmental Research
and Public Health*

Article

Removal of Emerging Contaminants and Estrogenic Activity from Wastewater Treatment Plant Effluent with UV/Chlorine and UV/H₂O₂ Advanced Oxidation Treatment at Pilot Scale

Eduard Rott *, Bertram Kuch, Claudia Lange, Philipp Richter, Amélie Kugele and Ralf Minke

Institute for Sanitary Engineering, Water Quality and Solid Waste Management, University of Stuttgart, Bandtäle 2, 70569 Stuttgart, Germany; bertram.kuch@iswa.uni-stuttgart.de (B.K.); claudia.lange@dekra.com (C.L.); philipp.richter@iswa.uni-stuttgart.de (P.R.); ask.kugele@t-online.de (A.K.); ralf.minke@iswa.uni-stuttgart.de (R.M.)
* Correspondence: eduard.rott@iswa.uni-stuttgart.de; Tel.: +49-711-685-60497

Abstract: Effluent of a municipal wastewater treatment plant (WWTP) was treated on-site with the UV/chlorine (UV/HOCl) advanced oxidation process (AOP) using a pilot plant equipped with a medium pressure UV lamp with an adjustable performance of up to 1 kW. Results obtained from parallel experiments with the same pilot plant, where the state of the art UV/H₂O₂ AOP was applied, were compared regarding the removal of emerging contaminants (EC) and the formation of adsorbable organohalogens (AOX). Furthermore, the total estrogenic activity was measured in samples treated with the UV/chlorine AOP. At an energy consumption of 0.4 kWh/m³ (0.4 kW, 1 m³/h) and in a range of oxidant concentrations from 1 to 6 mg/L, the UV/chlorine AOP had a significantly higher EC removal yield than the UV/H₂O₂ AOP. With free available chlorine concentrations (FAC) in the UV chamber influent of at least 5 mg/L (11 mg/L of dosed Cl₂), the total estrogenic activity could be reduced by at least 97%. To achieve a certain concentration of FAC in the UV chamber influent, double to triple the amount of dosed Cl₂ was needed, resulting in AOX concentrations of up to 520 µg/L.

Keywords: AOP; AOX; emerging contaminants; estrogenic activity; UV/chlorine; UV/H₂O₂

1. Introduction

Anthropogenic compounds have been detected in wastewater treatment effluent, surface water, and ground water over the last years [1–6]. Many compounds, referred to as emerging contaminants (ECs), are brought into the environment by the disposed effluent of municipal wastewater treatment plants (WWTP) due to their stability against biological decomposition. These compounds may endanger aquatic life forms and, ultimately, humans via the food chain. Some endocrine disrupting compounds (EDCs) interfere with the hormone system [7]. After exposure, some compounds may cause cancer in humans [8]. Therefore, an obligatory, additional treatment step in WWTPs will be required.

Today, numerous alternative treatment methods for the removal of these compounds have been considered, including activated carbon treatment [9,10] or membrane filtration [11,12]. The advanced oxidation process (AOP) is a modern solution for the reduction of EC concentrations in wastewater treatment plant effluent (WWTE) [13]. In UV/AOP, an oxidant is dosed to the WWTE and activated by UV radiation to form highly reactive and unselective hydroxyl radicals (•OH). Hence, organic pollutants can be oxidized to CO₂ and H₂O or at least rendered biodegradable for subsequent natural degradation. State of the art oxidants are hydrogen peroxide (H₂O₂) and ozone (O₃) [14,15].

UV/chlorine AOP is a promising alternative. In previous studies [16–20], good removal rates for specified ECs have been observed, with prospects of economic advantages and a better energy saving potential compared to state of the art UV/AOP implementations. The actual oxidizing effect of dissolved sodium hypochlorite (NaOCl) is based on the formation of hypochlorous acid (HOCl). The transformation of chlorine (Cl_2) in an aqueous solution into hypochlorous acid and hydrochloric acid (HCl) is shown in Equation (1) [21]. The dissociation of hypochlorous acid into hypochlorite anions (ClO^-) is pH dependent (Equation (2)) ($pK_a = 7.5$) [22]. ClO^- is a less effective oxidant, which is why at higher pH values the oxidation capability decreases [23]. In samples with pH values around 7, the predominant species is HOCl [23].

$$Cl_2 + H_2O \longrightarrow HOCl + Cl^- + H^+ \tag{1}$$

$$HOCl \longrightarrow ClO^- + H^+ \tag{2}$$

In the UV/chlorine process, inter alia, $\bullet OH$ and $Cl\bullet$ radicals are formed. Among others, the following reactions occur [24–26]:

$$HOCl + UV\ photons \longrightarrow \bullet OH + Cl\bullet \tag{3}$$

$$ClO^- + UV\ photons \longrightarrow \bullet O^- + Cl\bullet \tag{4}$$

$$\bullet O^- + H_2O \longrightarrow \bullet OH + OH^- \tag{5}$$

However, little research has been done in the study of UV/chlorine AOP treatment on wastewater at pilot scale. In this study, a continuous flow UV pilot plant was placed on the premises of a wastewater treatment plant. The goal was to examine the on-site feasibility of the UV/chlorine AOP applied to municipal WWTE.

2. Materials and Methods

2.1. Experimental Concept

The pilot plant, equipped with a medium pressure UV chamber with a maximum effective power of 1 kW, processed 1 m^3/h in all experiments. The gas pressure inside the UV lamp has a significant influence on the spectral emittance of mercury. In comparison to low pressure UV lamps, which emit at one single wavelength (254 nm), medium pressure UV lamps emit at a broader spectrum (200–400 nm). When free chlorine (HOCl and ClO^-) is dosed to wastewater, it can partially or fully react with wastewater components. The remaining active free Cl_2 is mostly referred to as "free available chlorine" (FAC). In UV/chlorine experiments of this work, NaOCl solution was dosed to obtain the desired FAC concentrations in the UV chamber influent. In the case of an incomplete reaction of FAC in the UV chamber, "residual free chlorine" (RFC) in the UV chamber effluent could occur. This RFC was quenched according to Equation (6) [27] and Equation (7) [28] by means of an additional dosage of H_2O_2 downstream of the UV chamber.

$$HOCl + H_2O_2 \longrightarrow Cl^- + H_2O + O_2 + H^+ \tag{6}$$

$$Cl\bullet + H_2O_2 \longrightarrow Cl^- + HO_2\bullet + H^+ \tag{7}$$

In parallel UV/H_2O_2 experiments, the oxidant dosed was H_2O_2 to allow for comparisons with the state of the art UV/H_2O_2 AOP. Two experiments were carried out, in which different experimental conditions were applied (no UV and no oxidant dosage; sole UV treatment; sole FAC or H_2O_2 treatment; and combinations out of UV, FAC and H_2O_2) as follows:

- Experiment 1: Variation of UV energy consumption (0.0, 0.4, 0.7, and 1.0 kWh/m^3) at 0 and 3 mg/L oxidant concentrations (FAC or H_2O_2).

- Experiment 2: Variation of oxidant concentration (1–6 mg/L FAC or H_2O_2) at 0.4 kWh/m^3 UV energy consumption.

Besides the removal of emerging contaminants, other impacts were examined. This was done with the aim to get a wide spectrum of the influence of the UV/chlorine and UV/H_2O_2 AOPs on WWTE using a continuous flow pilot plant. Cumulative effects on the WWTE were considered with the analysis of the bacterial count (in Supplementary Data), the total estrogenic activity, the formation of adsorbable organohalogens (AOX), combined Cl_2 and chlorine oxyanions.

2.2. Chemicals and Reagents

NaOCl solution (14% active chlorine) was purchased from VWR International (Radnor, PA, USA) and H_2O_2 solution (35% technical grade) was received from Siemens Water Technologies (Günzburg, Germany). Sodium thiosulfate ($Na_2S_2O_3 \cdot 5H_2O$, $\geq 99\%$) was purchased from Carl Roth (Karlsruhe, Germany) and nitric acid solution (65%, p.a.) was purchased from Merck (Darmstadt, Germany). *N,N*-diethyl-*p*-phenylenediamine (DPD) was contained in powder pillows obtained by Hach (Berlin, Germany).

2.3. Wastewater Treatment Plant Effluent (WWTE) and Emerging Contaminants (ECs)

The state of the art municipal Treatment Plant for Education and Research (LFKW, Lehr- und Forschungsklärwerk) (primary clarifier, activated sludge treatment: denitrification/nitrification, P precipitation, secondary clarifier, micro sieves) lies on the premises of ISWA next to Büsnau, a district in Stuttgart, Germany. The average amount of treated wastewater in a year is about 900,000 m^3 with a capacity of 30 L/s (9000 population equivalents). The raw wastewater of the LFKW is a mixture of domestic wastewater with a relatively high organic load and wastewater which is less concentrated from the university grounds. With the last treatment step of the LFKW, the water is filtered via micro sieves (15–20 µm pore size).

In Table 1, the initial parameter values c_0 of the UV pilot plant influent for both AOP experiments are shown. The UV/H_2O_2 experiments were performed two months after the UV/chlorine experiments. Thus, both WWTEs differed slightly from each other. Nevertheless, apart from the temperature most of the parameters were very similar.

Table 1. Initial parameter values c_0 measured in WWTE reference samples collected in both UV/chlorine AOP and UV/H_2O_2 AOP experiments (n. m.: not measured; COD: chemical oxygen demand; DOC: dissolved organic carbon; EEQ: 17β-estradiol equivalent; AOX: adsorbable organohalogens; * only one determination).

Parameter	Variation of UV Energy Consumption between 0 and 1 kWh/m^3 (Experiment 1)		Variation of Oxidant Concentration at 0.4 kWh/m^3 (Experiment 2)		
	0 and 3 mg/L FAC	0 and 3 mg/L H_2O_2	1–4 mg/L FAC	5–6 mg/L FAC	1–6 mg/L H_2O_2
Temperature (°C)	14.9	18.9	14.6	14.8	19.5
pH	7.0	7.0	7.0	7.0	7.0
COD (mg/L)	17.8 ± 1.3	20.4 ± 1.8	23.6 ± 0.3	23.2 ± 0.1	21.3 ± 0.8
DOC (mg/L)	5.8 ± 0.9	5.9 ± 0.3	6.0 ± 1.1	6.9 ± 0.2	5.5 ± 0.1
NH_4^+-N (mg/L)	<0.15	<0.11	<0.1	<0.1	<0.1
EEQ (ng/L)	1.83 ± 0.28	n. m.	3.92 ± 0.31	1.77 ± 0.25	n. m.
AOX (µg/L)	12 *	25 ± 1	21 ± 5	25 ± 5	22 ± 8
ClO_2^- (mg/L)	n. m.	n. m.	n. m.	<0.20	n. m.
ClO_3^- (mg/L)	n. m.	n. m.	n. m.	<0.06	n. m.
ClO_4^- (mg/L)	n. m.	n. m.	n. m.	<0.13	n. m.

In Figure 1, an overview of the ECs analyzed in this study is given. ECs can be found in WWTE in various concentrations depending on the sampling time. In this study, the initial concentrations varied from 0.04 to 2.6 µg/L (Table 2). Most of the initial concentrations did not differ significantly between the experiments. Exceptions were the insect repellent DEET (0.04–1.99 µg/L) and the organophosphorous compound TCEP (0.35–1.76 µg/L).

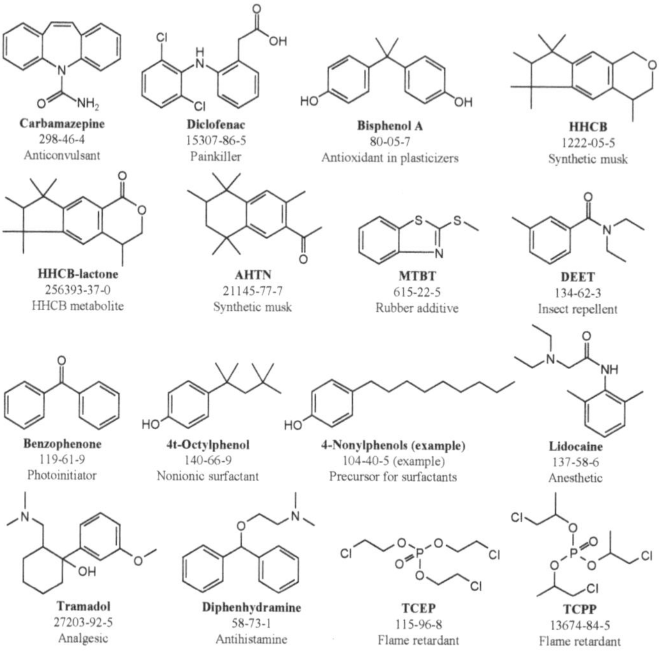

Figure 1. Overview of analyzed emerging contaminants with CAS numbers (based on [29]).

Table 2. Initial EC concentrations c_0 measured in WWTE reference samples collected in both UV/chlorine AOP and UV/H$_2$O$_2$ AOP experiments (three samples with single determination).

Emerging Contaminant (µg/L)	Variation of UV Energy Consumption between 0 and 1 kWh/m³ (Experiment 1)		Variation of Oxidant Concentration at 0.4 kWh/m³ (Experiment 2)		
	0 and 3 mg/L FAC	0 and 3 mg/L H$_2$O$_2$	1–4 mg/L FAC	5–6 mg/L FAC	1–6 mg/L H$_2$O$_2$
Carbamazepine	0.47 ± 0.00	0.48 ± 0.02	0.75 ± 0.02	0.83 ± 0.02	0.43 ± 0.02
Diclofenac	1.16 ± 0.02	2.28 ± 0.09	2.15 ± 0.08	2.55 ± 0.18	1.83 ± 0.14
Bisphenol A	0.77 ± 0.00	0.61 ± 0.04	0.85 ± 0.02	0.57 ± 0.30	0.53 ± 0.12
HHCB	1.20 ± 0.01	1.14 ± 0.02	1.24 ± 0.03	1.19 ± 0.06	1.12 ± 0.02
HHCB-lactone	1.33 ± 0.02	1.21 ± 0.03	1.61 ± 0.05	1.56 ± 0.09	1.20 ± 0.08
AHTN	0.16 ± 0.00	0.14 ± 0.00	0.18 ± 0.01	0.18 ± 0.01	0.14 ± 0.01
MTBT	0.24 ± 0.00	0.29 ± 0.01	0.24 ± 0.01	0.22 ± 0.03	0.28 ± 0.01
DEET	0.08 ± 0.00	1.99 ± 0.01	0.05 ± 0.00	0.04 ± 0.01	0.28 ± 0.02
Benzophenone	0.14 ± 0.00	0.20 ± 0.00	0.12 ± 0.00	0.13 ± 0.02	0.21 ± 0.01
4t-Octylphenol	0.04 ± 0.00	0.04 ± 0.00	0.04 ± 0.00	0.05 ± 0.01	0.03 ± 0.00
4-Nonylphenols	1.93 ± 0.03	1.67 ± 0.06	1.65 ± 0.13	1.55 ± 0.24	2.01 ± 0.12
Lidocaine	0.12 ± 0.01	0.22 ± 0.00	0.27 ± 0.00	0.23 ± 0.01	0.17 ± 0.01
Tramadol	0.11 ± 0.01	0.17 ± 0.01	0.16 ± 0.01	0.09 ± 0.02	0.21 ± 0.02
Diphenhydramine	0.25 ± 0.01	0.25 ± 0.01	0.22 ± 0.01	0.27 ± 0.02	0.24 ± 0.01
TCEP	1.06 ± 0.02	0.69 ± 0.02	1.76 ± 0.03	0.35 ± 0.04	1.27 ± 0.16
TCPP	0.91 ± 0.05	1.56 ± 0.05	1.42 ± 0.02	1.66 ± 0.22	1.17 ± 0.09

2.4. UV Pilot Plant

The pilot plant (Figure 2) was placed in a hall where the micro sieves of the LFKW are situated. It was fed with the effluent of the micro sieves using an eccentric screw pump (Moineau pump) with the flow rate of 1 m³/h. The untreated reference sample (c_0) could be collected from a tap behind the variable area flowmeter. In UV/chlorine AOP experiments, chlorine was dosed from a NaOCl stock solution (5–10 g/L free Cl$_2$) with a peristaltic pump (0.08–4 L/h). In UV/H$_2$O$_2$ AOP experiments, this peristaltic pump was used for the dosage of H$_2$O$_2$ stock solution (5–10 g/L H$_2$O$_2$). The NaOCl (or H$_2$O$_2$) dosed water passed a static mixer to guarantee an extensive mixing through turbulence. The temperature, the pH (single junction, combination electrode sensor) and the FAC concentration (potentiostatic electrode amperometry sensor) of the UV chamber influent were determined by two membrane sensors (Wallace & Tiernan, Günzburg, Germany). At a flow rate of 1 m^3/h, the contact time of chlorine until reaching the UV chamber was about 4.6–6.4 s. The brand-new immersion UV lamp (Wallace & Tiernan Barrier M35, type: WTL 1000, Siemens Water Technologies, Günzburg, Germany), protected by a quartz sleeve with a thickness of 1 mm and cut-off at 200 nm wavelength, was installed in a stainless steel chamber. The quartz sleeve could be cleaned by pushing an attached rubber ring back and forth. The irradiance could be controlled by a UV signal visualized on the cabinet determined by a 4–20 mA UV sensor (signal in W/m^2). The approximate contact time in the UV chamber was 6–10 s. Downstream of the UV chamber, H$_2$O$_2$ could be dosed to the water to quench RFC in the pilot plant effluent and make it thus less harmful. In UV/H$_2$O$_2$ experiments, no H$_2$O$_2$ dosage was performed here. The focus of this study was on the technical feasibility of the UV/chlorine process by applying a continuous flow pilot plant. Therefore, the peristaltic pump for quenching agent dosage was mainly operated in automatic mode. This H$_2$O$_2$ dosage was automatically controlled by means of a chemical feed analyzer (for the RFC concentration) and process controller (MFC Analyzer/Controller) from Wallace & Tiernan (the RFC concentration of the pilot plant effluent was determined downstream of two further static mixers in a measuring cell with a potentiostatic electrode amperometry sensor). The FAC concentration could vary during an experiment while the H$_2$O$_2$ dosage was running and the experiments were limited in time. It was therefore not possible to determine the RFC concentration on a regular basis in case of missing H$_2$O$_2$ dosage. This aspect is therefore not addressed in this article. The contact time of the quenching agent from its dosage point to the effluent of the pilot plant was approximately 4.8–6.7 s. The treated sample (c) was collected from a second sampling tap. A control valve at the effluent was used to adjust the pressure in the system to enable a uniform distribution of water into all outgoing branches of the measuring cells of the pilot plant. At a flow rate of 1 m^3/h, the approximate flow time from the pilot plant influent to the pilot plant effluent was 25–29 s.

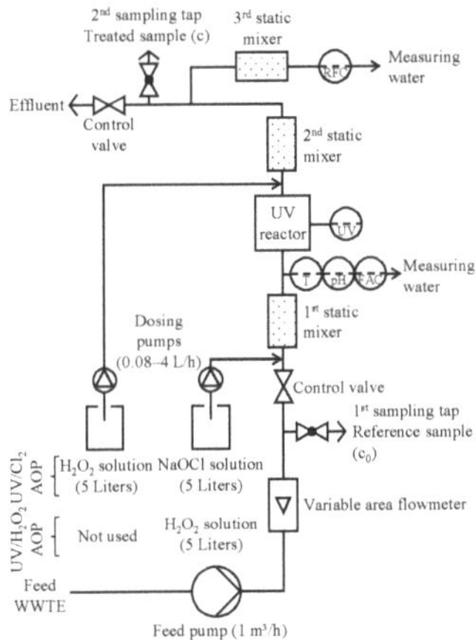

Figure 2. Technical scheme of the UV/chlorine AOP and UV/H$_2$O$_2$ AOP pilot plant used in the experiments.

2.5. Experimental Procedure

2.5.1. Variation of UV Energy Consumption at 0 and 3 mg/L Oxidant Concentrations (Experiment 1)

The UV/chlorine AOP and the UV/H$_2$O$_2$ AOP were compared in this experiment. At first, the flow rate was adjusted to 1 m^3/h. The oxidant concentration of 3 mg/L FAC or 3 mg/L H$_2$O$_2$ as well as the quenching agent dosage (not in UV/H$_2$O$_2$ AOP) was set. As soon as the pilot plant had reached a state of equilibrium (desired oxidant concentration achieved; no RFC measured in the pilot plant effluent), three reference samples (c_0) and the first three treated samples (c) (3 mg/L oxidant, 0.0 kWh/m^3) were taken from the sampling taps (separate bottles per analysis parameter). Then, the UV lamp was switched on and set to 0.4 kW. When the UV signal had stabilized, the next three treated samples were taken (3 mg/L oxidant, 0.4 kWh/m^3, 90 $\pm$ 10 W/m^2). Next, the UV lamp was set to 0.7 kW, and after 5 min, the next three treated samples were taken (3 mg/L oxidant, 0.7 kWh/m^3, 140 $\pm$ 10 W/m^2). This procedure was repeated with 1.0 kW (3 mg/L oxidant, 1.0 kWh/m^3, 185 $\pm$ 15 W/m^2). Then, the dosage of oxidant and quenching agent was switched off while the UV lamp was still running at 1.0 kW. After 10 min, the next three treated samples were collected (0 mg/L oxidant, 1.0 kWh/m^3, 185 $\pm$ 15 W/m^2). Subsequently, the UV lamp was set to 0.7 kW, and after 10 min, the next three treated samples were taken (0 mg/L oxidant, 0.7 kWh/m^3, 140 $\pm$ 10 W/m^2). This procedure was repeated with 0.4 kW (0 mg/L oxidant, 0.4 kWh/m^3, 90 $\pm$ 10 W/m^2). Finally, the UV lamp was switched off. After a further 10 min, three control samples (0 mg/L oxidant, 0.0 kWh/m^3) could be taken. During all settings, only small pH drifts with a maximum pH of 7.6 were measured in the pilot plant effluent.

2.5.2. Variation of Oxidant Concentration at 0.4 kWh/m^3 UV Energy Consumption (Experiment 2)

In this experiment, the UV/chlorine AOP and the UV/H$_2$O$_2$ AOP were compared at 0.4 kWh/m^3 UV energy consumption and varying oxidant concentrations (1–6 mg/L). At first, the flow rate was adjusted to 1 m^3/h. Next, the UV lamp was switched on and set to 0.4 kW while the desired oxidant concentration (1 mg/L) and quenching agent dosage (not in UV/H$_2$O$_2$ AOP) was set. After at least 5 min, the pilot plant had reached a state of equilibrium (desired oxidant concentration achieved; no RFC measured in the pilot plant effluent; stable UV signal in the range of 90 $\pm$ 10 W/m^2). Samples were then taken: For each analysis parameter, first, three sample bottles were filled with reference sample (c_0) consecutively from a sampling tap. Subsequently, for each analysis parameter, three sample bottles were filled with treated sample (c) from a different sampling tap at the effluent of the pilot plant. Then, the next desired oxidant concentration of 2 mg/L was set while the UV lamp was still running. When the pilot plant had reached a state of equilibrium, the treated samples could be collected in the same way as described above. These steps were repeated for the oxidant concentrations of 3, 4, 5 and 6 mg/L. During all settings, only small pH drifts with a maximum pH of 7.6 were measured in the pilot plant effluent.

2.6. Analytical Methods

2.6.1. Free Cl$_2$, Combined Cl$_2$, Total Cl$_2$

Cl$_2$, HOCl and OCl$^-$ are referred to as free Cl$_2$. Free Cl$_2$ becomes combined Cl$_2$ (CC, e.g., organic and inorganic chloramines) when it reacts with compounds in the water sample. Free Cl$_2$ and combined Cl$_2$ are summed up as total Cl$_2$. In this study, a Hach DPD powder pillow method (photometer: Merck SQ 118) was used for the measurement of free Cl$_2$ and total Cl$_2$ equivalent concentrations and the calibration of the free Cl$_2$ sensors on-site. In the following, there is also discussion of dosed Cl$_2$ (Equations (8) and (9)). Not all halogenated products can be determined by the total Cl$_2$ DPD method. Therefore, the dosed concentration of free Cl$_2$ from the NaOCl stock solution is presented (c_{dos}), which was calculated by means of Equation (10). c_{sol} is the free Cl$_2$ concentration of the dosed NaOCl stock solution. Q_{dos} is the flow rate of the dosing pump, Q the flow rate of the pilot plant (1 m^3/h). The term "Other Cl-containing reaction products" (OCRP) describes all substitution or oxidation/reduction products of the dosed Cl$_2$ containing the element chlorine which cannot be detected as total Cl$_2$ (e.g., chloride).

$$\text{dosed Cl}_2\ (c_{dos}) = \text{FAC/RFC} + \text{CC} + \text{OCRP} \tag{8}$$

$$\text{dosed Cl}_2\ (c_{dos}) = \text{total Cl}_2 + \text{OCRP} \tag{9}$$

$$c_{dos} = \frac{c_{sol} \times Q_{dos}}{Q} \tag{10}$$

During all experiments, on-site free Cl_2 and total Cl_2 measurements were carried out almost at about every process setting when a certain state of equilibrium was reached.

2.6.2. Chlorite (ClO_2^-), Chlorate (ClO_3^-), Perchlorate (ClO_4^-)

Chlorine oxyanions were determined only in the samples treated with 5 and 6 mg/L FAC at 0.4 kW UV power and their related reference sample in the 1st UV/chlorine AOP experiment using the standardized ISO 10304 method [30]. Each sample was filtered using C18 solid phase extraction cartridges and subsequent nylon filters with a pore size of 0.45 µm. The anions were detected by means of the Dionex ion chromatography system ICS-1000 (Waltham, MA, USA). An AS19a column (length: 25 cm, diameter: 2 mm) with a precolumn with anion self-regenerating suppressor was used. The gradient program was applied via a reagent-free controller. Each sample was measured twice. The limit of detection (LOD) for ClO_2^- was 0.2 mg/L; for ClO_3^-, it was 0.06 mg/L; and, for ClO_4^-, it was 0.13 mg/L.

2.6.3. Emerging Contaminants (ECs)

One-liter samples were quenched with 15 mg sodium thiosulfate ($Na_2S_2O_3$). The determination of ECs was performed via gas chromatography directly coupled with a mass selective spectrometer (GC Hewlett Packard 5890N Series II, Hewlett Packard 5972 Series detector, column: Varian VF-Xms, length: 30 m, diameter: 0.25 mm, film thickness: 0.25 µm). After the addition of internal standards, the samples were liquid–liquid extracted (dichloromethane, 2×40 mL) and evaporated to 100 µL. Quantification was done using the isotope dilution method and external calibration. The limit of quantification (LOQ) was 1 ng/L.

2.6.4. Total Estrogenic Activity (TEA)

For the determination of the total estrogenic activity, 1 L samples were collected without pretreating them prior to the analysis. The extracts obtained by solid-phase extraction were examined using an in vitro test system (E-screen assay) developed by Soto et al. [31] based on the instructions of Körner et al. [32] with modifications [33]. Thereby, the estrogenic activity reflects a sum parameter over all hormonal active compounds present in the samples expressed in concentration units of the reference compound 17β-estradiol. The LOQ of this method was 0.1 ng/L EEQ (17β-estradiol equivalent).

2.6.5. Adsorbable Organohalogens (AOX)

All samples were filled into 300 mL BOD bottles and acidified with 3 drops of 35% nitric acid (HNO_3) considering that the samples were free of headspace when the bottles were closed with glass stoppers. The determination of AOX concentrations was carried out using a standardized method [34]. In this method, the adsorbed and acidified sample is burned. Subsequently, the halogenide ions are determined via argentometry by means of microcoulometry (multi X 2000, Analytik Jena, Jena, Germany). The LOQ was 10 µg/L.

2.6.6. Number of Measurements

The given values in diagrams or tables are mean values calculated from determinations of three equivalent samples taken consecutively. Error bars in diagrams and numbers after the "±" symbol in tables correspond to the calculated standard deviation.

3. Results and Discussion

3.1. Chlorine Species and Adsorbable Organohalogens (AOX)

In Figure 3, for both UV/chlorine AOP experiments, the left columns depict the measured concentrations of FAC, combined Cl_2 and OCRP in the UV chamber influent. The right columns show the concentrations of these chlorine species in the pilot plant effluent after quenching. In the upper diagram (Experiment 1), it can be seen that, during the entire UV/chlorine AOP experiment, a dosage of about 7 mg/L was required to obtain a concentration of 3 mg/L FAC in the UV chamber influent. RFC could be eliminated successfully. However, about 1.9 mg/L of added free Cl_2 reacted to form compounds that could be measured as combined Cl_2 via the DPD method. The removal extent of combined Cl_2 was 10–20% with no significant effect of varying the UV performance (0.4, 0.7, and 1.0 kWh/m^3). With the UV lamp switched off, no removal of combined Cl_2 occurred. In Experiment 2 (lower diagram), to achieve a particular FAC concentration in WWTE, regardless of the desired FAC concentration, double to triple the amount of dosed Cl_2 was needed. In this experiment, the removal extent of combined Cl_2 varied between 10% and 50%, showing no linear correlation with the FAC concentration.

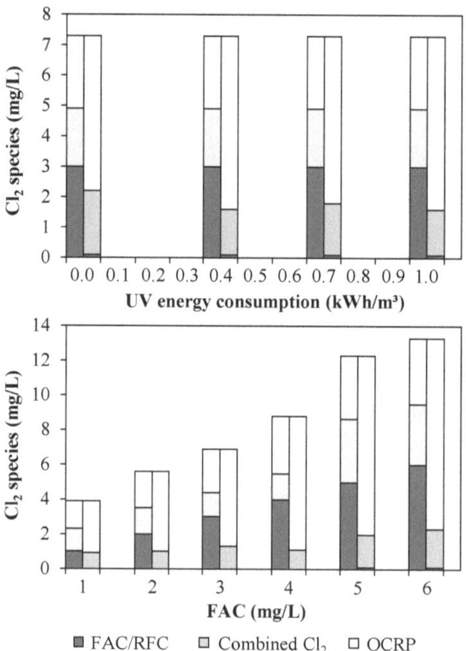

Figure 3. Chlorine species in UV chamber influent (1 m^3/h WWTE) (left columns) and in effluent of the pilot plant after UV treatment and subsequent quenching with H_2O_2 (right columns): (**Top**) Experiment 1 (0–1 kW, 3 mg/L FAC in UV chamber influent); and (**Bottom**) Experiment 2 (0.4 kW, 1–6 mg/L FAC in UV chamber influent).

Figure 4 sums up the measured concentrations of adsorbable organohalogens (AOX) in WWTE treated by the UV/chlorine AOP and the UV/H_2O_2 AOP. In all experiments, the AOX concentrations in the reference samples of the WWTE were very low, not exceeding 30 μg/L. In two control samples (sample of 2nd sampling tap, no oxidant dosage and no UV light), slightly higher AOX concentrations compared to the reference samples could be observed. However, the difference was insignificant since in all of these samples the AOX concentration was very low with <22 μg/L. In Experiment 1, both approaches with sole UV treatment (Figure 4, left and middle) led to AOX concentrations of <10–50 µg/L. However, considering the error susceptibility of the AOX determination method, the measured AOX concentrations were in such a small range that no significant AOX formation solely by UV treatment should be deduced. Furthermore, while the UV/H_2O_2 AOP had no significant effect on the AOX formation, already the sole dosage of FAC resulted in an AOX concentration of 314 ± 6 µg/L. Since only a slightly higher AOX concentration of up to 336 ± 6 µg/L was found with the UV/chlorine AOP at 0.4 kWh/m^3 and even lower values of down to 276 ± 6 µg/L were found at 1.0 kWh/m^3, it can be concluded that the formation of AOX is more due to chlorination and less to radical reaction. Much more, a higher UV power seemed to contribute to a reduction in AOX concentration. However, this reduction was very small. In Experiment 2 (right diagram), the UV/H_2O_2 AOP treatment did not increase the AOX concentration of the WWTE significantly as well (<30 µg/L AOX). In contrast, in WWTE samples treated with the UV/chlorine AOP, AOX could be measured up to 520 µg/L at FAC concentrations between 3 and 6 mg/L. With doses higher than 3 mg/L FAC, it seems as if a maximum AOX concentration was reached. This suggests that with 3 mg/L FAC most of the compounds in the WWTE that were available for chlorination were chlorinated. Furthermore, the two AOX concentrations at 0.4 kWh/m^3 and 3 mg/L FAC from Experiment 1 (336 ± 6 µg/L) and Experiment 2 (480 ± 32 µg/L) were different. Since both experiments were performed on different days, the slightly different wastewater composition (different amounts of chlorinable compounds) may have contributed to the deviation in results. This deviation, however, is not critical since the found AOX concentrations are in a similar range. In conclusion, although the pilot plant design did not allow to take samples directly in front of the UV chamber to analyze the AOX formation between the Cl_2 dosing point and UV chamber influent, there is considerable evidence that, despite the short contact time of 4.6–6.4 s, the AOX was formed before the UV chamber was reached, and this AOX formation was largely independent of the UV power.

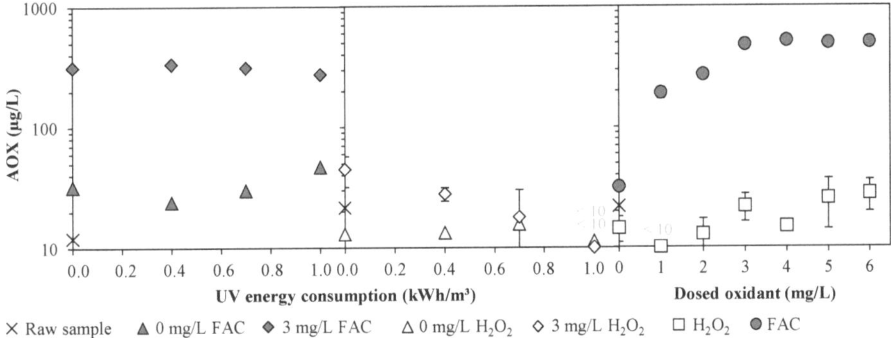

Figure 4. (**Left**, **Middle**) (Experiment 1) Influence of UV/chlorine AOP and UV/H_2O_2 AOP at 0.0, 0.4, 0.7, and 1.0 kWh/m^3 UV energy consumption (1 m^3/h, 0–1 kW) on AOX concentration in WWTE at oxidant concentrations of 0 and 3 mg/L; and (**Right**) (Experiment 2) influence of UV/chlorine AOP and UV/H_2O_2 AOP at 0.4 kWh/m^3 UV energy consumption (1 m^3/h, 0.4 kW) on AOX concentration in WWTE as a function of oxidant concentration.

The measured AOX concentrations were not high enough to fully explain the measured combined Cl_2 concentrations in the UV chamber influent and pilot plant effluent. In addition, the very low ammonium concentration (<0.15 mg/L NH_4^+-N) in the WWTE does not justify the conclusion that the measured combined Cl_2 consisted solely of inorganic chloramines. Furthermore, urea, a common precursor for inorganic chloramines, should not be found in WWTE due to its good removal in the wastewater treatment plant mainly based on hydrolysis [35]. The method for the determination of total Cl_2 is based on the principle that chloramines are capable of oxidizing iodide ions, dosed parallel to the DPD, to iodine, which also forms a red dye in reaction with DPD. Thus, it is useful to consider further compounds having an oxidative effect as a possible cause for the detection of combined Cl_2, i.e., also non-Cl-containing compounds [36]. To comprehend some of these oxidative by-products detected as combined Cl_2, the concentrations of the conjugated bases of oxyacids of chlorine (CBO) were determined in samples treated with 5 and 6 mg/L FAC at 0.4 kWh/m^3 (Experiment 2) (<0.2 mg/L ClO_2^-, 1.03 $\pm$ 0.02 mg/L ClO_3^-, <0.26 mg/L ClO_4^- at 5 mg/L FAC and <0.2 mg/L ClO_2^-, 1.11 $\pm$ 0.02 mg/L ClO_3^-, and <0.26 mg/L ClO_4^- at 6 mg/L FAC). In these samples, the measured concentrations of combined Cl_2 in the UV chamber influent were 3.7 $\pm$ 0.6 mg/L (at 5 mg/L FAC) and 3.5 $\pm$ 0.7 mg/L (at 6 mg/L FAC). In the corresponding samples of the UV pilot plant effluent, the combined Cl_2 concentrations were 1.9 $\pm$ 0.1 mg/L (at 5 mg/L FAC) and 2.2 $\pm$ 0.1 mg/L (at 6 mg/L FAC). From the experiment, it cannot be said whether chlorate was formed by chlorination, photolysis or radicals since this compound was only analyzed in the pilot plant effluent at parallel UV and chlorine dosage and not in the UV chamber influent. According to the literature, however, it can be assumed that UV light contributes significantly to its formation [24]. This indicates chlorate was surely hardly present in the UV chamber influent, where despite of that relatively high combined Cl_2 concentrations were present. Furthermore, it is obvious that also H_2O_2, which was dosed to the UV chamber effluent to quench RFC, contributed its part to increase the total Cl_2 concentration value. Approximately 9.1–9.2 mg/L H_2O_2 (about 270 μmol/L) were added at FAC dosages of 5 and 6 mg/L (70–85 μmol/L) at 0.4 kWh/m^3. According to Equation (6), this was 3–4 times the concentration that would be stoichiometrically required to quench 5–6 mg/L free Cl_2. Thus, surplus H_2O_2 must have been present in the pilot plant effluent, which must have contributed to a slightly falsified value of total Cl_2. In an experiment (Figure S1), this falsification was quantified to be 0.0388 mg total Cl_2/mg H_2O_2, so that at 9.1–9.2 mg/L H_2O_2 only a maximum interference of about 0.35 mg/L CC could be present. In the pilot plant effluent samples, which were treated with 5 and 6 mg/L FAC at 0.4 kWh/m^3, 1.9–2.2 mg/L combined Cl_2 was measured. With less than 20%, the falsification was therefore not large enough to fully explain the found concentration of combined Cl_2 in the pilot plant effluent and especially not in the UV chamber influent, where no H_2O_2 was present in the UV/chlorine AOP.

The relatively high concentration of combined Cl_2 in both the UV chamber influent and the pilot plant effluent can be attributed to a wide variety of other degradation products. Such disinfection by-products (DBP) resulting from the chlorination process are described in great detail in the literature. Trihalomethane (THM) formation is more pronounced in WWTEs with very low ammonium concentrations than in those with high ammonium concentrations [37]. The study of this work is based on WWTE with a very low NH_4^+-N concentration (<0.15 mg/L), so it can be assumed that THMs were prominently represented. It is also known that many dissolved organic nitrogen compounds (DON) in WWTE are essential precursors for N-DBPs [38,39]. In the effluent of the WWTP examined here, the annual average DON concentration was 1.6 mg/L (monthly average values varied between 0.9 and 2.4 mg/L N) and was thus in a range similar to the concentration of combined Cl_2 found. Extensive research by Pehlivanoglu-Mantas and Sedlak [40] showed that up to 10–20% of the DON concentration in WWTE can be attributed to amino acids and thus constitute a considerable precursor pool for the formation of N-DBPs such as dihaloacetonitriles [37–39]. This is relevant because amino acids have a relatively high reactivity with HOCl ($k_{app} > 1 \times 10^4$ M^{-1} s^{-1}) [23]. Other important known DBPs that may have been formed during the chlorination of WWTE are trichloronitromethanes, haloketones and chloral hydrates [17].

3.2. Emerging Contaminants

Figure 5 sums up the relative residual concentrations of ECs found in WWTE after the treatment with the UV/chlorine AOP (Figure 5, top) and the UV/H_2O_2 AOP (Figure 5, bottom) with different UV performances (0.0, 0.4, 0.7, and 1.0 kWh/m^3). Except for DEET, no compound could be eliminated more than 20% when no oxidant dosage was applied and the UV lamp was off. This indicates that only small reductions of the initial concentrations could be caused by adsorption processes in the static mixers. Some compounds could already be degraded by treating WWTE only with 3 mg/L FAC, such as diclofenac, bisphenol A, MTBT, 4t-octylphenol, 4-nonylphenols, tramadol and diphenhydramine. These are chemicals with electron-rich moieties (phenols, anilines, amines) that are preferably attacked by the selective chlorine molecule [41]. In the case of sole H_2O_2 dosage, no such effect was observed. Sole UV treatment reduced the concentration of many compounds to a certain degree. Exceptions were 4t-octylphenol, benzophenone, TCEP and TCPP. Bisphenol A, AHTN, 4-nonylphenols, tramadol and especially diclofenac were very susceptible to sole UV radiation. In comparison to sole UV treatment, the additional dosage of FAC paralleled with UV radiation was highly effective for almost all compounds except DEET. The UV/H_2O_2 AOP was only partly effective for ECs like HHCB, HHCB-lactone, benzophenone and 4t-octylphenol (only with high energy consumption), but not effective for any other ECs analyzed. Furthermore, the removal extents resulting from 3 mg/L oxidant and 0.4 kWh/m^3 were not increased more than 15 percentage points by increasing the energy consumption up to 1 kWh/m^3.

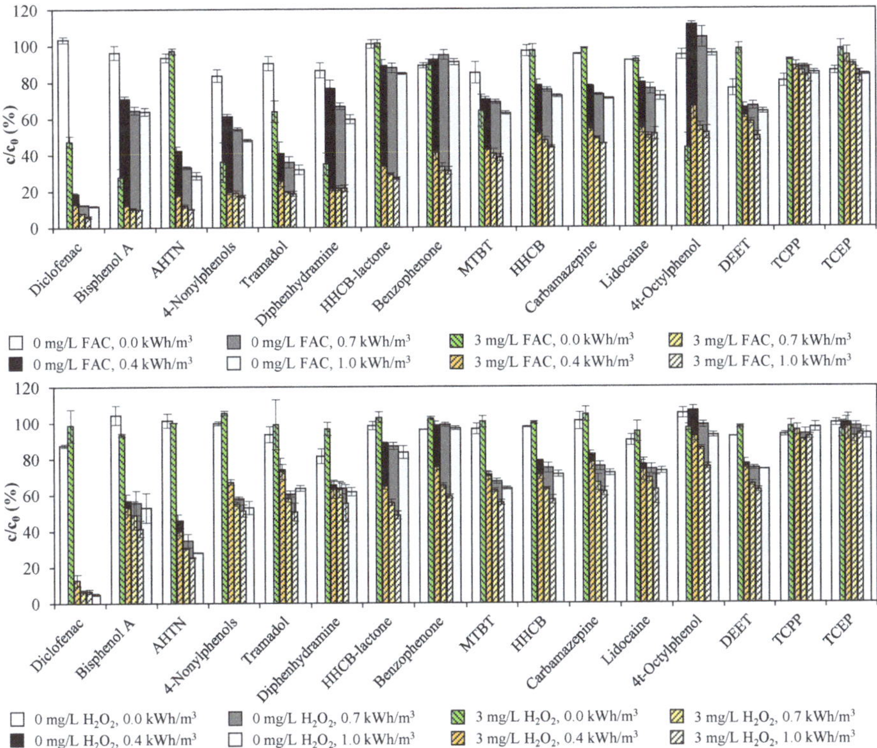

Figure 5. Removal of ECs found in WWTE by means of: UV/chlorine AOP (**top**); and UV/H_2O_2 AOP (**bottom**), at 0 and 3 mg/L oxidant concentrations depending on UV energy consumption at 1 m^3/h (Experiment 1).

In Figure 6, the removal of the same ECs found in WWTE due to the UV/chlorine AOP and UV/H_2O_2 AOP treatment at 0.4 kWh/m^3 energy consumption (1 m^3/h, 0.4 kW) is shown as a function of the oxidant concentration from 0 to 6 mg/L. The organophosphorous compounds TCEP and TCPP could not be eliminated with both AOPs. Both latter ECs are nevertheless designed to be resistant against oxidation [10]. Most of the partially eliminated compounds were degraded more effectively by the UV/chlorine AOP (exceptions were diclofenac and DEET). While 4t-octylphenol was not affected by the UV/H$_2$O$_2$ AOP, the UV/chlorine AOP treatment resulted in a 65% removal of that compound. The xenoestrogens bisphenol A and 4-nonylphenols could be removed by means of the UV/chlorine AOP by up to almost 90%. For most of the ECs, significantly higher removal extents could be achieved with higher oxidant concentrations. However, there were also compounds that did not seem to be affected by the variation of the oxidant concentration. Such compounds such as AHTN and diclofenac showed the same reaction for both the UV/chlorine AOP and the UV/H$_2$O$_2$ AOP. Additionally, these compounds also underwent degradation due to sole UV exposure (0 mg/L oxidant). The results obtained in Experiment 2 for 3 mg/L FAC or 3 mg/L H$_2$O$_2$ at 0.4 kWh/m^3 UV energy consumption did not differ more than 10% from the results of Experiment 1 in Figure 5. Both experiments were carried out on different days with slightly different wastewater compositions, which indicates a good reliability of the results.

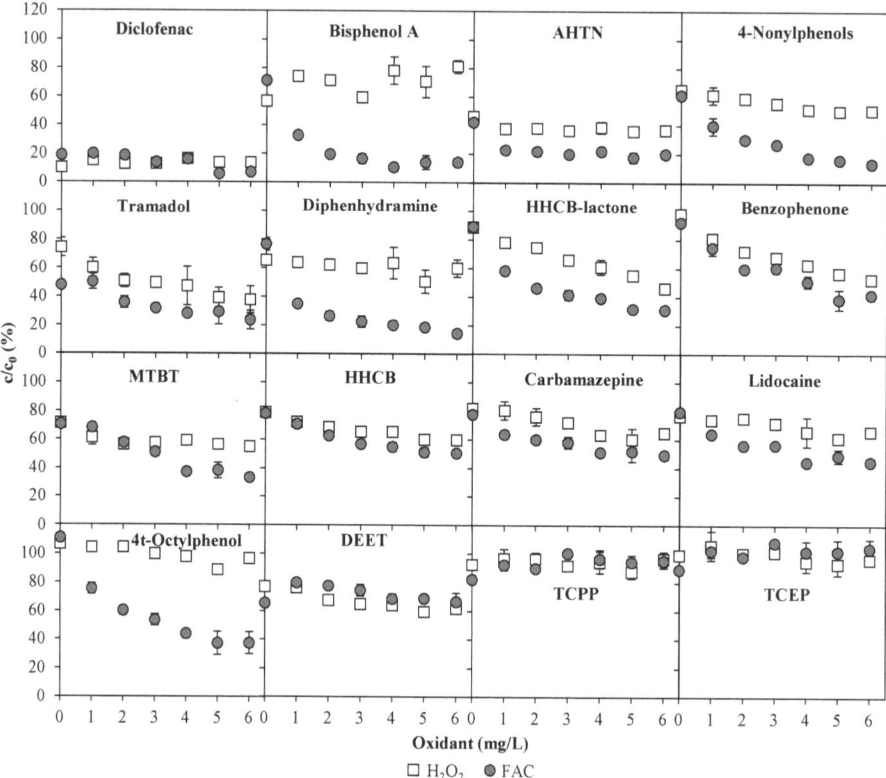

Figure 6. Removal of emerging contaminants found in WWTE by means of UV/chlorine AOP and UV/H$_2$O$_2$ AOP at 0.4 kWh/m^3 UV energy consumption (1 m^3/h, 0.4 kW) as a function of oxidant concentration (Experiment 2). At 0 mg/L oxidant: sole UV treatment.

In several publications, the very good EC elimination potential of the UV/chlorine AOP observed in this study could be seen as well (see Table S1 in Supplementary Materials summarizing results from other studies investigating the removal of ECs by the UV/chlorine AOP) [16–20]. However, none of these studies examined the UV/chlorine AOP on WWTE with such a high variation of analyzed ECs without spiking them and also applying the AOP at pilot scale. In this study, the initial concentration of diclofenac could be reduced up to 90% with both AOPs (most of the elimination can be attributed to photolysis). This pharmaceutical proved to be very susceptible to sole UV exposure, which can be ascribed to photoactive chromophores contained in the molecule [42]. As Sichel et al. [16] and Zhou et al. [43] had already observed, carbamazepine spiked in tap water or pure water could not significantly be eliminated with sole Cl_2 treatment (no UV exposure) even within a reaction time of 60 min. This poor degradation could be seen in this study as well, where contact times lower than 1 min were present. Here, carbamazepine could be eliminated by a maximum of approximately 50% with the application of the UV/chlorine AOP at the highest tested FAC concentration of 6 mg/L. A complete elimination of carbamazepine by the UV/chlorine AOP was only detected by Wang et al. [19]. However, their experiments were carried out with pure water and longer contact times.

The compounds diclofenac (DCF), bisphenol A (BPA) and 4-nonylphenols (4-NPh) were already eliminated at 50–70% with a dosage of 3 mg/L FAC. This happened at a contact time of about 9–16 s. The rate constants found in the literature for chlorination of these compounds at pH 7 were all determined at initial concentrations of ECs that were more than a hundred times higher than in the WWTE of this study ($k_{app,DCF}$ = 3.5 M^{-1} s^{-1} [44], $k_{app,BPA}$ = 62 M^{-1} s^{-1} [45], $k_{app,4\text{-}NPh}$ = 12.6 M^{-1} s^{-1} [46]). The resulting half-lives are >4 min and cannot be compared with the results gained in this study describing very small EC concentrations. Some investigations are available that examine the kinetics of the removal of ECs by the UV/chlorine AOP, with distinctions into chlorination, photolysis and radical reaction [18,19,43]. These investigations were usually carried out with much weaker UV lamps and simpler matrices than WWTE. Wang et al. [19], e.g., found a rate constant of k_{obs} = 0.78 min^{-1} for the degradation of 2 mg/L carbamazepine (8.5 µM) by 280 µM Cl_2 and 1.48 mW/cm^2 (41 W) in pure water matrix. At a contact time of 6–10 s, as in the UV chamber of this study, they found less than 15% degradation of carbamazepine, whereas in this study even at 42 µM FAC an elimination of 46% occurred. A direct transferability of the rate constants available in the literature is therefore not possible here either.

At a UV energy consumption of 0.4 kWh/m^3 and in a range of oxidant concentrations from 1 to 6 mg/L, the UV/chlorine AOP had a much better EC removal yield than the UV/H_2O_2 AOP for most of the analyzed compounds. This especially occurred with xenoestrogens like bisphenol A and 4-nonylphenols, which could be degraded very effectively. The more pronounced degradation yield by the UV/chlorine AOP compared to the UV/H_2O_2 AOP even at lower molar concentrations of FAC compared to H_2O_2 was also observed by Sichel et al. [16], Yang et al. [17] and Xiang et al. [18] (Table S1 in Supplementary Materials). Some possible reasons for this could be: The more efficient •OH radical yield due to different quantum yields at a wavelength of 254 nm and lower scavenger rates in the UV/chlorine AOP compared to the UV/H_2O_2 AOP [16,47]. Furthermore, other studies applying the UV/H_2O_2 AOP to real wastewater also showed that far higher concentrations of H_2O_2 were required for successful degradation yields than the 6 mg/L H_2O_2 used in this study [48,49].

3.3. Total Estrogenic Activity in the UV/Chlorine AOP Experiments

The total estrogenic activity (TEA) was analyzed in samples from UV/chlorine AOP experiments (Figure 7). With 65% EEQ removal, the sole dosage of 3 mg/L FAC had a stronger elimination effect than sole UV radiation at 0.4 kWh/m^3 (40% EEQ removal). This shows that xenoestrogens lose their estrogenic activity when they become chlorinated [50]. The UV/chlorine AOP could reduce the TEA of WWTE from 40% to at least 97% in the FAC concentration range of 0–6 mg/L at 0.4 kWh/m^3 (initial concentrations of 3.92 ± 0.31 and 1.77 ± 0.25 ng/L EEQ, respectively).

It is striking that those compounds that are known for their estrogenic activity were among those compounds that were removed best (60–90%) with the UV/chlorine AOP. These are in particular the endocrine disrupting compounds (EDCs) bisphenol A (BPA), 4-nonylphenols (4-NPh), and 4t-octylphenol (4t-OctPh) [51]. Similar to most estrogenic compounds, these EDCs bear phenolic hydroxyl groups. Lee et al. [52] suspected that the phenolic ring is oxidized by chlorination (it is likely that phenolic rings are preferably oxidized by chlorine [52]) in such a way that the TEA decreases. Wu et al. [50] also suspected that the EDCs are converted by chlorination into less estrogenic by-products leading to a decrease in TEA. Furthermore, Li et al. [20] had shown that the UV/chlorine AOP was the most efficient method compared to sole Cl_2 dosage and sole UV exposure in reducing estrogenic activity, even in the presence of NH_3 and wastewater matrix (Table S1 in Supplementary Materials). The reaction was very fast: the majority of the reaction was completed within less than 1 min. Similar contact times were observed for the pilot plant discussed here. Thus, in combination with the study of Li et al. [20], this work showed that even with complex matrixes such as WWTE and applied with a continuous mode pilot plant, the UV/chlorine AOP is an effective method for reducing the TEA.

Rosenfeldt and Linden [53], Rosenfeldt et al. [54] and Cédat et al. [48] showed with the EDCs 17β-estradiol, 17α-ethinylestradiol, bisphenol A and estrone that also the UV/H_2O_2 AOP, with sufficient dosage of H_2O_2, significantly decreases the estrogenic activity compared to sole UV treatment. However, Rosenfeldt et al. [54] also found that the reduction of estrogenic activity in wastewater matrix is significantly weaker compared to pure water matrix. In this study, the dosage of 3 mg/L H_2O_2 was obviously too low even at 1 kWh/m^3 UV exposure to significantly degrade the aforementioned EDCs (BPA, 4-NPh, 4t-OctPh). If one takes into account the statement of Lee et al. [52] that the elimination of estrogenic chemicals correlates with the elimination of TEA, it can thus be assumed that the UV/chlorine AOP reduces the TEA stronger than the UV/H_2O_2 AOP.

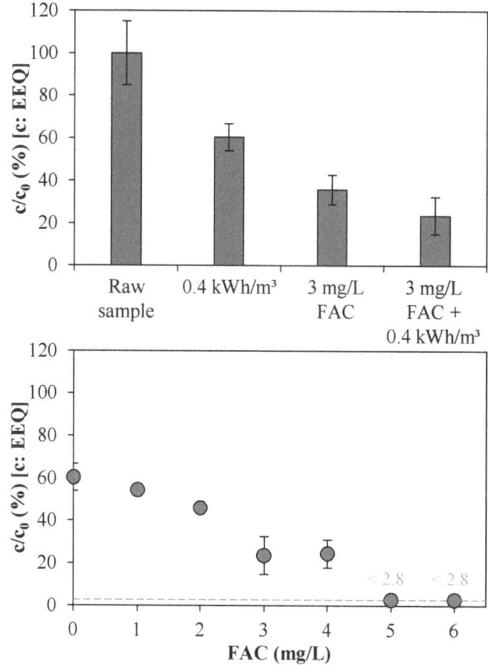

Figure 7. Effect of UV/chlorine AOP on total estrogenic activity of WWTE: measured in 17β-estradiol equivalents (EEQ) at different process settings at 1 m^3/h (**Top**) (Experiment 1); and as a function of FAC concentration in UV chamber influent at 0.4 kWh/m^3 energy consumption (1 m^3/h, 0.4 kW) (**Bottom**) (Experiment 2). Dashed line: LOQ of 0.1 ng/L EEQ (here 2.8%).

4. Conclusions

Effluent of a municipal wastewater treatment plant was treated with the UV/chlorine AOP on a technical scale on-site using a medium pressure UV lamp with an adjustable performance of up to 1 kW. In parallel experiments with the same pilot plant, the UV/H_2O_2 AOP was applied for comparison. The UV/chlorine AOP proved to be a highly effective method regarding the removal of bacteria (see Figure S2) and the removal of the estrogenic activity and thus endocrine disrupting compounds from WWTE. Compared to the UV/H_2O_2 AOP, most of the analyzed emerging contaminants were removed more efficiently with the UV/chlorine AOP. By-products in the form of AOX (most likely mainly by chlorination) and chlorate (most likely mainly by photolysis) occurred. Metabolites are of great concern regarding methods based on the oxidation of ECs [55], therefore, treatment of WWTE solely by the UV/chlorine AOP must be considered critically. Since AOX have a high tendency towards adsorption on activated carbon, an activated carbon treatment subsequent to the UV/chlorine AOP is recommended. Furthermore, with such a combination, compounds such as TCEP and TCPP can be eliminated as well [56], despite their high stability against UV/chlorine oxidation. Chlorine is known to react quickly with ammonium ions [57]. Especially in cold seasons, high ammonium concentrations can occur in WWTE. However, formed chloramines also have an oxidizing potential and can be transformed to radicals under UV exposure [47,58]. Thus, the effect of high ammonium concentrations on the effectiveness of the UV/chlorine AOP pilot plant still needs more research.

Supplementary Materials: Supplementary Data regarding influence of H_2O_2 on total Cl_2 analysis, results of UV/chlorine AOP in literature and elimination of bacterial count are available online at http://www.mdpi.com/1660-4601/15/5/935/s1. Table S1: Comparison of the results of different studies regarding the removal of important ECs (in %) by the UV/chlorine AOP and UV/H_2O_2 AOP (all studies except for this study and the study of Sichel et al. [16] were conducting batch experiments). Table S2: Initial bacterial count measured in WWTE reference samples collected in both UV/chlorine AOP and UV/H_2O_2 AOP experiments (CFU: colony forming units). Figure S1: Detected total Cl_2 concentrations in samples with different H_2O_2 concentrations without chlorine compounds. Figure S2: (Left, Middle) (Experiment 1) Influence of UV/chlorine AOP and UV/H_2O_2 AOP at 0.0, 0.4, 0.7, and 1.0 kWh/m^3 UV energy consumption (1 m^3/h, 0–1 kW) on bacterial count in WWTE at oxidant concentrations of 0 and 3 mg/L; and (Right) (Experiment 2) influence of UV/chlorine AOP and UV/H_2O_2 AOP at 0.4 kWh/m^3 UV energy consumption (1 m^3/h, 0.4 kW) on bacterial count in WWTE as a function of oxidant concentration.

Author Contributions: E.R. and B.K. conceived and designed the experiments; E.R. performed the experiments; E.R., B.K., and C.L. conducted the analyses; E.R., B.K., P.R., A.K., and R.M. analyzed the data; and E.R., P.R., and A.K. wrote the paper.

Conflicts of Interest: The authors declare no conflict of interest.

References

1. Kolpin, D.W.; Furlong, E.T.; Meyer, M.T.; Thurman, E.M.; Zaugg, S.D.; Barber, L.B.; Buxton, H.T. Pharmaceuticals, Hormones, and Other Organic Wastewater Contaminants in U.S. Streams, 1999–2000: A National Reconnaissance. *Environ. Sci. Technol.* **2002**, *36*, 1202–1211. [CrossRef] [PubMed]

2. Reemtsma, T.; Weiss, S.; Mueller, J.; Petrovic, M.; González, S.; Barcelo, D.; Ventura, F.; Knepper, T.P. Polar Pollutants Entry into the Water Cycle by Municipal Wastewater: A European Perspective. *Environ. Sci. Technol.* **2006**, *40*, 5451–5458. [CrossRef] [PubMed]

3. Stuart, M.; Lapworth, D.; Crane, E.; Hart, A. Review of risk from potential emerging contaminants in UK groundwater. *Sci. Total Environ.* **2012**, *416*, 1–21. [CrossRef] [PubMed]

4. Loos, R.; Carvalho, R.; António, D.C.; Comero, S.; Locoro, G.; Tavazzi, S.; Paracchini, B.; Ghiani, M.; Lettieri, T.; Blaha, L.; et al. EU-wide monitoring survey on emerging polar organic contaminants in wastewater treatment plant effluents. *Water Res.* **2013**, *47*, 6475–6487. [CrossRef] [PubMed]

5. Launay, M.A.; Dittmer, U.; Steinmetz, H. Organic micropollutants discharged by combined sewer overflows—Characterisation of pollutant sources and stormwater-related processes. *Water Res.* **2016**, *104*, 82–92. [CrossRef] [PubMed]

6. Ternes, T.A. Occurrence of drugs in German sewage treatment plants and rivers. *Water Res.* **1998**, *32*, 3245–3260. [CrossRef]

7. Auriol, M.; Filali-Meknassi, Y.; Tyagi, R.D.; Adams, C.D.; Surampalli, R.Y. Endocrine disrupting compounds removal from wastewater, a new challenge. *Process. Biochem.* **2006**, *41*, 525–539. [CrossRef]

8. Bolong, N.; Ismail, A.F.; Salim, M.R.; Matsuura, T. A review of the effects of emerging contaminants in wastewater and options for their removal. *Desalination* **2009**, *239*, 229–246. [CrossRef]

9. Reungoat, J.; Escher, B.I.; Macova, M.; Argaud, F.X.; Gernjak, W.; Keller, J. Ozonation and biological activated carbon filtration of wastewater treatment plant effluents. *Water Res.* **2012**, *46*, 863–872. [CrossRef] [PubMed]

10. Gerrity, D.; Gamage, S.; Holady, J.C.; Mawhinney, D.B.; Quiñones, O.; Trenholm, R.A.; Snyder, S.A. Pilot-scale evaluation of ozone and biological activated carbon for trace organic contaminant mitigation and disinfection. *Water Res.* **2011**, *45*, 2155–2165. [CrossRef] [PubMed]

11. Snyder, S.A.; Adham, S.; Redding, A.M.; Cannon, F.S.; DeCarolis, J.; Oppenheimer, J.; Wert, E.C.; Yoon, Y. Role of membranes and activated carbon in the removal of endocrine disruptors and pharmaceuticals. *Desalination* **2007**, *202*, 156–181. [CrossRef]

12. Dolar, D.; Gros, M.; Rodriguez-Mozaz, S.; Moreno, J.; Comas, J.; Rodriguez-Roda, I.; Barceló, D. Removal of emerging contaminants from municipal wastewater with an integrated membrane system, MBR-RO. *J. Hazard. Mater.* **2012**, *239–240*, 64–69. [CrossRef] [PubMed]

13. Legrini, O.; Oliveros, E.; Braun, A.M. Photochemical processes for water treatment. *Chem. Rev.* **1993**, *93*, 671–698. [CrossRef]

14. Esplugas, S.; Bila, D.M.; Krause, L.G.T.; Dezotti, M. Ozonation and advanced oxidation technologies to remove endocrine disrupting chemicals (EDCs) and pharmaceuticals and personal care products (PPCPs) in water effluents. *J. Hazard. Mater.* **2007**, *149*, 631–642. [CrossRef] [PubMed]

15. Ibáñez, M.; Gracia-Lor, E.; Bijlsma, L.; Morales, E.; Pastor, L.; Hernández, F. Removal of emerging contaminants in sewage water subjected to advanced oxidation with ozone. *J. Hazard. Mater.* **2013**, *260*, 389–398. [CrossRef] [PubMed]

16. Sichel, C.; Garcia, C.; Andre, K. Feasibility studies: UV/chlorine advanced oxidation treatment for the removal of emerging contaminants. *Water Res.* **2011**, *45*, 6371–6380. [CrossRef] [PubMed]

17. Yang, X.; Sun, J.; Fu, W.; Shang, C.; Li, Y.; Chen, Y.; Gan, W.; Fang, J. PPCP degradation by UV/chlorine treatment and its impact on DBP formation potential in real waters. *Water Res.* **2016**, *98*, 309–318. [CrossRef] [PubMed]

18. Xiang, Y.; Fang, J.; Shang, C. Kinetics and pathways of ibuprofen degradation by the UV/chlorine advanced oxidation process. *Water Res.* **2016**, *90*, 301–308. [CrossRef] [PubMed]

19. Wang, W.-L.; Wu, Q.-Y.; Huang, N.; Wang, T.; Hu, H.-Y. Synergistic effect between UV and chlorine (UV/chlorine) on the degradation of carbamazepine: Influence factors and radical species. *Water Res.* **2016**, *98*, 190–198. [CrossRef] [PubMed]

20. Li, M.; Xu, B.; Liungai, Z.; Hu, H.-Y.; Chen, C.; Qiao, J.; Lu, Y. The removal of estrogenic activity with UV/chlorine technology and identification of novel estrogenic disinfection by-products. *J. Hazard. Mater.* **2016**, *307*, 119–126. [CrossRef] [PubMed]

21. Wang, T.X.; Margerum, D.W. Kinetics of Reversible Chlorine Hydrolysis: Temperature Dependence and General-Acid/Base-Assisted Mechanisms. *Inorg. Chem.* **1994**, *33*, 1050–1055. [CrossRef]

22. Morris, J.C. The Acid Ionization Constant of HOCl from 5 to 35°. *J. Phys. Chem.* **1966**, *70*, 3798–3805. [CrossRef]

23. Deborde, M.; von Gunten, U. Reactions of chlorine with inorganic and organic compounds during water treatment—Kinetics and mechanisms: A critical review. *Water Res.* **2008**, *42*, 13–51. [CrossRef] [PubMed]

24. Buxton, G.V.; Subhani, M.S. Radiation chemistry and photochemistry of oxychlorine ions. Part 2—Photodecomposition of aqueous solutions of hypochlorite ions. *J. Chem. Soc. Faraday Trans. 1* **1972**, *68*, 958–969. [CrossRef]

25. Feng, Y.; Smith, D.W.; Bolton, J.R. Photolysis of aqueous free chlorine species (HOCl and OCl$^-$) with 254 nm ultraviolet light. *J. Environ. Eng. Sci.* **2007**, *6*, 277–284. [CrossRef]

26. Jin, J.; El-Din, M.G.; Bolton, J.R. Assessment of the UV/chlorine process as an advanced oxidation process. *Water Res.* **2011**, *45*, 1890–1896. [CrossRef] [PubMed]

27. Held, A.M.; Halko, D.J.; Hurst, J.K. Mechanisms of chlorine oxidation of hydrogen peroxide. *J. Am. Chem. Soc.* **1978**, *100*, 5732–5740. [CrossRef]

28. Graedel, T.E.; Goldberg, K.I. Kinetic studies of raindrop chemistry: 1. Inorganic and organic processes. *J. Geophys. Res.* **1983**, *88*, 10865. [CrossRef]

29. American Chemical Society. Database of SciFinder. Available online: https://scifinder.cas.org (accessed on 28 April 2016).

30. International Organization for Standardization. *Water Quality—Determination of Dissolved Anions by Liquid Chromatography of Ions—Part 1: Determination of Bromide, Chloride, Fluoride, Nitrate, Nitrite, Phosphate and Sulfate*; ISO 10304-1:2007; International Organization for Standardization: Geneva, Switzerland, 2007.

31. Soto, A.M.; Sonnenschein, C.; Chung, K.L.; Fernandez, M.F.; Olea, N.; Serrano, F.O. The E-SCREEN Assay as a Tool to Identify Estrogens: An Update on Estrogenic Environmental Pollutants. *Environ. Health Perspect.* **1995**, *103*, 113–122. [CrossRef] [PubMed]

32. Körner, W.; Hanf, V.; Schuller, W.; Kempter, C.; Metzger, J.; Hagenmaier, H. Development of a sensitive E-screen assay for quantitative analysis of estrogenic activity in municipal sewage plant effluents. *Sci. Total Environ.* **1999**, *225*, 33–48. [CrossRef]

33. Schultis, T. Erfassung der Estrogenen Wirksamkeit von Umweltproben und Reinsubstanzen durch Biologische Testsysteme—Entwicklung und Vergleich von In Vitro-Assays. Ph.D. Thesis, University of Stuttgart, Stuttgart, Germany, 2005.

34. International Organization for Standardization. *Water Quality—Determination of Adsorbable Organically Bound Halogens (AOX)*; ISO 9562:2004; International Organization for Standardization: Geneva, Switzerland, 2007.

35. Placak, O.R.; Ruchhoft, C.C. Studies of Sewage Purification: XVII. The Utilization of Organic Substrates by Activated Sludge. *Public Health Rep.* **1947**, *62*, 697–716. [CrossRef] [PubMed]

36. National Research Council. *Drinking Water and Health, Volume 7: Disinfectants and Disinfectant By-Products*; National Academies Press: Washington, DC, USA, 1987.

37. Krasner, S.W.; Westerhoff, P.; Chen, B.; Rittmann, B.E.; Amy, G. Occurrence of disinfection byproducts in United States wastewater treatment plant effluents. *Environ. Sci. Technol.* **2009**, *43*, 8320–8325. [CrossRef] [PubMed]

38. Bond, T.; Huang, J.; Templeton, M.R.; Graham, N. Occurrence and control of nitrogenous disinfection by-products in drinking water—A review. *Water Res.* **2011**, *45*, 4341–4354. [CrossRef] [PubMed]

39. Hong, H.C.; Wong, M.H.; Liang, Y. Amino Acids as Precursors of Trihalomethane and Haloacetic Acid Formation During Chlorination. *Arch. Environ. Contam. Toxicol.* **2009**, *56*, 638–645. [CrossRef] [PubMed]

40. Pehlivanoglu-Mantas, E.; Sedlak, D.L. Measurement of dissolved organic nitrogen forms in wastewater effluents: Concentrations, size distribution and NDMA formation potential. *Water Res.* **2008**, *42*, 3890–3898. [CrossRef] [PubMed]

41. Lee, Y.; von Gunten, U. Oxidative transformation of micropollutants during municipal wastewater treatment: Comparison of kinetic aspects of selective (chlorine, chlorine dioxide, ferrate VI, and ozone) and non-selective oxidants (hydroxyl radical). *Water Res.* **2010**, *44*, 555–566. [CrossRef] [PubMed]

42. Encinas, S.; Bosca, F.; Miranda, M.A. Photochemistry of 2,6-Dichlorodiphenylamine and 1-Chlorocarbazole, the Photoactive Chromophores of Diclofenac, Meclofenamic Acid and Their Major Photoproducts. *Photochem. Photobiol.* **1998**, *68*, 640–645. [CrossRef]

43. Zhou, S.; Xia, Y.; Li, T.; Yao, T.; Shi, Z.; Zhu, S.; Gao, N. Degradation of carbamazepine by UV/chlorine advanced oxidation process and formation of disinfection by-products. *Environ. Sci. Pollut. Res. Int.* **2016**, *23*, 16448–16455. [CrossRef] [PubMed]

44. Soufan, M.; Deborde, M.; Legube, B. Aqueous chlorination of diclofenac: Kinetic study and transformation products identification. *Water Res.* **2012**, *46*, 3377–3386. [CrossRef] [PubMed]

45. Gallard, H.; Leclercq, A.; Croué, J.-P. Chlorination of bisphenol A: Kinetics and by-products formation. *Chemosphere* **2004**, *56*, 465–473. [CrossRef] [PubMed]

46. Deborde, M.; Rabouan, S.; Gallard, H.; Legube, B. Aqueous Chlorination Kinetics of Some Endocrine Disruptors. *Environ. Sci. Technol.* **2004**, *38*, 5577–5583. [CrossRef] [PubMed]

47. Watts, M.J.; Linden, K.G. Chlorine photolysis and subsequent OH radical production during UV treatment of chlorinated water. *Water Res.* **2007**, *41*, 2871–2878. [CrossRef] [PubMed]

48. Cédat, B.; de Brauer, C.; Métivier, H.; Dumont, N.; Tutundjan, R. Are UV photolysis and UV/H_2O_2 process efficient to treat estrogens in waters? Chemical and biological assessment at pilot scale. *Water Res.* **2016**, *100*, 357–366. [CrossRef] [PubMed]

49. Rosario-Ortiz, F.L.; Wert, E.C.; Snyder, S.A. Evaluation of UV/H_2O_2 treatment for the oxidation of pharmaceuticals in wastewater. *Water Res.* **2010**, *44*, 1440–1448. [CrossRef] [PubMed]

50. Wu, Q.-Y.; Hu, H.-Y.; Zhao, X.; Sun, Y.-X. Effect of Chlorination on the Estrogenic/Antiestrogenic Activities of Biologically Treated Wastewater. *Environ. Sci. Technol.* **2009**, *43*, 4940–4945. [CrossRef] [PubMed]

51. Campbell, C.G.; Borglin, S.E.; Green, F.B.; Grayson, A.; Wozei, E.; Stringfellow, W.T. Biologically directed environmental monitoring, fate, and transport of estrogenic endocrine disrupting compounds in water: A review. *Chemosphere* **2006**, *65*, 1265–1280. [CrossRef] [PubMed]

52. Lee, B.-C.; Kamata, M.; Akatsuka, Y.; Takeda, M.; Ohno, K.; Kamei, T.; Magara, Y. Effects of chlorine on the decrease of estrogenic chemicals. *Water Res.* **2004**, *38*, 733–739. [CrossRef] [PubMed]
53. Rosenfeldt, E.J.; Linden, K.G. Degradation of Endocrine Disrupting Chemicals Bisphenol A, Ethinyl Estradiol, and Estradiol during UV Photolysis and Advanced Oxidation Processes. *Environ. Sci. Technol.* **2004**, *38*, 5476–5483. [CrossRef] [PubMed]
54. Rosenfeldt, E.J.; Chen, P.J.; Kullman, S.; Linden, K.G. Destruction of estrogenic activity in water using UV advanced oxidation. *Sci. Total Environ.* **2007**, *377*, 105–113. [CrossRef] [PubMed]
55. Wang, D.; Bolton, J.R.; Andrews, S.A.; Hofmann, R. Formation of disinfection by-products in the ultraviolet/chlorine advanced oxidation process. *Sci. Total Environ.* **2015**, *518–519*, 49–57. [CrossRef] [PubMed]
56. Wang, W.; Deng, S.; Li, D.; Ren, L.; Shan, D.; Wang, B.; Huang, J.; Wang, Y.; Yu, G. Sorption behavior and mechanism of organophosphate flame retardants on activated carbons. *Chem. Eng. J.* **2018**, *332*, 286–292. [CrossRef]
57. Lee, W.; Westerhoff, P. Formation of organic chloramines during water disinfection: Chlorination versus chloramination. *Water Res.* **2009**, *43*, 2233–2239. [CrossRef] [PubMed]
58. Zhang, X.; Li, W.; Blatchley, E.R.; Wang, X.; Ren, P. UV/chlorine process for ammonia removal and disinfection by-product reduction: Comparison with chlorination. *Water Res.* **2015**, *68*, 804–811. [CrossRef] [PubMed]

International Journal of
Environmental Research and Public Health

Article

Effect of Potassium Chlorate on the Treatment of Domestic Sewage by Achieving Shortcut Nitrification in a Constructed Rapid Infiltration System

Qinglin Fang [1], Wenlai Xu [1,2,*], Zhijiao Yan [2] and Lei Qian [2]

[1] State Key Laboratory of Geohazard Prevention and Geoenvironment Protection, Chengdu University of Technology, Chengdu 610059, China; txgsfy@163.com
[2] State Environmental Protection Key Laboratory of Synergetic Control and Joint Remediation for Soil and Water Pollution, Chengdu University of Technology, Chengdu 610059, China; yanzhijiaocdut@126.com (Z.Y.); qianleicdut@yeah.net (L.Q.)
* Correspondence: xuwenlai2012@cdut.cn; Tel.: +86-135-5102-9646

Abstract: A constructed rapid infiltration (CRI) system is a new type of sewage biofilm treatment technology, but due to its anaerobic zone it lacks the carbon sources and the conditions for nitrate retention, and its nitrogen removal performance is very poor. However, a shortcut nitrification–denitrification process presents distinctive advantages, as it saves oxygen, requires less organic matter, and requires less time for denitrification compared to conventional nitrogen removal methods. Thus, if the shortcut nitrification–denitrification process could be applied to the CRI system properly, a simpler, more economic, and efficient nitrogen removal method will be obtained. However, as its reaction process shows that the first and the most important step of achieving shortcut nitrification–denitrification is to achieve shortcut nitrification, in this study we explored the feasibility to achieve shortcut nitrification, which produces nitrite as the dominant nitrogen species in effluent, by the addition of potassium chlorate ($KClO_3$) to the influent. In an experimental CRI test system, the effects on nitrogen removal, nitrate inhibition, and nitrite accumulation were studied, and the advantages of achieving a shortcut nitrification–denitrification process were also analysed. The results showed that shortcut nitrification was successfully achieved and maintained in a CRI system by adding 5 mM $KClO_3$ to the influent at a constant pH of 8.4. Under these conditions, the nitrite accumulation percentage was increased, while a lower concentration of 3 mM $KClO_3$ had no obvious effect. The addition of 5mM $KClO_3$ in influent presumably inhibited the activity of ammonia-oxidizing bacteria (AOB) and nitrite-oxidizing bacteria (NOB), but inhibition of nitrite-oxidizing bacteria (NOB) was so strong that it resulted in a maximum nitrite accumulation percentage of up to over 80%. As a result, nitrite became the dominant nitrogen product in the effluent. Moreover, if the shortcut denitrification process will be achieved in the subsequent research, it could save 60.27 mg CH_3OH per litre of sewage in the CRI system compared with the full denitrification process.

Keywords: shortcut nitrification; constructed rapid infiltration system; potassium chlorate inhibition; domestic sewage

1. Introduction

Sewage treatment technology for domestic sewage and polluted surface water treatment in small towns—a constructed rapid infiltration (CRI) system—is a new sewage biofilm treatment technology put forward by Zhong Zuoshen et al. [1]. It presents both advantages of a sewage rapid infiltration land treatment system and a constructed wetland system [2]. A CRI system is mainly composed of a feeding tank, grill, preliminary sedimentation tank, rapid infiltration tank, and outlet system. A CRI system adopts the dry-wet (alternate running of feeding and drying in the CRI system) alternating operation mode and uses natural river sand, coal gangue, natural gravel, etc., to replace natural soil as the filling medium to improve the hydraulic load to 1.0–1.5 m/day [3]. Pictures of a practical example of a CRI system are shown in Figure 1. The removal mechanism of the CRI system is to use the filling medium and microorganisms grown on the filling medium to adsorb, intercept, and decompose the pollutants in sewage [4]. Especially, since the CRI system has the unique structure and feeding mode, its filling medium has the aerobic, facultative, and anaerobic environment to grow abundant microorganism to allow for efficient sewage treatment [5]. As the previous practice showed, a CRI system has a significant effect on the treatment of domestic sewage in small towns [6], whose removal rates of CODcr (chemical oxygen demand determined by potassium dichromate method), NH_4^+-N, suspended solid (SS), and linear alkylbenzene sulfonates (LAS) could reach above 85%, 90%, 95%, and 95%, respectively, and has the advantages of being less energy-intensive, more environmentally-friendly, and has a remarkable economic benefit compared with the conventional treatment systems [7]. Although a CRI system has a good removal effect of NH_4^+-N, due to its anaerobic zone it lacks the carbon sources for denitrification and the condition for nitrate retention [8], the concentration of nitrate in effluent is so high that the total nitrogen (TN) removal rate can only reach upwards of 10–30% [9].

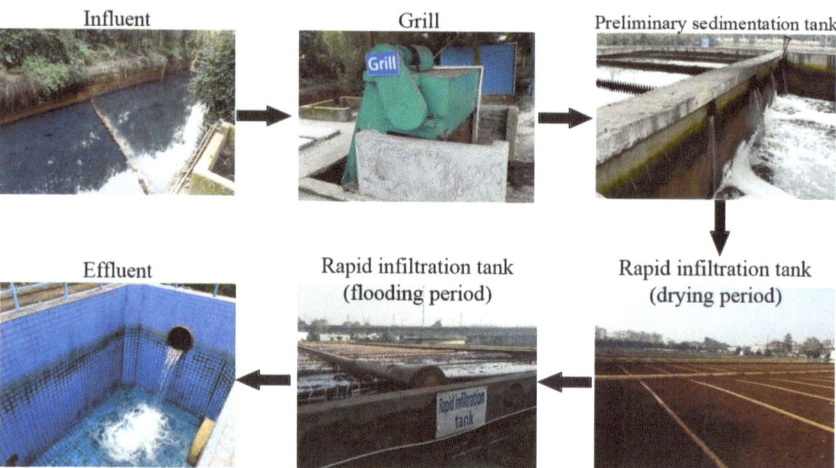

Figure 1. Practical engineering of the Phoniex River constructed rapid infiltration (CRI) system operated successfully for 12 years in Chengdu, China.

To enhance the nitrogen removal performance of the CRI system, the methods of adding external carbon sources, optimizing the packing structure [10], and changing the water feeding patterns [1] were adopted. However, those methods were all based on the full nitrification–denitrification process, making it difficult to overcome the problem of carbon source consumption and reduction of denitrifying bacteria activity during long-term operation, and were also difficult to popularize in the actual engineering due to their complex operating process.

Shortcut nitrification–denitrification is a novel biofilm nitrogen removal process which allows oxidation of ammonia to nitrite, but no further oxidation to nitrate and reduces nitrite into nitrogen gas directly to achieve nitrogen removal in the system. As Figure 2 shows, compared to the full nitrification–denitrification process, the shortcut nitrification–denitrification process reduces two reaction steps, which are "$NO_2^- \rightarrow NO_3^-$" and "$NO_3^- \rightarrow NO_2^-$". Thus, it will present the advantages of saving oxygen and requiring less organic matters. However, it can also be seen from Figure 2, for shortcut nitrification–denitrification to be employed, the key point is to achieve shortcut nitrification.

In other words, the system must accumulate and maintain enough nitrite, which is produced by ammonium-oxidizing bacteria (AOB) and, at the same time, inhibit or wash out nitrite-oxidizing bacteria (NOB), which would oxidize the produced nitrite to nitrate [11]. The conditions required to inhibit nitrite oxidization can be established with high concentrations of ammonium, a low concentration of dissolved oxygen, a high concentration of free nitrous acid, a relatively high temperature (30–35 °C) and a high pH (8–9). So far, shortcut nitrification has been achieved in various systems, such as aerated constructed wetlands [12], a sequencing batch reactor (SBR) [13] and submerged biofilters [14], all of which resulted in high nitrite accumulation percentages. The use of specific inhibitors can further improve shortcut nitrification. For example, Xu et al. [13] studied the effect of hydroxylamine addition on shortcut nitrification in SBR, and Chen et al. [15] used this same inhibitor in a CRI; both found nitrite accumulation percentages reaching more than 90%. Sukru and Erdal [14] and Cui et al. [16] found that increasing salinity could further promote the accumulation of nitrite. Moreover, Ge et al. [17] showed that low concentrations (4 mg/L) of chlorine could improve the nitrite accumulation percentage to reach 60–70%. Already in 1957 chlorate was described as a specific inhibitor of NOB: chlorate could inhibit the growth of autotrophic nitrite oxidizers at low concentration (4.2×10^{-3} M) and completely inhibit nitrite oxidation at high concentrations (1.7×10^{-2} M) [18]. Subsequent studies reported that the addition of chlorate could result in nitrite to become the dominant product of NO_x in the effluent, by allowing AOB activity while inhibiting NOB. For instance, Xu et al. [11] showed that the addition of chlorate to aerobic granules resulted in a 90% increase of nitrite accumulation in the effluent. Other studies showed that chlorate inhibited the oxidation of nitrite to nitrate, but it did not affect the oxidation of NH_4^+ to NO_2^- [19]; likewise, Xu et al. [11] found that oxidation of NH_4^+ to NO_2^- was not severely inhibited by chlorate. Such studies showed that shortcut nitrification can be achieved effectively by the addition of specific inhibitors, including chlorate, but the effect of adding potassium chlorate ($KClO_3$) in CRI system has not yet been studied in detail.

In this study, we tested whether potassium chlorate could improve the performance of shortcut nitrification and removal efficiency of pollutants in a CRI system under experimental conditions and prospected the benefits of achieving shortcut denitrification.

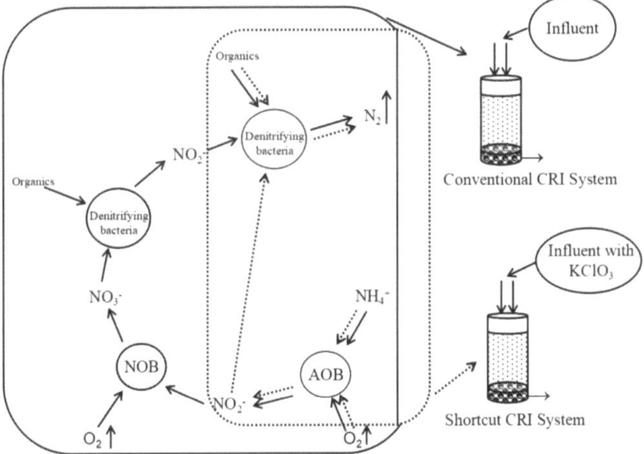

Figure 2. Comparison of the full nitrification–denitrification process and shortcut nitrification–denitrification process (→ represents the process of full nitrification–denitrification; --→ represents the process of shortcut nitrification–denitrification).

2. Materials and Methods

2.1. Experimental Design

Four separate CRI columns were constructed using PVC (polyvinyl chloride) (diameter 8 cm, height 30 cm) in the laboratory under controlled conditions. The temperature was kept constant at 34.2 ± 0.64 °C by constructing a temperature-controlling box around the CRI columns (Figure 3). The filling medium of the columns consisted of two functional layers: a 5 cm deep supporting layer consisting of pebbles (5.0–10.0 mm) and gravel (3.0–4.0 mm) at the bottom, a 20 cm deep treatment layer filled with 90% river sand (0.25–0.30 mm), 5% marble sand (1.0–2.0 mm), and 5% zeolite sand (1.5–1.7 mm) on the top of the supporting layer. The influent sewage was lifted by a peristaltic pump so that it entered at the top of the column, moved through the packing medium vertically, and left by the outlet where water quality was measured.

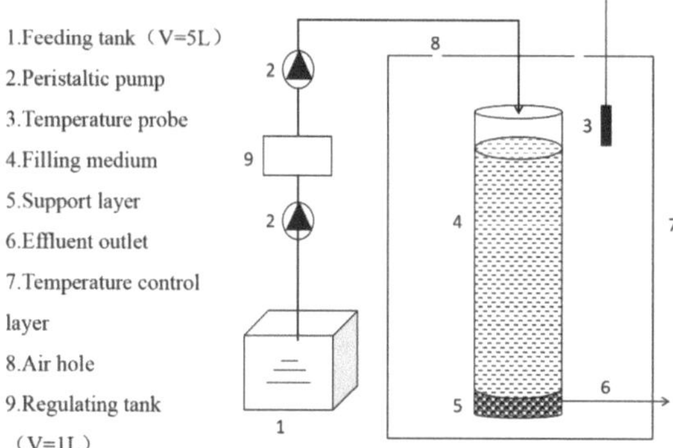

Figure 3. Experimental CRI system. CRI: constructed rapid infiltration.

2.2. Sewage and Operational Conditions

The influent sewage used in this study was a mixture of synthetic sewage and domestic sewage, the synthetic sewage was mainly made up of glucose, CH_3COONa, $(NH_4)_2SO_4$, KH_2PO_4, Na_2CO_3, and peptone and was refilled every three days. The water quality parameters are shown in Table 1. The whole experiment lasted for 110 days. The sewage was fed into the system by a dry-wet alternating operation mode as follows: water feeding was allowed twice daily with a hydraulic load of 0.6 m/day, each feeding time would last for 1.5 h, each drying time would last for 10.5 h, and the water flow was 200 mL/h. The system was operated for 70 days until the removal percentages of ammonium nitrogen (NH_4^+-N) in effluent of all columns reached to 88%, which indicated the biofilms had formed successfully in the CRI system.

In order to investigate the effect of potassium chlorate inhibition and pH control, the experimental columns were used as individual Tests. Test 1 was the control treatment not receiving additions, the pH of influent of Tests 2–4 was adjusted to 8.4 by addition of NaOH solution. Moreover, in Test 3, $KClO_3$ was added to the influent at a final concentration of 5 mM while, in Test 4, a concentration of 3 mM $KClO_3$ was used. Both the NaOH solution and $KClO_3$ were added and mixed in the regulating tank after it loaded with 600 mL sewage from feeding tank. Moreover, the scanning electron microscope (SEM) pictures (Figure 4) of the filling medium (sand) were taken on day 70, which could further show the situation of biofilm formation on the filling medium of tests 1–4. As we can see from Figure 4, the blank filling medium (picture a) which was not fed sewage, can hardly investigate the microbial flora attachment. However, the filling medium of Tests 1–4 (pictures b–e) which were fed sewage for 70 days had an obvious microbiological attachment, which indicated that biofilms were formed successfully in the filling medium of Tests 1–4.

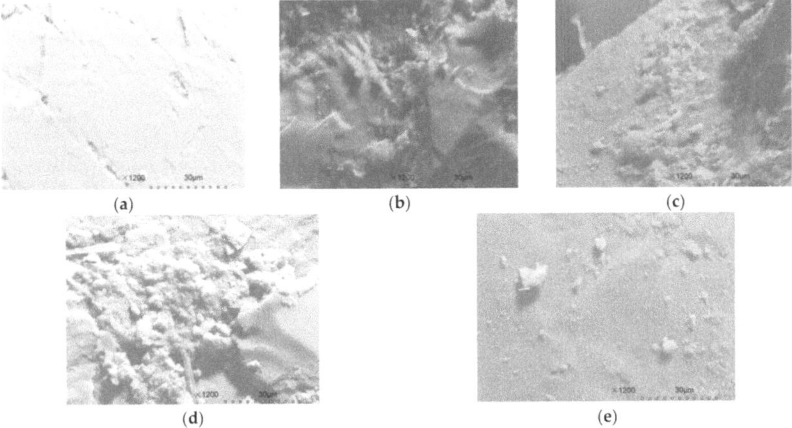

Figure 4. Scanning electron microscope (SEM) images of filling medium (sand) in the CRI columns after 70 days of operation. (**a**) blank filling medium; (**b–e**) filling medium (formed with biofilm) of Tests 1–4.

Table 1. Water quality parameters of influent.

Water Quality Parameters	Mean Concentration (mg/L)
Chemical Oxygen Demand (COD)	245.22 ± 27.11
NH_4^+-N	53.93 ± 3.81
NO_3^--N	1.15 ± 0.67
NO_2^--N	0.14 ± 0.09
Total Nitrogen (TN)	55.35 ± 6.01
pH	7.3 ± 0.14 (control), 8.4 (Tests 2–4)
Temperature (°C)	34.2 ± 0.64

2.3. Analytical Methods

Water samples from influent and effluent were collected every two days, filling medium samples were collected after biofilm formed successfully (on day 70). Concentration of COD in the sewage was determined using the potassium dichromate method (in a strong acid solution, a certain amount of potassium dichromate is used to oxidize the reducing substances in the water sample, then, ferroin (indicator) is added to the excess potassium dichromate before it is titrated with ammonium ferrous sulfate solution, and the oxygen consumption of the reductive substance in the water sample is calculated according to the amount of ammonium ferrous sulphate); the concentration of nitrogen in the form of ammonium was determined by Nessler's reagent colorimetric method (an alkaline solution made of mercuric iodide and potassium iodide reacted with ammonium nitrogen would generate reddish brown complex, the absorbance of the complex which was measured at 420 nm (visible light) and is proportional to the content of ammonium nitrogen), nitrate (NO_3^--N) by UV spectrometry (the concentration of nitrate nitrogen can be quantified by the absorption value of nitrate ion at a wavelength of 220 nm, however the dissolved organic matter was absorbed at both 220 nm and 275 nm, while the nitrate ion was not absorbed at 275 nm, therefore another measurement is made at 275 nm to correct the absorption of nitrate nitrogen), nitrite (NO_2^--N) by molecular absorption spectrophotometry (in phosphoric acid medium, nitrite is reacted with para-aminobenzene sulfonamide to produce diazonium salt, then, coupling with N-(1-naphthyl) ethylenediamine to produce red dye, finally, determining the absorbance of production at 540 nm (visible light)), and total nitrogen (TN) by UV spectrometry (at 120–124 °C basic potassium persulfate solution is used to convert nitrogen-containing compounds into nitrate in the water sample, then, the ultraviolet spectrophotometry method is used to determine the absorbency of the sample at 220 nm and 275 nm respectively, the corrected absorbance (A) is calculated according to the formula (A = $A_{220} - 2A_{275}$) and is proportional to the total nitrogen content), using standard procedures [20]. The nitrite accumulation percentage was calculated as the ratio of $NO_2^-/(NO_2^- + NO_3^-) \times 100\%$ [12].

2.4. Scanning Electron Microscope Detection

The biofilm of the filling medium was prepared by the glutaraldehyde fixation method [21] and observed by using a scanning electron microscope (SEM) (S-3000N, Hitachi Limited, Tokyo, Japan). The filling medium samples were fixed with 2.5% glutaraldehyde for 15 h and then rinsed in distilled water three times. Subsequently, the samples were dehydrated with series of ethanol (30%, 50%, 70%, 85%, 95%) for one time, and 100% ethanol for two times (20 min/time). After rinsing twice (20 min/time) with isoamyl acetate, the prepared samples were natural dried for 12 h. Finally, the dewatered samples were sputter-coated with gold and observed with SEM.

3. Results and Discussion

3.1. Effect of Potassium Chlorate on Removal Efficiency of Ammonium Nitrogen

The removal efficiency of nitrogen in the form of ammonium in the CRI system was compared between the controls (with and without pH adjustment) and after the addition of two concentrations of $KClO_3$ to the influent. Removal efficiency was calculated as the difference in concentration between influent and effluent (influent concentration minus effluent concentration) divided by the concentration in influent.

Adjustment of the influent pH to 8.4 of Test 2 only had a minor effect on ammonium nitrogen removal during the first 10 days (Figure 5), the reason may be that the AOB need time to adapt the new pH environment in the system. There was no difference in removal efficiency between Test 4 (pH 8.4, 3 mM $KClO_3$) and Test 2 (pH 8.4), as both reached approximately 87% removal on average (Figure 5). However, in presence of 5 mM $KClO_3$ (Test 3), the NH_4^+-N removal efficiency was reduced, though it still reached 66% on average. Xu et al. [11] also found that oxidation NH_4^+ to NO_2^- was slightly inhibited by chlorate. This is most likely the chlorate has a slight inhibition of AOB activity, as a result of which NH_4^+-N oxidation efficiency was less efficient.

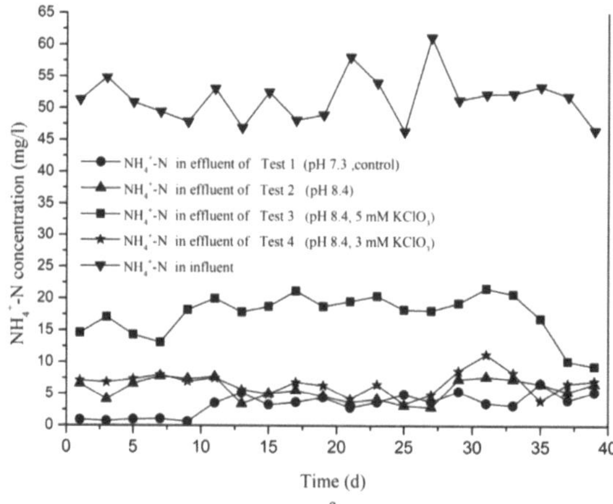

Figure 5. *Cont.*

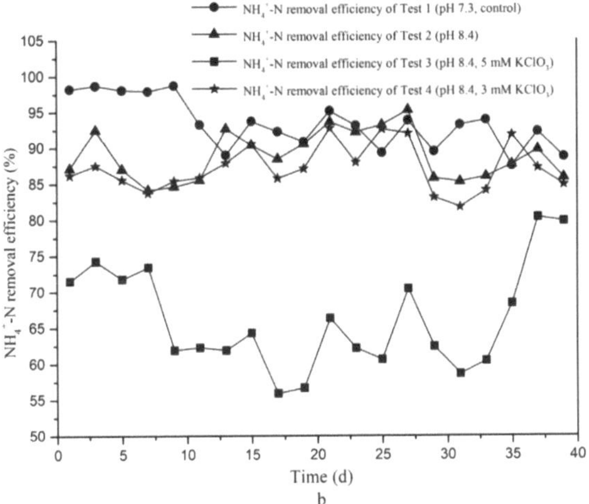

Figure 5. Ammonium-nitrogen removal. (**a**) Absolute concentrations of NH_4^+-N in effluent and influent; (**b**) Ammonium-nitrogen removal efficiency (in %) of the four experimental tests of CRI.

3.2. Effect of Potassium Chlorate on Nitrate Accumulation in a CRI System

As can be seen in Figure 6, there was no significant difference between Test 2 (pH 8.4), resulting in a nitrate concentration of on average 36.24 mg/L, and Test 4 (pH 8.4, 3 mM $KClO_3$), resulting in 34.51 mg/L. Very similar results were obtained for the control in which the pH of the influent had not been adjusted (Test 1, pH 7.3). In contrast, Test 3 (pH 8.4, 5 mM $KClO_3$) resulted in much lower nitrate concentrations of approximately 7.39 mg/L on average, which represented an 80% reduction compared to the control. As shown, the nitrate concentration in effluent of Test 3 was reduced within 48 h after addition of 5 mM $KClO_3$ and reached a minimum of 2.92 mg/L on day 13. This result shows that addition of 5 mM $KClO_3$ to the influent was able to strongly prevent the oxidation of nitrite, a condition that favours the accumulation of nitrite and is desired for shortcut nitrification achievement.

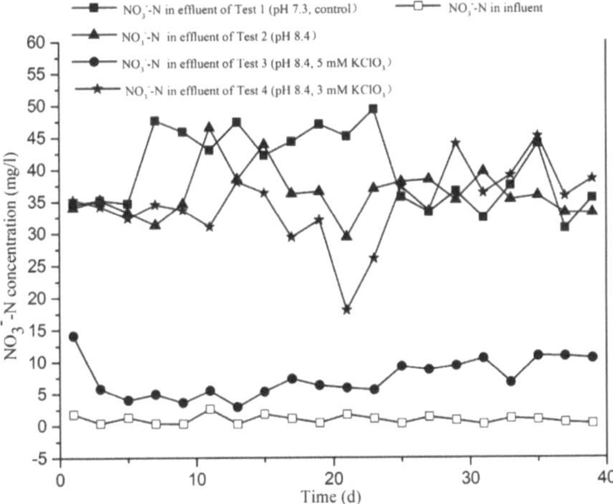

Figure 6. The nitrate-nitrogen concentration in influent and effluent in the four experimental tests of CRI system.

3.3. Effect of Potassium Chlorate and pH on Nitrite Accumulation in a CRI System

Previous studies have described that the pH of the influent is a decisive factor for inhibiting NOB activity. For instance, Banashri [19] described that nitrite accumulation can be improved at high pH (8–9). Glass and Silverstein [22] observed a significant increase of nitrite accumulation (250, 500 mg/L) in sequencing batch reactors when wastewater pH was increased during nitrification (pH 7.5, 8.5, respectively). Thus, we adjusted the influent sewage pH to 8.4 of Test 2 and observed (Figure 7a) that the average nitrite accumulation percentage of Test 2 (pH 8.4) was 1.5%, which was slightly higher than that of Test 1 (0.50%, pH 7.3). Nevertheless, this increase was too weak to support shortcut nitrification. Thus, a pH of 8.4 is, by itself, insufficient to enable effective shortcut nitrification in a CRI system.

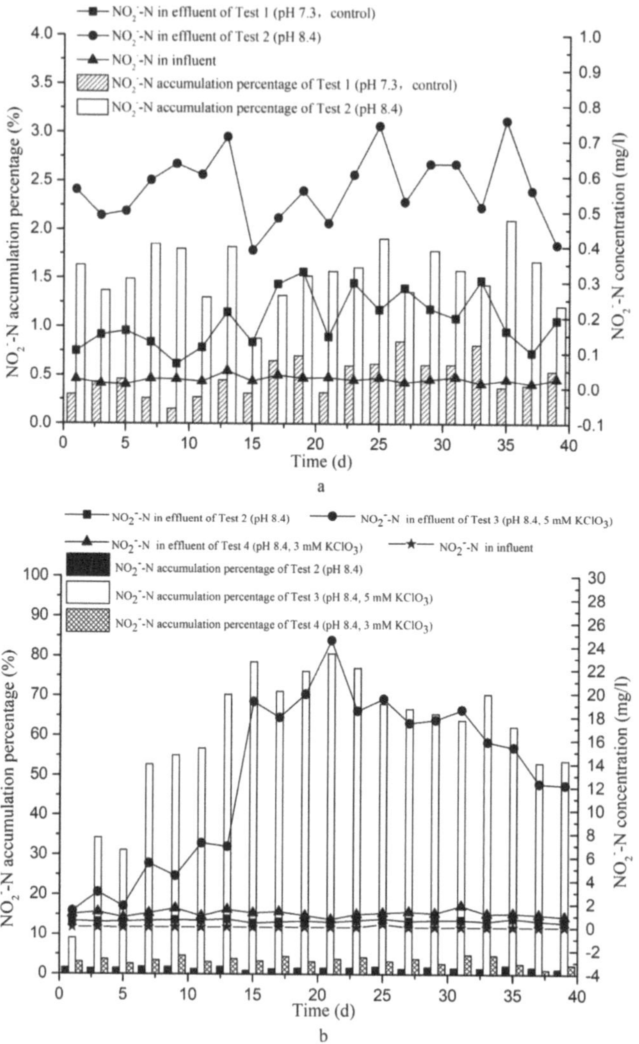

Figure 7. Nitrite accumulation in effluent. (**a**) Effect of pH on nitrite accumulation percentage (bars) and concentration (curves) in the effluent of Test 1 (pH, 7.3) and Test 2 (pH 8.4); (**b**) The nitrite accumulation percentage (bars) and concentration (curves) in effluent of Tests 2–4.

The average nitrite accumulation percentage in our tests are shown as bars in Figure 7b. As can be seen, these percentages were very low in Test 2 (1.56% on average) and Test 4 (3.43% on average), but much increased in Test 3, resulting in 59.80% accumulation percentages on average. Thus, the addition of 5 mM $KClO_3$ strongly supported accumulation of nitrite in the test CRI system. Combined with the data presented in Figures 5–7, it can be concluded that, whereas nitrate was the dominant product in effluent of Tests 1, 2, and 4, nitrite was the dominant nitrogen product of Test 3, as a result of effective nitrite oxidation inhibition.

As apparent in Figure 7b, the nitrite accumulation percentage in effluent of Test 3 increased sharply during the first seven days (from, initially, 9.02% to 52.76%) and further increased to reach a plateau of up to 80% during days 15–23. The nitrite concentration peaked at day 21 at 24.54 mg/L. After this, the nitrite accumulation percentage slightly decreased, but still reached 53% at day 39. The reason may be that the long-term addition of the chlorate will cause a slight inhibition of AOB activity. However, the result also indicates that shortcut nitrification can be not only be achieved, but also maintained in the tested CRI system by the addition of 5 mM $KClO_3$ in the influent at a pH of 8.4.

3.4. Prospects for the Achievement of Shortcut Nitrification–Denitrification in a CRI System

Xu et al. [11] mentioned that the shortcut nitrification–denitrification process could save 40% of carbon source consumption, compared with the full nitrification–denitrification process. Chen [23] observed that an increased COD concentration (51.3, 69.3, 73.3 mg/L) in the effluent in biological filters when the nitrite percentage ($NO_2^-/(NO_2^- + NO_3^-) \times 100\%$) in the influent was increased (0%, 50%, 80%, respectively). As we can see from Figure 7, the shortcut nitrification could be achieved successfully in the CRI system by adding 5 mM $KClO_3$ in Test 3, and its mean concentration and accumulation percentage of nitrite could reach to 12.98 mg/L and 59.80%, respectively. Thus, the accumulation of nitrite in Test 3 could provide electron acceptor for the subsequent shortcut denitrification. However, Chen et al. [15] found that the removal efficiency of COD could reach 90% during shortcut nitrification in the CRI system. Wang et al. [8] mentioned that carbon source, nitrate/nitrite and anaerobic environment are essential for denitrification in the CRI system since most of the denitrifying bacteria are facultative anaerobic and use organic matters as carbon sources under the anoxic condition to provide energy. However, in this study, we calculated that the removal percentage of CODcr of four tests all reached more than 91%, and the mean concentration of residual CODcr of Test 3 was only around 7.28 mg/L, which was too low to support the subsequent shortcut denitrification. Therefore, if a shortcut denitrification experiment will be conducted, an external carbon source is required to be added to the influent. Methanol, ethanol, acetic acid, and cellulose were all studied as external carbon sources for denitrification in previous studies. Yan et al. [24] found that methanol is easily biodegraded and used by denitrifying bacteria and the denitrification rate of methanol is very high. Zhang [25] found that the complete removal of nitrogen in effluent can be achieved by adding sufficient methanol in denitrification process. Gómez et al. [26] found that methanol is an ideal carbon source for denitrification. Thus, if methanol (CH_3OH) is chosen as an external carbon source for subsequent shortcut denitrification, the equations of full denitrification (Equation (1)) and shortcut denitrification (Equation (2)) are shown as follows:

$$NO_3^- + 1.08CH_3OH + 0.24H_2CO_3 \rightarrow 0.056C_3H_7O_2N + 0.47N_2 + 1.68H_2O + HCO_3^- \tag{1}$$

$$NO_2^- + 0.67CH_3OH + 0.53H_2CO_3 \rightarrow 0.004C_3H_7O_2N + 0.48N_2 + 1.23H_2O + HCO_3^- \tag{2}$$

According to Equations (1) and (2) and the data from Figures 6 and 7, the mean nitrate concentration of Test 2 (pH 8.4) and Test 3 (pH 8.4, 5 mM $KClO_3$) are about 36.24 mg/L and 7.39 mg/L, respectively, the mean nitrite concentration of Test 2 and Test 3 are about 0.57 mg/L and 12.98 mg/L, respectively. If the subsequent shortcut denitrification will be achieved, the dosage of CH_3OH used for Test 2 denitrification will consume 98.38 mg CH_3OH per litre of sewage during the operating period, but, Test 3 only needs 38.11 mg CH_3OH per litre of sewage for denitrification and shortcut denitrification, which was only 38.73% of the consumption of Test 2. Moreover, an anaerobic environment is also important for the denitrification in the CRI system. Fan et al. [1] added a sub-section intake and overflow pool in the CRI system simulated columns and found this method will increase the total nitrogen removal efficiency to 64.8%. Therefore, if the shortcut nitrification–denitrification process in the CRI system is implemented in the subsequent research, not only will the external carbon source be added in the denitrification section, but a saturated water layer will also be constructed in the bottom of the denitrification section to improve the total nitrogen removal performance of the CRI system. Furthermore, although, the chlorate is easily biodegraded by nitrate reductase in an organic-rich environment, the appropriate amounts of reductant also need to be added into the reactor to fully eliminate potential pollution when the shortcut nitrification process ended [11]. Thus, although achievement of shortcut nitrification–denitrification process in the CRI system will present many advantages, such as improving the denitrification rate, simplifying the reaction process, and saving carbon source consumption, there is still much research work needed to be done towards applying this new technology in a practical project.

4. Conclusions

(1) The addition of 3 mM $KClO_3$ to influent at a constant pH of 8.4 is not sufficient to inhibit that of NOB so that shortcut nitrification does not take place in the CRI system.

(2) Adjusting the pH of influent to 8.4 alone did not contribute much to establish shortcut nitrification in CRI.

(3) Although, the addition of 5 mM $KClO_3$ in influent could both inhibit the activity of ammonia-oxidizing bacteria (AOB) and nitrite-oxidizing bacteria (NOB), the inhibition of NOB was so strong that made the $NO_2{}^--N$ to be the dominant product of total oxidized nitrogen in effluent for a long period, showing that shortcut nitrification could be achieved and maintained successfully in a CRI system.

(4) According to the data of nitrate and nitrite in Figures 6 and 7, the consumption of external carbon source (CH_3OH) for subsequent denitrification was calculated and analysed by using Equations (1) and (2), the results showed that the consumption of carbon source (CH_3OH) of Test 3 (pH 8.4, 5 mM $KClO_3$) was only 38.73% of the consumption of Test 2 (pH 8.4). Therefore, compared with conventional sewage treatment methods, achievement of the shortcut nitrification–denitrification process in the CRI system will take both the advantages of the CRI system and shortcut nitrification–denitrification process; it will not only have a unique structure and feeding mode to construct aerobic, facultative, and anaerobic environments for microorganism enriching in the filling medium, but also improve the denitrification rate and save the carbon source consumption during the reaction process.

Acknowledgments: This research is funded by Chinese National Natural Science Foundations (41502333), Sichuan science and technology support project (2017JY0141), China Postdoctoral Science Foundation (2017M610598), and the State Key Laboratory of Geohazard Prevention and Geoenvironment Protection Foundation (SKLGP2016Z019). We received above funds for covering the costs to publish in open access.

Author Contributions: Wenlai Xu conceived and designed the experiments; Qinglin Fang performed the experiments; Zhijiao Yan analyzed the data; Lei Qian contributed reagents and analysis tools; and Wenlai Xu wrote the paper.

Conflicts of Interest: The authors declare no conflict of interest.

References

1. Fan, X.J.; Fu, Y.S.; Liu, F.; Xue, D.; Xu, W. Total nitrogen removal efficiency of improved constructed rapid infiltration system. *Technol. Water Treat.* **2009**, *10*, 021.

2. Ronald, W.C.; Sherwood, C.R.; Robert, K.B. Applying treatment wastewater to land. *Bio. Cycle* **2001**, *4*, 32–35.

3. He, J.T.; Zhong, Z.S.; Tang, M.G. New method of solving contradiction of rapid infiltration system land using. *Geoscience* **2001**, *15*, 339–345.

4. Xu, W.L.; Zhang, W.; Jian, Y. Analysis of nitrogen removal performance of constructed rapid infiltration system (CRIS). *Appl. Ecol. Environ. Res.* **2017**, *15*, 199–206. [CrossRef]

5. Xu, W.L.; Yang, Y.N.; Cheng, C. Treat Phoenix River water by constructed rapid Infiltration system. *J. Coast. Res.* **2015**, *73*, 386–390. [CrossRef]

6. Liu, G.Y.; Zhang, H.Z.; Zhang, X.; Li, W. Development of total nitrogen removing technology in constructed rapid infiltration systems. *Ind. Water Treat.* **2013**, *33*, 1–4.

7. Ling, Y.; Fan, L.K.; Min, X.; Yue, L.; Sen, W. Environmental economic value calculation and sustainability assessment for constructed rapid infiltration system based on emergy analysis. *J. Clean Prod.* **2017**, *167*, 582–588.

8. Wang, L.; Yu, Z.P.; Zhao, Z.J. The removal mechanism of ammoniac nitrogen in constructed rapid infiltration system. *China Environ. Sci.* **2006**, *26*, 500–504.

9. Zhang, J.B. Study on Constructed Rapid Infiltration for Wastewater Treatment. Ph.D. Thesis, University of Geosciences, Beijing, China, 2002.

10. Song, Z.X.; Zhang, H.Z.; Wang, Z.L.; Ping, Y.H.; Liu, G.Y.; Zhao, Q. Treating sewage by strengthened constructed rapid infiltration system. *Chin. J. Environ. Eng.* **2016**, *10*, 3491–3495.

11. Xu, G.J.; Xu, X.C.; Yang, F.L.; Liu, S.T. Selective inhibition of nitrite oxidation by chlorate dosing in aerobic granules. *J. Hazard. Mater.* **2011**, *185*, 249–254. [CrossRef] [PubMed]

12. Hou, L.; Xia, L.; Ma, T.; Zhang, Y.Q.; Zhou, Y.Y.; He, X.G. Achieving short-cut nitrification and denitrification in modified intermittently aerated constructed wetland. *Bioresour. Technol.* **2017**, *232*, 10–17. [CrossRef] [PubMed]

13. Xu, G.J.; Xu, X.C.; Yang, F.L.; Liu, S.T.; Gao, Y. Partial nitrification adjusted by hydroxylamine in aerobic granules under high DO and ambient temperature and subsequent Anammox for low C/N wastewater treatment. *Chem. Eng. J.* **2012**, *213*, 338–345. [CrossRef]

14. Sukru, A.; Erdal, S. Influence of salinity on partial nitrification in a submerged biofilter. *Bioresour. Technol.* **2012**, *118*, 24–29.

15. Chen, J.; Zhang, J.Q.; Wen, H.Y.; Zhang, Q.; Yang, X.; Li, J. The effect of hydroxylamine inhibition and pH control on achieving shortcut nitrification in constructed rapid infiltration system. *Acta Sci. Circumst.* **2016**, *36*, 3728–3735.

16. Cui, Y.W.; Peng, Y.Z.; Gan, X.Q.; Ye, L.; Wang, Y.Y. Achieving and maintaining biological nitrogen removal via nitrite under normal conditions. *J. Environ. Sci.* **2005**, *17*, 794–798.

17. Ge, L.P.; Qiu, L.P.; Liu, Y.Z.; Zhang, S.B. Effect of Free Chlorine on Shortcut Nitrification in Biological Aerated Filter. *J. Univ. Jinan Sci. Technol.* **2011**, *25*, 336–339.

18. Lees, H.; Simpson, J.R. The biochemistry of the nitrifying organisms. 5. Nitrite oxidation by *Nitrobacter*. *Biochem. J.* **1957**, *65*, 297–305. [CrossRef] [PubMed]

19. Banashri, S.A.P.A. Partial nitrification—Operational parameters and microorganisms involved. *Rev. Environ. Sci. Biotechnol.* **2007**, *6*, 285–313.

20. Wei, F.S. *The Standard Methods for the Examination of Water and Wastewater*, 4th ed.; China Environmental Science Press: Beijing, China, 2002; pp. 211, 254–279. ISBN 9787801634009.

21. Ni, H.; Xiong, Z.; Zhang, S.; Zeng, S.Q.; Li, L. Effect of porous ceramic on the immobilized microorganisms and scaning electron microscopy. *J. Hubei Univ. Nat. Sci.* **2011**, *33*, 182–186.

22. Glass, C.; Silverstein, J. Denitrification kinetics of high nitrate concentration water: pH effect on inhibition and nitrite accumulation. *Water Res.* **1998**, *32*, 831–839. [CrossRef]

23. Chen, J.W. Study on Nitrogen Removal in Partial Nitrification Denitrification Biological Filters. Master's Thesis, Harbin Institute of Technology, Harbin, China, 2017.

24. Yan, N.; Jin, X.B.; Zang, J.Q. A comparison between the processes of denitrification with glucose and methanol as carbon sourse. *J. Shanghai Teach. Univ. Nat. Sci.* **2002**, *31*, 41–44.

25. Zhang, Z.L. Selection of External Carbon Sources for Denitrification. Master's Thesis, Harbin Institute of Technology, Harbin, China, 2009.

26. Gómez, M.A.; Gonzálezlópez, J.; Hontoria-García, E. Influence of carbon source on nitrate removal of contaminated ground-water in a denitrifying submerged filter. *J. Hazard. Mater.* **2000**, *80*, 69–80. [CrossRef]

International Journal of
Environmental Research and Public Health

Article

Pilot-Scale Hydrolysis-Aerobic Treatment for Actual Municipal Wastewater: Performance and Microbial Community Analysis

Xiao Bian, Hui Gong * and Kaijun Wang *

State Key Joint Laboratory of Environment Simulation and Pollution Control, School of Environment, Tsinghua University, Beijing 100084, China; bianxiao8@126.com
* Correspondence: gongh14@tsinghua.org.cn (H.G.); wkj@tsinghua.edu.cn (K.W.);
  Tel.: +86-010-6278-9411 (H.G.); +86-010-6277-3065 (K.W.)

Received: 30 January 2018; Accepted: 27 February 2018; Published: 9 March 2018

Abstract: Low-energy cost wastewater treatment is required to change its current energy-intensive status. Although promising, the direct anaerobic digestion of municipal wastewater treatment faces challenges such as low organic content and low temperature, which require further development. The hydrolysis-aerobic system investigated in this study utilized the two well-proven processes of hydrolysis and aerobic oxidation. These have the advantages of efficient COD removal and biodegradability improvement with limited energy cost due to their avoidance of aeration. A pilot-scale hydrolysis-aerobic system was built for performance evaluation with actual municipal wastewater as feed. Results indicated that as high as 39–47% COD removal was achieved with a maximum COD load of 1.10 kg/m^3·d. The dominant bacteria phyla included *Proteobacteria* (36.0%), *Planctomycetes* (15.4%), *Chloroflexi* (9.7%), *Bacteroidetes* (7.7%), *Firmicutes* (4.4%), *Acidobacteria* (2.5%), *Actinobacteria* (1.8%) and *Synergistetes* (1.3%), while the dominant genera included *Thauera* (3.42%) and *Dechloromonas* (3.04%). The absence of methanogens indicates that the microbial community was perfectly retained in the hydrolysis stage instead of in the methane-producing stage.

Keywords: hydrolysis-aerobic; municipal wastewater treatment; microbial community; high-throughput sequencing

1. Introduction

Wastewater treatment accounts for a large amount of the energy load in society as a whole. It is estimated that 3–4% of total U.S. electricity is consumed for the movement and treatment of water and wastewater [1]. A similar situation could be observed in other developed and developing countries. High energy consumption also creates a large carbon footprint, accelerating global warming. As wastewater production increases due to population and economic growth, the energy-water burden becomes heavier for the future, undermining social sustainability. Thus, it is of great importance to reduce energy costs during the treatment of wastewater.

Wastewater treatment, which is mainly based on the conventional active sludge process (CAS), is criticized as energy-intensive for its heavy aeration. Based on a full-scale plant energy audit, energy as high as 50% was used to supply air for aeration tanks during the CAS process. The anaerobic digestion (AD) process, which avoids aeration and meanwhile recovers energy in the form of CH$_4$ from organics, is a perfect alternative choice for wastewater treatment [2]. Based on energetic calculation, the energy potential in domestic wastewater is higher than that consumed during treatment [3]. Namely, it is even possible to turn wastewater treatment (WWTP) into a net energy producer using a carefully designed process [4]. However, full-scale direct anaerobic treatment of domestic wastewater is mainly applied in developing countries in (sub)tropical regions such as Brazil, Mexico, Egypt, and

India. Its performance is generally unsatisfactory. Challenges exist for its wide application, including a low concentration of organics and low temperature in winter in temperate regions, both of which usually lead to biomass loss, inefficient organic biodegradation, and poor effluent quality.

Multiple measures have been considered to solve this problem, such as enhancing the anaerobic methane-producing process. The most viable idea is the use of the anaerobic membrane bioreactor (AnMBR), which has been thoroughly investigated in past decades [5]. The AnMBR utilizes the membrane with a high separation efficiency to retain enough biomass to maintain a sufficiently high solids retention time (SRT). The quality of effluent is also guaranteed with the use of membrane filtration. Recently another idea, which has drawn much attention, is to separate and concentrate organics in limited volume. First, this is performed through coagulation, adsorption or membrane filtration, and then it anaerobically transforms the concentrated organics into CH_4 in a more conventional way for energy recovery [6]. Although all these developments are promising, further efforts are still needed for their scale-up and full-scale application.

The hydrolysis-aerobic hybrid treatment investigated in this study combined the two well-proven processes of hydrolysis and aerobic oxidation, which was expected to achieve the advantages of the two processes. Due to the limitations of anaerobic methane production mentioned above, part of AD—namely hydrolysis—was applied and combined with aerobic post-treatment. Based on the individual bacteria species, the anaerobic process can be divided into three stages, including hydrolysis, acidogenesis and methanogenesis [7]. Hydrolysis is the initial stage when complex organic polymers are hydrolyzed into simple molecules by hydrolytic enzymes of fermenting bacteria. During the acidogenesis and methanogenesis processes, intermediate products including volatile fatty acids (VFA), such as acetate and H_2, were generated and finally converted into CH_4. For the first stage of AD, hydrolysis has been widely used as a biological pre-treatment. Hydrolysis can not only partially remove COD by fermentation and maybe also respiration of some of the hydrolysis products, but biodegradability can also be improved. Hydrolysis has been used for a lot of refractory feed, including both solid and liquid waste such as lignocellulosic biomass (e.g., sugarcane bagasse and rice straw) [8], excess sludge [9], food waste [10] and refractory/particulate-rich wastewater from tanneries [11] and the petrochemical industry [12]. The inhibiting issue of the antibiotic (e.g., tetracycline) production of wastewater for the following biological process could be avoided through the use of hydrolysis with a BOD/COD ratio increase [13]. Moreover, hydrolysis is based on anaerobes with no aeration and no additional heating requirements, indicating its low-energy cost for large-volume municipal wastewater treatment. Combined with aerobic post-treatment, effluent quality could be improved and also the energy requirement for post-aeration could be reduced due to a lower COD load and high biodegradability. Therefore, the hydrolysis-aerobic hybrid treatment process was expected to be a potential low-cost approach for the treatment of municipal wastewater. However, the performance of the hydrolysis-aerobic process still requires optimization through an investigation on the effects of operation parameters such as the COD load. The structure of the bacterial community also needs to be revealed.

In this study, a pilot-scale hydrolysis-aerobic system was built for performance evaluation with actual municipal wastewater as feed. The influence of the COD load on hydrolysis was investigated. The bacterial community structure of the hydrolysis system was analyzed using high-throughput sequencing.

2. Materials and Methods

2.1. Pilot Plant Experiments

The performance of hydrolysis-aerobic treatment was evaluated using a pilot-scale hydrolysis reactor (Figure 1) and aerobic post-treatment. All the pilot facilities were installed in Caoxian WWTP in Shandong province, China. Pilot-scale hydrolysis reactor with a vertical acrylic column (2 m diameter × 6 m length) was used, with double loop inside. It was composed of four main parts: the cone on the bottom, the cylinder, the tri-phase separator and the sludge circulation system. The sludge circulation pump controlled the rate of internal sludge circulation and provided suitable shear force for sludge at different growth stages. Sludge flowed from top to the bottom and sewage moved up-flow, with the down-flow/up-flow ratio more than 10:1 at the start-up stage.

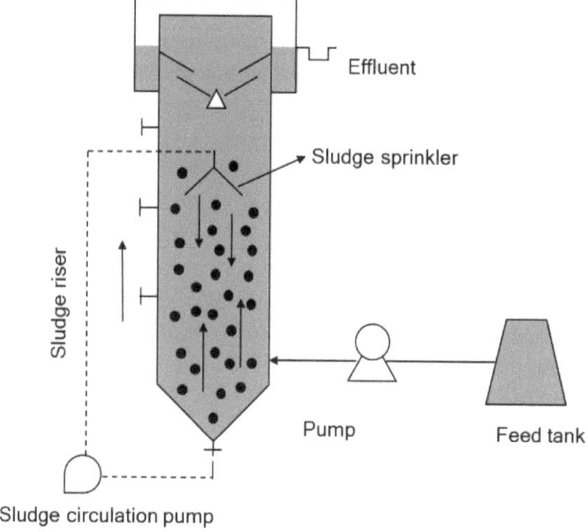

Figure 1. A schematic view of the hydrolysis reactor.

Actual municipal wastewater after the coarse filter of Caoxian WWTP was taken as feed. The BOD/COD of municipal wastewater feed was as low as about 0.2, due to some industrial wastewater flowing into WWTP. Sludge from the thickening tank was inoculated into the hydrolysis reactor with a final concentration of 25 g/L. The effective volume of the hydrolysis reactor is 16.5 m^3. Pilot-scale experiments lasted for approximately 45 days and had four stages. During the pilot experiment, temperature was maintained around 25 °C. The first stage was a start-up with approximately 50 h of hydraulic retention time (HRT). After 16 days of operation, HRT was reduced to 21 and 12 h in stage II and stage III with COD increasing, respectively. To evaluate the limitation, HRT was reduced further to 5 h in stage IV until the operation performance collapsed.

2.2. Analytical Methods

The chemical oxygen demand (COD) and NH$_4$-N were determined using colourimetric techniques with a HACH spectrophotometer (DR 5000, HACH, Loveland, CO, USA). Illumina HiSeq (Santiago, MN, USA) sequencing was applied to analyze the microbial consortium of the last stage sludge from the hydrolysis reactor. The procedures used are as follows: (1) The total genomic DNA was extracted using a DNA extraction kit (Mo Bio Laboratories, Carlsbad, CA, USA) according to the manufacturer's instructions; (2) PCR amplification was performed using specifically synthesized primers with a barcode (Forward: GTGCCAGCMGCCGCGGTAA, Reverse: GGACTACHVGGGTWTCTAAT) in the ABI GeneAmp® 9700 system (Bio-Rad, Hercules, CA, USA) using the following programme: 5 min of denaturation at 95 °C followed by 30 cycles of 30 s at 95 °C (denaturation), 30 s for annealing at 58 °C and 25 s at 72 °C (elongation), with a final extension at 72 °C for 7 min, and polymerase chain reaction (PCR) products were saved for Illumina HiSeq sequencing; and (3) high-throughput sequencing was performed using the Illumina HiSeq 2500 platform (Illmumina, Santiago, MN, USA).

3. Results and Discussion

3.1. Effects of the Organic Loading Rate (OLR) on Hydrolysis

As shown in Figure 1, the performance of a pilot-scale hydrolysis reactor was evaluated from start-up to stable stages with an increasing COD load. Stage I was a start-up with an average COD load of 0.14 kg/m^3·d. At the start, there was little to even negative COD removal due to inoculum sludge cell lysis and the release of organics. After 8 days of acclimatization, hydrolysis took over the main reaction and COD removal reached 40~50%. At the start-up stage, the total average COD removal was 25.1%. During stages II and III, the COD load increased to 0.34 kg/m^3·d and 0.71 kg/m^3·d. Meanwhile, the COD removal also increased to averages of 39.3% and 47.7%. The influence and effluence COD during pilot experiments is shown in Table 1. It should be noted that although the average COD removal efficiencies increased comparing Phases II and III, it was not significantly different due to fluctuating influent COD and limited measurements.

Table 1. COD removal during different stages of hydrolysis.

	Influent COD (mg/L)	Effluent COD (mg/L)	Removal Efficiency (%)	HRT (h)
Phase I	244.8 ± 138.4	183.4 ± 70.2	25.1	51.68
Phase II	270.3 ± 53.1	164.1 ± 44.4	39.3	21.29
Phase III	244.3 ± 70.7	127.8 ± 40.4	47.7	12.78
Phase IV	242.0 ± 122.8	201.5 ± 88.4	16.7	4.97

In the last stage, stage IV, the COD load increased to 1.10 kg/m^3·d with a very short HRT of 4.97 h. It should be noted that the hydrolysis performance was negatively impacted in this condition and the COD removal was only 20.2%. The main reason for the collapse of COD removal is probably the loss of microbial sludge in hydrolysis reactor due to the short HRT and high up-flow velocity. Recent research reporting granular hydrolysis sludge with much higher settleability was potential solution to improve performance of hydrolysis reactor, which could be investigated in further research.

3.2. Performance of the Hydrolysis-Aerobic System

The performance of the hydrolysis-aerobic system (stages III and IV) is illustrated in Figure 2. The COD of influent, after hydrolysis and the final effluent during stage III, was 244.3 ± 70.7 mg/L, 127.8 ± 40.4 mg/L and 59.6 ± 14.1 mg/L. The removal efficiencies of hydrolysis and the total hydrolysis-aerobic system were 44.8 ± 9.7% and 71.9 ± 13.1%, respectively. As high as 44.8% COD removal was achieved during hydrolysis, indicating its high efficiency. Additionally, the fluctuation was reduced for COD after hydrolysis, indicating its anti-shock loading capability. In other words, the hydrolysis system is a pre-treatment process with an anti-shock loading capability. The performance of the aeration process was not optimized and the effluent COD was still higher than 50 mg/L. It is expected that effluent COD can be reduced through modified aeration.

In comparison, at stage IV after hydrolysis and the final effluent, the COD of influent was 242.0 ± 122.8 mg/L, 201.5 ± 88.4 mg/L and 48.0 ± 8.2 mg/L. The removal ratios of hydrolysis and the total hydrolysis-aerobic system were 15.1 ± 5.1% and 75.8 ± 12.0%, respectively. The COD removal achieved through hydrolysis was remarkably limited due to the high COD load and the very short HRT over-reactor limits. Although the total COD removal of stage IV is close to that of stage III, its aeration energy consumption was probably higher than that of stage III. Furthermore, little nitrogen was removed by hydrolysis. No obvious NH$_4$-N reduction was observed (20.3 ± 4.8 mg/L after hydrolysis with influent of 19.3 ± 6.1 mg/L). Nitrification occurred in subsequent aeration process and NH$_4$-N decreased to 2.9 ± 1.0 mg/L in final effluent.

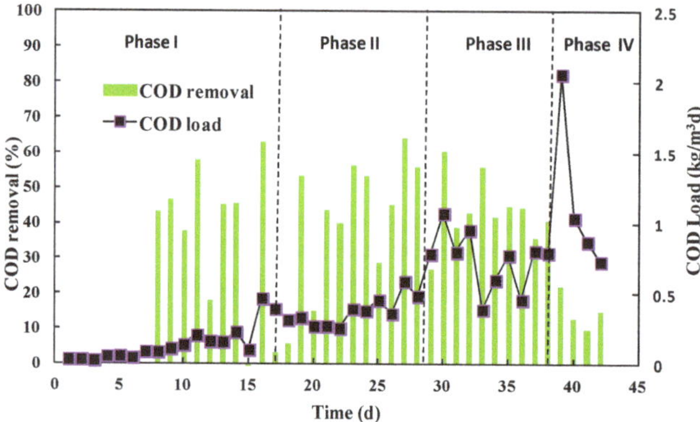

Figure 2. Effects of the organic loading rate (OLR) on the COD removal (based on TCOD) of the hydrolysis reactor.

From the perspective of cost saving, the hydrolysis-aeration hybrid system demonstrates its potentials on two advantages. The first is energy saving by degrading part of organics during hydrolysis process, which otherwise would be removal by energy-intensive aeration process e.g., about 105~120 mg/L COD was removal during hydrolysis (stages II and III). Here we assume the energy for aeration was providing 1 kg O_2/kWh energy input. Based the roughly estimation, the energy saving by COD removal of hydrolysis process could be as high as 0.11~0.12 kWh/m^3, indicating its advantage over conventional process. The other potential advantage is to improve biodegradability by degrade complex organic compounds into small molecules during hydrolysis.

3.3. Microbial Community Analysis

The structure and diversity of bacterial communities in the hydrolysis reactor were revealed using high-throughput sequencing (HiSeq). Analysis of both the phylum and genus levels of the bacterial community is shown in Figure 3. Bacteria phyla with a relative abundance above 1% included *Proteobacteria* (36.0%), *Planctomycetes* (15.4%), *Chloroflexi* (9.7%), *Bacteroidetes* (7.7%), *Firmicutes* (4.4%), *Acidobacteria* (2.5%), *Actinobacteria* (1.8%) and *Synergistetes* (1.3%). These phyla were also identified by other researchers as the dominant bacteria phyla for wastewater treatment [14]. It is reported that Proteobacteria, Bacteroidetes and Firmicutes are dominant bacteria phyla in a conventional active sludge process and also in a hydrolysis system. In this study, they account for approximately 48.1%. Proteobacteria has been widely reported in the wastewater treatment process for its multiple functions [15]. It was revealed that Bacteroidetes has the capacity to convert organic carbon to CO_2 through aerobic oxidation [16]. Firmicutes, as reported previously, was connected with hydrolyzing long-chain organics into small molecules [17]. The hydrolysis reactor exhibited a strong performance in removing and degrading the complex organic compounds in the municipal wastewater.

The hydrolysis processes required the presence of a diverse group of bacteria, which were probably closely dependent on each other. There were as many as approximately 200 genera identified in the hydrolysis system. The utilisation of actual wastewater increased the complexity of the microbial niche and led to high diversity. As high as 87.4% of the bacteria were unclassified and even unknown based on genus-level analysis. As shown in Figure 4b, the top ten genera included *Thauera* (3.42%), *Dechloromonas* (3.04%), *Rubinisphaera* (1.05%), *Nitrosomonas* (0.98%), *Gemmata* (0.66%), *Desulfobacter* (0.65%), *Blastopirellula* (0.64%), *Ornatilinea* (0.63%), *Reyranella* (0.56%), *Thiobacillus* (0.48%), *Longilinea* (0.48%). The genus *Thauera* belongs to the *Betaproteobacteria* class and the Proteobacteria phylum. It had a denitrification capacity and is widely distributed in wastewater treatment plants, rivers and polluted pools [18]. It is also reported to be dominant in the denitrification process, such as in the

DEAMOX (Denitrifying Ammonium OXidation) system. *Dechloromonas* frequently demonstrated a capacity to degrade aromatic compounds such as benzene, and this was related to nitrogen removal [19]. Furthermore, it is widely found in WWTP effluence [19]. *Nitrosomonas* was strictly of the ammonia-oxidizing bacterial (AOB), indicating the potential for nitrosification [20]. *Desulfobacter* is a genus of sulfate-reducing bacteria (SRB) that utilise sulfate as an electron accepter and degrade aromatic compounds. Together with *Desulfobacula*, it was classified as part of the Proteobacteria phylum, *Deltaproteobacteria* class, *Desulfobacterales* order and *Desulfobacteraceae* family. During hydrolysis, no aeration reduced the SO_4^{2-} reduction and inhibited SRB abundance and diversity [12]. *Longilinea* of the *Anaerolineae* class and *Thiobacillus* of the *Betaproteobacteria* class were reported to degrade refractory organics including naphthaline and SCN^-.

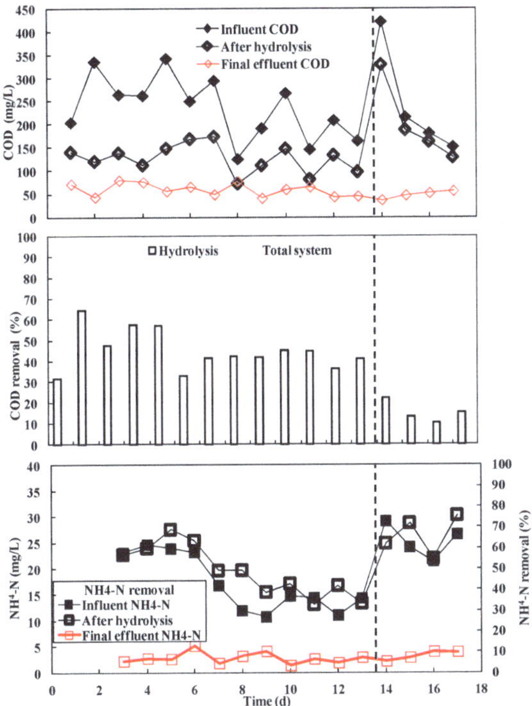

Figure 3. COD and NH_4^+-N concentration of the influent and effluent and their removal efficiencies during the hydrolysis-aerobic process (stages III and IV).

Archaeal diversities and communities were also investigated. No methanogens were found as dominant and only *Methanomassiliicoccus* was identified with a very low abundance of 0.0096%. Both the absence of methanogens and the process performances indicated that the microbial community was perfectly retained within the hydrolysis stage instead of entering into the methane-producing stage. Moreover, it should be noted that microbial community structure analysis by HiSeq was performed only for the last stage with single sampling. The evolving of microbial community structure could be revealed by further research with various sampling over different stages and it would be more informative approach to provide microbial information.

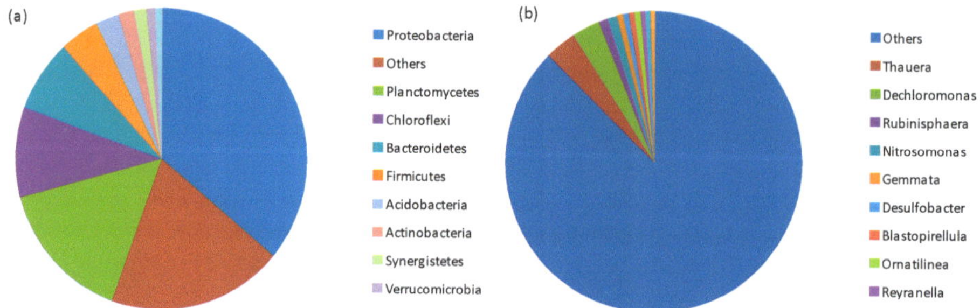

Figure 4. Microbial community structures in the hydrolysis reactor at phylum level (**a**) and genus level (**b**).

4. Conclusions

The hydrolysis achieved effective pre-treatment with 39–47% COD removal in the hybrid hydrolysis-aerobic process for the low-cost treatment of municipal wastewater. The performance of the pilot-scale system was inhibited when the COD load increased to 1.10 kg/m^3·d. The microbial community in the hydrolysis reactor was analyzed. The dominant bacteria phyla included *Proteobacteria* (36.0%), *Planctomycetes* (15.4%), *Chloroflexi* (9.7%), *Bacteroidetes* (7.7%), *Firmicutes* (4.4%), *Acidobacteria* (2.5%), *Actinobacteria* (1.8%) and *Synergistetes* (1.3%). The dominant bacteria genera included *Thauera* (3.42%) and *Dechloromonas* (3.04%). High microbial diversity was observed with actual wastewater as feed, indicating a large proportion of genera that needs to be investigated further. The absence of methanogens indicates that the microbial community was perfectly retained in the hydrolysis stage instead of in the methane-producing stage.

Author Contributions: Hui Gong and Kaijun Wang conceived and designed the experiments; Xiao Bian performed the experiments; Xiao Bian and Hui Gong analyzed the data and wrote the paper.

Conflicts of Interest: The authors declare no conflict of interest. The founding sponsors had no role in the design of the study; in the collection, analyses, or interpretation of data; in the writing of the manuscript, and in the decision to publish the results.

References

1. U.S. Environmental Protection Agency. *Wastewater Management Fact Sheet, Energy Conservation*; U.S. Environmental Protection Agency: Washington, DC, USA, 2006; p. 7.
2. Schaum, C.; Lensch, D.; Bolle, P.Y.; Cornel, P. Sewage sludge treatment: Evaluation of the energy potential and methane emissions with COD balancing. *J. Water Reuse Desalination* **2015**, *5*, 437–445. [CrossRef]
3. Mccarty, P.L.; Jaeho, B.; Jeonghwan, K. Domestic wastewater treatment as a net energy producer-can this be achieved? *Environ. Sci. Technol.* **2011**, *45*, 7100–7106. [CrossRef] [PubMed]
4. Wett, B.; Buchauer, K.; Fimml, C. Energy self-sufficiency as a feasible concept for wastewater treatment systems. In *IWA Leading Edge Technology Conference*; Asian Water Singapore: Singapore, 2007.
5. Stuckey, D.C. Recent developments in anaerobic membrane reactors. *Bioresour. Technol.* **2012**, *122*, 137–148. [CrossRef] [PubMed]
6. Gong, H.; Wang, Z.; Zhang, X.; Jin, Z.; Wang, C.; Zhang, L.; Wang, K. Organics and nitrogen recovery from sewage via membrane-based pre-concentration combined with ion exchange process. *Chem. Eng. J.* **2016**, *311*, 13–19. [CrossRef]
7. Mccarty, P.L.; Smith, D.P. Anaerobic wastewater treatment. *Environ. Sci. Technol.* **1986**, *20*, 1200–1206. [CrossRef]
8. Santos, D.A.; Oliveira, M.M.; Curvelo, A.A.S.; Fonseca, L.P.; Porto, A.L.M. Hydrolysis of cellulose from sugarcane bagasse by cellulases from marine-derived fungi strains. *Int. Biodeterior. Biodegrad.* **2017**, *121*, 66–78. [CrossRef]

9.   Buntner, D.; Spanjers, H.; Lier, J.B.V. The influence of hydrolysis induced biopolymers from recycled aerobic sludge on specific methanogenic activity and sludge filterability in an anaerobic membrane bioreactor. *Water Res.* **2014**, *51*, 284–292. [CrossRef] [PubMed]

10. Hafid, H.S.; Nor'Aini, A.R.; Mokhtar, M.N.; Talib, A.T.; Baharuddin, A.S.; Kalsom, M.S.U. Over production of fermentable sugar for bioethanol production from carbohydrate-rich Malaysian food waste via sequential acid-enzymatic hydrolysis pretreatment. *Waste Manag.* **2017**, *67*, 95–105. [CrossRef] [PubMed]

11. Wang, K.; Li, W.; Gong, X.; Li, X.; Liu, W.; He, C.; Wang, Z.; Minh, Q.N.; Chen, C.L.; Wang, J.Y. Biological pretreatment of tannery wastewater using a full-scale hydrolysis acidification system. *Int. Biodeterior. Biodegrad.* **2014**, *95*, 41–45. [CrossRef]

12. Yang, Q.; Xiong, P.; Ding, P.; Chu, L.; Wang, J. Treatment of petrochemical wastewater by microaerobic hydrolysis and anoxic/oxic processes and analysis of bacterial diversity. *Bioresour. Technol.* **2015**, *196*, 169–175. [CrossRef] [PubMed]

13. Yi, Q.; Gao, Y.; Zhang, H.; Zhang, H.; Zhang, Y.; Yang, M. Establishment of a pretreatment method for tetracycline production wastewater using enhanced hydrolysis. *Chem. Eng. J.* **2016**, *300*, 139–145. [CrossRef]

14. Bakare, B.F.; Mtsweni, S.; Rathilal, S. Characteristics of greywater from different sources within households in a community in Durban, South Africa. *J. Water Reuse Desalination* **2017**, *7*, 520–528. [CrossRef]

15. Niu, T.; Zhou, Z.; Shen, X.; Qiao, W.; Jiang, L.M.; Pan, W.; Zhou, J. Effects of dissolved oxygen on performance and microbial community structure in a micro-aerobic hydrolysis sludge in situ reduction process. *Water Res.* **2016**, *90*, 369–377. [CrossRef] [PubMed]

16. Xie, X.; Liu, N.; Yang, B.; Yu, C.; Zhang, Q.; Zheng, X.; Xu, L.; Li, R.; Liu, J. Comparison of microbial community in hydrolysis acidification reactor depending on different structure dyes by Illumina MiSeq sequencing. *Int. Biodeterior. Biodegrad.* **2016**, *111*, 14–21. [CrossRef]

17. Chen, H.; Chang, S. Impact of temperatures on microbial community structures of sewage sludge biological hydrolysis. *Bioresour. Technol.* **2017**, *245*, 502–510. [CrossRef] [PubMed]

18. Podosokorskaya, O.A.; Podosokorskaya, O.A.; Bonch-Osmolovskaya, E.A.; Novikov, A.A.; Kolganova, T.V.; Kublanov, I.V. *Ornatilinea apprima* gen. nov., sp. nov., a cellulolytic representative of the class Anaerolineae. *Int. J. Syst. Evolut. Microbiol.* **2013**, *63*, 86–92. [CrossRef] [PubMed]

19. Carosia, M.F.; Okada, D.Y.; Sakamoto, I.K.; Silva, E.L.; Varesche, M.B.A. Microbial characterization and degradation of linear alkylbenzene sulfonate in an anaerobic reactor treating wastewater containing soap powder. *Bioresour. Technol.* **2014**, *167*, 316–323. [CrossRef] [PubMed]

20. Bin, M.; Wang, S.; Cao, S.; Miao, Y.; Jia, F.; Du, R.; Peng, Y. Biological nitrogen removal from sewage via anammox: Recent advances. *Bioresour. Technol.* **2015**, *200*, 981–990.

International Journal of
*Environmental Research
and Public Health*

Article

Processing Technology Selection for Municipal Sewage Treatment Based on a Multi-Objective Decision Model under Uncertainty

Xudong Chen [1], Zhongwen Xu [2], Liming Yao [2,*] and Ning Ma [2]

[1] College of Management Science, Chengdu University of Technology, Chengdu 610059, China; chenxudong198401@163.com

[2] Business School, Sichuan University, Chengdu 610065, China; xuzhongwen1996@163.com (Z.X.); ningma_610@163.com (N.M.)

* Correspondence: lmyao@scu.edu.cn; Tel.: +86-28-8541-7897

Received: 6 January 2018; Accepted: 26 February 2018; Published: 5 March 2018

Abstract: This study considers the two factors of environmental protection and economic benefits to address municipal sewage treatment. Based on considerations regarding the sewage treatment plant construction site, processing technology, capital investment, operation costs, water pollutant emissions, water quality and other indicators, we establish a general multi-objective decision model for optimizing municipal sewage treatment plant construction. Using the construction of a sewage treatment plant in a suburb of Chengdu as an example, this paper tests the general model of multi-objective decision-making for the sewage treatment plant construction by implementing a genetic algorithm. The results show the applicability and effectiveness of the multi-objective decision model for the sewage treatment plant. This paper provides decision and technical support for the optimization of municipal sewage treatment.

Keywords: processing technology selection; municipal sewage treatment; multi-objective decision

1. Introduction

With the continuous advancement of China's economic construction and the enhancement of an overall social consciousness regarding environmental protection, municipal sewage treatment, known as an important measure to protect water resources and control water pollution, has received increasing attention. Many studies have shown that municipal sewage contains large amounts of contaminants, such as nitrogen and phosphorus [1], resistance genes and hormones [2,3] and trace contaminants [4,5]. Several countries in Europe and the United States have constructed many municipal sewage treatment plants, which have effectively controlled environmental water pollution [6,7]. However, in some of the sewage treatment plants, especially the county-level sewage treatment plants, there is a lack of funds for their operation [8]. In addition, with the annual increase in municipal sewage treatment capacity, a dramatic increase in municipal sewage sludge production has occurred, with concomitant concerns regarding improper sludge disposal and retention of organic contaminants [9]; therefore, secondary pollution to the water and the atmosphere will occur if the sludge enters the environment without the appropriate disposal measures [10]. Hence, sludge should be considered from a comprehensive perspective. In China, water resources per capita are scarce, especially in the densely populated cities, and water pollution is more serious [11]. Unfortunately, there are also some existing backwards for municipal sewage treatment, such as high operation costs and low removal efficiency [12]. Therefore, in order to align with the sustainable development of society, it is necessary to analyze the overarching factors, such as technology, economy, environment and management in relation to sewage treatment and put forward scientific suggestions for the construction of sewage treatment plants.

Processing technology is the most important factor affecting sewage treatment, and mainstream sewage treatment processes rely on biological treatment methods because of technical and cost efficiencies [13,14], which mainly include the activated sludge method and the biofilm method. In addition, the activated sludge method is more widely used because of its sewage removal capacity [15], including Intermittent Cycle Extended Aeration System (ICEAS) process, Anaerobic Anoxic Oxic (AAO) method, and oxidation ditch. Bhatnagar and Sillanpää, [16] analyzed the application of adsorbent in sewage treatment. Kartal et al. [17] claimed that the use of anoxic ammonium-oxidizing (anammox) bacteria can facilitate effective removal of ammonia and nitrite. Plakas and Karabelas [18] argued that nanofiltration (NF) and low-pressure reverse osmosis (LPRO) can be used to treat pesticides in sewage. Xu et al. [19] and Fu et al. [20] summarized the application of zero-valent iron (ZVI), including nanoscale zero-valent iron (nZVI)), in wastewater treatment.

In addition to the processing technology, there are many factors that will affect the sewage treatment plant construction program design, such as construction scale and site selection of the sewage treatment plant [21,22]. Hernandez-Sancho et al. [21] considered the impact of environmental and economic benefits on the design of wastewater treatment plants. Cornel and Schaum [23], Bresler [24] and Dodane [25] focused on the impact of costs in the construction of wastewater treatment plants, which was related to a number of factors [26]. Molinos-Senante et al. [27] considered a cost-benefit analysis of wastewater treatment plants from an economical perspective. In addition, to consider the impact of the size of the sewage treatment plant on the cost, Hernandez-Sancho et al. [28] also considered the impact of the pollutant removal rate and the equipment age. Mojahed et al. [29] analyzed the planning of wastewater treatment plants under environmental control. There are also some studies that analyzed the construction of sewage treatment plants from the life cycle assessment (LCA) perspective [30–32]. In all, there are many factors that affect the design of municipal sewage treatment, such as processing technology, site location, environmental protection, investment costs and operational efficiency. Through a review of the above literature, it can be deduced that existing municipal sewage treatment facilities lack an important single factor, namely, a systematic theoretical research approach; moreover, the research conclusion is often not very practical. Hence, faced with these five indicators in terms of environmental protection and economic benefit, a multi-objective programming approach is more suitable [33].

This paper aims to put forward some scientific insights regarding the construction of sewage treatment plants by developing a multi-objective decision model that is subject to some relevant constraints, such as pollutant control capacity, etc. In addition, the contributions of this paper are summarized as follows:

1. By combining multiple factors influencing the construction of sewage treatment plants, this study considers the trade-offs between environmental protection and economic benefit.
2. A general model of multi-objective decision-making under uncertainty conditions is proposed for optimizing the problem of municipal sewage treatment plant construction.

The structure of this paper is as follows: The introduction provides reviews of previous academic research in the area of urban sewage treatment planning. A multi-objective decision model to solve the urban sewage treatment plan problem is established in Section 2. Section 3 is dedicated to application research. Section 4 discusses conclusions drawn from the research.

2. Description of Multi-Objective Decision Model Building

2.1. Key Problem Statement

The treatment of municipal sewage has become one of the essential measures to protect water resources and to control water pollution. Nevertheless, the operation of current municipal sewage treatment plants is not promising. A large number of the sewage treatment plants are not open throughout the entire year because of high operation costs, weak technical strength, inadequate staff ability and so on. Therefore, some plants fail to operate at full capacity; moreover, China's urban sewage treatment facilities seriously lag more robust implementations and are inadequate [34], leading to a poor return on this portion of the national investment. Hence, a comprehensive analysis of the factors impacting the municipal sewage treatment plants from the perspective of systems engineering is needed, which can deliver more acceptable solutions that address the perspectives of the stakeholders [35].

According to the reports of relevant departments, there are multiple reasons for these phenomena, for instance, unclear operational costs, weak technical strengths, inadequately trained staff, poor operations management and design problems.

2.1.1. Principal Factors

Sewage treatment technology is the most important factor affecting the results of the sewage treatment. A suitable processing technology is important and should be chosen according to the actual local water requirements. Given the different qualities of municipal sewage and domestic sewage, different processing technologies are required. The primary technologies used for municipal sewage treatment in China are the Sequencing Batch Reactor (SBR) process, the oxidation ditch process and the traditional activated sludge process. Each technology for sewage treatment has its advantages, characteristics, applicable conditions, and deficiencies.

Additionally, the site selection has influences on the sewage treatment plants. The optimal site of a sewage treatment plant is critical because an optimally established wastewater treatment plant can efficiently protect the environment and enable the sustainability of economic and urban development. Moreover, the control target of pollutant discharge, the urban geographic and geological environment, the function and flow volume of the receiving water and land type and quality are factors that require consideration.

2.1.2. Environmental Impacts and Economic Benefits

Environmental impacts and economic benefits are of equal importance. From the perspective of environmental protection, the construction of the sewage treatment plants generally should not cause irreparable damage to the surrounding environment. Therefore, unsuitable locations include upwind of a city, upstream of an urban water source or in close proximity to residential areas. After the sewage treatment plant is put into operation, its effects should be minimized on the environmentally sensitive areas, the downstream areas for water conservation and, in particular, the aquaculture at the towns. Moreover, the impact should not exceed the local environmental capacity.

From the perspective of economic benefits, the reuse of wastewater is becoming an essential approach to addressing water shortages. After the treated sewage reaches the prescribed standard, it can be used as irrigation water, industrial water, and domestic water for toilet flushing, garden watering and road cleaning. Therefore, the benefits of return water utilization can be considered when optimizing the construction scheme.

2.1.3. Uncertainty Problem

Uncertainty widely exists in various social phenomena, natural phenomena and engineering practice. As a complex system, the municipal sewage treatment system includes three kinds of elements: human, matter and environment, and has three key processes: input, inner system and output, so it includes various kinds of uncertainty (see Figure 1).

First, inherent uncertainties in nature include hydrology, geography, temperature, precipitation, and amount of solar radiation, and the environmental parameters can constantly change with the changes in the weather and other conditions. In a sewage treatment system, the condition of the original sewage input, the treatment effectiveness and the environmental influence of the treated sewage output are subject to uncertainty due to changes in such parameters as temperature, humidity, and radiation exposure. For example, the input volume is highly related to the temperature and rainfall; the treatment effectiveness is related to the temperature, light and humidity. Second, the uncertainties caused by human activity in a sewage treatment system are mainly reflected in the input and treatment processes. For, example, the releasing of water saving policies could decrease the original sewage volume; domestic sewage is greater on non-working days and industrial sewage is greater on working days. Meanwhile, possible sudden emergencies such as a pipeline leakage may increase the pollutant level in the sewage. Additionally, the serious working attitude of the staff in the treatment plant has a positive influence on the effectiveness of sewage treatment. Third, the uncertainties caused by an engineering system can be reflected in the input and treatment processes. In the input process, the pollutant level of the original sewage may decrease with technological improvements adopted by the factories. In the treatment process, operational mistakes in the sewage treatment or monitoring deviations in sewage detection can negatively impact sewage treatment. Therefore, using a systems perspective, the sewage treatment problem has uncertainty in many aspects of its subsystems, and these characteristics facilitate using an uncertainty framework to solve this problem within the context of its environment.

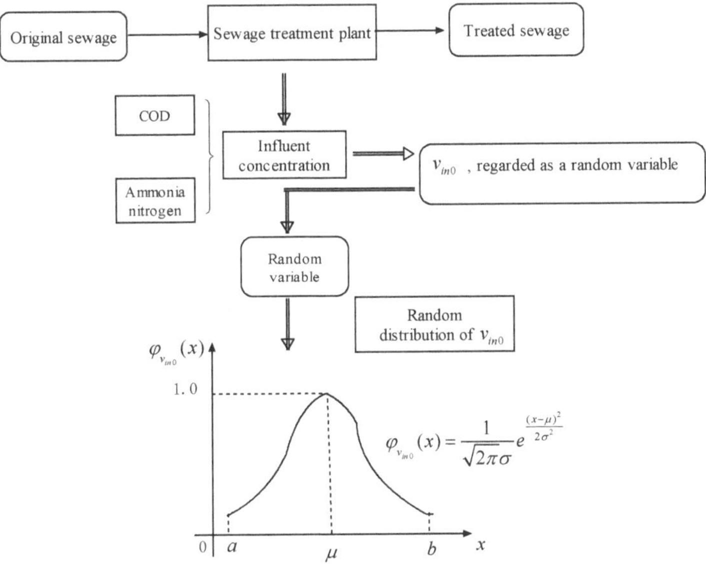

Figure 1. Uncertainty of urban sewage treatment problem.

The sewage concentration is the most important element that is related to the decision of each process step and the final goals. The sewage influent concentration may influence the site and technology selections of a treatment plant and further influence the economic and ecological goals. The influent concentration involves the rate of the pollutants quantity and sewage volume, which is affected by various uncertain factors. Therefore, the influent concentration is also uncertain and difficult to define with crisp values. In this paper, the influent concentration is valued by two indices: Chemical Oxygen Demand (COD) v_{in0}^{α} and ammonia nitrogen v_{in0}^{β} [12,36]. The corresponding managers can only predict the most possible value of influent concentration v_{in0} which approximately follows a normal distribution, i.e., $v_{in0} \sim N\left(\mu, \delta^2\right)$. Therefore, it is appropriate to consider a sewage treatment system as a random environment.

Using the background of uncertainty, factors such as the site selection, selection of wastewater treatment technology, environmental impacts and economic benefits allow for the development of a scientific construction plan for the sewage treatment plant. The uncertainty approach ensures the quality of sewage treatment, guarantees the stable operation of the plant and meets the treatment standards. Hence, the construction scheme is the focus of this paper (see Figure 2).

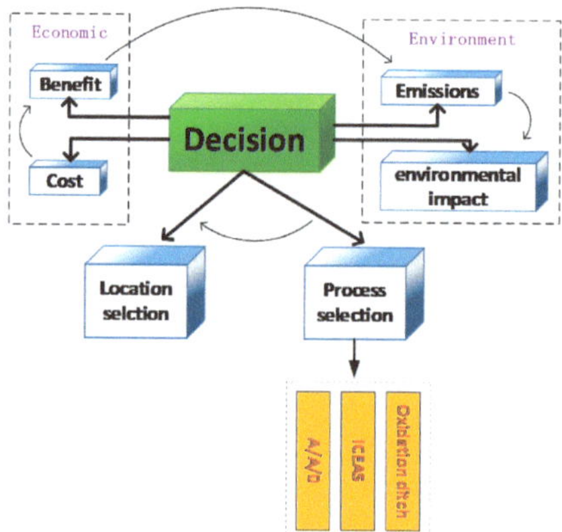

Figure 2. Economic-environment trade-off problem.

2.2. Question Assumption

1. All construction schemes are aligned with the overall land utilization planning, urban master planning, special planning of a sewage treatment, as well as such regulations as relevant laws, specifications, and procedures.
2. Due to the constraints of factors such as the floor space, the paper assumes that in terms of decision-making, the number of sewage treatment plants to be built is less than N.
3. It is assumed that the distributions of the main sewage source at the construction site of a treatment plant is determined and assumed that the influent concentrations are the same.
4. Apart from the infrastructure investments affected by the design treatment capacity and efficiency, the construction and investment costs of a sewage treatment plant also include the purchase and installation costs of different pieces of treatment equipment required by different processes.
5. The management and operation costs mainly consist of, for instance, pipeline maintenance, energy consumption, and equipment repair, which is different from sewage treatment engineering.

In terms of the impact of a sewage treatment plant on the environment, the impact of the noise and odor of that plant on the surrounding residential land and sensitive spots are principally considered. Taking all of the environmental impact factors into account was accomplished by utilizing the fuzzy multi-attribute method; the larger the value, the greater the impacts of the site on the environment.

2.3. Modeling

Based on the general multi-objective, decision-making model of a sewage treatment plant and considering the specific conditions of this case, this section establishes the objective functions and constraint conditions from the perspective of a cost model, eventually forming the multi-objective programming model which is suitable for this case.

2.3.1. Objective Function

There are multiple factors affecting the construction of a sewage treatment plant, for instance, the overall consideration of the economic, social and environmental benefits, etc. This paper sets four objectives, namely, minimum cost, minimum environmental impact, maximum benefit of recycled water and maximum sewage treatment efficiency.

x_i represents whether the optional site A_i can be selected to build a sewage treatment plant: if $x_i = 1$, then, site A_i can be chosen to build a treatment plant. If $x_i = 0$, then, site A_i is not suitable to build such a plant. y_{ij} denotes whether the j technology T_j can be used in treatment plant i: if $y_{ij} = 1$, then, T_j can be used in the treatment plant A_i. If $y_{ij} = 0$, then, T_j is not suitable for plant A_i. The treatment capacity of technology j is denoted t_j.

According to Assumption 4, the initial capital cost comprises construction cost of a treatment plant, cost of off-site sewage interception pipelines, land expropriation cost and household demolition compensation cost. Hence, the initial capital cost is:

$$\sum_{i=1}^{n} C_i(Q_i, y_{ij})x_i + \sum_{i=1}^{n} L_i(l_i)x_i + \sum_{i=1}^{n} x_i D_i(s_i) \tag{1}$$

According to Assumption 5, the management operation cost is:

$$\sum_{i=1}^{n} x_i F_i(y_{ij}) \tag{2}$$

Then, the objective function I pertains to the minimum investment cost, which can be written as:

$$\min f_1(x_i, y_{ij}) = \sum_{j=1}^{m} \left\{ \sum_{i=1}^{n} C_i(Q_i, y_{ij})x_i + \sum_{i=1}^{n} L_i(l_i)x_i + \sum_{i=1}^{n} x_i D_i(s_i) + \sum_{i=1}^{n} x_i F_i(y_{ij}) \right\} \tag{3}$$

The objective function II pertains to the minimum environmental impact, which can be written as:

$$\min f_2(x_i) = \sum_{i=1}^{n} \delta_i x_i \tag{4}$$

Accordingly, the reclaimed sewage benefit of treatment plant i is the recycled water income $B_i \lambda_i Q_i$; therefore, the objective function III pertains to the maximum benefit, which can be written as:

$$\max f_3(x_i) = \sum_{i=1}^{n} B_i \lambda_i Q_i x_i \tag{5}$$

Upon completion of a treatment plant, the sewage treatment efficiency is expected to be maximized; the objective functions IV and V, therefore, are the minimum discharge of COD and ammonia nitrogen, respectively:

$$\min f_4(x_i, y_{ij}) = E\left[\sum_{i=1}^{n} Q_i x_i (1 - \lambda_i) \widetilde{v}_{outi}^{\alpha} \right] \tag{6}$$

$$\min f_5(x_i, y_{ij}) = E\left[\sum_{i=1}^{n} Q_i x_i (1 - \lambda_i) \widetilde{v}_{outi}^{\beta} \right] \tag{7}$$

where $v_{outi}^{\alpha} = v_{in0}^{\alpha} - t_j y_{ij}$; $v_{outi}^{\beta} = v_{in0}^{\beta} - t_j y_{ij}$.

2.3.2. Constraints

1. Pollutant processing capacity. The pollutant processing capacity of a sewage treatment plant should not be less than the sewage quantity, and the actual processing capacity should be more than 60% of the design capacity. In other words:

$$0.6\sum_{i=1}^{n} Q_i x_i \leq \sum_{i=1}^{m} q_i \leq \sum_{i=1}^{n} Q_i x_i \tag{8}$$

2. Construction investment limitation. Before constructing a sewage treatment plant, the planning department sets a certain total investment; in other words, the construction investment funds should not exceed the upper limit of the total investment. It can be written as:

$$\sum_{i=1}^{n} C_i(Q_i, y_{ij})x_i + \sum_{i=1}^{n} L_i(l_i)x_i + \sum_{i=1}^{n} x_i D_i(s_i) \leq M \tag{9}$$

where M is the prescribed investment limit.

3. Emissions standards. According to the total pollutant control target, after completion of a sewage treatment plant, effluent must meet certain standards; in other words, the main pollutant content should be less than the standard content. It can be written as:

$$\text{COD}: E\left[\sum_{i=1}^{n} Q_i x_i (1 - \lambda_i)\left(\tilde{v}_{in0}^{\alpha} - t_j y_{ij}\right)\right] < W_{\text{COD}}, \tag{10}$$

$$\text{NH}_3 - \text{NE}\left[\sum_{i=1}^{n} Q_i x_i (1 - \lambda_i)\left(\tilde{v}_{in0}^{\beta} - t_j y_{ij}\right)\right] < W_{\text{NH}_3-\text{N}}, \tag{11}$$

where W_{COD} and $W_{\text{NH}_3-\text{N}}$ represent COD and ammonia nitrogen emissions control standards of that region, respectively.

4. Effluent concentration limitation. To meet discharge standards, effluent concentration should be less than the upper limit of the stated standard. That is:

$$E\left[v_{in0}^{\alpha} - t_j y_{ij}\right] < \eta_0^{\alpha}, i = 1, 2, \ldots, n \tag{12}$$

$$E\left[v_{in0}^{\beta} - t_j y_{ij}\right] < \eta_0^{\beta}, i = 1, 2, \ldots, n \tag{13}$$

where η_0^{α} and η_0^{β} denote the stated upper limits of wastewater COD and ammonia nitrogen concentration, respectively. Pollutant discharge standards are generally divided into three types: Primary standard A (COD, 50 mg/L; NH$_3$-N, 8 mg/L), Primary standard B (COD, 60 mg/L; NH$_3$-N, 15 mg/L) and secondary standard (COD, 100 mg/L; NH$_3$-N, 25 mg/L).

5. Pollutant treatment rate limitation. The pollutant treatment rate should be more than the lowest limit:

$$E\left[\left(v_{in0}^{\alpha} - t_j y_{ij}\right)/v_{in0}^{\alpha}\right] > \theta_0^{\alpha}, i = 1, 2, \ldots, n \tag{14}$$

$$E\left[\left(v_{in0}^{\beta} - t_j y_{ij}\right)/v_{in0}^{\beta}\right] > \theta_0^{\beta}, i = 1, 2, \ldots, n \tag{15}$$

where θ_0 is the stated lower limit of the pollutant treatment rate.

6. Plant quantity limitation. The number of sewage treatment plants to be constructed should not exceed a certain limit:

$$\sum_{i=1}^{n} x_i \leq N \tag{16}$$

7. Processing technology selection limitation. Each wastewater treatment plant can only choose to use one kind of disposal process:

$$\sum_{j=1}^{m} y_{ij} = 1, i = 1, 2, \ldots, n \tag{17}$$

Based on the analysis above, the paper establishes the following multi-objective programming model subject to constraints (8)–(17):

$$\min f_1(x_i, y_{ij}) = \sum_{i=1}^{n} C_i(Q_i, y_{ij})x_i + \sum_{i=1}^{n} L_i(l_i)x_i + \sum_{i=1}^{n} x_i D_i(s_i) + \sum_{i=1}^{n} x_i F_i(y_{ij})$$

$$\min f_2(x_i) = \sum_{i=1}^{n} \delta_i x_i$$

$$\max f_3(x_i,) = \sum_{i=1}^{n} B_i \lambda_i Q_i x_i$$

$$\min f_4(x_i, y_{ij}) = E\left[\sum_{i=1}^{n} Q_i x_i (1 - \lambda_i)(\tilde{v}^{\alpha}_{in0} - t_j y_{ij})\right]$$

$$\min f_5(x_i, y_{ij}) = E\left[\sum_{i=1}^{n} Q_i x_i (1 - \lambda_i)(\tilde{v}^{\beta}_{in0} - t_j y_{ij})\right]$$

$$\text{s.t.} \begin{cases} 0.6 \sum_{i=1}^{n} Q_i x_i \leq \sum_{i=1}^{m} q_i \leq \sum_{i=1}^{n} Q_i x_i \\[6pt] \sum_{i=1}^{n} C_i(Q_i, y_{ij})x_i + \sum_{i=1}^{n} L_i(l_i)x_i + \sum_{i=1}^{n} x_i D_i(s_i) \leq M \\[6pt] E\left[\sum_{i=1}^{n} Q_i x_i (1 - \lambda_i)(\tilde{v}^{\alpha}_{in0} - t_j y_{ij})\right] < W_{\text{COD}} \\[6pt] E\left[\sum_{i=1}^{n} Q_i x_i (1 - \lambda_i)(\tilde{v}^{\beta}_{in0} - t_j y_{ij})\right] < W_{\text{NH}_3-\text{N}} \\[6pt] E\left[\tilde{v}^{\alpha}_{in0} - t_j y_{ij}\right] < \eta^{\alpha}_0, \ i = 1, 2, \ldots, n \\[6pt] E\left[\tilde{v}^{\beta}_{in0} - t_j y_{ij}\right] < \eta^{\beta}_0, \ i = 1, 2, \ldots, n \\[6pt] E\left[(\tilde{v}^{\alpha}_{in0} - v^{\alpha}_{outi})/\tilde{v}^{\alpha}_{in0}\right] > \theta^{\alpha}_0, \ i = 1, 2, \ldots, n \\[6pt] E\left[(\tilde{v}^{\beta}_{in0} - v^{\beta}_{outi})/\tilde{v}^{\beta}_{in0}\right] > \theta^{\beta}_0, \ i = 1, 2, \ldots, n \\[6pt] \sum_{i=1}^{n} x_i \leq N \\[6pt] \sum_{j=1}^{m} y_{ij} = 1, \ i = 1, 2, \ldots, n \\[6pt] x_i, \ y_{ij} = 0 \text{ or } 1 \end{cases} \qquad (18)$$

3. Application Research

The resident population of the urban area of X county, a Chengdu suburb in Sichuan Province, is 112,000. According to the county development plan, it will reach 150,000 by 2015. By reference to per capita household water consumption in Sichuan Province, which is 128.5 L/day, the total domestic water consumption of that area is approximately 1.9×10^4 m^3/day. By calculating on the basis of 90% of the predicted total domestic water consumption, household sewage discharge will be 1.7×10^4 m^3/day. In addition to sanitary sewage, due to a large number of textile mills and garment factories in the county, the industrial water consumption is 1.5×10^4 m^3/day. By calculating on the basis of 0.7 of the design coefficient, the predicted industrial wastewater discharge is approximately 1.1×10^4 m^3/day. Therefore, the recent overall sewage discharge of that county will be 2.8×10^4 m^3/day, 60.7% of which is sanitary sewage.

Because of rapid industrial development in the county, industrial wastewater discharge increases annually; therefore, a sewage treatment plant needs to be established so that the wastewater can be processed and reach the discharge standard. Three optional sites around the county are available, from which one or two sites can be selected. Since the county is located in a flood-prone area, the issue of flood control should not be ignored when building a sewage treatment plant. In addition, some other actual situations of the county should be taken into account, such as, the planned area, the overall layout of principal pipelines, the matched technical process, etc.

3.1. Schemes for Site Selection

Minjiang River runs through the county, which is a resort for holidays, exploration, and recuperation, with abundant tourism resources and national nature reserves. According to the overall sewage quantity of the county, principles of site selection and the specific situation of the county industrial development, the designed process capability of a sewage treatment plant is between 20,000 and 50,000 m^3/day. Considering both the construction scale and the surrounding areas, the planning area of the plant is approximately 10,000–50,000 m^3. Through a preliminary survey conducted by relevant personnel in the planning process and a solicitation of opinions from involved parties, three relatively satisfactory plans for the location were selected in accordance with the principles and requirements of constructing a sewage plant. The details are as follows:

Scheme I: A sewage plant can be built on a mountain slope with a gradient of 40°. Hence, the land requires leveling; thus, the capital cost of earthwork is evaluated as 190 thousand USD.

Scheme II: A sewage plant can be built on a river bank with a floodwall, which is used for resolving a flood issue. Due to the reinforced concrete structure of the floodwall, a total of 4800 m^3 of reinforced concrete is needed. Calculating the comprehensive cost of the reinforced concrete at 63.49 USD/m^3, the required investment is 304.76 thousand USD.

Scheme III: A sewage plant can be built on a river bank without a floodwall. In this situation, the processing equipment needs to be adjusted. It is worth noting that this scheme adopts an ICEAS process (an improved SBR technique). Adopting this process can completely avoid the flood control drawbacks of the aeration equipment. Before the flood, we need only to move the ordinary movable elements without worrying about the key equipment in the sewage plant.

All of the capital expenditures of a sewage treatment plant in the schemes above are illustrated in Table 1. In addition, the proposed reuse rate of the reclaimed water of all schemes is 30%, and the economic benefit of the reclaimed water per unit is 0.19 USD/m^3.

Table 1. Cost schedule of all schemes.

Cost Items	Scheme I	Scheme II	Scheme III
Occupied land (10,000 m^2)	2.4	4.6	3.5
Land expropriation cost (10,000 USD)	28.57	38.10	34.92
Demolition cost (10,000 USD)	14.29	23.81	20.63
Initial capital cost (10,000 USD)	50.79	47.62	57.14
Sewage pipeline construction cost (10,000 USD)	15.87	31.75	23.81
Sewage pump station construction cost (10,000 USD)	23.81	7.94	15.87
pipeline maintenance cost (10,000 USD)	0.63	0.95	0.63
Pump station operation cost (10,000 USD)	4.13	2.86	3.81
Other operation cost (10,000 USD)	2.54	1.59	1.90
Double circuit return pipe installation cost (10,000 USD)	19.05	15.87	14.29

3.2. Influent and Effluent

For the convenience of calculation, the paper assumes that the influent concentration of the sewage source is consistent in this area. Because domestic wastewater accounts for a large percent of the county sewage, and the industrial wastewater belongs to general industrial sewage, COD and NH_3-N are considered the major pollutants. The influent qualities of sewage water pouring into the treatment plant in this area are determined as:

$$COD: 252 \text{ mg/L} \quad NH_3\text{-}N: 35 \text{ mg/L}$$

According to the river system distribution and the sewage treatment plant, the treated water ultimately flows into Minjiang River and its tributaries. In accordance with GB18918-2002 "Pollutant

Discharge Standard for Municipal Sewage Treatment Plants" primary standard B, the effluent qualities of the treatment plant are required as follows:

$$COD: \leq 60 \text{ mg/L} \quad NH_3\text{-N}: \leq 15 \text{ mg/L}$$

The proposed influent and effluent quality and the pollutant removal rate of the treatment plant are shown in Table 2.

Table 2. The designed influent and effluent quality and pollutant removal rate of the sewage plant.

Item	pH	COD	NH$_3$-N
Influent quality (mg/L)	6–9	252	35
Effluent quality (mg/L)	6–9	60	15
Process rate (%)	—	76	57

COD: Chemical Oxygen Demand.

3.3. Processing Technology

Based on the designed processing scale of a sewage plant, the water characteristics, environmental function and the local actual situation and requirements, the sewage treatment process can be selected subject to comprehensive technical and economic constraints. According to the requirements of the influent and effluent quality in this county, as well as the actual situation of Chengdu area, this research considers the following three optional schemes, namely, the oxidation ditch, ICEAS and A/A/O method as the alternatives (see Figure 3).

Table 3 lists the structures of oxidation ditch, ICEAS, and A/A/O and shows that some structures are consistent, while some are different. Table 4 presents the required equipment of the different structures.

Table 3. Comparison between three types of sewage treatment process structures.

Treatment Process	Oxidation Ditch	ICEAS *	A/A/O
Same structures	Coarse screen wells and pumping station, fine screen and grit chamber, blower room, sludge tank, dewatering room, instruments and center control room		
Different structures	Oxidation ditch biological reaction tank, return sludge pump room	ICEAS reaction tank	A/A/O biological reaction tank, secondary sedimentation tank, return sludge pump room

* ICEAS: Intermittent Cycle Extended Aeration System.

Table 4. Comparison between three types of sewage treatment process equipment.

Treatment Process	Oxidation Ditch	ICEAS	A/A/O
Equipment	Surface aerator, rotating disc aerator, underwater agitator, submersible axial pump	Micro porous aeration device, plug-flow agitator, water decanter, ICEAS submersible sewage pump	Underwater agitator, underwater propeller, aerator, rotating door, submersible sewage pump, mud scraper, electric hoist, excess sludge pump

To summarize, the difference in the equipment costs of the same-size sewage plants primarily lies in the diverse pieces of equipment of a biological treatment unit structure. Therefore, we integrate the cost in terms of equipment when establishing the model.

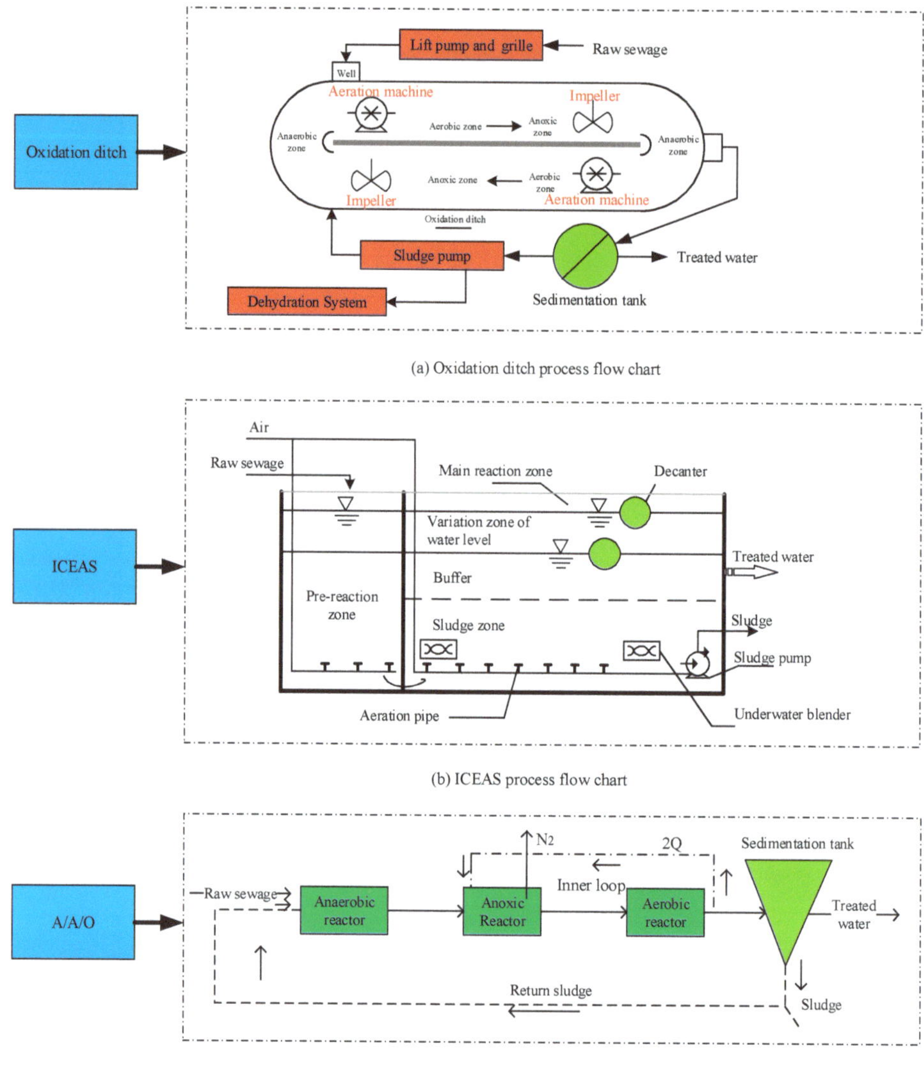

(a) Oxidation ditch process flow chart

(b) ICEAS process flow chart

(c) A/A/O process flow chart

Figure 3. Three types of sewage treatment process flow charts.

3.4. Total Costs and Solution Procedure

The total cost of a sewage treatment plant consists of an initial investment cost and an operation cost in this study. In addition, the specific forms of the initial investment cost and the operation cost in practical calculation are different.

The initial investment cost of a sewage treatment plant in this case comprises costs mentioned in Table 1 and the diverse process investment cost in terms of different techniques. For the convenience of calculation, initial investment cost models of three kinds of techniques were borrowed from other literature and were utilized in this case, as shown in Table 5.

Table 5. Initial investment cost of three types of sewage treatment processes (lv 2009 [37]).

Treatment Process	Oxidation Ditch	ICEAS	A/A/O
Initial Investment Cost	$C_1' = -0.0036Q_i^2 + 819.13Q_i + 10^6$;	$C_2' = -0.0009Q_i^2 + 692.12Q_i + 2 \times 10^6$;	$C_3' = -0.0079Q_i^2 + 1085.9Q_i + 2 \times 10^6$

In addition, the operation cost of a sewage plant refers to the expenditure for sewage treatment after the completion of the project. In the operation of a sewage plant, the operating cost mainly focuses on pipeline maintenance, energy consumption, equipment repair and so forth. In addition to the operation cost listed in Table 6, the sewage treatment costs from different processes are also included.

Table 6. Operation costs of three types of sewage treatment process (lv 2009 [37]).

Treatment Process	Oxidation Ditch	ICEAS	A/A/O
Operation cost	$F_1 = 1.8177Q_i^{1-0.0534}$	$F_2 = 1.4113Q_i^{1-0.0253}$	$F_3 = -9E - 12Q_i^3 - 1E - 06Q_i^2 + 1.1188Q_i$

In this paper, the proposed model is solved by a heuristic algorithm, the Genetic Algorithm (GA), because it is challenging to solve directly. Figure 4 and Table A1 illustrate the basic process of the applied genetic algorithm.

3.5. Result Analysis

The parameters for the GA in this problem are as follows: crossover rate 0.2, mutation rate is 0.4, population size is 25 and maximum generation is 500.

Assign weight to the four objective functions in turn:

$$\omega_1 = 0.7, \ \omega_2 = 0.1, \ \omega_3 = 0.1, \ \omega_4 = 0.1$$

Through calculation, the optimal result of the multi-objective model in this situation is:

$$x_1 = x_2 = 0, \ x_3 = 1, \ Q_3 = 3.7 \times 10^4, \ v_{ou3}^a = 35.63, \ v_{ou3}^b = 2.39$$

Accordingly, under the optimal solution, the minimum cost is 4.19×10^6 USD; the maximum economic benefit is 2.17×10^4 USD; and the removal rates of COD and ammonia nitrogen are approximately 85.1% and 93.2%, respectively. Because $x_1 = x_2 = 0$, which means that schemes I and II are not considered, any arbitrary value assigned to neither the design capacity nor the removal rate of COD and ammonia nitrogen in the model is meaningless.

The results above show that within the planned area of 10,000–50,000 m^2, to achieve such goals as reaching the designed capacity of 20,000–50,000 m^3/day and meeting the requirements of minimum investment cost, Scheme III should be selected. In Scheme I a method of building a highland sewage treatment plant is adopted, leading to larger investment in earth work and annual operation cost. Through analysis, the paper concludes that Scheme I is not economically feasible. In Scheme II, the sewage plant is surrounded by a flood wall, resulting in a series of problems, such as a large investment and landscape affects. Scheme III adopts provisions to install submersible sewage pumps and jet aeration devices, plan a control room and install the electric transformation and distribution equipment with localized elevation in the building of the sewage plant. Although the investment increases to a certain extent, the overall increase is not too large. Scheme III, therefore, is technically feasible and economically rational.

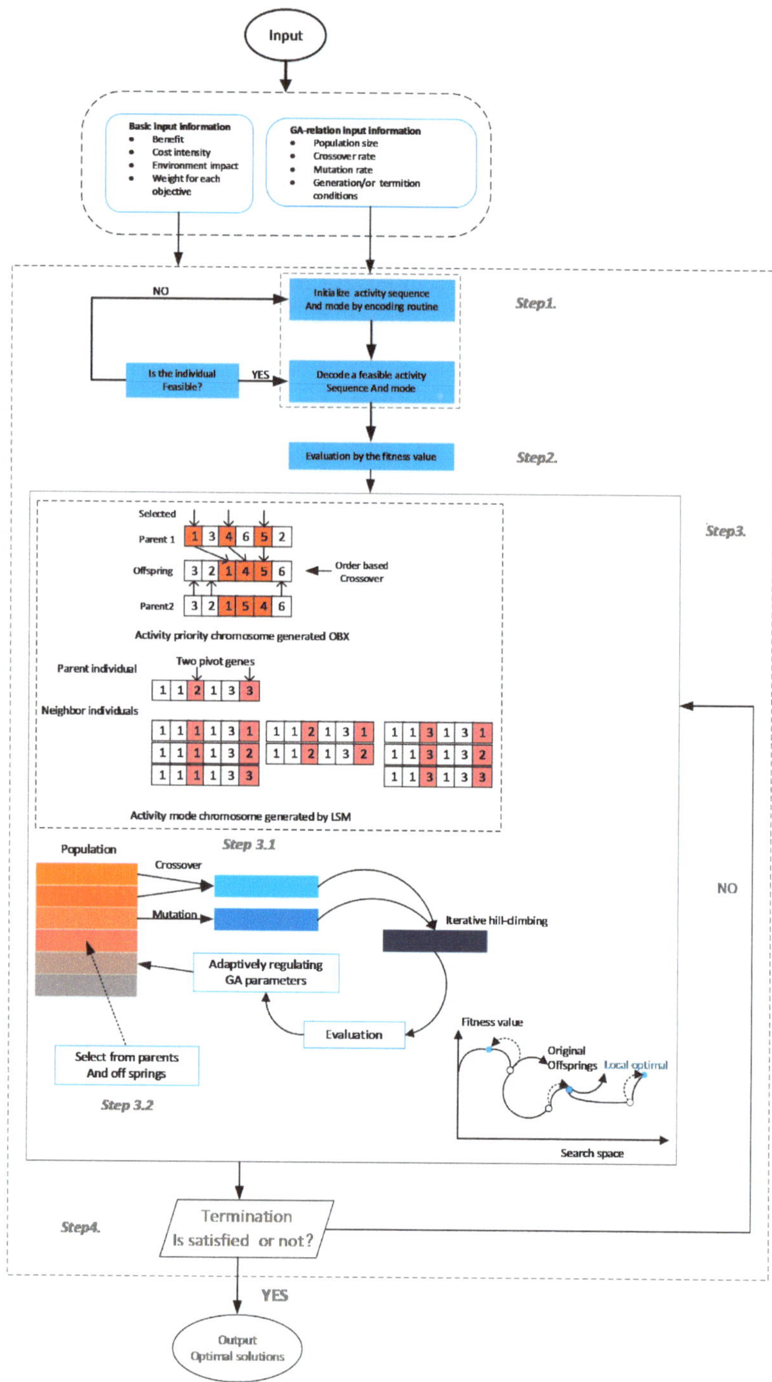

Figure 4. Basic Steps of the Genetic Algorithm.

In terms of techniques, although the improved SBR technique with an ICEAS process is only adopted in Scheme III, it can effectively dispose of pollutants in wastewater and reduce their current content, thereby meeting the requirements of the total discharge control of pollutants. This is attributable to the principal sewage in this area being composed of domestic wastewater, the biodegradability of which is relatively good; meanwhile, the ICEAS process is the secondary biochemical treatment, enabling the COD and ammonia nitrogen removal rate to reach a higher level at 85.1% and 93.2%, respectively. In contrast, even though the optional processes in schemes I and II can achieve the desired effect discussed in the preceding, they are restricted economically due to their higher construction and operation costs.

From the perspective of environmental protection, Scheme III is also based on realistic considerations. The COD and ammonia nitrogen concentrations of the treated sewage are calculated as 35.63 mg/L and 2.39 mg/L, respectively, completely meeting GB18918-2002 "Pollutant Discharge Standard for Municipal Sewage Treatment Plants" primary standard B. Additionally, this scheme locates the treatment plant on the river bank without a flood wall surrounding it. In accordance with the normal requirements, the structural foundation is laid without causing irreversible damage to the surrounding environment (including water, groundwater, cultivated land, forests, aquatic products, landscape, scenic spots, nature reserves, etc.). Moreover, neither is it located in the upper wind zone of urban or residential areas, nor upstream of an urban water resource; therefore, the treatment plant will not affect the residents' normal life.

According to the importance attached to each target by the decision makers, different weights were assigned to each objective function. Table 7 illustrates the value of each objective function for the conditions and the optimal decision results.

Table 7. Optimal decision results under different weight conditions (variable unit in the table is the same as that in this paper).

λ_1	λ_2	λ_3	λ_4	f_1	f_2	f_3	f_4	x_i	y_{ij}	Q	v_{outi}^{ff}	v_{outi}^{fi}
0.7	0.1	0.1	0.1	0.42	0.22	0.15	0.01	$x_3 = 1$	$y_{32} = 1$	3.7	35.63	2.39
0.4	0.4	0.1	0.1	0.46	0.29	0.14	0.01	$x_1 = 1$	$y_{11} = 1$	4.2	29.58	2.31
0.4	0.1	0.4	0.1	0.42	0.21	0.12	0.01	$x_3 = 1$	$y_{32} = 1$	3.9	27.39	2.44
0.3	0.1	0.2	0.4	0.43	0.22	0.14	0.01	$x_3 = 1$	$y_{32} = 1$	3.7	33.22	2.31

Construction cost is a primary factor considered when building a treatment plant; therefore, it is generally assigned a greater weight value. In accordance with the importance attached to reclaimed water benefit and pollutant discharge amount the weight is adjusted. It is manifested from Table 7 that the weight adjustment of the cost and the reclaimed water benefit will significantly affect the decision result, whereas the weight changes of pollutant treatment efficiency or pollutant emission will not substantially affect the decision result.

Based on the actual situation and the development plan of an area we can adjust the parameters, such as modifying the sewage quantity and the reuse rate of reclaimed water. Through such adjustments, we can set up a new situation, establish different multi-objective decision models for constructing a treatment plant and obtain various corresponding results by calculation. Since some parameters in this case were calculated based on those of other towns with the same levels, the relevant data might not precisely match the corresponding data; therefore, a situation could arise where the adjustment of a certain index will not significantly affect the decision result, in contradiction with the real situation. Therefore, decision makers should fully understand the specific characteristics of their practical application because the accuracy of the data determines the accuracy of the model. In the practical applications, however, researchers can adjust this model to a certain extent and use this flexibility with actual conditions that can better reflect the real conditions of a given problem domain.

4. Conclusions

Considering multiple factors affecting the construction of sewage treatment plants, this paper developed a general multi-objective, decision-based optimizing model for the sewage treatment plant construction plan. In the proposed model, COD and ammonia nitrogen were characterized by random variables because their inherent uncertainty precluded defining them as crisp values or members of only one set. In the case study, the results showed that within the planning area of 10,000–50,000 m^3, the ICEAS process that was adopted was able to reach the design capacity of 20,000–50,000 m^3/day and meet the requirements of minimum investment cost, etc. The main conclusions of this study were as follows:

1. This paper introduced the sewage treatment plant construction program's related issues, specifically describing the existing problems of the sewage treatment plant construction and combining an actual analysis of the sewage treatment plant construction factors. This process identified the key link (the sewage treatment process) in the construction of a sewage treatment plant.

2. Using the framework of uncertainty, random variables were used to characterize COD and ammonia nitrogen, which reduced information losses or distortions while relieving the decision makers' burden.

3. With the construction of a sewage treatment plant in a suburb of Chengdu as an example, this paper empirically tested the general multi-objective decision model for sewage treatment plant construction. The results verified the applicability and effectiveness of the proposed model and provided decision makers with technical support for the optimization of a sewage treatment plant construction plan.

Last but not least, the weights of each objective were adjusted to account for more information on the construction of the sewage treatment plant and treatment technology selection, and the results showed that changing the weights of the cost and the reclaimed water benefit will significantly affect the decision result. There were many non-modeling activities in designing a solution for the construction of sewage treatment plants, such as the decision-making problem identification and the analysis and assessment of the results, which required the decision-makers' experience and knowledge and highlights limitations in articulating this practical knowledge; therefore, additional uncertainty in this area can be expected. However, further research can take into account the various complex uncertainties in the decision-making process, and enable the construction of uncertain variables that can be implemented in an uncertainty-based, multi-objective decision-making model.

Acknowledgments: This research is supported by the National Natural Science Foundation of China (Grant No. 71771157, 71301109), Soft Science Program of Sichuan Province (Grant No. 2017ZR0154), Funding of Sichuan University (Grant No. skqx201726), China Postdoctoral Science Foundation Funded Project (Grant No. 2017M610609), Funding of Research Center for Systems Science & Enterprise Development (Xq17c06) and Sichuan Province Mineral Resources Research Center (SCKCZY2017-YB055).

Author Contributions: Xudong Chen: Research idea and design, grant holder of research financing, participation in related article writing. Zhongwen Xu: Literature review, participation in model construction, model solution, writing and formatting. Liming Yao: Grant holder of research financing, model construction, model solution, the results analysis, writing and formatting. Ning Ma: Literature review, article writing and formatting.

Conflicts of Interest: The authors declare no conflict of interest.

Nomenclature

$T_{j,j=1,2\ldots m}$	The sewage treatment process
$A_{i,i=1,2\ldots n}$	The sites of sewage treatment plant
$Q_{i,i=1,2\ldots n}$	The sewage treatment capacity
$v_{in0}^{\alpha} v_{in0}^{\beta}$	Influent concentrations (COD and ammonia nitrogen)
$v_{outi,i=1,2\ldots n}^{\alpha} v_{outi,i=1,2\ldots n}^{\beta}$	Effluent concentrations (COD and ammonia nitrogen)
$B_{i,i=1,2\ldots k}$	The main distributions of sewage source around the treatment plant
$q_{i,i=1,2\ldots k}$	The sewage volumes of each sewage source
$C_i(Q_i, T_j), i = 1, 2 \ldots n; j = 1, 2 \ldots m$	Investment costs of the plant construction
$L_i(l_i), i = 1, 2 \ldots n$	The cost of off-site pipe network
$l_{i,i=1,2\ldots n}$	The length of the pipeline
$D_i(s_i), i = 1, 2 \ldots n$	The costs of land expropriation and household demolition compensation
$s_{i,i=1,2\ldots n}$	The occupied area of the sewage treatment plant
$\lambda_{i,i=1,2\ldots n}$	The reuse rates of reclaimed water
$E_{i,i=1,2\ldots n}$	The economic benefits per unit reclaimed water
$F_i(T_j), i = 1, 2 \ldots n; j = 1, 2 \ldots m$	The management operation cost
$\delta_{i,i=1,2\ldots n}$	The score value of the optional site i for the decision makers

Appendix A

Table A1. The GA algorithm for the sewage treatment selection.

Procedure: The GA Algorithm for the Sewage Treatment Selection
Input: Initial data and GA parameters
Output: Best solutions and objective values
Step 1. Initialization. Encode the decision variables and randomly generate an initial population;
Step 2. Evaluation. Decode a feasible activity sequence and calculate the fitness function;
Step 3. Upgrade. Select the best set of solution from the current generation population and offspring populations;
Step 3.1 Offspring generation. Fulfill crossover and mutation * and produce offspring solutions;
Step 3.2 Evolution. Compare the value of newly obtained individual fitness function with those contained in the previous set. Replace the solutions with the new solutions if their value is higher;
Step 4 Stopping criterion. The algorithm ends when a maximal number of generations is reached, and the best solution, together with the corresponding objective functions, are given as output.

* The mutation operator maintains the diversity of the population and increases the possibility of not losing any potential solution while finding the global optimal solution. The crossover operator is a technique used for rapid exploration of the search space.

References

1. Oakley, S.M.; Gold, A.J.; Oczkowski, A.J. Nitrogen control through decentralized wastewater treatment: Process performance and alternative management strategies. *Ecol. Eng.* **2010**, *36*, 1520–1531. [CrossRef]

2. Xu, J.; Xu, Y.; Wang, H.M.; Guo, C.S.; Qiu, H.Y.; He, Y.; Zhang, Y.; Li, X.C.; Meng, W. Occurrence of antibiotics and antibiotic resistance genes in a sewage treatment plant and its effluent-receiving river. *Chemosphere* **2015**, *119*, 1379–1385. [CrossRef] [PubMed]

3. Belhaj, D.; Athmouni, K.; Jerbi, B.; Kallel, M.; Ayadi, H.; Zhou, L. Estrogenic compounds in Tunisian urban sewage treatment plant: Occurrence, removal and ecotoxicological impact of sewage discharge and sludge disposal. *Ecotoxicology* **2016**, *25*, 1–9. [CrossRef] [PubMed]

4. Pan, C.G.; Liu, Y.S.; Ying, G.G. Perfluoroalkyl substances (PFASs) in wastewater treatment plants and drinking water treatment plants: Removal efficiency and exposure risk. *Water Res.* **2016**, *106*, 562–570. [CrossRef] [PubMed]

5. Gros, M.; Blum, K.M.; Jernstedt, H.; Renman, G.; Rodríguez-Mozaz, S.; Haglund, P.; Andersson, P.L.; Wiberg, K.; Ahrens, L. Screening and prioritization of micropollutants in wastewaters from on-site sewage treatment facilities. *J. Hazard. Mater.* **2017**, *328*, 37–45. [CrossRef] [PubMed]

6. Smith, B.R. Re-thinking wastewater landscapes: Combining innovative strategies to address tomorrow's urban wastewater treatment challenges. *Water Sci. Technol.* **2009**, *60*, 1465–1473. [CrossRef] [PubMed]

7.	Libralato, G.; Ghirardini, A.; Avezzù, F. To centralise or to decentralise: An overview of the most recent trends in wastewater treatment management. *J. Environ. Manag.* **2012**, *94*, 61–68. [CrossRef] [PubMed]
8.	Murray, A.; Ray, I. Wastewater for agriculture: A reuse-oriented planning model and its application in Peri-Urban China. *Water Res.* **2010**, *44*, 1667–1679. [CrossRef] [PubMed]
9.	Cai, Q.Y.; Mo, C.H.; Wu, Q.T.; Zeng, Q.Y.; Katsoyiannis, A. Occurrence of organic contaminants in sewage sludges from eleven wastewater treatment plants, China. *Chemosphere* **2007**, *68*, 1751–1762. [CrossRef] [PubMed]
10.	Kelessidis, A.; Stasinakis, A.S. Comparative study of the methods used for treatment and final disposal of sewage sludge in European countries. *Waste Manag.* **2012**, *32*, 1186–1195. [CrossRef] [PubMed]
11.	Gleick, P.H.; Palaniappan, M.; Morikawa, M.; Morrison, J. The world's water 2008~2009: The biennial report on freshwater resources. *Environ. Conserv.* **2009**, *36*, 171–175.
12.	Shen, W.Q. Study on Ammonia nitrogen in municipal sewage adopted by actived coal-serial kaolin. *J. Anhui Agric. Sci.* **2010**, *38*, 10845–10847.
13.	Hernández-Sancho, F.; Sala-Garrido, R. Technical efficiency and cost analysis in wastewater treatment processes: A DEA approach. *Desalination* **2009**, *249*, 230–234. [CrossRef]
14.	Rizzoa, L.; Manaiab, C.; Merlinc, C.; Schwartzd, T.; Dagote, C.; Ployf, M.C.; Michaelg, I.; Fatta-Kassinosg, D. Urban wastewater treatment plants as hotspots for antibiotic resistant bacteria and genes spread into the environment: A review. *Sci. Total Environ.* **2013**, *447*, 345–360. [CrossRef] [PubMed]
15.	Ostoich, M.; Serena, F.; Tomiato, L. Environmental controls for wastewater treatment plants: Hierarchical planning, integrated approach, and functionality assessment. *J. Integr. Environ. Sci.* **2010**, *7*, 251–270. [CrossRef]
16.	Bhatnagar, A.; Sillanpää, M. Utilization of agro-industrial and municipal waste materials as potential adsorbents for water treatment. *Chem. Eng. J.* **2010**, *157*, 277–296. [CrossRef]
17.	Kartal, B.; Kuenen, J.G.; Loosdrecht, M.C.M.V. Sewage treatment with anammox. *Science* **2010**, *328*, 702–703. [CrossRef] [PubMed]
18.	Plakas, K.V.; Karabelas, A.J. Removal of pesticides from water by NF and RO Membranes—A review. *Desalination* **2012**, *287*, 255–265. [CrossRef]
19.	Xu, P.; Zeng, G.M.; Huang, D.L.; Feng, C.L.; Hu, S.; Zhao, M.H.; Lai, C.; Wei, Z.; Huang, C.; Xie, G.X.; et al. Use of iron oxide nanomaterials in wastewater treatment: A review. *Sci. Total Environ.* **2012**, *424*, 1–10. [CrossRef] [PubMed]
20.	Fu, F.L.; Dionysios, D.; Liu, H. The use of zero-valent iron for groundwater remediation and wastewater treatment: A review. *J. Hazard. Mater.* **2014**, *267*, 194–205. [CrossRef] [PubMed]
21.	Garnier, J.; Brion, N.; Callens, J.; Passy, P. Modeling historical changes in nutrient delivery and water quality of the Zenne River (1790s–2010): The role of land use, waterscape and urban wastewater management. *J. Mar. Syst.* **2013**, *128*, 62–76. [CrossRef]
22.	Molinos-Senante, M.; Garrido-Baserba, M.; Reif, R.; Hernández-Sancho, F.; Poch, M. Assessment of wastewater treatment plant design for small communities: Environmental and economic aspects. *Sci. Total Environ.* **2012**, *42*, 11–18. [CrossRef] [PubMed]
23.	Cornel, P.; Schaum, C. Phosphorus recovery from wastewater: Needs, technologies and costs. *Water Sci. Technol.* **2009**, *59*, 1069–1076. [CrossRef] [PubMed]
24.	Bresler, S.E. Policy recommendations for reducing reactive nitrogen from wastewater treatment in the great bay estuary, NH. *Environ. Sci. Policy* **2012**, *19*, 69–77. [CrossRef]
25.	Dodane, P.H.; Mbéguéré, M.; Sow, O.; Strande, L. Capital and operating costs of full-scale fecal sludge management and wastewater treatment systems in Dakar, Senegal. *Environ. Sci. Technol.* **2012**, *46*, 3705–3711. [CrossRef] [PubMed]
26.	Ostace, G.S.; Cristea, V.M.; Agachi, P.Ş. Cost reduction of the wastewater treatment plant operation by MPC based on modified ASM1 with two-step nitrification/denitrification model. *Comput. Chem. Eng.* **2011**, *35*, 2469–2479. [CrossRef]
27.	Molinos-Senante, M.; Hernández-Sancho, F.; Sala-Garrido, R. Economic feasibility study for wastewater treatment: A cost-benefit analysis. *Sci. Total Environ.* **2010**, *408*, 4396–4402. [CrossRef] [PubMed]
28.	Hernandez-Sancho, F.; Molinos-Senante, M.; Sala-Garrido, R. Cost modelling for wastewater treatment processes. *Desalination* **2011**, *268*, 1–5. [CrossRef]
29.	Mojahed, S.; Aghazadeh, F. Major factors influencing productivity of water and wastewater treatment plant construction: Evidence from the deep south USA. *Int. J. Proj. Manag.* **2008**, *26*, 195–202. [CrossRef]

30. Renou, S.; Thomas, J.S.; Aoustin, E.; Pons, M.N. Influence of impact assessment methods in wastewater treatment LCA. *J. Clean. Prod.* **2008**, *16*, 1098–1105. [CrossRef]

31. Hong, J.L.; Hong, J.M.; Otaki, M.; Jolliet, O. Environmental and economic life cycle assessment for sewage sludge treatment processes in Japan. *Waste Manag.* **2009**, *29*, 696–703. [CrossRef] [PubMed]

32. Robinson, K.G.; Robinso, C.H.; Raup, L.A.; Markum, T.R. Public attitudes and risk perception toward land application of biosolids within the south-eastern United States. *J. Environ. Manag.* **2012**, *98*, 29–36. [CrossRef] [PubMed]

33. Jung, W.; Koo, B.; Han, S.H. A multi-objective linear programming framework for evaluating the financial viability of supplementary facilities in build-transfer-lease projects in Korea. *KSCE J. Civ. Eng.* **2012**, *16*, 29–37. [CrossRef]

34. Di, Z. *Using GIS-Based Multi-Criteria Analysis for Optimal Site Selection for a Sewage Treatment Plant*; University Library in Gävle: Gävle, Sweden, 2015.

35. Baxter, G.; Sommerville, I. Socio-technical systems: From design methods to systems engineering. *Interact. Comput.* **2011**, *23*, 4–17. [CrossRef]

36. Anthony, M.; Mathieu, S.; Corinne, C. Comparison of sludge characteristics and performance of a submerged membrane bioreactor and an activated sludge process at high solids retention time. *Water Res.* **2006**, *40*, 2405–2415.

37. Lv, Y. *The Typical City Secondary Sewage Plant Cost Models and the Optimization Design*; Kunming University of Science: Kunming, China, 2009.

International Journal of
*Environmental Research
and Public Health*

Article

The Performance of a Self-Flocculating Microalga *Chlorococcum* sp. GD in Wastewater with Different Ammonia Concentrations

Junping Lv [1], Xuechun Wang [1], Wei Liu [1], Jia Feng [1], Qi Liu [1], Fangru Nan [1], Xiaoyan Jiao [2] and Shulian Xie [1,*]

[1]   School of Life Science, Shanxi University, Taiyuan 030006, China; lvjunping024@sxu.edu.cn (J.L.);
      201623101009@email.sxu.edu.cn (X.W.); 201523103002@email.sxu.edu.cn (W.L.); fengj@sxu.edu.cn (J.F.);
      liuqi@sxu.edu.cn (Q.L.); Nangfr@sxu.edu.cn (F.N.)
[2]   Institute of Agricultural Environment and Resource, Shanxi Academy of Agricultural Sciences,
      Taiyuan 030031, China; jiaoxiaoyan@sxagri.ac.cn
*   Correspondence: xiesl@sxu.edu.cn; Tel.: +86-351-7018121

Received: 22 January 2018; Accepted: 27 February 2018; Published: 2 March 2018

Abstract: The performance of a self-flocculating microalga *Chlorococcum* sp. GD on the flocculation, growth, and lipid accumulation in wastewater with different ammonia nitrogen concentrations was investigated. It was revealed that relative high ammonia nitrogen concentration (20–50 mg·L^{-1}) was beneficial to the flocculation of *Chlorococcum* sp. GD, and the highest flocculating efficiency was up to 84.4%. It was also found that the highest flocculating efficiency occurred in the middle of the culture (4–5 days) regardless of initial ammonia concentration in wastewater. It was speculated that high flocculating efficiency was likely related to the production of extracellular proteins. 20 mg·L^{-1} of ammonia was found to be a preferred concentration for both biomass production and lipid accumulation. 92.8% COD, 98.8% ammonia, and 69.4% phosphorus were removed when *Chlorococcum* sp. GD was cultivated in wastewater with 20 mg·L^{-1} ammonia. The novelty and significance of the investigation was the integration of flocculation, biomass production, wastewater treatment, and lipid accumulation, simultaneously, which made *Chlorococcum* sp. GD a potential candidate for wastewater treatment and biodiesel production if harvested in wastewater with suitable ammonia nitrogen concentration.

Keywords: ammonia nitrogen; biomass production; lipid accumulation; wastewater treatment; self-flocculation

1. Introduction

The integration of microalgae-based wastewater treatment and lipid production has major advantages for both industries [1,2]. Nevertheless, there are some challenges on microalgal harvesting from wastewater due to small cell size, low cell density, and homogeneous distribution of cell in culture systems [3]. The contribution of the harvesting cost to the total cost is reportedly in the range of 20–30% or beyond [4]. Therefore, improving the efficiency of microalgal harvesting is very important for the process of microalgae-based wastewater treatment and lipid production.

Currently, microalgal harvesting mainly involves physical-, chemical-, and biological-based methods. For example, efficiencies higher than 90% are reached by applying ultrasound waves to harvest microalgae [5]. Xu et al. [6] reports that chitosan is an effective flocculant for concentrating *Chlorella sorokiniana* with more than 99% efficiency. Bioflocculant production from *Solibacillus silvestris* W01 shows 90% flocculating efficiency on *Nannochloropsis oceanica* [7]. Some other flocculants, such as aluminum nitrate sulfate, cationic starch and poly γ-glutamic acid also exhibit excellent flocculating efficiency (>79%) on microalgae [8]. However, high energy consumption, production cost, and potential biological toxicity become a bottleneck for the further application of above harvesting technologies.

Self-flocculating microalgae are a kind of microalgae that can aggregate together by themselves. The aggregate can be easily harvested by sedimentation. Currently, some self-flocculating microalgae, such as *Ankistrodesmus falcatus*, *C. vulgaris* JSC-7, *Ettlia texensis*, *Scenedesmus obliquus* AS-6-1, *Scenedesmus* sp. BH, and *Tetraselmis suecica*, have been screened [9–15]. It has been reported that the flocculating efficiency of *Scenedesmus* sp. BH is up to 92.3% [14]. Because the self-flocculating process of microalgae is a spontaneous behavior, it does not incur extra cost for microalgal harvesting compared to the high cost of separation and purification of flocculants from microorganisms for biological flocculation [16]. Compared to physical and chemical flocculation [17], there is no biological toxicity and almost no energy consumption in the self-flocculating process. The flocculating efficiency of self-flocculating microalgae is also comparable to that of traditional flocculation technologies. Therefore, flocculation technology based on cultivation of self-flocculating microalgae exhibits some potential advantages of high efficiency, environment-friendliness, and low cost; it represents the future research direction of microalgal harvesting.

For microalgal cultivation in wastewaters, ammonia is one important form of nitrogen and can influence microalgal growth and nutrient uptake [18,19]. The response of flocculating capability to ammonia concentration is worthy of attention for the cultivation of self-flocculating microalgae. It is hoped that self-flocculating microalgae can maintain excellent flocculating ability to facilitate the efficient recovery of cells from wastewater, in addition to keeping the high growth potential and nutrient uptake efficiency.

At present, the flocculating ability of self-flocculating microalgae under different growth phasees, temperatures, and pHs has been investigated. It has been demonstrated that the suitable growth phase, temperature, and pH are beneficial to flocculation of self-flocculating microalgae [12,20]. Nevertheless, the flocculating property of self-flocculating microalgae at different ammonia concentrations is not clear, and there are no papers published so far. The effect of ammonia nitrogen on the growth of self-flocculating microalgae is also unknown.

Chlorococcum sp. GD belonging to green microalgae has been reported to exhibit excellent self-flocculating capability in secondary effluent [21], which was comparable to that of *C. vulgaris* JSC-7, *E. texensis*, and *S. obliquus* AS-6-1 [11–13]. In the present study, the performance of *Chlorococcum* sp. GD on the flocculation with different ammonia nitrogen concentrations is investigated. The response of growth to varied ammonia concentrations is also analyzed. The novelty and significance of the study is to evaluate the possibility of the integration of sewage treatment, biomass production, and microalgal harvesting of *Chlorococcum* sp. GD, simultaneously, by altering ammonia nitrogen concentration in wastewater.

2. Materials and Methods

2.1. Microalgal Strain

A strain of *Chlorococcum* sp. GD was used in the study. It was isolated from the moss *Entodon obtusatus* of Shanxi Province, China [22] and exhibited excellent self-flocculating property as described by Lv et al. [21].

2.2. Synthetic Wastewater

In the present study, synthetic wastewater was formulated to simulate municipal wastewater. The exact concentration of COD and phosphorus of the synthetic wastewater was derived from the average concentration of municipal wastewater in Shanxi Province, China. The composition of macro element of the synthetic wastewater was provided by Aslan and Kapdan [23] with some modifications. The composition and concentration of trace elements in the synthetic wastewater was consistent to those in Blue-Green Medium (BG11). The exact compositions of the synthetic wastewater were as follows (mg/L): glucose 350, NH_4Cl 38.21–191.04, KH_2PO_4 21.97, $NaHCO_3$ 100, NaCl 64, $MgSO_4$ 90, $FeSO_4$ 5, $CaCl_2$ 25, H_3BO_3 2.86, $MnCl_2 \cdot 4H_2O$ 1.86, $ZnSO_4 \cdot 7H_2O$ 0.22, $Na_2MoO_4 \cdot 2H_2O$ 0.39, $CuSO_4 \cdot 5H_2O$ 0.08, and $Co(NO_3)_2 \cdot 6H_2O$ 0.05. The COD of the synthetic wastewater was around 325 mg·L^{-1} and total phosphorus concentration was 5 mg·L^{-1}. The ammonia concentration varied from 10 to 50 mg·L^{-1}. In the study, synthetic wastewater was sterilized (121 °C for 20 min) before experiments.

2.3. Experimental Set-Up

In the study, *Chlorococcum* sp. GD was firstly cultivated in BG11 for 13 days. After that, microalgal suspension was centrifuged at 5000 rpm for 5 min. The pellet was then washed with deionized water and centrifuged at 5000 rpm for 5 min again. The washed pellet was distributed in synthetic wastewater and inoculated into 500 mL transparent conical flasks with an initial concentration of 20 mg·L^{-1}. In order to investigate the effect of different ammonia nitrogen concentrations, four samples with initial ammonia nitrogen of 10, 20, 30, and 50 mg·L^{-1} were prepared. Cultures were grown at a constant temperature of 25 °C in a shaker with a shaking rate of 160 revolutions per minute (rpm). The cultures were illuminated with fluorescent lamps, providing an incident light intensity of 3000 lux under 14 h:10 h light:dark cycle. After daily sampling, pH of cultures was adjusted to 7.5 by supplementing HCl or NaOH solution. All these experiments were performed in triplicate.

2.4. Analytical Methods

2.4.1. Flocculating Ability Test

During the process of microalgal cultivation, 25 mL culture was harvested and distributed in 25 mL cylindrical glass tubes. The culture was gently mixed for 1 min at room temperature. After that, an aliquot of the culture was withdrawn at a height of two-thirds from the bottle when the culture had settled for 3 h. The optical density of the aliquots was then measured at 680 nm. The flocculating ability was calculated using the following Equation (1):

$$\text{Flocculating ability} = \frac{A - B}{A} \times 100\% \tag{1}$$

in which A and B are the optical density (OD_{680}) of the aliquot before and after flocculation.

2.4.2. Extraction and Analysis of Extracellular Polymeric Substances (EPS)

EPS mainly consisted of proteins and polysaccharides. The EPS were extracted according to procedures of Yang and Li [24] with some modifications. Microalgal suspension was dewatered by centrifugation at 5000 rpm for 5 min. The pellet was washed with deionized water and centrifuged at 5000 rpm for 5 min again. The washed pellet was diluted with deionized water and was heated to 80 °C for 30 min. The mixture was then centrifuged at 10,000 rpm for 10 min. After that, the supernatant was filtered with 0.45 µm acetate cellulose membranes, and the filtrate was regarded as EPS fraction. Protein concentration was analyzed by the coomassic brilliant blue method [25] using albumin from bovine serum (BSA) as the standard. Polysaccharides concentration was analyzed by the Anthrone method [26] with glucose as the standard.

2.4.3. Determination of Microalgal Growth

Microalgal biomass concentration was measured according to the method of suspended solid (SS) measurement [27]. The specific growth rate, mean microalgal biomass production, and double time during the cultivation period were calculated by Equations (2)–(4).

$$\text{Specific growth rate } (\mu, d^{-1}) = \ln(DW_{t2} - DW_{t1})/(t_2 - t_1) \tag{2}$$

$$\text{Mean biomass productivity } (mgDW \cdot L^{-1} \cdot d^{-1}) = (DW_{t2} - DW_{t1})/t \tag{3}$$

$$\text{Double time (day)} = \ln 2/\mu \tag{4}$$

In Equations (2) and (3), DW_{t1} and DW_{t2} represent the microalgal dry weight on day t_1 and t_2, $mg \cdot L^{-1}$; t_1 and t_2 are the cultivation time, d.

2.4.4. Determination of Water Quality

For the measurement of water quality, the microalgal suspension was centrifuged at 6000 rpm for 5 min, and the supernatant was filtered through 0.22 μm membranes. The filtered supernatant was then used for the determination of chemical oxygen demand (COD), ammonia nitrogen, and total phosphorous. COD, ammonia, and total phosphorus were analyzed by dichromate method, Nessler's reagent spectrophotometry, and ammonium molybdate spectrophotometric method, respectively [27].

Michaelis–Menten kinetic can be used to evaluate the ammonia nitrogen removal capability of microalgae [23].

$$R = R_{max} \times S / (K_m + S) \tag{5}$$

in which R is the ammonia nitrogen removal rate, R_{max} is the maximal ammonia nitrogen removal rate, S is the ammonia nitrogen concentration, and K_m is the ammonia nitrogen concentration at which ammonia removal rate reaches half-maximum.

Equation (5) can be linearized in double reciprocal form as in Equation (6) to determine the ammonia nitrogen removal kinetic coefficients K_m and V_{max} [23].

$$1/R_{X_i} = 1/V_{max} + K_m / (V_{max} \times S_0) \tag{6}$$

in which R_{X_i} ($mg\ N \cdot g^{-1}\ DW \cdot d^{-1}$) is the specific ammonia nitrogen removal rate per unit microalgal biomass; S_0 ($mg \cdot L^{-1}$) is the initial ammonia nitrogen concentration; and V_{max} ($mg\ N \cdot g^{-1}\ DW \cdot d^{-1}$) is the maximal ammonia nitrogen removal rate.

2.4.5. Determination of Microalgal Lipid

During the cultivation of microalgae, a suspension of microalgae (240 μL) was stained with 0.5 $mg \cdot mL^{-1}$ Nile red (9-diethylamino-5H-benzo(α)phenoxazine-5-one, dissolved in DMSO, 1 μL), and the mixture was incubated for 10 min at 37 °C. The fluorescence of the mixture was then determined using a microplate reader Tecan Infinite 200 Pro (Tecan, Switzerland) with a 96-well plate. The fluorescence of microalgae alone was measured. The fluorescence of Nile red alone was also measured. The fluorescence intensity of microalgal lipid was obtained when the autofluorescence of microalgae and Nile red was subtracted. The excitation and emission wavelengths were 543 and 598 nm, respectively. Each experiment was performed in triplicate.

The specific Nile Red fluorescence intensity was calculated as following:

$$sFI = FI \times DW^{-1} \tag{7}$$

in which sFI is the specific fluorescence intensity ($a.u. \cdot mg^{-1}$ DW biomass), FI is the total Nile Red fluorescence intensity of 240 μL microalgal suspension (a.u.), and DW is biomass dry weight in 240 μL microalgal suspension.

At the end of the culture, the total lipid was extracted with a chloroform/methanol solution (1/1, v/v) and was quantified gravimetrically [28].

2.4.6. Statistical Analysis

The measured value was expressed as the mean ± standard deviation. If necessary, data were analyzed by analysis of variance (ANOVA) conducting by Statistical Product and Service Solutions (SPSS) software (version 19.0, IBM Corporation, Armonk, NY, USA). There was a statistically significant difference when $p < 0.05$. Pearson correlation analysis was also conducted by SPSS software (version 19.0), and there was a significant relationship when $p < 0.05$.

3. Results

3.1. The Flocculating Property of Chlorococcum sp. GD in Wastewater with Different Ammonia Nitrogen Concentrations

In Figure 1, the flocculating ability of *Chlorococcum* sp. GD cultivated with different ammonia nitrogen concentrations is presented. Throughout the culture period, the flocculating ability of *Chlorococcum* sp. GD ranged from 54.3% to 84.4%. More specifically, the flocculating ability of *Chlorococcum* sp. GD cultivated with 10 mg·L^{-1} ammonia nitrogen decreased sharply from 74.6% to 54.3% after 3 days of culture. It then increased to 79% on the 4th day. Afterwards, flocculation efficiency gradually decreased to 62.7%. For *Chlorococcum* sp. GD cultivated with 20 mg·L^{-1} ammonia nitrogen, the highest flocculating ability was 79.4% on the 4th day. Then, the flocculating ability slightly decreased and maintained at around 72% until the end of cultivation. For *Chlorococcum* sp. GD cultivated with 30 mg·L^{-1} ammonia nitrogen, the flocculating ability remained at about 70%, except for a significant decrease to 60.6% on the 2nd day. The flocculating ability of *Chlorococcum* sp. GD cultivated with 50 mg·L^{-1} ammonia nitrogen was highest and maintained at about 80% at the end of the culture. Generally speaking, the flocculating ability of *Chlorococcum* sp. GD was affected by ammonia nitrogen concentration in wastewater. *Chlorococcum* sp. GD had an excellent and stable flocculating performance when cultivated in wastewater with 50 mg·L^{-1} ammonia nitrogen. *Chlorococcum* sp. GD cultivated with 10 mg·L^{-1} ammonia nitrogen exhibited the worst flocculation ability. It was also found that the flocculating ability of *Chlorococcum* sp. GD was dynamic when cultivated in wastewater with different ammonia nitrogen concentration. The highest flocculating efficiency occurred in the middle of the culture (4–5 days) regardless of initial ammonia nitrogen concentration in wastewater.

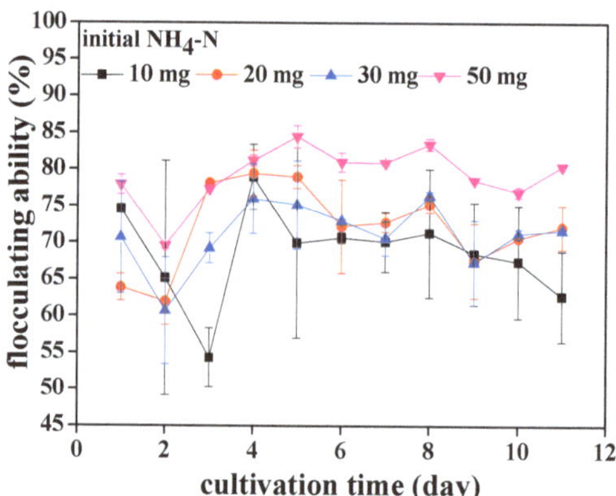

Figure 1. The flocculating ability of *Chlorococcum* sp. GD for 3 h settling under different ammonia nitrogen concentrations.

In the present study, the concentration of proteins and polysaccharides from EPS attached to the cell surface of *Chlorococcum* sp. GD cultivated in different ammonia nitrogen concentrations is measured (Figure 2a,b). The proteins concentration significantly increased from 31.4 mg·g·DW^{-1} to 83.9, 67.3, 81.5, and 79 mg·g·DW^{-1}, respectively, after 4 days of cultivation of *Chlorococcum* sp. GD in wastewater with 10, 20, 30, and 50 mg·L^{-1} ammonia nitrogen. The proteins concentration then decreased to 28.6, 38.8, 43.9, and 35.1 mg·g·DW^{-1}, respectively, at the end of cultivation. The polysaccharides concentration of *Chlorococcum* sp. GD in wastewater with 10 and 20 mg·L^{-1} ammonia nitrogen slightly increased from 21.7 mg·g·DW^{-1} to 26 and 24 mg·g·DW^{-1}, respectively, after 4 days of cultivation. It then decreased significantly to 6.0 and 7.8 mg·g·DW^{-1}, respectively, at the end of cultivation. The polysaccharides concentration of *Chlorococcum* sp. GD in wastewater with 30 and

50 mg·L^{-1} ammonia nitrogen significantly decreased from 21.7 mg·g·DW^{-1} to 5.4 and 5.9 mg·g·DW^{-1}, respectively, throughout the culture period. Based on data above, it indicated that both proteins and polysaccharides content were dynamic regardless of ammonia nitrogen concentrations in wastewater throughout the culture period. It was obvious that the proteins content from EPS was higher than the polysaccharides content at each growth stage of *Chlorococcum* sp. GD, and the peak of extracellular protein content of *Chlorococcum* sp. GD appeared in day 4 for all treatments. It was also found that *Chlorococcum* sp. GD in wastewater with 10 mg·L^{-1} ammonia nitrogen had the lowest concentration of proteins at the end of the culture.

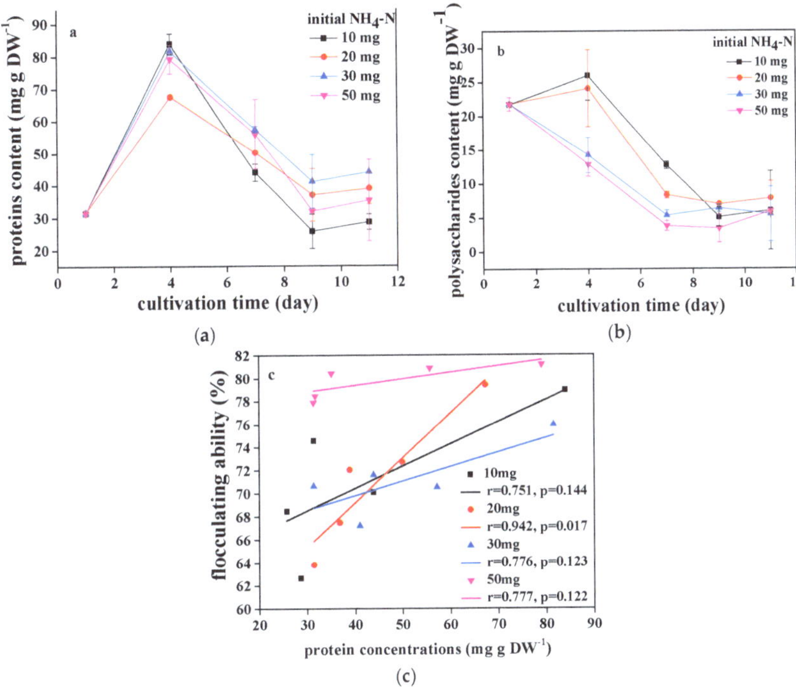

Figure 2. The extracellular proteins (**a**), polysaccharides concentration (**b**) and correlation analysis between flocculating ability and extracellular proteins (**c**) of *Chlorococcum* sp. GD under different ammonia nitrogen concentrations.

As shown in Figure 2c, the relationship between extracellular protein content and flocculation efficiency in each treatment is analyzed. When the flocculation efficiency was compared with the proteins content, it was seen that there was a definite trend. In practice, there was a good degree of correlation ($r = 0.9$, $p < 0.05$) when *Chlorococcum* sp. GD was cultivated in wastewater with 20 mg·L^{-1} ammonia nitrogen. That is to say, flocculation might be positively related to extracellular protein content in EPS.

3.2. The Growth and Biomass Production of Chlorococcum sp. GD in Wastewater with Different Ammonia Nitrogen Concentrations

The growth potential of *Chlorococcum* sp. GD with different ammonia nitrogen concentrations is shown in Figure 3. There was no lag phase of growth curves, which indicated that *Chlorococcum* sp. GD could well adapt in synthetic wastewater with different ammonia nitrogen concentrations. After 11 days of cultivation, the biomass production increased and ranged from 138 to 190.1 mg·L^{-1}.

The biomass production was highest when ammonia nitrogen concentration in synthetic municipal wastewater was 20 mg·L^{-1}. The specific growth rate of *Chlorococcum* sp. GD cultivated in wastewater with different ammonia nitrogen concentrations is also calculated as shown in Table 1. The specific growth rate was significantly affected by ammonia nitrogen concentration ($p < 0.05$). *Chlorococcum* sp. GD cultivated in 20 mg·L^{-1} ammonia nitrogen had higher specific growth rate (0.24 d^{-1}) than other ammonia nitrogen concentrations (0.19–0.21 d^{-1}) (Table 1). The mean biomass productivity was also the highest when *Chlorococcum* sp. GD was cultivated in 20 mg·L^{-1} ammonia nitrogen. Similarly, when *Chlorococcum* sp. GD grew in wastewater with 20 mg·L^{-1} ammonia nitrogen, the double time was the shortest (Table 1). All results above demonstrated that the growth and biomass production of *Chlorococcum* sp. GD were significantly affected by ammonia nitrogen, and the most suitable ammonia nitrogen concentration for the growth of *Chlorococcum* sp. GD was 20 mg·L^{-1}.

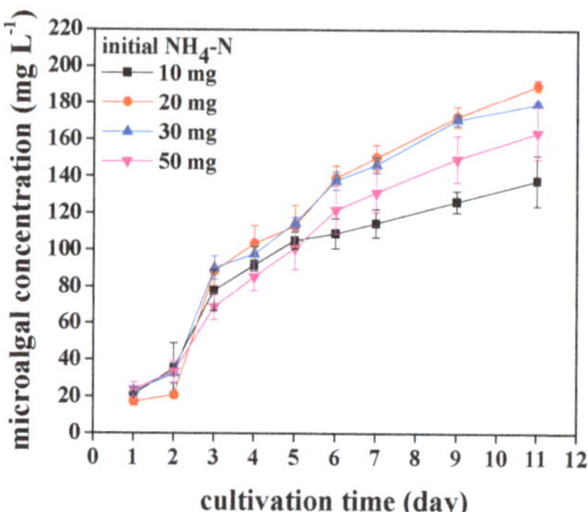

Figure 3. Biomass production of *Chlorococcum* sp. GD under different ammonia nitrogen concentrations.

Table 1. Growth parameters of *Chlorococcum* sp. GD under different ammonia nitrogen concentrations.

Ammonia Concentration (mg·L^{-1})	N/P Ratio	Specific Growth Rate (μ, d^{-1})	Mean Biomass Productivity (mg DW·L^{-1}·d^{-1})	Double Time (d^{-1})
10	2	0.19 ± 0.02 [a]	10.64 ± 1.35 [a]	2.37 ± 0.08 [a]
20	4	0.24 ± 0.01 [b]	15.70 ± 0.48 [b]	2.12 ± 0.06 [b]
30	6	0.21 ± 0.01 [a]	14.26 ± 0.06 [b,c]	2.27 ± 0.03 [a,b]
50	10	0.19 ± 0.03 [a]	12.74 ± 1.61 [c]	2.34 ± 0.1 [a]

Different letters (a, b and c) followed by values on columns indicated that they were significantly different at a probability level of 0.05 according to ANOVA test. DW: dry weight.

3.3. Pollutants Removal of Chlorococcum sp. GD in Wastewater with Different Ammonia Nitrogen Concentrations

As shown in Figure 4a, the organic pollutant concentration decreased at the end of the culture in every ammonia nitrogen concentration group. The COD removal efficiencies and rates varied from 87.6% to 92.8% and from 25.3 to 27.7 mg COD·L^{-1}·d^{-1}, respectively. *Chlorococcum* sp. GD showed excellent COD removal performance, and COD removal did not seem to be affected by ammonia nitrogen concentration. Conversely, the ammonia nitrogen removal efficiency was related to the ammonia nitrogen concentration. The ammonia nitrogen removal efficiency was around 98% when the ammonia nitrogen concentration of wastewater was 10 and 20 mg·L^{-1}. The removal efficiency decreased to 43.9–78.0% when the initial ammonia nitrogen concentration was more than 20 mg·L^{-1} (Figure 4b). Although the phosphorus removal efficiency was less than 70% in every ammonia nitrogen concentration group (Figure 4c), the phosphorus removal efficiency under 20–50 mg·L^{-1} ammonia nitrogen was superior to that under 10 mg·L^{-1} ammonia nitrogen. According to the above results, pollutants removal performance of *Chlorococcum* sp. GD was significantly affected by ammonia nitrogen concentration, and pollutants removal efficiency was greatest when the ammonia nitrogen concentration was 20 mg·L^{-1}. It was also found that most of pollutants were removed from wastewater after 4–5 days of cultivation of *Chlorococcum* sp. GD in each treatment.

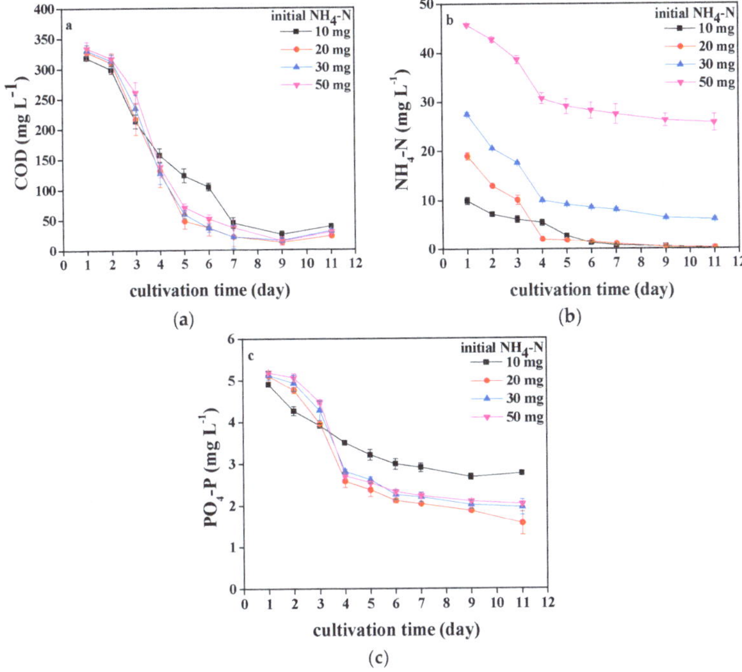

Figure 4. COD (**a**), ammonia nitrogen (**b**), and phosphorous (**c**) removal of *Chlorococcum* sp. GD under different ammonia nitrogen concentrations.

The data on ammonia nitrogen removal of *Chlorococcum* sp. GD given in Figure 4b is plotted in form of $1/R_{X_i}$ versus $1/(NH_4\text{-}N)_0$ as shown in Figure 5. From the slope and intercept of best fit line of this plot, kinetic coefficients of ammonia nitrogen removal by *Chlorococcum* sp. GD were V_{max} = 15.2 mg NH$_4$-N·g^{-1} DW·d^{-1} and K_m = 13.4 mg NH$_4$-N·L^{-1}.

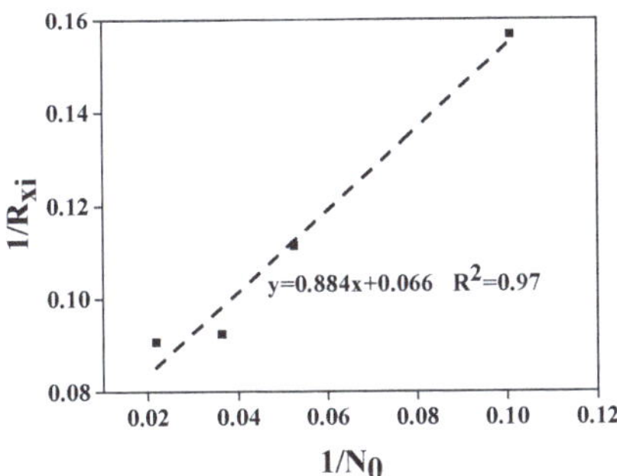

Figure 5. Determination of kinetic coefficients V_{max} and K_m for NH$_4$-N removal of *Chlorococcum* sp. GD.

3.4. Lipid Accumulation of Chlorococcum sp. GD in Wastewater with Different Ammonia Nitrogen Concentrations

The lipid accumulation of *Chlorococcum* sp. GD cultivated with different ammonia nitrogen concentrations is shown in Figure 6a. The lipid accumulation was affected by the initial ammonia nitrogen concentration. When *Chlorococcum* sp. GD was cultivated with 10 and 20 mg·L^{-1} ammonia nitrogen, the lipid accumulation firstly decreased and then increased. The difference was that the turning point for *Chlorococcum* sp. GD cultivated with 10 mg·L^{-1} ammonia nitrogen was in the 7th day, whereas the turning point for *Chlorococcum* sp. GD cultivated with 20 mg·L^{-1} ammonia nitrogen was in the 4th day. When *Chlorococcum* sp. GD was cultivated with 30 and 50 mg·L^{-1} ammonia nitrogen, the lipid accumulation decreased during the whole culture period. In conclusion, after 11 days of culture, the highest lipid accumulation of *Chlorococcum* sp. GD occurred when cultivated with 10 mg·L^{-1} ammonia nitrogen, followed by 20 and 50 mg·L^{-1} ammonia nitrogen. *Chlorococcum* sp. GD had lowest lipid accumulation when cultivated with 30 mg·L^{-1} ammonia nitrogen. The lipid content of *Chlorococcum* sp. GD cultivated with different ammonia nitrogen concentrations is shown in Figure 6b, according to the gravimetric method. After 11 days of culture with wastewater of 10 mg·L^{-1} ammonia nitrogen, *Chlorococcum* sp. GD had the highest lipid content at 47.5%, which was significantly higher than that of *Chlorococcum* sp. GD cultivated with wastewater of 20, 30, and 50 mg·L^{-1} ammonia nitrogen ($p < 0.05$). The lipid content of *Chlorococcum* sp. GD cultivated with 20, 30, and 50 mg·L^{-1} ammonia nitrogen was of the same order of magnitude.

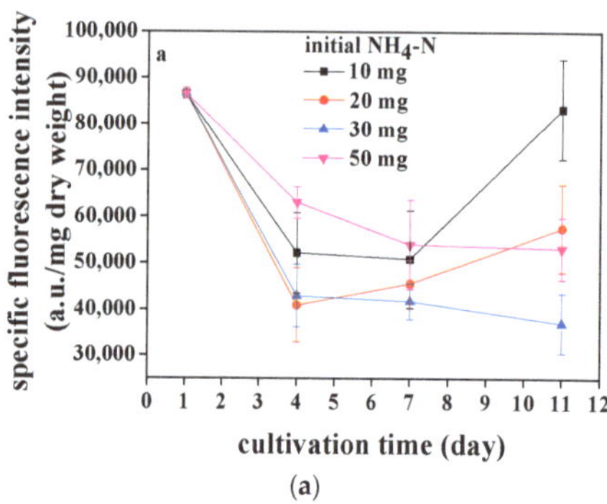

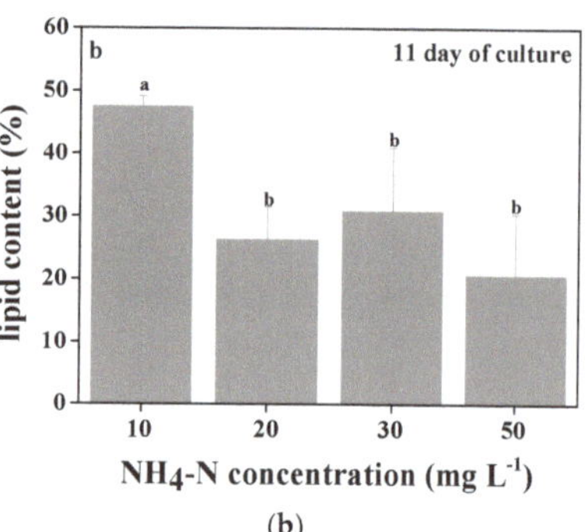

Figure 6. The specific fluorescence intensity (**a**) and lipid content (**b**) of *Chlorococcum* sp. GD under different ammonia nitrogen concentrations. In (**b**), different letters on columns indicated that they were significantly different at a probability level of 0.05 according to ANOVA test.

4. Discussion

In the present study, *Chlorococcum* sp. GD cultivated in wastewater with high ammonia nitrogen concentration (20–50 mg·L^{-1}) had high concentration of extracellular proteins at the end of the culture. Nitrogen is an essential nutrient that is required in microalgal growth and is used to synthesize a variety of biological substances, such as peptides, proteins, etc. [29]. Durmaz and Sanin [30] found that EPS of activated sludge had high protein and low carbohydrate content at a C/N ratio of 5. When the C/N ratio increased to 40, the proteins content decreased sharply, whereas the carbohydrates content increased. Thus, it could be seen that nitrogen deficiency in high C/N ratio was not beneficial to the synthesis of extracellular proteins, which was reconfirmed in this study. It was also found that there was a positive trend between flocculation efficiency of *Chlorococcum* sp. GD and protein content. Salim et al. [15] found that autoflocculation of *E. texensis* was due to extracellular proteins patched to the cell surface. As reported by Díaz-Santos et al. [31], a soluble cell wall protein similar to glucanases was isolated from the yeast *Saccharomyces bayanus* var. *uvarum* and could effectively induce flocculation of *Chlamydomonas reinhardtii* and *Picochlorum* sp. HM1. Díaz-Santos et al. [32] expressed the *FLO5* gene (flocculation gene, specific cell surface lectin-like glycoproteins relating to the flocculation process in yeasts) from *S. bayanus* by cotransformation in *C. reinhardtii*. *C. reinhardtii* transformants exhibited self-flocculation abilities between 2- and 3.5-fold higher than the control untransformed strain. All these cases demonstrated the importance of extracellular proteins in microalgal flocculation, which was proved again in this study. Based on the analysis and discussion above, it was speculated that nitrogen indeed affected the flocculation efficiency of *Chlorococcum* sp. GD, which was likely related to the production of extracellular proteins. It was found that the flocculating ability of *Chlorococcum* sp. GD was affected by ammonia nitrogen. *Chlorococcum* sp. GD had an excellent flocculating performance when ammonia nitrogen concentration of wastewater was more than 10 mg·L^{-1}. *Chlorococcum* sp. GD had the highest flocculating efficiency in the middle of the culture (4–5 days) regardless of initial ammonia nitrogen concentration in wastewater. The phenomenon likely came from the maximum extracellular protein content and high degradation efficiency of pollutants of *Chlorococcum* sp. GD synchronously in the middle of the culture (4–5 days). Of course, the extracellular proteins content of *Chlorococcum* sp. GD was obviously higher than the extracellular polysaccharides content, which was similar to self-flocculating microalga *E. texensis*, since it had high extracellular proteins content [12,15].

It was found that ammonia nitrogen affected the growth of *Chlorococcum* sp. GD. The optimal ammonia nitrogen concentration for its growth was 20 mg·L^{-1}, which was consistent with when Tam and Wong [18] showed that *C. vulgaris* grew well in cultures containing 20 mg·L^{-1} ammonia nitrogen. Of course, many microalgae could grow well in cultures with more than 20 mg·L^{-1} ammonia nitrogen [19,33], which might be species-dependent or related to other factors, such as light intensity, immobilization, operation modes, etc. [34,35]. In this study, the 190.1 mg·L^{-1} of microalgal biomass was low compared to the common biomass levels of *Chlorococcum* sp. in BG11 (360–4140 mg·L^{-1}) [36]. The possible reason was the low initial inoculation concentration used in the study.

The pollutants removal was also affected by ammonia nitrogen, and pollutants removal efficiency was the best when the ammonia nitrogen concentration was 20 mg·L^{-1}. Commonly, the elementary composition ratio of microalgal cells gave a hint of the optimal N/P ratio in wastewaters. The empirical elementary composition for microalgae was $C_{106}H_{263}O_{110}N_{16}P$ (N/P ratio = 7.2:1). In our work, the optimal N/P ratio for pollutants removal was 4:1, which was relatively lower than the empirical value. Some investigations also found that *Klebsormidium* sp. and *Pseudanabaena* sp. had excellent nitrogen and phosphorus removal capability with high N/P ratio of 7–10 and 7–20 [37], which was higher than the empirical value. Therefore, the optimal N/P ratio for pollutants removal is likely to depend on the strain and growth conditions. Kinetics of ammonia nitrogen removal was calculated by Michaelis-Menten Kinetics. Kinetic coefficients of ammonia nitrogen removal by *Chlorococcum* sp. GD were V_{max} = 15.2 mg NH$_4$-N·g^{-1} DW·d^{-1} and K_m = 13.4 mg NH$_4$-N·L^{-1}, which was lower than those of *Klebsormidium* sp. and *Pseudanabaena* sp. [37]. Normally, a high V_{max} is an indicator of the high potential capability of removing pollutants, while a low K_m meant that the microalga could reach its highest pollutant removal rate at low pollutant concentrations [37]. Based on the analysis above, it is indicated that *Chlorococcum* sp. GD has a competitive advantage in wastewaters with relatively low ammonia nitrogen concentration.

In this work, it was found that the lipid content of *Chlorococcum* sp. GD was influenced by ammonia nitrogen. Nitrogen deprivation was commonly the main trigger for lipid accumulation of microalgae [38]. Because of nitrogen deprivation of initial inoculum cultivated in BG11 for 13 days, it had the highest lipid content. For *Chlorococcum* sp. GD cultivated with 30 and 50 mg·L^{-1} ammonia nitrogen, there was still a large amount of nitrogen not being degraded at the end of the culture. Microalgal cells consequently did not accumulate large amounts of lipids, and the lipid content was relatively low. For low nitrogen concentration groups (10 and 20 mg·L^{-1}), almost all ammonia nitrogen was degraded within 7 days of culture. Afterwards, *Chlorococcum* sp. GD was in the stage of nitrogen deprivation, and thus accumulated high content of lipid. Of course, the lipid content in 10 mg·L^{-1} nitrogen concentration group was significantly higher than that in 20 mg·L^{-1}

nitrogen concentration group. Nevertheless, the total lipid content in 20 mg·L^{-1} nitrogen concentration group (10,913,050 a.u. L^{-1}) was almost the same as that in 10 mg·L^{-1} nitrogen concentration group (11,498,226.9 a.u. L^{-1}) owing to high biomass production of *Chlorococcum* sp. GD cultivated with 20 mg·L^{-1} ammonia nitrogen. High pollutants removal efficiency also occurred in wastewater with 20 mg·L^{-1} ammonia nitrogen. Therefore, it was suggested that *Chlorococcum* sp. GD was cultivated with 20 mg·L^{-1} ammonia nitrogen for lipid production. In the study, 26.3% of lipid content was achieved when *Chlorococcum* sp. GD was cultured with 20 mg·L^{-1} ammonia nitrogen, which was comparable to lipid content of microalgae cultivated in wastewaters in former literatures [39,40]. Interestingly, for the cultivation of *Chlorococcum* sp. GD, high efficient accumulation of lipid occurred at the end of the culture, while optimum efficiency of pollutants degradation and microalgal harvesting occurred in the middle of the culture. Therefore, it was worth paying attention to how to solve this contradiction in the future.

5. Conclusions

In the study, it was found that the flocculating ability of *Chlorococcum* sp. GD was affected by ammonia nitrogen concentration in wastewater. Relative high ammonia nitrogen concentration (20–50 mg·L^{-1}) was beneficial to flocculation, and the highest flocculating efficiency was in the middle of the culture (4–5 days) regardless of ammonia nitrogen concentration in wastewater. It was speculated that high flocculating efficiency was likely related to the production of extracellular proteins. Relative high ammonia nitrogen concentration (20 mg·L^{-1}) was also beneficial to biomass production and lipid accumulation of *Chlorococcum* sp. GD. Pollutants removal efficiency was the best when the ammonia nitrogen concentration was 20 mg·L^{-1}. A better understanding of the response of *Chlorococcum* sp. GD to ammonia nitrogen is beneficial for cultivating and harvesting self-flocculating microalga from wastewater.

Acknowledgments: This research project was financed by the Natural Science Foundation of China (No. 31700310), the Key Scientific Development Project of Shanxi Province, China (No. FT-2014-01-15 and FT-2014-02-15), the Natural Science Foundation of Shanxi Province, China (No. 2015021159), the Social Development Foundation of Shanxi, China (No. 201603D321008), and 1331 Innovation team of Shanxi.

Author Contributions: Shulian Xie and Xiaoyan Jiao conceived and designed the experiments; Junping Lv, Xuechun Wang, and Wei Liu performed the experiments; Jia Feng, Qi Liu, and Fangru Nan analyzed the data; Junping Lv wrote the paper.

Conflicts of Interest: The authors declare no conflict of interest.

References

1. Li, Y.; Zhou, W.; Hu, B.; Min, M.; Chen, P.; Ruan, R.R. Integration of algae cultivation as biodiesel production feedstock with municipal wastewater treatment: Strains screening and significance evaluation of environmental factors. *Bioresour. Technol.* **2011**, *102*, 10861–10867. [CrossRef] [PubMed]
2. Zhu, L.; Wang, Z.; Shu, Q.; Takala, J.; Hiltunen, E.; Feng, P.; Yuan, Z. Nutrient removal and biodiesel production by integration of freshwater algae cultivation with piggery wastewater treatment. *Water Res.* **2013**, *47*, 4294–4302. [CrossRef] [PubMed]
3. Olguín, E.J. Dual purpose microalgae-bacteria-based systems that treat wastewater and produce biodiesel and chemical products within a Biorefinery. *Biotechnol. Adv.* **2012**, *30*, 1031–1046. [CrossRef] [PubMed]
4. Milledge, J.J.; Heaven, S. A review of the harvesting of micro-algae for biofuel production. *Rev. Environ. Sci. Biotechnol.* **2013**, *12*, 165–178. [CrossRef]
5. Bosma, R.; van Spronsen, W.A.; Tramper, J.; Wijffels, R.H. Ultrasound, a new separation technique to harvest microalgae. *J. Appl. Phycol.* **2003**, *15*, 143–153. [CrossRef]
6. Xu, Y.; Purton, S.; Baganz, F. Chitosan flocculation to aid the harvesting of the microalga *Chlorella sorokiniana*. *Bioresour. Technol.* **2013**, *129*, 296–301. [CrossRef] [PubMed]
7. Wan, C.; Zhao, X.Q.; Guo, S.L.; Alam, M.A.; Bai, F.W. Bioflocculant production from *Solibacillus silvestris* W01 and its application in cost-effective harvest of marine microalga *Nannochloropsis oceanica* by flocculation. *Bioresour. Technol.* **2013**, *135*, 207–212. [CrossRef] [PubMed]

8. Wan, C.; Alam, M.A.; Zhao, X.Q.; Zhang, X.Y.; Guo, S.L.; Ho, S.H.; Chang, J.S.; Bai, F.W. Current progress and future prospect of microalgal biomass harvest using various flocculation technologies. *Bioresour. Technol.* **2015**, *184*, 251–257. [CrossRef] [PubMed]

9. Salim, S.; Bosma, R.; Vermuë, M.H.; Wijffels, R.H. Harvesting of microalgae by bio-flocculation. *J. Appl. Phycol.* **2011**, *23*, 849–855. [CrossRef] [PubMed]

10. Salim, S.; Vermuë, M.H.; Wijffels, R.H. Ratio between autoflocculating and target microalgae affects the energy-efficient harvesting by bio-flocculation. *Bioresour. Technol.* **2012**, *118*, 49–55. [CrossRef] [PubMed]

11. Guo, S.L.; Zhao, X.Q.; Wan, C.; Huang, Z.Y.; Yang, Y.L.; Alam, M.A.; Ho, S.H.; Bai, F.W.; Chang, J.S. Characterization of flocculating agent from the self-flocculating microalga *Scenedesmus obliquus* AS-6-1 for efficient biomass harvest. *Bioresour. Technol.* **2013**, *145*, 285–289. [CrossRef] [PubMed]

12. Salim, S.; Shi, Z.; Vermuë, M.H.; Wijffels, R.H. Effect of growth phase on harvesting characteristics, autoflocculation and lipid content of *Ettlia texensis* for microalgal biodiesel production. *Bioresour. Technol.* **2013**, *138*, 214–221. [CrossRef] [PubMed]

13. Alam, M.A.; Wan, C.; Guo, S.L.; Zhao, X.Q.; Huang, Z.Y.; Yang, Y.L.; Chang, J.S.; Bai, F.W. Characterization of the flocculating agent from the spontaneously flocculating microalga *Chlorella vulgaris* JSC-7. *J. Biosci. Bioeng.* **2014**, *118*, 29–33. [CrossRef] [PubMed]

14. He, Y.; Zhou, L.; Xu, X.; Wang, S.; Wang, C.; Dai, B. Uniform design for optimizing biomass and intracellular polysaccharide production from self-flocculating *Scenedesmus* sp. BH. *Ann. Microbiol.* **2014**, *64*, 1779–1787. [CrossRef]

15. Salim, S.; Kosterink, N.R.; Wacka, N.T.; Vermuë, M.H.; Wijffels, R.H. Mechanism behind autoflocculation of unicellular green microalgae *Ettlia Texensis*. *J. Biotechnol.* **2014**, *174*, 34–38. [CrossRef] [PubMed]

16. Wan, C.; Zhang, X.Y.; Zhao, X.Q.; Bai, F.W. Harvesting microalgae via flocculation: A review. *Chin. J. Biotech.* **2015**, *31*, 161–171. (In Chinese)

17. González-Fernández, C.; Ballesteros, M. Microalgae autoflocculation: An alternative to high-energy consuming harvesting methods. *J. Appl. Phycol.* **2013**, *25*, 991–999. [CrossRef]

18. Tam, N.F.Y.; Wong, Y.S. Effect of ammonia concentrations on growth of *Chlorella vulgaris* and nitrogen removal from media. *Bioresour. Technol.* **1996**, *57*, 45–50. [CrossRef]

19. Kim, S.; Lee, Y.; Hwang, S.J. Removal of nitrogen and phosphorus by *Chlorella sorokiniana* cultured heterotrophically in ammonia and nitrate. *Int. Biodeter. Biodegr.* **2013**, *85*, 511–516. [CrossRef]

20. Guo, S.L. Development of Transgenic Flocculation *Scenedesmus obliquus* and Characterization of the Flocculation of *S. obliquus*. Ph.D. Thesis, Dalian University of Technology, Dalian, China, 2013; pp. 91–92. (In Chinese)

21. Lv, J.P.; Guo, J.Y.; Feng, J.; Liu, Q.; Xie, S.L. A comparative study on flocculating ability and growth potential of two microalgae in simulated secondary effluent. *Bioresour. Technol.* **2016**, *205*, 111–117. [CrossRef] [PubMed]

22. Feng, J.; Guo, Y.; Zhang, X.; Wang, G.; Lv, J.; Liu, Q.; Xie, S. Identification and characterization of a symbiotic alga from soil bryophyte for lipid profiles. *Biol. Open* **2016**, *5*, 1317–1323. [CrossRef] [PubMed]

23. Aslan, S.; Kapdan, I.K. Batch kinetics of nitrogen and phosphorus removal from synthetic wastewater by algae. *Ecol. Eng.* **2006**, *28*, 64–70. [CrossRef]

24. Yang, S.F.; Li, X.Y. Influences of extracellular polymeric substances (EPS) on the characteristics of activated sludge under non-steady-state conditions. *Process. Biochem.* **2009**, *44*, 91–96. [CrossRef]

25. Bradford, M.M. A rapid and sensitive method for quantification of microgram quantities of protein utilizing principle of protein-dye binding. *Anal. Biochem.* **1976**, *72*, 248–254. [CrossRef]

26. Gaudy, A.F. Colorimetric determination of protein and carbohydrate. *Ind. Water Wastes* **1962**, *7*, 17–22.

27. APHA-AWWA-WEF. *Standard Methods for the Examination of Water and Wastewater*, 21st ed.; American Public Health Association/American Water Works Association/Water Environment Federation: Washington, DC, USA, 2005.

28. Bligh, E.G.; Dyer, W.J. A rapid method of total lipid extraction and purification. *Can. J. Biochem. Physiol.* **1959**, *37*, 911–917. [CrossRef] [PubMed]

29. Cai, T.; Park, S.Y.; Li, Y. Nutrient recovery from wastewater streams by microalgae: Status and prospects. *Renew. Sustain. Energy Rev.* **2013**, *19*, 360–369. [CrossRef]

30. Durmaz, B.; Sanin, F.D. Effect of carbon to nitrogen ratio on the composition of microbial extracellular polymers in activated sludge. *Water Sci. Technol.* **2001**, *44*, 221–229. [PubMed]

31. Díaz-Santos, E.; Vila, M.; de la Vega, M.; León, R.; Vigara, J. Study of bioflocculation induced by *Saccharomyces bayanus* var. *uvarum* and flocculating protein factors in microalgae. *Algal Res.* **2015**, *8*, 23–29.

32. Díaz-Santos, E.; Vila, M.; Vigara, J.; León, R. A new approach to express transgenes in microalgae and its use to increase the flocculation ability of *Chlamydomonas reinhardtii*. *J. Appl. Phycol.* **2016**, *28*, 1611–1621. [CrossRef]

33. He, P.J.; Mao, B.; Shen, C.M.; Shao, L.M.; Lee, D.J.; Chang, J.S. Cultivation of *Chlorella vulgaris* on wastewater containing high levels of ammonia for biodiesel production. *Bioresour. Technol.* **2013**, *129*, 177–181. [CrossRef] [PubMed]

34. Ruiz-Marin, A.; Mendoza-Espinosa, L.G.; Stephenson, T. Growth and nutrient removal in free and immobilized green algae in batch and semi-continuous cultures treating real wastewater. *Bioresour. Technol.* **2010**, *101*, 58–64. [CrossRef] [PubMed]

35. Su, Y.; Mennerich, A.; Urban, B. Coupled nutrient removal and biomass production with mixed algal culture: Impact of biotic and abiotic factors. *Bioresour. Technol.* **2012**, *118*, 469–476. [CrossRef] [PubMed]

36. Feng, P.; Deng, Z.; Hu, Z.; Wang, Z.; Fan, L. Characterization of *Chlorococcum pamirum* as a potential biodiesel feedstock. *Bioresour. Technol.* **2014**, *162*, 115–122. [CrossRef] [PubMed]

37. Liu, J.; Vyverman, W. Differences in nutrient uptake capacity of the benthic filamentous algae *Cladophora* sp., *Klebsormidium* sp. and *Pseudanabaena* sp. under varying N/P conditions. *Bioresour. Technol.* **2015**, *179*, 234–242. [CrossRef] [PubMed]

38. Hu, Q.; Sommerfeld, M.; Jarvis, E.; Ghirardi, M.; Posewitz, M.; Seibert, M.; Darzins, A. Microalgal triacylglycerols as feedstocks for biofuel production: Perspectives and advances. *Plant J.* **2008**, *54*, 621–639. [CrossRef] [PubMed]

39. Jebali, A.; Acién, F.G.; Gómez, C.; Fernández-Sevilla, J.M.; Mhiri, N.; Karray, F.; Dhouib, F.; Molina-Grima, E.; Sayadi, S. Selection of native Tunisian microalgae for simultaneous wastewater treatment and biofuel production. *Bioresour. Technol.* **2015**, *198*, 424–430. [CrossRef] [PubMed]

40. Kuo, C.M.; Chen, T.Y.; Lin, T.H.; Kao, C.Y.; Lai, J.T.; Chang, J.S.; Lin, C.S. Cultivation of *Chlorella* sp. GD using piggery wastewater for biomass and lipid production. *Bioresour. Technol.* **2015**, *194*, 326–333. [CrossRef] [PubMed]

International Journal of
*Environmental Research
and Public Health*

Article

Mechanisms of Phosphorus Removal by Recycled Crushed Concrete

Yihuan Deng * and Andrew Wheatley

School of Architecture, Building and Civil Engineering, Loughborough University, Loughborough, LE113 TU, UK; a.d.wheatley@lboro.ac.uk
* Correspondence: y.deng@lboro.ac.uk; Tel.: +44-(0)-1509-222-222

Received: 9 January 2018; Accepted: 8 February 2018; Published: 17 February 2018

Abstract: Due to urbanisation, there are large amounts of waste concrete, particularly in rapidly industrialising countries. Currently, demolished concrete is mainly recycled as aggregate for reconstruction. This study has shown that larger sizes (2–5 mm) of recycled concrete aggregate (RCA) removed more than 90% of P from effluent when at pH 5. Analysis of the data, using equilibrium models, indicated a best fit with the Langmuir which predicated an adsorption capacity of 6.88 mg/g. Kinetic analysis indicated the equilibrium adsorption time was 12 h, with pseudo second-order as the best fit. The thermal dynamic tests showed that the adsorption was spontaneous and, together with the evidence from the sequential extraction and desorption experiments, indicated the initial mechanism was physical attraction to the surface followed by chemical reactions which prevented re-release. These results suggested that RCA could be used for both wastewater treatment and P recovery.

Keywords: desorption; recycled concrete; phosphorus removal and recovery; phosphorus speciation by sequential extraction; thermal dynamic; infinite focused microscope

---

1. Introduction

Concrete is a major construction material which includes significant embedded resources, these include 50% of all the raw materials used and 40% of the energy. After manufacture, the other 50% of the total waste was generated by the concrete industry [1]. Recently, large volumes of demolition concrete waste have been produced due to limited reuse of older concrete buildings. For example in 2014, China produced 1.5 billion tonnes of construction waste; concrete was 34% of this total and only about 5% was reused [2]. This compares with above 50% in developed countries. New and alternative uses for recycled concrete are a priority in China where, traditionally, construction waste has been disposed of in landfills or deposited on river banks. This disposal used large areas of land as well as causing environmental nuisance and complaints.

There has been previous work on using recycled concrete aggregate (RCA) for water treatment and one of the most researched applications has been as a filter media due to its surface roughness and desirable chemical content (e.g., Mg, Ca, Fe, Al). Li et al. (2007) [3], for example, used fly ash aerated crushed concrete at laboratory scale to achieve 95.6% total nitrogen removal from sewage and landfill leachate. Guo et al. (2009) [4] also used RCA as a wastewater filter and showed it could remove 37% of COD and 55% of total P. The mechanism has been reported to be based on the calcium, aluminium, and iron content which can bind with phosphate [5]. Phosphorus (P) is needed for optimum crop production and mineral phosphorus reserves are thought to be limited with no substitutes. Research in both the U.S. and Europe has shown mineral P resources may be depleted in a few hundred years and recovery of P is urgent to maintain agricultural productivity [6]. Furthermore, removal of P from wastewater to meet environmental quality standards is now a significant cost in chemicals and power, complementing the benefits of recovery, particularly in China [7]. RCA is a lime enriched material, widely available and potentially a sustainable method for P recovery. Other work on P adsorption from wastewater has been reported by Xiang et al. (2013) [5] who suggested RCA could be suitable and sustainable for adsorbing low phosphate concentrations from wastewater. They also reported that the RCA did not re-release P or metals during sludge treatment. RCA has also been used for pH correction of acid wastewaters because of its weakly alkaline properties [8,9]. Instead of using concrete particles, others have used reconstituted concrete blocks. Chen (2001) [10] designed a wastewater treatment system based on concrete blocks. He operated the media as a submerged aerated filter and reported removals of 50% of COD and BOD, 70% of TP, and 20% of TN. Yuan et al. (2006) [11] published results from wastewater using reconstituted concrete blocks, known as eco-blocks which removed 76% of COD and 94.9% of BOD_5. Japanese researchers have proposed that concrete blocks have greater porosity and provide a larger internal and external surface area for microorganisms [12].

Upon review of the previous literature, there are few well-controlled experiments on the detailed removal mechanisms or reproducibility of P adsorption by either RCA or blocks. Similarly there is little theoretical information on how to design P adsorption systems based on the common adsorption isotherms or reaction kinetics. The aim of this study was to determine the process of P adsorption by RCA. This was achieved by several batch tests with a mechanistic isotherm model. Also, the kinetics and thermodynamics studies were involved to further explain the adsorption mechanism. A novel study of P speciation was carried by sequential extraction. It is first time reported the type of P contained in RCA. This study could contribute to new usage of RCA.

2. Materials and Methods

2.1. Sorption Studies

Batch experiments were used to evaluate optimum operational conditions for phosphorus adsorption by RCA and data was analysed using adsorption isotherms, kinetics, and thermodynamics. The studies carried out are summarised in Table 1.

Table 1. Experimental schedule for the adsorption of P.

No.	Study	Parameters
1	Effect of pH of solution on sorption	The pH of the solution was varied between 6.0 and 8.0. Dose of media 2 g; initial P concentration 20 mg/L; contact time 24 h; agitator 180-rpm
2	Effect of dose of sorbent on phosphorus sorption	RCA was varied in the range 1–10 g, 20 mg/L P solution; contact time 24 h; agitator 180-rpm; pH-5
3	Effect of initial phosphorus concentration	Phosphate concentrations in the range 5–30 mg/L. Dose of media 2 g contact time 24 h; agitator 180-rpm; pH-5;
4	Equilibrium studies	Evaluation of maximum adsorption by isotherm models
5	Kinetic analysis	Initial P concentration 15 mg/L with 2 g of RCA, contact time 24 h at 298 K, agitator 180-rpm, pH 5
6	Effect of temperature on sorption	Initial P concentration 15 mg/L with 2 g of RCA and contact time 24 h at 298, 318 and 328 °C, pH 5

RCA: recycled concrete aggregate.

Before the experiments, all the glassware and plastic devices were soaked in 10% nitric acid solution for 1 day and then rinsed before each test with 250 mL deionised water. Sieved (2–5 mm) and manually hammered RCA was used for the experiments. RCA was also scanned by an infinite focus microscope (Alicona) (see Figure 1). An infinite focus microscope can quantitatively measure surface texture or deviations in a surface character. Figure 1 shows RCA has a crude surface, which indicates more adsorption sites. The specific surface area of RCA was also identified as 35 m^2/m^3. The RCA was examined by EDS (ED3000) and composed of 56% oxygen, 18% Si, 10% Carbon, 10% Ca, 2% Al, 2% K, and 1% Mg. Erhlmeyer flasks, with measured amounts of dried RCA media were used with 100 mL of a known P concentration as a solution of Potassium orthophosphate. The flasks were run in a thermostatically controlled orbital shaker for the defined periods (Table 1). By operating the shaker at various temperatures (Table 1), the data could be used to determine thermodynamic parameters. The pH of the solution was altered using a pH meter (HANNA HI9812-5) with either 0.1 M HCl or NaOH. The final solutions were filtered by a 0.45 μm membrane filter and dissolved concentrations of P were determined by Inductively Coupled Plasma analysis (Shimadzu ICP-9000) according to the international standard method [13]. All tests were conducted in triplicate. The amount adsorbed was calculated using the formula below:

$$q_e = \frac{V(C_o - C_e)}{W}$$

where q_e is the adsorption capacity (mg/g), C_o is the initial concentration of solution (mg/L), C_e is the concentration after adsorption (mL/L), W is the weight of adsorbent (g) and V is the volume of solution (L).

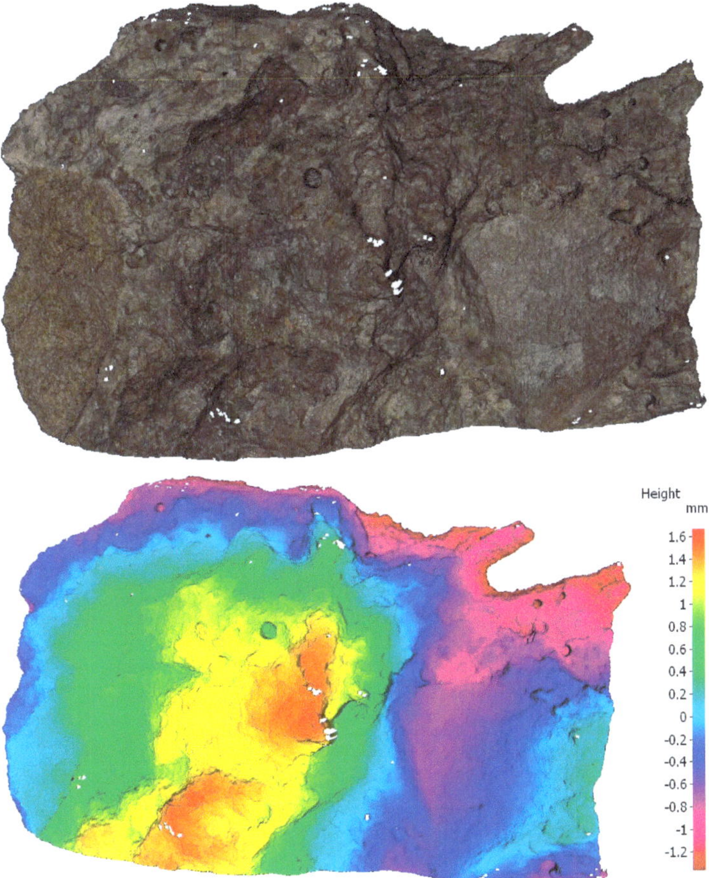

Figure 1. RCA scanning by infinite focused microscope.

2.2. Desorption of P

The P-treated RCAs from the previous experiments, with 2 g of RCA, were elutriated with 100 mL deionised water in Erhlmeyer flasks. Tests were conducted in triplicate for 24 h mixing and residual P concentration analysed as before.

2.3. Fractionation of Inorganic Phosphorus

Sequential extraction to analyse the various P-forms was used to identify the chemical reaction that occurred between the surface of the media and the P solution. Analysis of the P species was first reported by Dean (1938) [14] in soil complexes. Dean defined two types of P according to whether they were extracted by acid or alkali. Chang and Jackson (1957) [15] refined this by using four extraction solutions: loosely bound phosphate (LBP), aluminium phosphate (Al-P), iron phosphate (Fe-P), calcium phosphate (Ca-P), and occluded phosphate (O-P). Most procedures for P characterisation are based on the Chang and Jackson (1957) method with modification or improvements. Normally, Ca and Mg-P are extracted together by HCl [16–18]. The Ca-P amount was determined separately and the Mg-P by subtraction. The modification of the Chang and Jackson (1957) [15] and Hartikainen (1979) [19] method was used (see Table 2). After the standard adsorption test, 2 g RCA was used and the analysis was carried out in triplicate.

Table 2. Inorganic phosphorus by a sequential extraction for filter media [19,20].

Step	Inorganic P	Extraction Reagents	Concentration	Condition
I	LBP	NH_4Cl	1 mol/L	50 mL, shaking 0.5 h
II	Al-P	NH_4F (pH 8)	0.5 mol/L	50 mL, shaking 1 h
III	Fe-P	NaOH	0.1 mol/L	50 mL, shaking 2 h
IV	O-P	CDB (pH 7.6)	-	45 mL, shaking 0.5 h
V	Ca-P	H_2SO_4	0.5 mol/L	50 mL, shaking 1 h
VI	Mg-Ca-P	HCl	0.5 mol/L	50 mL, shaking 1 h

Note: CDB (Sodium citrate-sodium dithionite-sodium bicarbonate); $Na_3C_3H_6O_7$ (0.3 M); $NaHCO_3$ (1 M); $Na_2S_2O_4$ (1 g).

3. Results and Discussion

3.1. Sorption Studies

3.1.1. Effect of pH of Solution on Sorption

The best adsorption was observed at pH 5.0, when the adsorption capacity of P was 0.85 mg/g and removal reached 93.5% (Figure 2). The trend shows a decline in P adsorption with increasing pH to the lowest capacity at pH 8.0. Agyeia et al. [21] also pointed out acidic pH was best for P removal. There are two possible mechanisms, one has been attributed to the double layer effect whereby acidic H^+ was attracted to the concrete surface by Ca, Al, and Mg hydroxide content creating a secondary positive layer to bind the negative orthophosphate. The second possibility was that the phosphate ions could be converted to their acidic forms (e.g., $H_2PO_4^-$ and HPO_4^{2-}) binding to the positive surface. The amount of these phosphate ions generated would be proportional to the acidity of the solution. The acidic pH could also help the cement release more Ca^{2+} ions into the solution to react with the hydrogen phosphate causing deprotonation and precipitation of $Ca_3(PO_4)_2$. The increase in P removal above pH 9 could be due to a similar mechanism but from the formation of OH enriched complexes precipitating calcium phosphate.

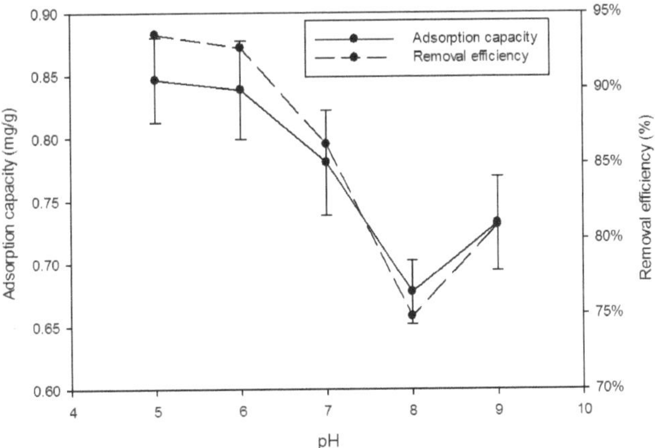

Figure 2. Effect of pH of solution on sorption of P. Dose of media 2 g; initial P concentration 2 mg/L; contact time 24 h; agitator 180-rpm.

3.1.2. Effect of Dose of Sorbent on Phosphorus Sorption

In this experiment, the optimum pH 5 was used and the results in Figure 3 show the amount of P adsorbed was proportional to the amount of RCA added. At the lowest dose (1 g), P removal was 55.5% but 2 g removed 95% of P, similar to the maximum achieved. The normalised adsorption capacity reduced with increasing RCA from 0.99 to 0.17 mg/g, as added RCA was increased from 2 to 10 g. Similar trends have been also reported by other authors [22,23]. Increasing the dose of sorbent increased the total removal of P by increasing the availability of sorption sites. Specific adsorption capacity is a measure of the amount of P bonded by a unit weight of sorbent. The competition by ions for the sites caused a decrease in the specific uptake once all the sites were filled and adsorption capacity decreased with increments in sorbent dose. The simpler models assume a fixed number of adsorption sites according to the molecular structure and the number of positive ions responsible for the binding.

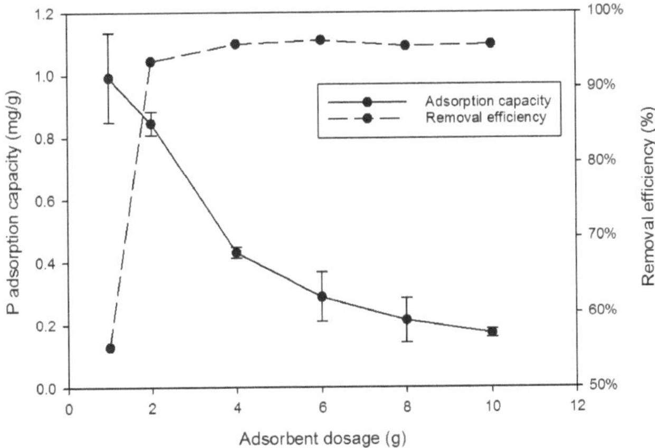

Figure 3. Effect of dose of media on sorption of P. Initial P concentration 20 mg/L; contact time 24 h; agitator 180-rpm; pH-5.

3.1.3. Effect of Initial Phosphorus Concentration

Figure 4 shows that the initial concentration of P in the solution influenced the equilibrium uptake of P achieved. The adsorption capacity increased linearly with the initial P concentration from 5 to 30 mg/L, but the proportion of P removed reached maximum at an initial concentration of 15 mg/L. This suggests the simple, single layer, fixed number of adsorption sites model may not be suitable at higher P concentrations. This is in line with previous equilibrium adsorption capacity experiments, which suggest that higher solute concentrations could encourage other mechanisms such as greater boundary concentrations which in turn lead to double layer adsorption and complex formation [23,24].

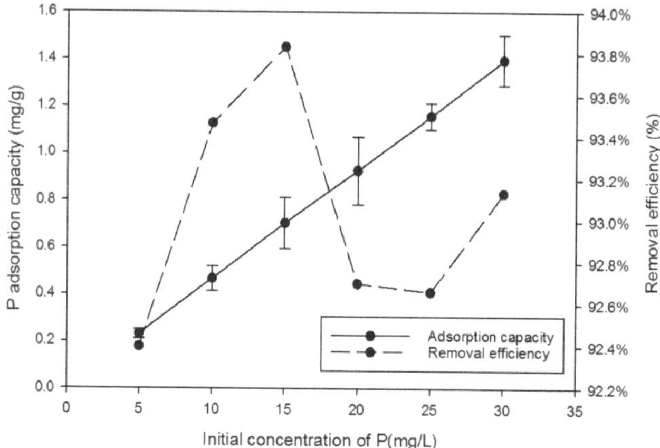

Figure 4. Effect of initial concentration of media on removal of P. Dose of media 2 g contact time 24 h; agitator 180-rpm; pH-5.

3.1.4. Equilibrium Studies

The equilibrium equations are shown in Table 3, and were used to determine the mechanisms of adsorption. The Langmuir and Freundlich isotherms gave the best fit (presented in Table 4), the Langmuir showing a maximum adsorption capacity of 6.88 mg/g. The Langmuir is the simplest isotherm, based on monolayer sorption onto a surface with a finite number of identical sites, homogeneously distributed over the sorbent surface. Others have also reported that the Langmuir isotherm was the more robust when modelling complex aqueous mediums. The Frumkin and BET equations were introduced as extensions to the Langmuir isotherm, the Frumkin to take account of solute–solute interaction at a non-ideal surface [25]. The BET was introduced to model multi-layer adsorption but, except for the first layer, it also assumes equal energies of adsorption for each layer and no interactions between layers [26]. The linear correlation of both the Frumkin and BET are weaker than the Langmuir (Table 4). The Freundlich equation included empirical constants as an acceptance that the adsorption sites were heterogeneous. The Temkin and the Dubinin-Radushkevich (D-R) adsorption equations are based on the thermodynamics of adsorption. The Tempkin is based on the principle that adsorption reduces the energy or heat in the adsorbed layer and the D–R on a Gaussian energy distribution as expected from a heterogeneous surface. The D–R model was reported to be suited to high and medium concentration solutions such as these P adsorption experiments [27]. The adsorption energy (E) derived from the D–R model can be used to predict adsorption mechanisms. Typical results are between 1–16 kJ/mol with E values lower than 8 kJ/mol indicating physical sorption [28]. In this study, E was calculated to be 2.24 kJ/mol, which would therefore be considered as physical adsorption.

Table 3. Equilibrium study models [25–27,29–31].

Isotherm Models	Linear Expression	Associated Equations	Parameters
Freundlich	$\log q_e = \log K_f + \frac{1}{n} \log C_e$	K and n are empirical constants	K_f (L/mg) = Langmuir equilibrium constant; n = dimensionless correction factor; qe = adsorption capacity (mg/g)
Langmuir	$\frac{1}{q_e} = \frac{1}{q_m} + \frac{1}{K_L q_m C_e}$	$R_L = \frac{1}{1 + K_L C_o}$	K_L = Isotherm constant (L/mg); C_O = initial concentration; q_m = mainxmum adsorption capacity (mg/g)
Tempkin	$q_e = B \ln A_T + B \ln C_e$	-	A_T = Tempkin isotherm equilibrium binding constant (L/g); B = constant related to heat of sorption (J/mol)
D-R	$\ln q_e = \ln(q_s) - (K_{ad}\varepsilon^2)$	$E = \frac{1}{\sqrt{2K_{ad}}}$	q_s = theoretical isotherm saturation capacity (mg/g), K_{ad} = Dubinin–Radushkevich isotherm constant (mol²/kJ²) and ε = Dubinin–Radushkevich isotherm constant
Frumkin	$\ln\left(\frac{\frac{\theta}{1-\theta}}{C_e}\right) = \ln K_{fr} + 2\alpha\theta$	-	θ = fractional surface coverage; α = lateral interaction coefficient; K_{fr} (L/g) = Frumkin equilibrium constant,
BET	$\frac{C_e}{q_e(C_s - C_e)} = \frac{1}{q_s C_{BET}} + \frac{(C_{BET}-1)}{q_s C_{BET}}\left(\frac{C_e}{C_s}\right)$	-	C_e = equilibrium concentration (mg/L); C_s = adsorbate monolayer saturation concentration (mg/L); C_{BET} = BET adsorption isotherm relating to the energy of surface interaction (L/mg)

Table 4. Adsorption Isotherm constants for P adsorption by concrete.

Langmuir Adsorption Isotherm				Freundlich Adsorption Isotherm		
q_m (mg/g)	K_L (L/mg)	R_L	R^2	n	K_f	R^2
6.88	0.089	0.281	0.984	0.996	0.669	0.983

Tempkin Adsorption Isotherm				Dubinin-Radushkevich Isotherm			
A_T (L/mg)	A_t	B	R^2	q_s (mg/g)	K_{ad} (mol²/kJ²)	ε (kJ/mol)	R^2
0.195	134.9	0.649	0.958	1.4	2×10^{-7}	2.24	0.968

Frumkin Adsorption Isotherm			BET Adsorption Isotherm		
α	K_{fr}	R^2	C_{BET}	q_s	R^2
2.959	0.00342	0.669	-0.798	0.553	0.898

Previous filtration studies using cement or concrete have been summarised in Table 5. The majority of these studies used finer particles (>1 mm) to adsorb P because of the greater surface area and consequent adsorption capacity available from smaller sized particles. The aerated concrete had the greatest surface area and gave the largest adsorption capacity. It was reported as too fragile for water treatment however, and small media would also be vulnerable to clogging and wash out, giving high operating costs. The results of this study suggest larger sizes could be used as a compromise for wastewater treatment. The cement studies—by Zheng et al. [32] and Renman and Renman [33]—used a similar size as used here (Table 5) but both materials were unstable. Most of these studies have operated under acidic pH (<3) to achieve maximum P adsorption capacity but these acidic conditions would be unrealistic for most wastewaters. This study has demonstrated that concrete demonstrated reliable P removal at a higher pH (pH 5) which could be achieved in the microenvironment of a wastewater filter.

Table 5. Adsorptive capacity of various studies by concrete.

Types	Media	Size (mm)	Time	Q_{max} (mg/g)	Reference
Empirical	Ordinary Portland cement	0.045–0.300	16 h	19.90	[21]
Empirical	Recycled Crushed Concrete	0.125–0.250	1 h	0.134	[22]
Empirical	Crushed concrete	0.125	40 days	19.6	[34]
Theoretical	Cement	0.425–0.85	24 h	1.185	[35]
Empirical	Cement	0.85	28 days	16.16	[36]
Empirical	Gas concrete	0.063–2	1 h	11.5	[37]
Theoretical	Recycled crushed concrete	0.3–2.3	24 h	6.1	[23]
Empirical	Crushed autoclaved aerated concrete	2–4	24 h	70.9	[33]
Theoretical	Recycled concrete	2–5	24 h	6.88	Present study
Theoretical	Cement	3–5	32 h	4.98	[32]

3.1.5. Kinetic Analysis

Four kinetic models were also used to analyse the data (Table 6). These were the standard first and second-order equations plus the Fractional power equation which is a modified form of Freundlich. Finally the Elovich equation was used since it had previously been used to study NH_4^+ removal by Chien and Clayton (1980) [38]. The results and correlation coefficients (R^2) are summarised in Table 7. The best fit was pseudo second-order followed by the first order equation. The experimentally measured adsorption capacity (0.749 mg/g) was between those calculated from the pseudo first- and second-order equations. The experimental results indicated that RCA adsorbed most when at high P concentrations; the calculated maximum capacity suggested adsorption could reach 0.9 mg/g.

Table 6. Kinetic study models [39,40].

Kinetic Study	Linear Expression	Parameters
The pseudo-first-order	$\log(Q_e - Q_t) = -\frac{k_1}{2.303}t + \log Q_e$	Q = the amount of adsorption time (min) (mg/g); k_1 = the rate, constant of pseudo first-order sorption (L/min); Q_e = adsorption capacities at equilibrium, Q_t = adsorption capacities at time t(min)
The pseudo second-order	$\frac{t}{Q_t} = \frac{1}{k_2 Q_e^2} + \frac{1}{Q_e}t$	k_2 = the rate constant of the second-order equation
Elovich model equation	$Q_t = \left(\frac{1}{\beta}\right)\ln(\alpha\beta) + \left(\frac{1}{\beta}\right)\ln t$	α and β are constants
Fractional power model	$\ln q_t = \ln a + b\ln t$	q_t = the amount of adsorbate sorbed by adsorbent at a time t; and b = constants with b < 1

Table 7. Kinetic parameters for P adsorption on concrete.

The Pseudo-First-Order			The Pseudo Second-Order		
k_1	Q_e	R^2	k_2	Q_e	R^2
0.211	0.657	0.9876	0.279	0.893	0.9916
Elovich Model Equation			Fractional Power Model		
α	β	R^2	a	b	R^2
0.476	5.061	0.9599	0.184	0.5	0.8986

Figure 5 shows the adsorption rate and removal efficiency increased until equilibrium was reached at 93% P adsorbed after 12 h and 99% at 24 h. The results were carried out at typical domestic wastewater P concentrations (around 10–15 mg/L) and demonstrated a rate of adsorption likely to be achievable in a wastewater treatment plant without excessive costs.

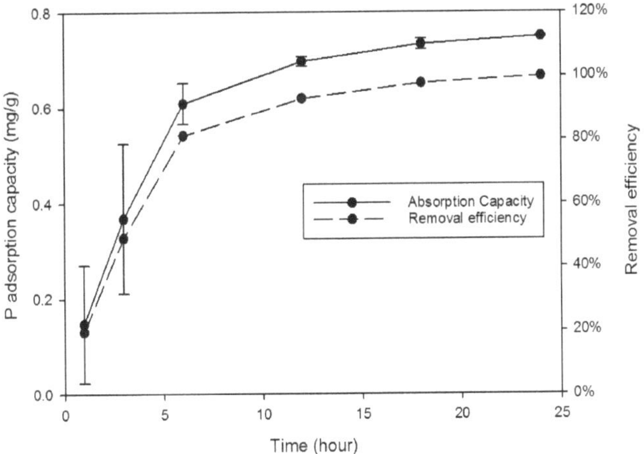

Figure 5. Kinetic studies of P adsorption; Initial P concentration 15 mg/L; agitator 180-rpm; pH-5; temperature = 27 °C.

3.1.6. Effect of Temperature on Sorption

The influence of temperature on the thermodynamics of adsorption (Gibbs free energy ($\Delta G°$), Enthalpy change ($\Delta H°$), and Entropy change ($\Delta S°$)) was calculated according to the procedure described by Tosun (2012) [28] as shown in Equations (1)–(3):

$$\Delta G° = -RTb \tag{1}$$

where R = gas constant (8.314 J/mol K); T = absolute temperature; b = the distribution coefficient, which was calculated by:

$$b = \frac{C_a}{C_e} \tag{2}$$

where C_a = equilibrium concentration of P on adsorbent (mg/L). The relation between ($\Delta H°$), ($\Delta G°$) and ($\Delta S°$) was given by:

$$\Delta G° = \Delta H° - T\Delta S° \tag{3}$$

Figure 6 shows that the adsorption of P increased with rises in temperature and that the adsorption process was endothermic. It is suggested that the higher temperature resulted in swelling of the RCA, creating more space for adsorption as well as accelerating molecular diffusion and the chemical reaction. The b value was obtained from $\ln(\frac{Q_e}{C_e})$ vs. Q_e and by extrapolating Q_e to zero to obtain the intercept value. Results for $\Delta G°$, $\Delta H°$, and $\Delta S°$ were calculated from the slope and intercept of a plot of ln K vs. (1/T) (see Figure 7).

The thermodynamic parameters are summarised in Table 8. The negative values of $\Delta G°$ indicate spontaneous adsorption of P on the RCA and the decreasing values of $\Delta G°$ confirm the experimental data showing acceleration at higher temperatures. The $\Delta H°$ is positive and less than 20 kJ/mol, while the value of $\Delta H°$ is less than 20 kJ/mol, indicating the adsorption process is physical [41]. It has been suggested that the positive value of $\Delta S°$ is due to the release of the water from phosphate as it is adsorbed onto the surface to increase the degrees of freedom and entropy of the system [42,43]. Other thermodynamic sorption studies for P are summarised in Table 9. The free energy range of RCA was similar to dolomite and activated carbon. Adsorption to these natural materials were also spontaneous, endothermic, and with increased entropy. The similarity to activated carbon also suggests non-specific adsorption sites corroborating the results from the isotherm and kinetic studies.

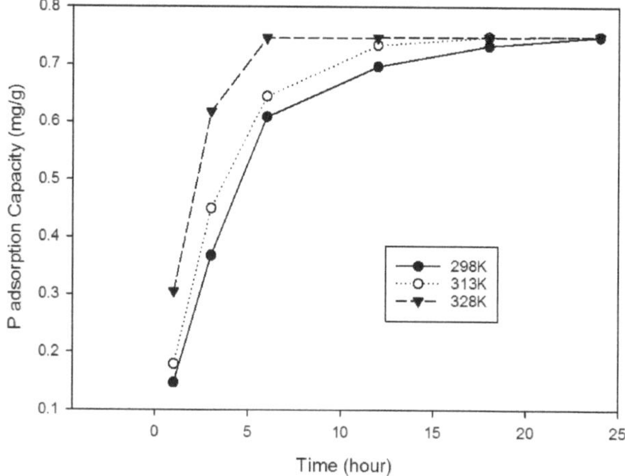

Figure 6. P sorption by concrete studies at various temperatures; Dose of media 2 g; initial P concentration 15 mg/L; contact time 24 h; agitator 180-rpm; pH 5.

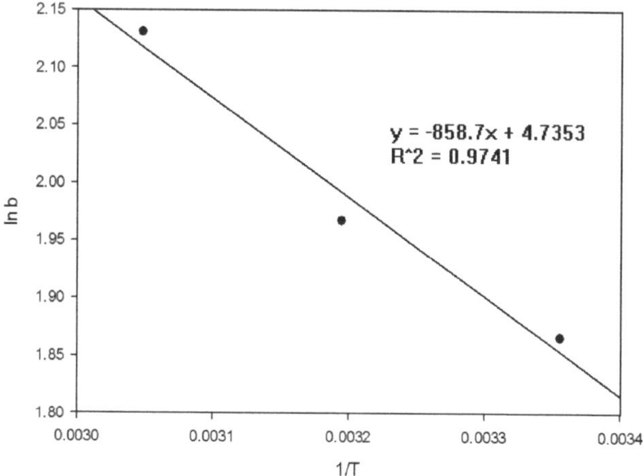

Figure 7. Plot of ln K vs. 1/T for estimation of thermodynamic parameters (ΔH° and ΔS°).

Table 8. Thermodynamic parameters for adsorption of P on concrete particles.

Thermodynamic Parameters	Temperature (K)		
	298	313	328
b	6.460	7.150	8.418
ΔG° (kJ/mol)	−4.623	−5.119	−5.808
ΔH° (kJ/mol)	7.139	-	-
ΔS° (J/mol)	39.336	-	-

Table 9. Thermodynamic studies for P sorption by other materials.

Adsorbent	$\Delta H°$ (kJ/mol)	$\Delta S°$ (J/mol)	Reference
Clinoptilolite rich tuff	20.8	100	[44]
Coir-pith activated carbon	3.88	21.88	[41]
Dolomite	−5.85	−10.17	[45]
Granulated ferric hydroxide	15.1	80	[44]
Iron hydroxide-eggshell waste	81.84	-	[46]
Bentonite	−5.3	10	[44]
RCA	7.139	39.336	Present study
Slovakite	104.9	300	[44]

3.2. Desorption of Phosphorus

Figure 8 shows the proportion of P re-released. The desorption was in the range of 4–7% and lowest at the higher initial P concentrations with maximum desorption at the initial P = 10 mg/L. The adsorption and desorption capacities were not equal, corroborating a previous hypothesis that different mechanisms were involved at different P concentrations and that some were irreversible. It is suggested that at the lower concentrations adsorption was due to just physical adsorption, which then enabled easier P release. Higher concentrations of P, on the other hand, could be sufficient to cause reactions with the metals such as complexation and precipitation; desorption would therefore be reduced [47]. A lower desorption rate at smaller P concentrations, was also reported by Yu et al. (2009) [35]. They tested desorption from mortar (cement mixed with sand) and found the release was 9% but they deduced that the adsorption was mainly due to chemical reaction.

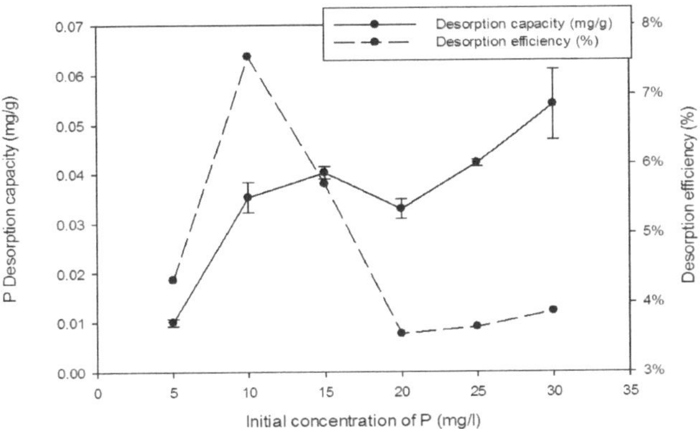

Figure 8. P desorption of concrete on different initial P concentration. Contact time 24 h; agitator 180-rpm; pH-5.

3.3. Fractionation of Inorganic Phosphorus

Concrete is composed of Ca, Mg, Si, and Al hydroxides [33,37,48]. The proportions of each metal bound P types were measured. The Si was assumed to be inactive and not analysed. The data is shown in Figure 9: in the raw concrete O-P was the predominant form of P (37%), the remainder was mostly Ca-P (29%) combined with the Al-P, Mg-P and labile P were less than 10%. There was no Fe-P observed in this study. In the used RCA, on the other hand, labile P (43.3%) was the highest proportion, O-P remained high (25%), Al-P and Ca-P were reduced but Mg-P increased, possibly reflecting variations in the mineral composition. Previous work indicates P reacts with calcium and magnesium, from the cement, to form insoluble precipitate (detail see Table 10).

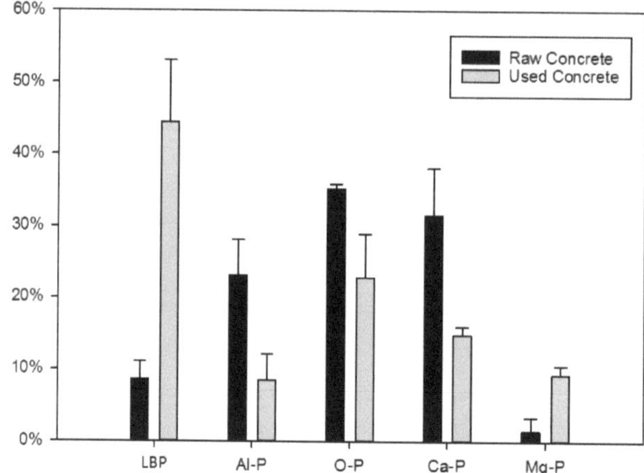

Figure 9. The comparison of different type of P before and after test.

Table 10. Formation of P precipitate.

Sequence	Chemical Formula
1	$Mg^{2+} + HPO_4^{2-} + 3H_2O \rightarrow MgHPO_4 \cdot 3H_2O \downarrow$
2	$Ca^{2+} + HCO_3^- + OH^- \rightarrow CaCO_3 \downarrow + H_2O$
3	$5Ca^{2+} + 4OH^- + 3HPO_4^{2-} \rightarrow Ca_5(OH)(PO_4)_3 \downarrow + 3H_2O$
4	$Al^{3+} + PO_4^{3-} + 2H_2O \rightarrow AlPO_4 \cdot 2H_2O$

In this study, Mg-P was low in the raw concrete and there are few studies of Mg-P due to its high water solubility. This solubility and reactivity of the Mg-ion compared to Ca, Fe, and Al means Mg-P forms rapidly, but is less crystalline and vulnerable to remobilisation by a wide range of conditions (including pH). Ca-P is more stable due to slower crystallisation [49]. Previous work has also indicated the form of P does vary according to the chemical composition of the filter media (Table 11). Oyster shell, for example, was more than 50% CaO which then formed the majority of the adsorbed P as Ca-P [50]. Zeolite was reported to be 70% SiO_2, 10% Al_2O_3, 2.5% CaO, and 1.5% Fe_2O_3 [51–53]. The inactivity of Si would explain its poor performance for P removal. In other materials—such as drinking water treatment residues, fly ash, and Bauxite residues—the total P amount was higher than RCA, but the soluble P and sizes available were much smaller, and were not convenient as wastewater filter media.

The used RCA contained 43% soluble P which is the second highest when compared to previous work and blast furnace slag contained most soluble P (see Table 11). Hylander (1999) [58] carried out loading scale pot experiments and stated that slag sorbed P was highly plant available. Therefore, RCA could be used as fertiliser. This is due to the fact that soluble P could be recovered through equilibrium soluble P-concentration, pH change, or redox potential [59,60]. It was confirmed in previous tests that soluble P could be re-released as it was physical attraction, corroborating the predictions of the isotherm, kinetic, and desorption tests. The reaction time was important because previous work has shown that the surface-attracted P might gradually become more permanently bound through complexation and precipitation reactions [54,61]. This was demonstrated in Table 11 which showed adsorption was followed by the formation of apatite complexes. O-P was the second highest indicating that between a third and a quarter of the P became fixed immediately, and would resist re-release, but it still could be re-converted into available P by soil microorganisms (e.g., Phosphate solubilising bacteria) [62].

Table 11. P specification of various saturated material.

Media	Total (mg/g)	Soluble P (mg/g)	Al-P (mg/g)	Fe-P (mg/g)	O-P (mg/g)	Ca-P (mg/g)	Reference
Blast furnace granulated slag	0.086	0.042 / 49%	- / -	0.007 / 8%	- / -	0.037 / 43%	[54]
Zeolite	0.448	0.024 / 5.4%	0.352 / 78.6%	0.041 / 9.2%	0.016 / 3.6%	0.0131 / 2.9%	[20]
Volcanic rock	0.516	0.066 / 12.8%	0.311 / 60.3%	0.027 / 5.2%	0.054 / 10.5%	0.058 / 11.2%	[20]
Crushed Bricks	0.956	0.068 / 7.1%	0.498 / 52.1%	0.277 / 29.0%	0.024 / 2.5%	0.089 / 9.3%	[20]
RCA	1.298	0.577 / 44.42%	0.110 / 8.50%	0 / 0%	0.297 / 22.86%	0.193 / 14.86%	Present study
Oyster shell	3.596	0.363 / 10.1%	0.08 / 2.2%	0.014 / 0.4%	0.589 / 16.4%	2.55 / 70.9%	[20]
Light-weight expanded clay	6.527	0.053 / 1%	1.641 / 25%	0.022 / <1%	- / -	4.811 / 74%	[55]
Gas desulfurization products	8.607	2.544 / 30%	1.452 / 17%	0.008 / 0%	0.9 / 10%	3.703 / 43%	[56]
Bauxite residual	19.487	0.568 / 3%	14.31 / 73%	2.203 / 11%	1.21 / 6%	1.196 / 6%	[56]
Fly ash	28.074	9.131 / 33%	16.631 / 59%	0.147 / 1%	1.316 / 5%	0.849 / 3%	[56]
Drinking Water treatment residual	30.031	0.367 / 1%	20.726 / 69%	4.66 / 16%	1.755 / 6%	2.523 / 8%	[56]
Electric arc fumace steel slag	-	- / 0.63%	- / 3.05%	- / 13.67%	- / -	- / 82.65%	[57]
Iron melter slag	-	- / 2.41%	- / 22.88%	- / 12.69%	- / -	- / 62%	[57]

4. Conclusions

This study tested recycled concrete for P removal from wastewater, using a larger sized (2–5 mm) and therefore more suitable media than previous studies in practice. More than 90% P removal was achievable using concentrations typical of sewage effluents ($p < 15$ mg/L) with an equilibrium time of 24 h. Model analysis predicted that capacity could reach 6.88 mg/g. The experiments using an analysis of the kinetics (pseudo second-order R = 0.9916) and thermodynamics indicated the adsorption process was spontaneous and, together with the differential extraction and desorption tests results, indicated the mechanism was surface physical electrostatic attraction followed by chemical precipitation. The results suggested that half the P adsorbed to the surface was labile and available for recovery, or, after further crushing, could be used directly as a fertilizer supplement. Further studies on the uptake and impact on plant growth from P enriched RCA would be useful.

The results offer a new usage for high cement RCA which is undesirable for aggregate, due to a lower density, high water adsorption, and reduced structural performance [63,64]. Further work is also suggested to measure how concrete adsorbs heavy metals and persistent organics and how these might interact with C, P, and N in effluents.

Author Contributions: Yihuan Deng and Andrew Wheatley conceived and designed the experiments; Yihuan Deng performed the experiments; Yihuan Deng analyzed the data; Andrew Wheatley contributed reagents/materials/analysis tools; Yihuan Deng and Andrew Wheatley wrote the paper.

Conflicts of Interest: The authors declare no conflict of interest.

References

1. Oikonomou, N.D. Recycled concrete aggregates. *Cem. Concr. Compos.* **2005**, *27*, 315–318. [CrossRef]

2.   National Development and Reform Commission. *The Annual Report on Comprehensive Utilization of Resources of China*; National Development and Reform Commission: Beijing, China, 2014. (In Chinese)
3.   Li, W.; Wu, Y.; Li, S. The Research of the Biologic Padding of Fly Ash Applied in the Sewage Transaction. *Modem Sci. Instrum.* **2007**, *4*, 63–66. (In Chinese)
4.   Guo, Y.; Li, Y.; Wang, Y.; Gao, K.; Yang, L.; Qin, X.; Du, J.; Wang, S. Study on pollutants removal performance of some biological carrier in wastewater land treatment system. *Environ. Sci. Technol.* **2009**, *22*, 33–35. (In Chinese)
5.   Xiang, H.; Han, Y.; Liu, Y. Substrate screening for phosphorus removal in low concentration phosphorus-containing water body. *Acta Sci. Circumst.* **2013**, *33*, 3227–3233. (In Chinese)
6.   Reijnders, L. Phosphorus resources, their depletion and conservation, a review. *Resour. Conserv. Recycl.* **2014**, *93*, 32–49. [CrossRef]
7.   Deng, Y.; Wheatley, A. Wastewater treatment in Chinese rural areas. *Asian J. Water Environ. Pollut.* **2016**, *13*, 1–11. [CrossRef]
8.   Sasaki, A.; Yoshikawa, E.; Mizuguchi, H.; Endo, M. The Improvement of water quality in an acidic river environment using waste concrete aggregates. *J. Water Environ. Technol.* **2013**, *11*, 235–247. [CrossRef]
9.   Zhou, L.; Huang, Z.; Li, T. The Application of Wasted Architecture Walling Materials Used as a constructed wetland media. In *Water Infrastructure for Sustainable Communities China and the World*, 1st ed.; Hao, X., Novotny, V., Nelson, V., Eds.; International Water Association: London, UK, 2010; pp. 481–489.
10. Chen, Z. Technology of Using Eco-concrete for Sewage Treatment. *J. Build. Mater.* **2001**, *4*, 60–64. (In Chinese)
11. Yuan, J.; Jin, L.; Zou, C.; Wan, Y.; Liu, H.; Lan, Y. Microorganisms and effectiveness of a low cost concrete biofilm reactor for sewage treatment. *Environ. Pollut. Control* **2006**, *28*, 568–571. (In Chinese)
12. Tamai, M.; Okinaka, T. Present and future views of environmentally-friendly concrete. In *Concrete Technology for a Sustainable Development in the 21st Century*, 1st ed.; Gjorv, O.E., Sakai, K., Eds.; E&FN: New York, NY, USA, 2000; pp. 264–273.
13. American Public Health Association. *Standard Methods for the Examination of Water and Wastewater*, 21st ed.; American Public Health Association: Washington, DC, USA, 2005.
14. Dean, L.A. An attempted fractionation of the soil phosphorus. *J. Agric. Sci.* **1938**, *28*, 234–246. [CrossRef]
15. Chang, S.C.; Jackson, M.L. Frationation of soil phosphorus. *Soil Sci.* **1957**, *84*, 133–144. [CrossRef]
16. Leader, J.W.; Dunn, E.J.; Reddy, K.R. Phosphorus Sorbing Materials: Sorption Dynamics and Physicochemical Characteristics. *J. Environ. Qual.* **2008**, *37*, 174–181. [CrossRef] [PubMed]
17. Reddy, K.R.; Wang, Y.; DeBusk, W.F.; Fisher, M.M.; Newman, S. Forms of Soil Phosphorus in Selected Hydrologic Units of the Florida Everglades. *Soil Sci. Soc. Am. J.* **1998**, *62*, 1134–1147. [CrossRef]
18. White, S.A.; Taylor, M.D.; Albano, J.P.; Whitwell, T.; Klaine, S.J. Phosphorus retention in lab and field-scale subsurface-flow wetlands treating plant nursery runoff. *Ecol. Eng.* **2011**, *37*, 1968–1976. [CrossRef]
19. Hartikainen, H. Phosphorus and its reactions in terrestrial soils and lake sediments. *J. Sci. Agric. Soc. Finl.* **1979**, *51*, 537–624.
20. Wang, Z.; Liu, C.; Dong, J.; Liu, L.; Li, P.; Zheng, J. Screening of phosphate-removing filter media for use in constructed wetlands and their phosphorus removal capacities. *China Environ. Sci.* **2013**, *33*, 227–233. (In Chinese)
21. Agyeia, N.M.; Strydomb, C.A.; Potgieter, J.H. The removal of phosphate ions from aqueous solution by fly ash, slag, ordinary Portland cement and related blends. *Cem. Concr. Res.* **2002**, *32*, 1889–1897. [CrossRef]
22. Burianek, P.; Skalicky, M.; Grunwald, A. Phosphates adsorption from water by recycled concrete. *GeoSci. Eng.* **2014**, *LX*, 1–8. [CrossRef]
23. Molle, P.; Lienard, A.; Grasmick, A.; Lwema, A. Phosphorus retention in subsurface constructed wetlands: Investigations focused on calcareous materials and their chemical reactions. *Water Sci. Technol.* **2003**, *48*, 75–83. [PubMed]
24. Cucarella, V.; Renman, G. Phosphorus sorption capacity of filter materials used for on-site wastewater treatment determined in batch experiments–a comparative study. *J. Environ. Qual.* **2009**, *3*, 381–392. [CrossRef] [PubMed]
25. Agrawal, S.G.; King, K.W.; Fischer, E.N.; Woner, D.N. PO4 3- removal by and permeability of industrial byproducts and minerals: Granulated blast furnace slag, cement kiln dust, coconut shell activated carbon, silica sand, and zeolite. *Water Air Soil Pollut.* **2011**, *219*, 91–101. [CrossRef]

26. Dawodu, F.A.; Akpomie, K.G. Kinetic, Equilibrium, and Thermodynamic Studies on the Adsorption of Cadmium (II) Ions using "Aloji Kaolinite" Mineral. *Pac. J. Sci. Technol.* **2014**, *15*, 268–276.

27. Dada, A.O.; Olalekan, A.P.; Olatunya, A.M.; DADA, O. Langmuir, Freundlich, Temkin and Dubinin-Radushkevich Isotherms Studies of Equilibrium Sorption of Zn^{2+} Unto Phosphoric Acid Modified Rice Husk. *J. Appl. Chem.* **2012**, *3*, 38–45.

28. Tosun, İ. Ammonium Removal from Aqueous Solutions by Clinoptilolite Determination of Isotherm and Thermodynamic Parameters and Comparison of Kinetics by the Double Exponential Model and Conventional Kinetic Models. *Int. J. Environ. Res. Public Health* **2012**, *9*, 970–984. [CrossRef] [PubMed]

29. Cheng, Y.; Yang, C.; He, H.; Zeng, G.; Zhao, K.; Yan, Z. Biosorption of Pb(II) Ions from Aqueous Solutions by Waste Biomass from Biotrickling Filters: Kinetics, Isotherms, and Thermodynamics. *J. Environ. Eng.* **2015**, *142*, C4015001. [CrossRef]

30. Gu, D.; Zhu, X.; Vongsay, T.; Huang, M.; Song, L.; He, Y. phosphorus and Nitrogen removal using novel porous bricks incorporated with wastes and minerals. *Pol. J. Environ. Stud.* **2013**, *22*, 1349–1356.

31. Tewaria, N.; Vasudevana, P.; Guhab, B.K. Study on biosorption of Cr(VI) by Mucor hiemalis. *Biochem. Eng. J.* **2005**, *23*, 185–192. [CrossRef]

32. Zheng, J.; Yin, X.; Liu, B.; Meng, G. Preparation of Porous Hardened Cement Paste Synthetic Filter Material and Phosphorus Removal from Wastewater by Adsorption. *J. Civ. Archit. Environ. Eng.* **2013**, *35*, 115–120. (In Chinese)

33. Renman, G.; Renman, A. Sustainable use of crushed autoclaved aerated concrete (CAAC) as a filter medium in wastewater purification. In *WASCON*; Arm, M., Ed.; ISCOWA and SGI: Gothenburg, Sweden, 2012.

34. Egemose, S.; Sønderup, M.J.; Beinthin, M.V.; Reitzel, K.; Hoffmann, C.C.; Flindt, M.R. Crushed concrete as a phosphate binding material: A potential new management tool. *J. Environ. Qual.* **2012**, *41*, 647–653. [CrossRef] [PubMed]

35. Yu, Z.; Ding, Z.; Zha, X.; Cheng, T.; Ding, Z.; Zha, X.; Cheng, T. Phosphorus Adsorption Characteristics of Different Substrates in Constructed Wetland. *China Water Wastewater* **2009**, *25*, 80–82. (In Chinese)

36. Hu, C.; Liu, P. Study on the effect of red mud in improving the phosphorus removal ability of eco-concrete. *Concrete* **2014**, *293*, 151–153. (In Chinese)

37. Oguz, E.; Gurses, A.; Yalcin, M. Removal of phosphate from waste waters by adsorption. *Water Air Soil Pollut.* **2003**, *148*, 279–287. [CrossRef]

38. Chien, S.H.; Clayton, W.R. Application of Elovich Equation to the Kinetics of Phosphate Release and Sorption in Soils. *Soil Sci. Soc. Am. J.* **1980**, *44*, 265–268. [CrossRef]

39. Gerente, C.; Lee, V.K.C.; Cloirec, P.L.; McKay, G. Application of chitosan for the removal of metals from wastewaters by adsorption—mechanisms and models review. *Crit. Rev. Environ. Sci. Technol.* **2007**, *37*, 41–127. [CrossRef]

40. Ho, Y.S.; McKay, G. Application of kinetic models to the sorption of copper (II) on to peat. *Adsorp. Sci. Technol.* **2002**, *20*, 797–815. [CrossRef]

41. Kumar, P.; Sudha, S.; Chand, S.; Srivastava, V.C. Phosphate Removal from Aqueous Solution Using Coir-Pith Activated Carbon. *Sep. Sci. Technol.* **2010**, *45*, 1463–1470. [CrossRef]

42. Kim, D.S. Adsorption characteristics of Fe(III) and Fe(III)–NTA complex on granular activated carbon. *J. Hazard. Mater.* **2004**, *106*, 67–84. [CrossRef]

43. Wang, Z.; Nie, E.; Li, J.; Yang, M.; Zhao, Y.; Luo, X.; Zheng, Z. Equilibrium and kinetics of adsorption of phosphate onto iron-doped activated carbon. *Environ. Sci. Pollut. Res.* **2012**, *19*, 2908–2917. [CrossRef] [PubMed]

44. Chmielewská, E.; Hodossyová, R.; Bujdoš, M. Kinetic and Thermodynamic Studies for Phosphate Removal Using Natural Adsorption Materials. *Pol. J. Environ. Stud.* **2013**, *22*, 1307–1316.

45. Yuan, X.; Xia, W.; An, J.; Yin, J.; Zhou, X.; Yang, W. Kinetic and Thermodynamic Studies on the Phosphate Adsorption Removal by Dolomite Mineral. *J. Chem.* **2015**, *2015*, 1–8. [CrossRef]

46. Mezenner, N.Y.; Bensmaili, A. Kinetics and thermodynamic study of phosphate adsorption on iron hydroxide-eggshell waste. *Chem. Eng. J.* **2009**, *147*, 87–96. [CrossRef]

47. Zhang, Y.; Zou, Y.; Huang, Y.; Wang, C.; Li, F. Phosphate adsorption and desorption characteristic of several fly ashes. *Chin. J. Appl. Ecol.* **2005**, *16*, 1756–1760. (In Chinese)

48. Berg, U.; Donnert, D.; Ehbrecht, A.; Bumiller, W.; Kusche, I.; Weidler, P.G.; Nüesch, R. "Active filtration" for the elimination and recovery of phosphorus from waste water. *Colloids Surf. A Physicochem. Eng. Asp.* **2005**, *265*, 141–148. [CrossRef]

49. Josan, M.S.; Nair, V.D.; Harris, W.G.; Herrera, D. Associated release of magnesium and phosphorus from active and abandoned dairy soils. *J. Environ. Qual.* **2005**, *34*, 184–191. [CrossRef] [PubMed]

50. Zhong, B.; Zhou, Q.; Chan, Y.; Yu, F. Structure and Property Characterization of Oyster Shell Cementing Material. *Chin. J. Struct. Chem.* **2012**, *31*, 85–92.

51. Erdem, E.; Karapinar, N.; Donat, R. The removal of heavy metal cations by natural zeolites. *J. Colloid Interface Sci.* **2004**, *280*, 309–314. [CrossRef] [PubMed]

52. Jiang, J.; Chen, Y.; Deng, Z.; Wang, W.; Zhao, Q. A research on ammonia treated by zeolite from landfill leachate. *Water Wastewater Eng.* **2003**, *29*, 6–9. (In Chinese)

53. Petrus, R.; Warchol, J.K. Heavy metal removal by clinoptilolite. An equilibrium study in multi-component systems. *Multi-Compon. Syst.* **2005**, *39*, 819–830. [CrossRef] [PubMed]

54. Korkusuz, E.A.; Beklioglub, M.; Demirer, G.N. Use of blast furnace granulated slag as a substrate in vertical flow reed beds Field application. *Bioresour. Technol.* **2007**, *98*, 2089–2101. [CrossRef] [PubMed]

55. Jenssen, P.D.; Krogstad, T.; Paruch, A.M.; Mæhlum, T.; Adam, K.; Arias, C.A.; Heistad, A.; Jonsson, L.; Hellström, D.; Brix, H.; et al. Filter bed systems treating domestic wastewater in the Nordic countries—Performance and reuse of filter media. *Ecol. Eng.* **2010**, *36*, 1651–1659. [CrossRef]

56. Penn, C.J.; Bryant, R.B.; Callahan, M.P.; McGrath, J.M. Use of Industrial By-products to Sorb and Retain Phosphorus. *Commun. Soil Sci. Plant Anal.* **2011**, *42*, 633–644. [CrossRef]

57. Drizo, A.; Cummings, J.; Weber, D.; Twohig, E.; Druschel, G.; Bourke, B. New Evidence for Rejuvenation of Phosphorus Retention Capacity in EAF Steel Slag. *Eviron. Sci. Technol.* **2008**, *42*, 6191–6197. [CrossRef]

58. Hylander, L.D.; Johansson, L.; Renman, G.; Ridderstolpe, P.; Siman, G. Phosphorus recycling from waste water by filter media used as fertilisers. *Nordisk Jordbrugsforsking* **1999**, *81*, 102–105.

59. Grobbelaar, J.U.; House, W.A. Phosphorus as a limiting resource in inland waters; interactions with nitrogen. In *Phosphorus in the Global Environment-Transfers, Cycles and Management*; Tiessen, H., Ed.; Wiley: Hoboken, NJ, USA, 1995; pp. 55–274.

60. Hillbricht-Ilkowska, A.; Ryszkowski, L.; Sharpley, A.N. Phosphorus Transfers and Landscape Structure: Riparian Sites and Diversified Land Use Patterns. In *Phosphorus in the Global Environment-Transfers, Cycles and Management*; Tiessen, H., Ed.; Wiley: Hoboken, NJ, USA, 1995; pp. 201–228.

61. Frossard, E.; Brossard, M.; Hedley, M.J.; Metherell, A. Reactions Controlling the Cycling of P in Soils. In *Phosphorus in the Global Environment-Transfers Cycles and Management*; Tiessen, H., Ed.; Wiley: Hoboken, NJ, USA, 1995; pp. 107–138.

62. Chen, Y.P.; Rekha, P.D.; Arun, A.B.; Shen, F.T.; Lai, W.-A.; Young, C.C. Phosphate solubilizing bacteria from subtropical soil and their tricalcium phosphate solubilizing abilities. *Appl. Soil Ecol.* **2006**, *34*, 33–41. [CrossRef]

63. Juan, M.S.D.; Gutiérrez, P.A. Study on the influence of attached mortar content on the properties of recycled concrete aggregate. *Construct. Build. Mater.* **2009**, *23*, 872–878. [CrossRef]

64. Sagoe-Crentsil, K.K.; Taylor, T.B. Performance of concrete made with commercially produced coarse recycled concrete aggregate. *Cem. Concr. Res.* **2001**, *31*, 707–712. [CrossRef]